Mebus A. Geyh    Helmut Schleicher

# Absolute Age Determination

## Physical and Chemical Dating Methods and Their Application

English by R. Clark Newcomb

With 146 Figures

Springer-Verlag Berlin Heidelberg New York
London Paris Tokyo Hong Kong Barcelona

Prof. Dr. MEBUS A. GEYH
Niedersächsisches Landesamt
für Bodenforschung
Alfred-Bentz-Haus
Postfach 51 01 53
D - 3000 Hannover

Prof. Dr. HELMUT SCHLEICHER
Mineralogisch-petro-
graphisches Institut
Universität Freiburg
Albertstraße 23b
D - 7800 Freiburg

ISBN 3-540-51276-4 Springer-Verlag Berlin  Heidelberg  New York
ISBN 0-387-51276-4 Springer-Verlag New York  Berlin  Heidelberg

Library of Congress Cataloging-in-Publication Data. Geyh, Mebus A. Absolute age determination: physical and chemical dating methods and their application / Mebus A. Geyh, Helmut Schleicher. p. cm. Includes bibliographical references. 1. Geological time――Measurement. 2. Radiocarbon dating. I. Schleicher, Helmut, 1947– . II. Title. QE508.G443 1990 551.7'01――dc20 90-9406

© Springer-Verlag Berlin  Heidelberg 1990
Printed in Germany

The use of registered names, trademarks, etc. in this publication does not imply, even in the absence of a specific statement, that such names are exempt from the relevant protective laws and regulations and therefore free for general use.

Typesetting: Thomson Press India Ltd., New Delhi, India
2132/3145(3011)-543210  –  Printed on acid-free paper

# Preface

With the growing recognition during the last two centuries that the Earth has an immense age and processes over long periods of time have changed the morphology and composition of the Earth's crust, geologists have become increasingly interested in determination of absolute ages. A relative geochronology was established on the basis of the lithostratigraphic and biostratigraphic principles developed during the last century. With the discovery of radioactivity, the basis for a new geoscientific discipline – geochronology – was established (Rutherford 1906). It is the study of geological time, based mainly on the time signatures provided by the isotopic composition in geologic materials. The isotopic signature in a rock yields more information than that provided by the geochemical signature alone because it reflects the origin and history of the element in the rock.

The aim of geochronology is to calibrate and standardize chronostratigraphic scales, to develop geological time scales that have a sensitive or at least useful resolution in order to place the geological events in the correct chronological order, and to assign their proper time spans. In practice, the application of geochronology is much wider because the data in the „natural archives" often provide information on the origin, genesis, and history of the materials. This, of course, requires an understanding of the geochemical behavior of the substances involved.

Geochronological and isotope geological studies, to an increasing degree, are solving or contributing to the solution of problems encountered in prospecting, research on ore deposits and hydrocarbon maturation, earthquake forecasting, geothermal exploration, ecological studies (environmental physics, e.g., the siting of waste repositories), and paleoclimatology (e.g., climatic jumps). Often, the results of geochronological studies yield unexpected insights into tectonic and other geological structures that had not been previously achieved by other geoscientific methods. In addition, attempts are being made to recognize geological trends (Doe 1983), the aim being not so much an understanding of the past as drawing conclusions about the future by comparing the long-term records of the past with the present.

Taylor (1987a) classifies dating methods according to the „physical and chemical, time-sensitive mechanism employed". Absolute ages are obtained with „fixed-rate processes such as nuclear decay". Variable-rate processes, such as chemical reactions, yield age estimates, but calibration with a fixed-rate method is possible. Then a „calibrated variable rate" is available „for time placement purposes". Each dating method „operates within the framework of a specific set of contextual and geochemical-geophysical assumptions and constraints". The reliability of the dates obtained is „directly tied to the degree to which each assumption is fulfilled for each sample or geochemical-geophysical environment. ... Each method can be characterized by its geographic or environmental restrictions, age range (defined by maximum and minimum ages), relevant sample types, and time zero processes".

The spectrum of physical and chemical dating methods now covers the entire range of Earth history. But there are so many methods that it is becoming more and more difficult to select those that are appropriate for solving a specific problem. It is also becoming increasingly difficult to assess the meaning of the data obtained; for example, the question may arise whether the determined age is the age of formation, early or late diagenesis, or some stage of metamorphism. Moreover, different components of a sample may yield different kinds of ages, depending on the method applied. In some cases, even no date at all may be obtained. Hence, the selection of the most appropriate method for a particular task remains a problem. The aim of further development of the physical and chemical dating methods is to widen the range of geological application and to increase the precision of the methods.

This book is addressed to everyone interested in the application of physical and chemical dating methods to the geosciences and archaeology. It should be especially valuable as a concise, but comprehensive reference for students and practitioners using these methods. The geochronologist may find that certain details of the methods and techniques are missing, for which the reader will have to refer to the extensively cited literature. Numerous publications deal with dating methods, but usually they discuss only the radiometric methods, the most well-known ones (e.g. Hamilton 1965; York and Farquar 1972; Michels 1973; Fitch et al. 1974; Aitken 1978; Jäger and Hunziker 1979; Tite 1881; Currie 1982; Mahaney 1984; Roth et al. 1985, 1989; Faure 1986). They do not, however, cover the whole spectrum of methods, which is one of the objectives of this book. In modern geochronological studies, radioactive dating methods are used as well as chronostratigraphic and chemical methods. Hence, a complete compilation is justified.

The information provided by the absolute dates alone is not sufficient to make chronological sense. This information must be supplemented by geological and petrological data, including lithostratigraphic and biostratigraphic data, paleoclimatic records, and information about the origin of the sample, geochemical behavior, and any diagenetic changes that may have occurred.

Hannover/Freiburg, Spring 1990                                    M. A. GEYH
                                                                  H. SCHLEICHER

# Contents

In Chapters 6 - 8 the following classification of dating methods has been used:
*** standard methods; ** routine methods; * individual case study methods;
without asterix: methods in development or obsolete.

**Foldout Table: Dating Methods, Ranges, and Materials**
(inside back cover)

# 1 Introduction

This book is meant to be both a textbook and a reference book of all methods of physical and chemical age determination. Owing to the profusion of material, a compressed presentation is necessary but with sufficient detail to give a comprehensive picture. The descriptions of the individual methods are, to a large extent, complete and understandable in themselves.

Those who are collecting samples for dating for the first time should first read the chapter on "Selection, Collection, Packing, Storage, Transport, and Description of the Samples" (Chap. 3). A general discussion of the "Treatment and Interpretation of the Raw Data" is given in Chap. 4. Aspects that apply only to one method are discussed in the individual chapters on the methods.

Terms commonly used in geochronology are explained in the glossary (App. A). Words that begin with Greek letters are alphabetized according to the name of that letter. The physical constants for the radionucleides that are important for geochronology are given in App. B. The most recently determined values for the half-lives and isotopic abundances, the commonly used ones, and those set by international convention are given (Steiger and Jäger 1977). Half-lives [Eq. (5.3)], rather than decay constants, are used in soft-rock dating because they better reflect the dating ranges. A selection of addresses of large, well-known, or multi-disciplinary geochronological laboratories is given in App. C. These laboratories are now so numerous that such a list can make no pretense of being complete, but it may be of help to users of this book to make initial contacts.

The methods, dating ranges, and the different kinds of materials that can be dated with each are compiled in the Foldout Table at the end of the book. Please keep in mind, however, that this table provides only general information. In each case, a geochronological laboratory should be consulted as to whether a specific sample is really suitable. The materials listed in the table are not necessarily of equal suitability for the dating range given. In some cases, only certain laboratories (see App. C) are capable of handling special samples and dating ranges.

The section number in which the method is discussed is given together with the name of the method in the Foldout Table. Due to the large number of datable materials, it was necessary to group them, which required compromises. The materials are classified according to their suitability for dating: very good (*), good (+), and limited (−).

The methods are marked in the headings according to their present-day state of the art and use: Standard methods are indicated by ***. These are employed worldwide and can be used for many different materials. The standard methods are distinguished from routine methods (**) that are widely applied but for which there is no international convention on how to evaluate the results. In some cases, fundamental technical problems have yet to be solved. Methods that have been used only in individual case studies are marked with *. Methods that are in development or are obsolete are not marked. The papers in which the principle or a case study is described for the first time are cited for each method.

The discussion of each method is organized according to a uniform outline: The scope of application is given in the first subsection of the section on a given method. This includes dating range, attainable precision, suitable materials, and the amount of sample required. Laboratories using the method are listed in App. C.

The second subsection "Basic Concept" gives the fundamentals of the method. The principles are described, together with the major equations. This subsection is followed by a brief description of "Sample Treatment and Measurement Techniques". The number of procedures that have been developed by the laboratories has become so large that only the frequently used or especially important ones are included. Special tips on the collection and storage of the samples that go beyond the general suggestions (Chap. 3) are also given in this subsection. The scope of the method is discussed in the subsection "Scope and Potential, Limitations, Representative Examples". Recurring problems and the attempts to solve them by modification of the model are described to illustrate the possibilities and limits of the method. References are given that reflect the spectrum of applications of the method and allow the reader to study a particular aspect in greater detail than is possible in this book, as well as aid him in the interpretation of the ages.

The last subsection outlines areas of "Non-Chronological Applications". It is impossible to give a representative selection of references. The citations are meant to stimulate the prospective user of the method to read the papers that have been published on studies applying the method. Comprehensive reviews of non-chronological applications of isotope techniques have been published, for example, by Fritz and Fontes (1980, 1986), Allègre (1987), and Wasserburg (1987).

# 2    Time Scales and Ages

The question as to whether physical and chemical dating methods yield absolute or relative ages is the subject of continuing debate. Fuel for this discussion has been supplied especially by those methods that, for example, utilize cosmogenic isotopes (Sect. 6.2), and have time scales which deviate systematically from the solar calendar (Fig. 2.1).

The differing opinions can be bridged if a distinction is made between the methods and the results. Most dating methods, especially the physical ones, are based on processes that are governed only by the passage of time (time-sensitive fixed-rate mechanisms). These methods yield, in principle, absolute ages. Radioactive decay is the most important of these processes. On the other hand, methods that are based, for example, on stratigraphic anomalies (Chap. 7) or utilize chemical reaction rates as a measure of time (Chap. 8) can generally yield only relative ages. This is because environmental factors, particularly temperature, which vary from place to place and from time to time, often have a decisive influence on these rates that cannot always be corrected by taking into account other chemical or physical information. In such cases calibration tables or curves on independent obtained dates must be established.

If the results of the dating methods are considered, one can talk about absolute ages in the narrower sense of the term only if they are valid worldwide and, thus, are comparable with each other and correspond to the siderial (absolute) time scale (Sect. 2.1). Some ages derived from the isotopic concentrations of a sample often do not meet this requirement. The exceptions to this are the ages obtained using the standard methods for which international agreements have been made (Sects. 6.1.1, 6.1.3, and 6.2.1). So-called conventional ages are obtained, which are comparable with each other worldwide. Because these conventional ages do not necessarily agree with the ages obtained with other methods, they are usually not called absolute ages, but rather terms such as "physical", "radiometric", "chemical" ages are used or they are given a prefix referring to the applied method. However, because they can be compared worldwide, it is not fundamentally incorrect, in a wider sense, to talk about absolute dates. But interpretation in terms of which event is being dated makes even conventional dates relative.

## 2.1    Absolute Time Scales

Absolute ages, for which the solar year (also called the siderial or astronomical year) is the unit of time, are based either on historical or astronomical events or on

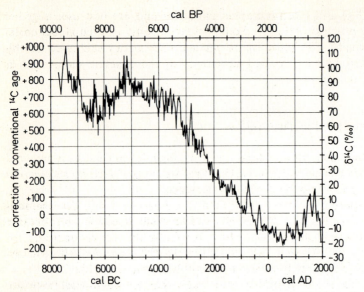

**Fig. 2.1.** Deviations of the conventional radiocarbon time-scale (Sect. 6.2.1) from the dendrochronological time-scale (and $\delta^{14}C$ values) plotted against the calibrated radiocarbon time-scale (after Stuiver and Kra 1986)

dendrochronological studies (Eckstein et al. 1984). They are valid for any site and period.

After the Gregorian calendar, the best recorded absolute chronology is the several-thousand-year dynastic succession of Egyptian pharaohs, although it is not completely free of error. Others are the Mayan calendar and the Scandinavian varve chronology. The latter is an absolute chronology for the late Pleistocene, for which corrections are still being made (e.g., Olsson 1970). After historically and astronomically verified dates, the dendrochronological dates of the Holocene master curves are the most exact and most reliable. They are derived from nonsystematic variations in the thickness of the annual rings of trees. These curves have been determined from the *Sequoia gigantea* and the bristle cone pine (*Pinus aristata*) in North America and from the oak (e.g., *Quercus robur* and *Q. petraea*) in Europe (e.g., Pilcher et al. 1984) and are being continually extended. The resolution of the dendrochronological dating method is about one year, obtained (with some exceptions) with samples containing at least 100 annual rings. The possibilities of this dating method are limited to certain tree species and regions. Very long sections of the master curves have been correlated between North America and Europe. Dendrochronological ages cannot yet be determined for the late Pleistocene or earlier periods (Stuiver and Kra 1986).

## 2.2  Relative Time Scales

Relative time scales are valid only for limited areas and time periods. The most important methods are based on lithostratigraphic (e.g., tephrochronology, see e.g.,

Paterne et al. 1986) and biostratigraphic principles and are best used together with absolute or physical dating methods so that regional or even global correlations can be made. The precision of relative time scales calibrated in this way is determined by the standard deviations of the absolute dates used as references.

## 2.3  Physical and Chemical Time Scales

The physical dating methods are based, as far as possible, on changes in material properties that are dependent only on time. Environmental factors, particularly temperature and pressure, should have no influence.

The most important process of this kind is radioactive decay. The methods based on this process, however, do not all yield absolute dates (Sect. 6.2) because geophysical or geochemical processes often complicate the model conditions for the age determination, which is why, for example, there are deviations of the radiocarbon time scale (Sect. 6.2.1) from the absolute time scale (Fig. 2.1).

Most of the chronostratigraphic time scales that are valid worldwide are based on ages obtained with physical methods (e.g., Harland et al. 1982). They are being continually improved. Chronologic time scales based on only one method are given the name of that method so that their limitations can be taken into consideration, e.g., the radiocarbon (Sect. 6.2.1) chronology.

The restriction that, strictly speaking, only dates obtained with the same physical or chemical method can be easily compared does not mean their use has to be limited (Sect. 4.2.2). Sometimes an absolute date for a geological event may be less informative than the complex of information on the genesis and history of the material that is inherent in dates obtained with different methods.

Other terms are used in the literature to narrow the meaning of the dates: For example, a distinction is made between true, model, apparent, and conventional ages. The word model indicates that the age is derived from special material properties within the "framework of a specific set of contextual and geochemical–geophysical assumptions and constraints" (Taylor 1987a). If these assumptions are fulfilled, the age is called a "true" model age. If not, "apparent" model ages are obtained. Conventional ages are determined according to international guidelines; these have been laid down for the $^{14}$C (Sect. 6.2.1), K/Ar (Sect. 6.1.1), Rb/Sr (Sect. 6.1.3), and U/Th/Pb (Sect. 6.1.9) methods (e.g., Steiger and Jäger 1977). Conventional ages are the most precise of all the dates determined by physical methods and can be compared better than any of the others.

This comparison of $^{14}$C dates, for example, shows that the published standard deviations are too small by as much as a factor of 2. As a consequence, it is recommended that the standard deviation be multiplied by a laboratory-specific error multiplier before any further statistical treatment of the data (International Study Group 1982; Stuiver and Kra 1986). However, this procedure does not increase the confidence probability as, for example, is done by increasing the value of the confidence coefficient $k$ (Sect. 4.2). The basis for the laboratory-specific error multiplier may change with time.

An example of the problems that may arise is that caused by contamination. The extent to which it affects the various methods can differ greatly. U/Th dates for speleothem that are less than 10 ka often prove to be too large by many thousands of years without any indication (e.g., the presence of $^{232}$Th) that such is the case. The cause can be clay that contains traces of "detritial" $^{230}$Th or it can be due to leaching of uranium. This effect has been demonstrated for slowly growing flowstone, for instance (Geyh and Hennig 1986). On the other hand, high $^{14}$C ages (> 30,000 years) are usually minimum ages for the sample being dated because even a very small proportion of recent material in the sample is sufficient to lower the $^{14}$C age considerably (Fig. 6.55, Sect. 6.2.1).

Comparison of conventional $^{14}$C dates with dates given in calender years (Figs. 2.1 and 6.52) is another example of the problems of working with different time scales. The two usually deviate widely from one another and require calibration tables or calibration curves (Sect. 6.2.1) for direct comparison of the results.

# 3 Selection, Collection, Packing, Storage, Transport, and Description of the Samples

Despite the great differences between the numerous methods of physical and chemical age determination and the large variety of datable materials, general recommendations can be made on the selection, collection, packing, storage, transport, and description of the samples. These recommendations are especially important for geologically young samples ("soft-rock dating"). Further recommendations are given in the chapters for the individual methods.

## 3.1 Selection and Collection of the Samples

Before sampling site and samples are selected, it must be decided whether more than one dating method should be used and whether different substances or fractions should be analyzed in order to increase the reliability of the results. The type and amount of sample depends on the method(s) to be applied. It is advisable to select the most suitable samples for dating from a small number of "pilot" samples analyzed first. The requirements placed on the pilot samples do not necessarily have to be the same as those for the samples finally used for dating. For whole-rock samples, for example, hand specimens are sufficient to roughly determine the range of geochemical variation needed for "isochron dating". But the samples that are finally used for the age determination usually require a very exact preliminary chemical and/or mineralogical and petrographic analysis.

It is very important that all aspects of the site that affect the age determination be taken into consideration (e.g., stratigraphy and geochemistry) and be mentioned in the sample description submitted to the analyst. An inadequate description can lead to the omission of necessary sample preparation procedures or corrections of the results. It is, therefore, highly recommended that a geochronologist participate in the initial sample collection and the setting up of the pilot study. When the results of the pilot analysis become available, the samples for the final analysis can be chosen by the submitter with little or no consultation.

If at all possible the samples should be taken from stratigraphically known positions. For organic samples, attention must be paid to whether they contain roots or whether bioturbation has occurred. In general, rocks must be unweathered (Sect. 4.1.2). The choice of tools is immaterial as long as no contamination or mixing (e.g., with certain kinds of drilling equipment) occurs through their use (e.g., Sect. 7.1) or a false stratigraphic assignment results (e.g., by dredging under water).

## 3.2   Packing, Storage, and Transport of the Samples

The rules given in the following sections are more for "soft-rock dating" (dating of Quaternary samples) and archeological samples (Sect. 4.1.1) than for "hard-rock dating" (i.e., dating of rock and mineral samples using the decay of long-lived radioactive isotopes; Sect. 4.1.2).

An important, but unfortunately not always fulfilled condition for satisfactory dating is that the samples arrive undamaged at the laboratory. The samples must be carefully packed and labeled at the sampling site. Strong boxes are recommended for rock samples and stable plastic bags for light material. Paper bags and cardboard boxes are not suitable because they can be affected by moisture and/or humic acid in the samples. The packing material must not be allowed to mix with the samples. Moreover, very small bags should be avoided as they may be lost or overlooked. Since the label written on plastic bags can be easily wiped off, it is recommended that the bag containing the sample be put in a second bag and that a paper label be placed between the two bags.

Materials that can decompose are either dried before storage or frozen. The latter is especially important for deep-sea sediment samples (Geyh et al. 1974). Water samples are kept in sealed glass bottles, which must be completely full. If the analysis is for traces of alkali or other chemical elements, polyethylene bottles are preferable because the samples can be contaminated by glass.

## 3.3   Sample Description

Rather extensive descriptions of the samples and sampling sites are needed by the geochronological laboratory in order to be able to optimally prepare the sample and correctly interpret the raw data. Some laboratories provide application forms for samples submitted to them. This form contains questions on the kind and condition (e.g., fresh, weathered, presence of roots) of the material to be dated, measures taken for its preservation after collection, the amount of sample (helpful for identification), about the stratigraphic location of the sample, the depth from which the sample was taken (possibly with reference to m.s.l.), and the geographic coordinates of the site. It is very important for the geochronologist to know which geological or archeological period the sample belongs to, as well as the results of any other age determinations and literature citations that are relevant. And last, but not least, a brief description of the problem to be solved with the individual samples and the relationship of the samples to one another must be included.

# 4  Treatment and Interpretation of the Raw Data

## 4.1  Suitability of a Sample for Dating and Reliability of the Dates

The evaluation, treatment, and interpretation of the results of an age determination require long experience and an extensive knowledge of the literature. The laboratory analysis for physical and chemical age determinations is often depreciated as routine work, but careful laboratory work requires great skill and the guidance of an experienced scientist.

There are no generally applicable procedures for interpreting dating results. Each case must be considered separately and requires a specific model (Jäger and Hunziker 1979) and the definition of limiting conditions. Often, complicated mathematical models must be developed. The conditions at the sampling site and the composition and origin of the sample, as well as the way the sample was prepared in the laboratory are decisive for the interpretation of the results. The success of this will always be aided by consultation with a geochronologist.

### 4.1.1  Soft-Rock Dating

A simple set of criteria – a modification of recommendations published by Waterbolk (1983) – can be used to assess the suitability of a Quaternary sample for dating. A sample that is otherwise appropriate for a particular method may be considered suitable if

1) its stratigraphic association is certain;
2) it is not contaminated and has not been diagenetically altered;
3) it is autochthonous; and
4) other sources of error for the dating can be exluded.

If one or more of these conditions is not fulfilled, the sample must be considered relatively unsuitable or even unusable for dating. If a sample is classified as being not very suitable, an attempt can be made by applying different dating methods, possibly to different fractions (e.g., humic acids and residual material). If the results agree, a reliable age can be assumed to have been obtained. But there is no certainty that this will be the case.

When dates from different laboratories are compared, the following criteria must be taken into consideration:

- dating method and sample treatment;
- origin and genesis of the sample;
- basis of the calculations (e.g., the half-life values used) and the kind of corrections (e.g., $\delta^{13}C$ correction, see Sect. 6.2.1); and
- reference year and standards.

Depending on the number of criteria for which the dates are in agreement, the dates can be classified according to their comparability: noncomparable, moderately comparable, good and excellent comparability.

Although dates that are noncomparable should not be used quantitatively, they can be sometimes interpreted as minimum or maximum ages. For a quantitative evaluation of age data (Sect. 4.2), e.g., the calculation of sedimentation rates or mean values, only dates that belong to the same group should be used together. The rule must be taken strictly into account especially when setting up a data bank that is accessible to anyone, since otherwise impermissible coupling of uncomparable dates will too easily occur.

### 4.1.2   Hard-Rock Dating

Determination of the age of formation or metamorphism of a rock on the basis of the radioactive decay of natural isotope has become an important instrument of geological and petrological research in the last several decades. To be of interest for geochronological research, a radioisotope must have a half-life [Eq. (5.3)] long enough that it has not yet fully decayed, but short enough that measurable changes in the isotopic ratios are produced within the geological time periods under investigation. To determine the age of a geological sample older than the Quaternary, the half-life should be between $10^8$ and $10^{12}$ years. The age range covered by a specific method depends primarily on the half-life of the parent isotope and the analytically attainable precision and sensitivity.

All classical dating methods based on the radioactive decay of natural isotopes with long half-lives are based on the following model assumptions:

1) The decay constant $\lambda$ must be known with sufficient accuracy.
2) The mineral or rock has formed a closed geochemical system (i.e., neither parent nor daughter element is added or taken away) since time $t_0$ (i.e., the starting point of the "radiometric clock").
3) There was none of the daughter element in the mineral or rock at time $t_0$, or the isotopic composition of the daughter element initially present can be determined reliably (e.g., by the isochron method) and corrected for.

These conditions are seldom completely fulfilled. This is especially true for the closed-system requirement (assumption 2). Differences in the decay constants (assumption 1) used by the various laboratories also lead to differences in the radiometric ages determined. Since the introduction of conventional decay constants proposed by the IUGS in 1976 (see App. B), assumption 1 is no longer a serious problem for the standard radiometric dating methods, e.g., K/Ar (Sect. 6.1.1), Rb/Sr (Sect. 6.1.3), and U/Th/Pb (Sect. 6.1.9).

Determination of the age of rocks and minerals by any of several physical methods is now "routine" for experienced multi-method laboratories. Skilled specialists, however, are necessary not only for the very complicated and lengthy analytical procedures, including sample preparation, but also for sample collection and interpretation of the analytical data. The geologist interested in age determinations should be aware that the collection and selection of the samples (Chapt. 3) are decisive for the success of the determination. An age determination can only be as good as the sample used for it! This is equally true for both hard-rock and soft-rock dating (e.g., Waterbolk 1983).

Advances in the analytical methods (e.g., mass spectrometry, Sect. 5.2.3.1) have increased the precision of the age determination; but the accuracy depends mainly on the nature of the sample. Errors arising from incorrect sample collection are often considerably larger than the analytical error of the age determination. This can result in incorrect geological interpretation. Therefore, problems of sample collection and data interpretation should be briefly discussed.

The first step of any geochronological study must be the determination which geological event is to be dated and which samples or fractions of them are representative for this event so that the most suitable dating method(s) can be chosen. A radiometric age determination always dates the termination of a physical or chemical process, e.g., the crystallization of a mineral (crystallization age) or its cooling below a certain threshold temperature (cooling age). (This threshold temperature is called a blocking or closure temperature.) When different methods are applied to the same rock or mineral, quite different geological events may be dated. On the other hand, the same method can date different events when applied to different minerals or a whole-rock sample.

The hypothetical case of a rhyolitic tuff can be discussed as an example: After the tuff was formed, it was subjected to medium-grade metamorphism at temperatures up to 500°C. This was followed by rather rapid uplift where it became subject to erosion. This is illustrated in schematic form in Fig. 4.1.

Conventional K/Ar age determinations (Sect. 6.1.1.1) on the minerals muscovite, biotite, and K-feldspar extracted from the tuff date various stages of cooling after the metamorphic event. If the cooling occurred rapidly, then the age values will be concordant. If the cooling was sufficiently slow, different ages will be obtained in accordance with the different closure temperatures of the minerals. Thus, with very precise dating (e.g., using the $^{39}$Ar/$^{40}$Ar method, Sect. 6.1.1.2), the rate of cooling can be estimated. The Rb/Sr age (Sect. 6.1.3) dates another stage of cooling than the K/Ar method carried out on the same mineral because the Rb/Sr and K/Ar systems have different closure temperatures for each mineral. The fission-track method (Sect. 6.4.7) also yields cooling ages for the different minerals. Apatite fission-track ages for our hypothetical metavolcanite would give the time of the last (tectonic) uplift above an isotherm of about 120°C.

The Rb/Sr mineral isochron for several minerals yields a mean cooling age. In contrast, whole-rock isochron ages reach beyond the metamorphic event and date the time when the original tuff material was expelled.

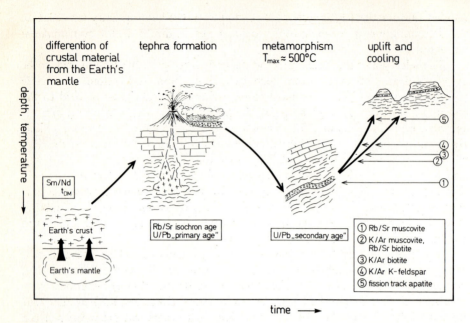

**Fig. 4.1.** Hypothetical case history of a rhyolitic tuff showing the different "ages" determined by different dating methods and materials. After the formation of crustal material by differentiation processes within the Earth's mantle, anatectic melting of this crustal matter produces a granitic magma that is expelled to form a rhyolitic tuff. Involvement of this material in orogenic processes results in metamorphic overprinting followed by either rapid or slow uplift and cooling

U/Pb discordia ages (Sect. 6.1.9) obtained from zircon usually yield two age values. In favorable cases, the so-called "primary" age corresponds to the time of the tuff eruption and the "secondary" age to the time of metamorphism. However, if the rhyolitic melt in our example was formed by anatectic melting of crustal matter, a significant proportion of radiogenic lead might have been included. This can, in connection with other factors, lead to "ages" that cannot be interpreted geologically.

The Sm/Nd age (Sect. 6.1.6) determined for our hypothetical rhyolitic tuff is a $t_{DM}$ model age (DM = depleted mantle), representing the time the source material of the rhyolitic melt (or the corresponding crustal segment) was differentiated from the Earth's mantle.

This example is highly simplified and does not include by far all of the possibilities of interpretation of the methods mentioned. However, it illustrates that each radiometric 'age' is never more than an analytically determined geochemical parameter (date) which can provide information about the time of a specific geological event only when all known geological, petrographic, and geochemical aspects are included in the interpretation. It must also be kept in mind that not all of the possible effects of the geological processes on the various dating methods can be completely understood or even recognized. For example, there are indications that the condition of isotopic homogeneity of a magmatic body at time $t_0$, prerequisite for isochron dating of magmatic rock, is not always

fulfilled. But for the Rb/Sr system, for example, initial heterogeneity woul[
the determination of a whole-rock isochron age in doubt, if not even [
impossible.

The selection of the dating method depends largely on the type of rock to be
dated, its degree of preservation, chemical composition, and content of datable
minerals, as well as its age. In most cases, serious attention must be paid to the
collection of especially fresh material. Rb/Sr and K/Ar whole-rock and mineral
age determinations require, for example, the best preserved material possible.
However, if biotite is slightly altered, it may still be suitable for the K/Ar method,
but not for the $^{39}$Ar/$^{40}$Ar method. Zircon for the U/Pb method can, if necessary,
also be taken from slightly weathered material. The Sm/Nd system is extremely
resistent to secondary alteration of the samples so that extremely fresh material is
not absolutely necessary. Due to the large number of requirements, the suitability
of each sample must be clarified by routine geological, petrographic, and
geochemical analysis before the isotopic analysis.

In addition, it must be certain that no younger igneous rock or hydrothermal
veins occur near the sampling site that could have caused allochemical overprinting,
thus putting the presence of a closed system in doubt. Younger thermal overprinting can reset or partially reset the radiometric clock, resulting in ages that
are too young. It must be kept in mind that such events can occur beyond the
immediate site, as systematic studies in the contact area of intrusive bodies have
established (e.g., Hart 1964). As the various dating systems react quite differently
to this kind of overprinting, they might be used to reconstruct a geological history.
Therefore, the person making the study must know the geological situation in the
study area very well. This is a prerequisite for successful dating.

## 4.1.3  Isotope Geochemistry

In addition to the determination of the ages of rocks and minerals, isotope
geochemistry studies utilizing natural radionuclides with long half-lives makes it
possible to reconstruct the chemical differentiation of the Earth's crust and mantle
during its history. The isotope geochemistry of these nuclides differs considerably
from the isotope geochemistry of the stable isotopes; the latter provides a view of
the fractionation processes of geological systems. The information provided by the
former is based on the fact that the chemical fractionation between parent and
daughter nuclides that occurred in the past have left time-dependent differences
in the isotopic composition of the daughter element. Thus, the isotope ratios of
the daughter elements can provide information on such fractionation processes
and permit a reconstruction of their occurrence during the history of the Earth.

A comprehensive presentation of isotope geochemistry cannot be the purpose
of this book. For this the reader is referred, for example, to Jäger and Hunziker
(1979), Ivanovich and Harmon (1982), Fritz and Fontes (1980, 1986). But geochronology and isotope geochemistry are very closely related fields; some methods are
not applied mainly for the dating of rocks but for the information provided about
the geochemical environment in which the rock was formed (e.g., the common-lead

method, Sect. 6.1.10). Such applications are mentioned in the last section of each chapter as "Non-Chronological Applications".

## 4.2  Mathematical Evaluation of Physical and Chemical Age Data

If a random event (for example, replicate dating of a sample) is repeated a large number of times, the results $t_i$ range about a mean value $\bar{t}$. The frequency of a specific value decreases with increasing distance from $\bar{t}$ (bell-shaped curve, normal or Gaussian distribution; Fig. 4.2). $\bar{t}$ is an estimate of $t^\varepsilon$, the expected value of the distribution. A measure of the data is given by the standard deviation $\sigma_t$, often incorrectly called the error. For a normal (Gaussian or bell-shaped) distribution, a measure of the range of the data is given by

$$t \pm k\sigma_t.$$

Any value taken at random will be within this range with a probability $P_i$. The value of $k$ is determined by the percentage coverage required. Analytical results

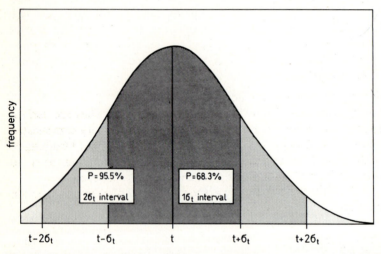

**Fig. 4.2.** Frequency distribution of random events (Gaussian or normal distribution, bell-shaped curve) showing the boundaries of the $1\sigma_t$ and $2\sigma_t$ confidence intervals with the probabilities $P$ for these intervals

**Table 4.1.** Width of the confidence interval as a function of the confidence coefficient $k$ (in %)

| Confidence interval | Example (years) | * $k$ | ** $P_i$ | Probability | Use |
|---|---|---|---|---|---|
| $t \pm 1\sigma_t$ | $2000 \pm 100$ | 68.3 | 31.7 | Probable | Usual expression |
| $t \pm 2\sigma_t$ | $2000 \pm 200$ | 95.5 | 4.5 | Very probable | For comparisons |
| $t \pm 3\sigma_t$ | $2000 \pm 300$ | 99.7 | 0.3 | Extremely probable | Rigorous comparisons |
| $t \pm 4\sigma_t$ | $2000 \pm 400$ | 99.9 | 0.1 | Very nearly certain | None |

*Confidence coefficient (in %); **probability of error (in %).

are generally given with a one-sigma interval ($k = 1$, $P_i = 68.3\%$) (Fig. 4.2). If the results of two or more laboratories or related samples are to be compared, a two-sigma interval ($k = 2$, $P_i = 95.5\%$) should be used. In a rigorous comparison, 3 of 1000 values are permitted to lie outside the interval ($k = 3$) (see Table 4.1). A four-sigma interval has no practical importance (see for example, Tsoulfanidis 1983).

### 4.2.1   Rules for Simple Calculations with the Dating Results; Statistical Tests

The results of physical and chemical age determinations consist of two dependent terms, the age $t$ and the standard deviation $\sigma_t$ (e.g., $2000 \pm 50$ a), which yield the confidence interval ($t - \sigma_t$ to $t + \sigma_t = 1950$ to $2050$ a) in which the true, but unknown value $t^e$ is to be expected with a probability $P$. Any calculation involving $t$ and $\sigma_t$ (addition, substraction, multiplication, division) must follow certain rules (e.g., Tsoulfanidis 1983). For example:

If

$$t = t_1 + t_2 \qquad \text{and} \qquad \sigma t_1 = \sigma t_2 = \sigma_t,$$

then

$$\sigma t = \sqrt{\sigma_t^2 + \sigma_t^2} = \sqrt{2\sigma_t^2}$$

thus, the usual rules of arithmetic are not being followed.

For a few simple cases, the following formulas can be used. They are derived from the error propagation law, which yields solutions even for very complicated cases. The application of this law, however, usually requires advanced training in error theory and should, therefore, be undertaken in cooperation with an experienced specialist from the cooperating geochronological laboratory.

According to the error propagation law, the standard deviation of a function $f(x_i)$ with independent variables is given by

$$\sigma_{f(x_i)}^2 = \sum_{i=1}^{n} \left( \frac{\partial f}{\partial x_i} \right)^2 \sigma_{x_i}^2 \tag{4.1}$$

#### Addition and Subtraction

Addition and subtraction of dates $t_i \pm \sigma t_i$ are carried out as follows:

$$t^* \pm \sigma_t^* = t_1 + t_2 - t_3 \pm \sqrt{\sigma_1^2 + \sigma_2^2 + \sigma_3^2} \tag{4.2}$$

Example:

$$t_1 = 2000 \pm 100$$
$$t_2 = 1000 \pm 50$$
$$t_1 - t_2 = 1000 \pm \sqrt{100^2 + 50^2} = 1000 \pm 110$$
$$t_1 + t_2 = 3000 \pm \sqrt{100^2 + 50^2} = 3000 \pm 110$$

## Comparison of Dates

The question whether two dates $t_1$ and $t_2$ may be viewed as equal is answered by a comparison of the absolute difference $\Delta$ between the two and their standard deviations. If $|\Delta t| \leqslant k\sigma_{\Delta t}$, then the two dates may be considered statistically the same.

## Multiplication and Division

Mixed products and quotients can be calculated as follows:

$$t^* \pm \sigma_t^* = \frac{t_1 \cdot t_2}{t_3} \pm t^* \sqrt{\frac{\sigma_{t_1}^2}{t_1^2} + \frac{\sigma_{t_2}^2}{t_2^2} + \frac{\sigma_{t_3}^2}{t_3^2}} \tag{4.3}$$

Multiplication with a constant K is a special case of Eq. (4.3):

$$t^* \pm \sigma_t^* = K \cdot t \pm k \cdot \sigma_t \tag{4.4}$$

## Weighted Mean

The weighted mean (and its standard deviation) of several dates with different standard deviations is calculated as follows:

$$\bar{t} \pm \sigma_{\bar{t}} = \frac{\sum\limits_{i=1}^{n} t_i w_i}{\sum\limits_{i=1}^{n} w_i} \pm \sqrt{\frac{1}{\sum\limits_{i=1}^{n} w_i}} \tag{4.5}$$

where the weight $w_i = 1/\sigma_{t_i}^2$.

Example:

The weighted mean of $1000 \pm 100$ and $1200 \pm 50$ is $1160 \pm 45$ (not $1150 \pm 150$!), the difference is $200 \pm 112$ [see Eq. (4.2)].

The standard deviation of a single date normalized with respect to the mean standard deviation is given by $\bar{\sigma}_t \sqrt{n}$, where $n$ is the number of dates used for the mean (sometimes called the standard error). The calculation of the mean and its standard deviation is valid only when it can be shown they have the same true value $t^e$ and a common standard deviation $\sigma_t$. A suitable test for this is the chi-square test:

## Chi-square test

A mean may be calculated for a series of dates only if they all belong to the same normal distribution. The $\chi^2$ test is used to determine whether this condition is fulfilled. If the calculated $\chi^2$ value [Eq. (4.6)] for a given degree of freedom (number of values minus 1) is smaller than the $\chi^2$ value given in the tables (e.g., Handbook of

**Table 4.2.** Values of $\chi^2$ for a 1% probability of error and various degrees of freedom $(n-1)$

| $n$ | 2 | 3 | 4 | 5 | 6 | 7 | 8 | 9 | 10 | 20 | 30 |
|-----|---|---|---|---|---|---|---|---|----|----|----|
| $\chi^2$ | 6.6 | 9.2 | 11.3 | 13.3 | 15.1 | 16.8 | 18.5 | 20.1 | 21.2 | 38.9 | 52.1 |

Chemistry and Physics), then we cannot reject the hypothesis that they belong to the same normal distribution.

$$\chi^2 = \sum_{i=1}^{n} \frac{(\bar{t} - t_i)^2}{\sigma_{t_i}^2} \tag{4.6}$$

Using a one percent probability of error which is sufficient for comparisons (Table 4.1), the assumption that the dates belong to the same frequency distribution will be false in only one of 100 cases if the calculated $\chi^2$ value is equal to or smaller than the listed value (Table 4.2). Outlaying values that do not meet the $\chi^2$ test must be considered individually before rejection (Tsoulfanidis 1983). The mean can be calculated only after outliers have been removed.

### 4.2.2   Comparison of Age Values

When several dates are available for a particular problem being studied, the following cases should be distinguished:

- dates obtained with the same method in the same laboratory;
- dates obtained with the same method in the same laboratory but at quite different times (i.e., years apart);
- dates obtained with the same method in different laboratorties; and
- dates obtained with different methods.

Considering these four cases, it might be expected that two dates for the same sample determined using the same method in the same laboratory would be the most likely to agree within their standard deviations. Unfortunately, the results of international comparison tests (e.g., International Study Group 1982; Wehmiller 1984b; Hennig et al. 1985) have shown that this expectation is not always fulfilled.

This comparison of $^{14}$C dates, for example, shows that the published standard deviations are too small by as much as a factor of 2. As a consequence, it is recommended that the standard deviation be multiplied by a laboratory-specific error multiplier before any further statistical treatment of the data (International Study Group 1982; Stuiver and Kra 1986). However, this procedure does not increase in the confidence interval as, for example, is done by increasing the value of $k$. The basis for the laboratory-specific error multiplier may change with time.

Suitable statistical tests (e.g., the chi-square test, Sect. 4.2.1) should be used if possible to determine whether the dates that are available for a particular problem belong to the same frequency distribution. Sometimes it can be helpful to construct a histogram of the dates (Sect. 4.2.3).

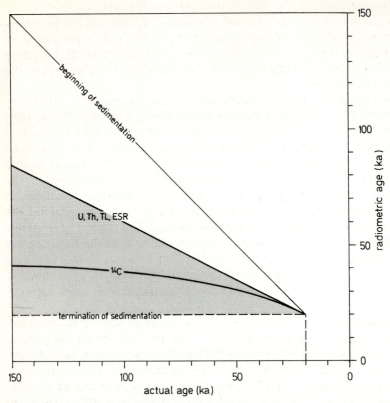

**Fig. 4.3.** Hypothetical case: Comparison of conventional $^{14}$C and U/Th ages of coeval samples that slowly formed over the time-span between 150,000 and 20,000 years ago

The most difficult dates to compare are those obtained using different geochronological methods. The least success is with dates that lie at the limits of a method (Foldout Table). This is illustrated by the following example (Geyh 1983): Samples that consist of mixtures of material of differing ages or material formed relatively slowly may represent an age span of tens of thousands of years or even hundreds of thousands of years. The $^{14}$C (Sect. 6.2.1) and U/Th (Sect. 6.3) ages of such samples differ greatly (Fig. 4.3). The reason for this is that the $^{14}$C concentration decreases exponentially during aging and the conventional age calculated from the average $^{14}$C concentration is always smaller than the actual age (Geyh et al. 1971a). In contrast, the $^{230}$Th concentration increases quasi-linearly in the dating range of the $^{14}$C method so that the average age is approximately equal to the U/Th age.

### 4.2.3  Numerical and Graphical Evaluation of Age Values

When a large number of dates covering a relatively long period of time is available for a specific problem from different, often unknown sources, the question arises whether time boundaries can be derived from this data. A prerequisite for this is that the samples must have been collected at random, i.e., collected independently by several researchers under different aspects and possibly at different sites. If all of the the samples were taken for a specific purpose, e.g., to determine the beginning and/or end of an event, the frequency distribution will show a double peak owing to the manner of selection.

Data for a specific problem can be evaluated either numerically by calculating a weighted mean [Eq. (4.5); e.g., Rhodes et al. 1980] or graphically (Geyh 1980a; Geyh and de Maret 1982). Usually, it is recommendable to use both methods.

For a numerical evaluation, a weighted mean is calculated for dates that all lie within a narrow range or for samples that have certain characteristics. Statistical tests (Sect. 4.2.1) are to be applied to determine whether the dates belong to the same frequency distribution, a prerequisite for such an evaluation.

Graphical representation of the dates in the form of a histogram can make it easier to recognize the information content of the data. If a number of the dates are incorrect, a numerical evaluation may yield misleading results, but a histogram will at least show the trends in the data and may help pinpoint erroneous results (Geyh and Rhode 1972). When this is done, however, the standard deviations must be multiplied by a factor $< 1$ to compensate for the resulting artificial widening of the histogram peaks.

The class interval of the histogram should be chosen as short as possible, but not shorter than the average confidence interval of the dates so that the greatest possible resolution with respect to time is obtained. On the other hand, the class intervals should be wide enough to ensure that the number of dates assigned to each class is large enough to keep the statistical uncertainty as low as possible (rarely wider than twice the average confidence interval) (Geyh 1980a). The scale of the $y$-axis depends on the manner of representation of the histogram.

The simplest way is to assign each date to a class and represent the classes by rectangles with areas proportional to the number of dates assigned to them (Fig. 4.4). The width of the base of each rectangle can, for example, be the average confidence interval. The line connecting the tops of these superposed rectangles represents the frequency distribution. The $y$-axis gives the number of dates per class. This procedure is recommended for large numbers of dates (more than five per class), but not for a limited number of dates of differing precision. A refinement is the representation in a dispersion diagram (Ottaway 1973). As the standard deviations are not taken into account by this procedure, it can be used only if they are about the same size or if the number of dates is large.

A more complicated way is to assign each date and its standard deviation to a polygon or bell-shaped curve with a given area and superimpose them at the position of their class (Fig. 4.4). Dates with large standard deviations give relatively short, broad polygons that may stretch over many classes; precise dates yield tall, narrow peaks. The advantage of this procedure, which can be used for a small number of dates, is that the precision of the dates is taken into consideration. A disadvantage is that peaks may be obtained which have little or no meaning because they are formed by a few, very precise dates. In this case, the scale of the $y$-axis is arbitrary and does not correspond to the class frequency. The histograms

a)

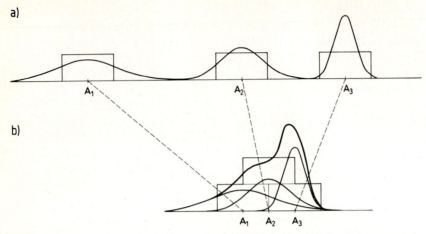

b)

**Fig. 4.4.** Two types of representation of histograms for three dates of differing precision: **a** All of the dates are represented by *rectangles* of the same size without consideration of the confidence intervals. **b** All of the dates are represented by *bell-shaped curves* whose widths are determined by the respective standard deviation, but the areas below the curves are all the same size. When these areas are superimposed the enveloping curve represents the histogram for the three dates. Both representations yield the same curve when a large number of dates is used

obtained with this method correspond fully to those of the simple class frequency presentation if the number of dates is large enough.

The following criteria must be taken into consideration when interpreting a histogram:

1) If the time scale of the dating method deviates systematically from calendar years, as is the case, for example, with the $^{14}$C method (Fig. 2.1 and Sect. 6.2.1), the histogram is highly distorted and the mean value of the uncorrected dates cannot be converted to an absolute age. If a time period of 100 calendar years, for example, corresponds to thirty $^{14}$C years (or vice versa), the peaks of the histogram (Fig. 4.5) are shifted up (or down). Histograms for which the units of the x-axis are calendar years do not show this error, but cannot be made unambiguously from $^{14}$C dates (Geyh 1980a).

2) There are phenomena in the history of the Earth, e.g., eustatic sea-level fluctuations (Geyh 1980a), for which $^{14}$C dates correlate with lithostratigraphic units. But if the x-axis is converted to calendar years, the correlation disappears. In such cases, it may be assumed that the cause of the phenomenon being studied is extraterrestrial, like that responsible for the medium-term changes in $^{14}$C production. In our example, solar activity may have caused both the sea-level changes and the variations in radiocarbon production (Sect. 6.2.1).

3) The number of dates in a histogram is finite and the number of individual peaks is almost always small. Thus, the amplitude of the peaks has an uncertainty that is given roughly by the square root of the number of dates in the class. For example, a peak formed by ten dates has an uncertainty of $\pm 33\%$! Apparent peaks and minima are therefore not uncommon (Geyh and de Maret 1982).

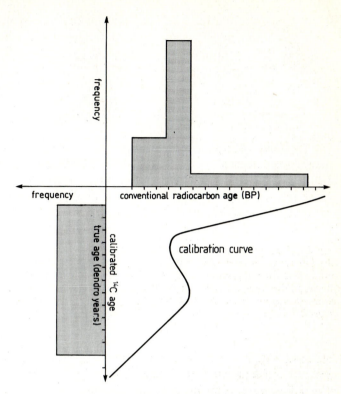

**Fig. 4.5.** Example of the conversion of a histogram for the distribution of dates on the conventional $^{14}$C time-scale to one for the calibrated $^{14}$C time-scale. It can be seen that the uniform distribution of dates along the calibrated $^{14}$C time-scale corresponds to a distinctly distorted distribution of dates along the conventional $^{14}$C time-scale and vice versa

4) Only the existence of a peak, not its amplitude, is significant. Time spans that contain no $^{14}$C dates, however, may be interpreted with more certainty, e.g., as intervals in which no datable material was formed (e.g., wood in glacial or dry periods) or it was eroded away or decomposed (Geyh and Rohde 1972). Thus, only one question can be unambiguously answered with a histogram: In which periods of time are no dates to be expected? (see Fig. 6.56)
5) Additional information can be included in a histogram if the areas for dates with characteristics in common are marked by color or in some other way (Waterbolk 1983).

## 4.3    Publication of the Age Values

When age data are published, it is important that the following accessory information by included:

– age and standard deviation;
– for isochron dates: initial isotope ratios and standard deviations;

- dating method and the constants used for calculating the age, e.g., half-life or decay constant, initial activity or initial isotope ratios, atomic ratios; [if the ages are calculated according to international convention, e.g., IUGS (Steiger and Jäger 1977), it is sufficient to only mention the fact];
- any corrections that have been made for systematic error;
- address or international code of the laboratory;
- the sample number assigned by the laboratory.

   To avoid publication of improperly interpreted data, large amounts of data should be published together with an experienced geochronologist, possibly a staff member of the laboratory that carried out the analysis. If this is not possible, the geochronological sections of any manuscript should at least be read through and corrected by a specialist (App. C).

# 5 Physical Dating Methods

## 5.1 Principles

Most physical dating methods are based on processes that are strictly a function of time or that contain a prominent time-dependent component. This is especially true for the radiometric methods. Even with these methods, however, derivations from the siderial time-scale sometimes occur as a result of processes not taken into consideration by the assumptions of the model (Fig. 2.1). Such processes can be geochemical or geophysical, e.g., diagenetic mobilization of parent or daughter nuclides in a mineral or rock system, the occurrence or annealing of radiation damage, isotope fractionation, or long-term fluctuations in the production of cosmogenic radionuclides.

Other physical dating methods are based on chronologies based on time markers present globally (Chap. 7), e.g., temperature-controlled changes in the isotope ratios of oxygen in ice or pelagic sediments (Sect. 7.2), and secular variations in the dipole moment of the Earth's magnetic field (Sect. 7.1) preserved, for instance, in magmatic rocks.

The radiometric methods of age determination are among the most important of the absolute dating methods (e.g., Roth et al. 1985, 1989; Faure 1986). The basic principles of these methods are discussed in this chapter.

It has been known since 1907 that the atoms of a chemical element can have different masses. Atoms of the same element with different masses are called isotopes. For example, carbon has three natural isotopes with mass numbers 12, 13, and 14. Carbon-12 (also written $^{12}C$) and carbon-13 (or $^{13}C$) are stable, i.e., they do not change with time. Carbon-14 ($^{14}C$) is radioactive; such isotopes are called radionuclides, which decay at an isotope-specific rate (expressed by a decay constant $\lambda$) to other isotopes or to a lower energy level. For example, radiocarbon decays to a stable isotope of nitrogen.

Particle or electromagnetic radiation is emitted during the radioactive decay of radionuclides. The decay of each radionuclide is characterized by the kind(s) of radiation they emit, the energy(ies), and the half-life [Eq. (5.3)]. The decay of a nucleus in more than one way is called dual or branched decay. The branching coefficient $\xi$ is defined as the ratio of the frequencies in which the parent nuclide decays into two or more daughter nuclides (Fig. 6.2). Partial decay constants are obtained from the half-life and the branching ratio.

There are four main modes of radioactive decay (Fig. 5.1):

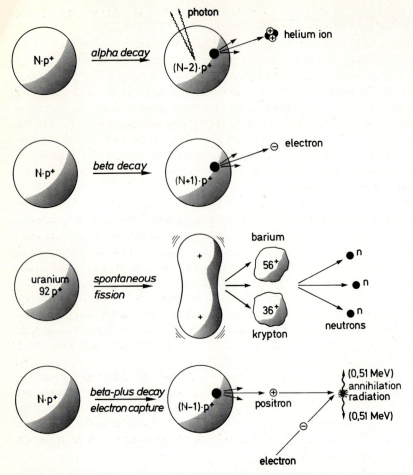

**Fig. 5.1.** Schematic representation of alpha- and beta-decay, spontaneous fission, and $\beta^+$-decay or electron capture; the disintegrating atoms are assumed to have N protons. In the case of $\alpha$-decay, the disintegration product has N-2 protons; one helium atom and one photon is emitted. In $\beta^-$-decay, a neutron disintegrates into a proton and an electron, which is ejected. Only the heaviest elements undergo spontaneous fission (see glossary). In the case of K-capture, an orbital electron is captured by the nucleus, where it combines with a neutron with the emission of X-rays

*Alpha-decay:* emission of helium-4 nuclei ($^4$He). Owing to their size and charge, alpha-particles ionize other atoms when passing through matter, rapidly losing energy in the process. The range of $\alpha$-radiation is therefore short and is only a few micrometers in solid matter. Only elements with a relatively high atomic number decay by emission of $\alpha$-particles, which is almost always accompanied by gamma-radiation, i.e., short-wave photons which have a practically infinite range. The emission of an $\alpha$-particle results in a product nuclide with an atomic number two less and a mass number four less than the decaying nuclide. Examples of $\alpha$-decay are found in the natural radioactive decay series (Sect. 6.3) beginning with

$^{238}$U, $^{235}$U, and $^{232}$Th (Fig. 6.64), e.g. $^{226}$Ra to $^{222}$Rn; another example is decay of $^{147}$Sm to $^{143}$Nd (Sect. 6.1.6).

*Beta-decay:* conversion of a neutron to a proton with the emission of an electron (e$^-$). These electrons have a continuous energy distribution with a maximum ($E_{max}$) that is characteristic for each isotope (App. B). Since electrons have a mass that is about 1/7300 that of an α-particle, they interact with the matter they pass through less than α-particles do and their range is accordingly greater. The emission of a $\beta^-$-particle results in an increase in atomic number by one, the mass number remains constant. An example of $\beta^-$-decay is provided by radiocarbon (Sect. 6.2.1), which decays ($E_{max} = 156$ keV) to the stable nuclide $^{14}$N. The Rb/Sr (Sect. 6.1.3), Lu/Hf (Sect. 6.1.7), and Re/Os (Sect. 6.1.8) dating methods are also based on this mode of decay.

*$\beta^+$-Decay and orbital-electron capture (EC):* $\beta^+$-decay and orbital-electron capture are similar in their effects. In both processes, the atomic number decreases by one and the mass number remains constant. In orbital-electron capture (also called K-electron capture or simply K-capture), an electron (usually from the K shell, from which the usual name is derived) is captured by the nucleus with the formation of a neutron from a proton. In $\beta^+$-decay, a proton is converted into a neutron with the emission of a positron (e$^+$), which soon combines with an electron with the release of two γ-photons (0.51 MeV) radiated in diametrically opposed directions (annihilation radiation). The decay of the $^{40}$K nucleus is a good example. Potassium-40 decays to $^{40}$Ar by K-capture and to $^{40}$Ca by emission of a $\beta^-$-particle (Fig. 6.2), an example of dual decay.

*Spontaneous nuclear fission:* the spontaneous disintegration of a heavy nucleus, e.g., $^{238}$U, into fragments of about the same size. The half-lives for spontaneous fission are on the order of $10^{15}$ a. Owing to their size and velocity, the recoiling fragments cause radiation damage (fission tracks) in solid matter in the immediate neighborhood of formation, which after etching of the sample can be observed under the microscope (Sect. 6.4.7).

The decay processes are not influenced by factors such as temperature, pressure, or chemical environment. The isotope-specific decay constant ($\lambda$) is therefore defined as the statistical probability that a specific nucleus will decay within a specific unit of time. The number of nuclei that decay within a time unit $dN/dt'$ (called the activity) is thus proportional to the number of radioactive nuclei $N$ that are present:

$$dN/dt' = -\lambda N \tag{5.1}$$

Integration yields the law of radioactive decay:

$$\ln(N/N_0) = -\lambda t; \qquad N = N_0 e^{-\lambda t}; \qquad t = -\lambda \ln N/N_0 \tag{5.2}$$

where $N_0$ is the number of parent nuclides at time zero (the time the system commenced, i.e., the start of the "radioactive clock") and $N$ is the number of parent nuclides remaining at time $t$ (Fig. 5.2).

In geochronology, the easy-to-understand parameter half-life $\tau$ is used in preference to the decay constant $\lambda$. The half-life is the time in which half of the originally present nuclei decay. Setting $t' = \tau$ and $N = N_0/2$ in the logarithmic form

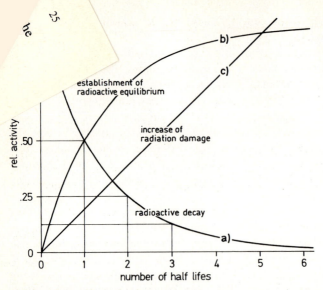

establishment of
radioactive equilibrium

increase of
radiation damage

radioactive decay

rel. activity

.50

.25

0

b)

c)

a)

0    1    2    3    4    5    6
number of half lifes

**Fig. 5.2.** The three main possibilities for radiometric dating: **a** the decrease in isotope concentration resulting from radioactive decay of the parent isotope as a function of time (the radioactive activity decreases by a factor of two with each half-life; **b** the increase in the concentration of the radiogenic daughter nuclide; the approach of saturation conditions in a closed system when the radionuclide with a decay constant $\lambda$ is reproduced at a constant rate P; and **c** the rather linear increase in radiation damage due to the decay of long-lived isotopes

of Eq. (5.2) yields

$$\tau = \ln 2/\lambda = 0.693/\lambda \tag{5.3}$$

Radioactive decay is excellently suited for dating methods because it is not influenced by temperature, pressure, or chemical environment. The individual radiometric dating methods utilize the characteristics of various radionuclides (App. B). The methods based on the ratio of atoms of radioactive parent isotope to stable daughter isotope assume that at the time of the formation of the rock, mineral or meteorite, the parent and daughter nuclides were either completely separated from one another or the amount of daughter isotope originally present can be estimated. This condition is fulfilled in nature only in exceptional cases. The time and manner in which the separation (fractionation) occurred determine the information provided by the method, e.g., the time of solidification, mineralization (or crystallization), or gas retention. After its formation, the object being studied must have formed a closed system, i.e., no transport of isotopes between the system and its surroundings (e.g., by diffusion, chemical exchange, weathering, evaporation, condensation, and contamination). This condition is also not always fulfilled; many rocks have been proven to represent an open system.

The following assumptions are made for the derivation of the basic equations used for the age calculations: If a certain number $N_0$ of atoms of the parent nuclide were present at the time the radioactive clock was started ($t = 0$, e.g., the formation of the rock or mineral) and if $N_d^*$ atoms (the * is used to indicate radiogenic products) of the daughter nuclides are formed during time $t$ and $N_p$ parent nuclides remain, then Eq. (5.4) follows from Eq. (5.2):

$$N_d^* = N_p(e^{\lambda t} - 1) \tag{5.4}$$

where $N_p = N$ and $N_d^* = N_0 - N_p$. Thus,

$$t = \frac{1}{\lambda} \ln\left(\frac{N_d^*}{N_p} + 1\right) \tag{5.5}$$

To obtain a radiometric age, therefore, the present atom ratio of the parent and daughter nuclides ($N_p$ and $N_d^*$, respectively) must be determined. Equation (5.5) must be expanded by a term if both parent and daughter nuclides are present at time zero:

$$t = \frac{1}{\lambda} \ln\left(\frac{N_d - N_{d_0}}{N_p} + 1\right) \tag{5.6}$$

where $N_d = N_{d_0} + N_d^*$.

The larger the initial concentration $N_{d_0}$ of the daughter nuclide and the lower the age of the sample, the more inaccurate the age that is calculated, because under these conditions the proportion of radiogenic daughter nuclide is rather small. $N_d^*$ and $N_d$ are determined chemically, radiometrically, or mass spectrometrically (Sect. 5.2), depending on the kind and abundance of the nuclides to be analyzed. $N_{d_0}$ must be determined or estimated separately or it must be eliminated mathematically (for example, by the isochron method, see Nicolaysen 1961).

The use of cosmogenic radionuclides for age determinations (Sect. 6.2) is based on the assumption that these are produced by cosmic radiation at roughly a constant rate $P$ (atoms/time unit) and that their specific activities in the various georeservoirs are, thus, also constant. Adding the term $P$ to Eq. (5.1), the following equation is obtained after integration:

$$N = \frac{P}{\lambda}(1 - e^{-\lambda t'}) \tag{5.7}$$

When $t'$ approaches infinity (Fig. 5.2), the system attains steady-state equilibrium, a basic assumption for radiometric age determinations. Under these conditions, $N$ remains constant:

$$N_\infty = P/\lambda \tag{5.8}$$

Because the chronometer begins at a time when the steady-state has already been reached, the value of $N_\infty$ is set to $N_0$. $N_0$ can differ in the various georeservoirs of the earth. For example, the atmosphere, the biosphere, and the hydrosphere are reservoirs for radiocarbon. In addition to production rate and half-life, $N_\infty$ is controlled by the rate of exchange of the nuclide between the various georeservoirs (see, for example, Oeschger et al. 1975).

When Eq. (5.2) is used to calculate an age, $t$, the specific activity of the standard corresponding to an age of zero is taken as the initial activity $N_0$. The absolute activity in the object being dated and that in the standard need not be measured since only their ratio occurs in the equation. However, owing to the very low activities of cosmogenic isotopes in natural matter, special low-level techniques must be used for their determination (e.g., Oeschger and Wahlen 1975).

One group of radiometric dating methods is based on disturbances in the radioactive equilibrium between members of the natural radioactive decay series (Sect. 6.3). The isotopes $^{238}U$, $^{235}U$, and $^{232}Th$ decay via a number of radioactive descendent nuclides to stable lead isotopes (Fig. 6.64). The half-lives of the various members of the series differ greatly. In rocks or minerals forming a closed system, radioactive equilibrium is attained in which the activities of all members of the series are the same and the isotopic abundances are proportional to the half-lives.

Since the members of the decay series are different chemical elements, geochemical processes can easily disturb the radioactive equilibrium, for instance, the leaching of uranium or radium and the loss of radon from rocks by diffusion. Fractionation of members of the radioactive decay series also occurs in aqueous environments, for example, owing to the lower solubility of thorium than that of uranium. This leads, in this case, to a relative enrichment of thorium in pelagic sediments. The excess of the short-lived (geologically speaking) $^{230}Th$ disappears via radioactive decay within a geologically relatively short time and a new radioactive equilibrium is approached. The extent to which equilibrium has been reattained can be used as a measure of the time since the formation of the sediment on rock (commonly termed age).

Age determinations based on radiation damage (Sect. 6.4) depend on changes in matter, e.g., alterations in the lattice spacing, caused by the interaction of the matter with radiation passing through it. If the radiation damage increases at a rather constant rate (Fig. 5.2), the number of damages can be used as a measure of the time (radiation age) the sample was subjected to the radiation. The most important parameters for these methods are the dose rate of the natural source of radiation, the sample-specific generation of radiation damage, and the density of the damages.

## 5.2  Sample Treatment and Measurement Techniques

Isotope abundances or ratios are measured for most of the physical dating methods utilizing isotopes. Two principal methods are used:

– decay counting of radionuclides and
– atom counting of stable or long-lived radioisotopes.

The traditional methods of measuring radioactivity (decay counting techniques) use various detectors, such as proportional counters, scintillation counters, and germanium and silicon solid-state counters. The detector is installed in a more or less complicated shielding of lead or iron designed to shield it from environmental (background) radiation. The electronics of the instrument is designed to optimize the sensitivity of the detector to a specific type of radiation and a specific energy range (Oeschger and Wahlen 1975; Tsoulfanidis 1983). When absolute measurements are to be made rather than relative ones, more sophisticated devices are needed (Hoppes 1984).

The atom-counting methods cover a significantly larger range of isotope abundances. Mass spectrometers are used for measuring atom ratios down to $10^{-5}$, which is suitable for many of the stable isotopes and long-lived natural radionuclides. Accelerator mass spectrometers (AMS) are used to measure radionuclides with very long half-lives and atom ratios down to $10^{-17}$ (Wölfli et al. 1984; Gove et al. 1987). Resonance ionization spectrometry (RIS), with its ability to detect individual atoms, has the greatest sensitivity (Lehmann et al. 1985).

Elemental analyses often need to be made in addition to isotope measurements. AA, RIS, NAA, ICP, and XRF are examples of the analytical methods used for this purpose (Sect. 5.2.4). The reader is referred to Tite (1981) for a comprehensive description of these techniques.

### 5.2.1 Sample Treatment

#### 5.2.1.1 Hard-Rock Samples

After the field work, the second most important step in the dating of rocks is sample preparation (separation of minerals and/or whole-rock preparation). An age determination can only be as good as the mineral fraction on which it is carried out: Even minute amounts of impurities increase the danger of error in the date obtained. The mineral fractions should have a purity of at least 95%, if possible more than 99%. Sample preparation is best done in a geochronology laboratory or at least supervised by a specialist, since it requires some experience to avoid contamination and memory effects.

A generalized scheme for the separation of the usual minerals is shown in Fig. 5.3. Weathered parts are carefully removed from the fresh rock (1–50 kg), which is then broken up into pieces with a size of 5–10 mm. An aliquot is held in reserve for a whole-rock analysis. The sample is further ground to a grain size of less than about 500 μm and sieved into various grain-size fractions. To avoid contamination, it is best to use a new plastic screen for each sample for grain sizes under 150 μm. The question whether to screen wet or dry must be decided from case to case. Wet sieving separates the grain sizes more effectively but can lead to contamination.

The individual grain-size fractions are separated further on a shaking table, with high-density liquids and/or a magnetic separator. The initial enrichment of the heavy minerals is best done on a wet shaking table; when the accessory minerals are desired, this is almost always necessary. The separation according to density (heavy liquids) and according to magnetic susceptibility (magnetic separator) is shown schematically in Fig. 5.4 and Table 5.1.

Dry shaking tables are very effective for the initial enrichment of phyllosilicates (e.g., mica). Special separation problems can be solved utilizing differences in the dielectrical behavior of the minerals. Mica fractions must, as a final step, be ground in alcohol to remove mineral inclusions (e.g., apatite). All mineral fractions should finally be washed in an ultrasonic bath using double-distilled water. A purity of >99% can be attained, almost always, only by handpicking.

**Fig. 5.3.** Scheme for the extraction of the common minerals from hard-rock samples

**Table 5.1.** Mineral separation using a magnetic divider

| Ferromagnetic | Paramagnetic | | | | |
|---|---|---|---|---|---|
| Current<br>Inclination<br>Property | 0.4 A<br>$S = 20°$<br>Magnetic | 0.8 A<br>$S = 20°$<br>Magnetic | 1.2 A<br>$S = 20°$<br>Magnetic | 1.2 A<br>$S = 5°$<br>Magnetic | 1.2 A<br>$S = 5°$<br>Non-magnetic |
| Magnetite<br>Pyrrhotite | Olivine<br>Garnet<br>Allanite<br>Perovskite<br>Hematite<br>Cordierite | Biotite<br>Chlorite<br>Glaucophane<br>Hornblende<br>Allanite<br>Staurolite<br>Augite | Muscovite<br>Diopside<br>Staurolite | Apatite<br>Leucoxene<br>Sphene<br>Monazite<br>Xenotime | Dolomite<br>Fluorite<br>Apatite<br>Kyanite<br>Barite<br>Zircon<br>Pyrite<br>Feldspar<br>Quartz |

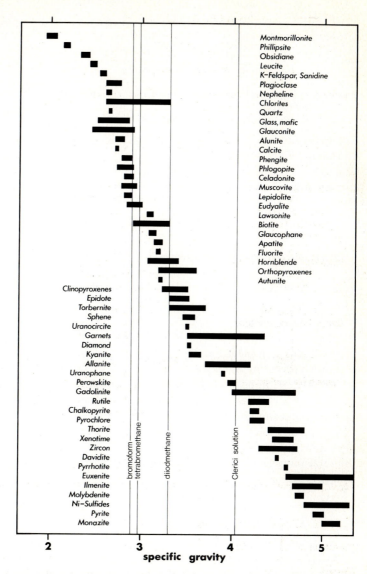

**Fig. 5.4.** Specific gravities of selected minerals used for geochronologic purposes and some frequently used heavy liquids. Mineral separation is done with heavy liquids on the basis of specific gravity

For the preparation the whole-rock samples, the aliquot mentioned above is divided and a split (50–100 g) is ground in an agate mortar. The pulverized rock is divided again until the aliquots have the weight necessary for each analysis.

### 5.2.1.2  Soft-Rock Samples

The variety of samples to be dated by soft-rock dating methods (Sects. 6.2, 6.3, 6.4 and Chaps. 7 and 8) is so great that the description of the pretreatment techniques is restricted to the individual chapters. In the case of groundwater sampling, special care must be taken to avoid or at least recognize mixing and contamination during pumping for sample collection (e.g., Egboka 1985).

### 5.2.2  Radioactivity Measurements: Decay Counting Methods

Radioactivity measurements are based on the effects of the interaction of radiation with matter. The most important effects for this purpose are ionization and scintillation. In the first case, electrically neutral materials become conducting; in the latter case, a substance emits fluorescent light.

The activity is measured by counting the number of atoms that decay within a certain length of time (i.e., measurement of the decay rate). The accuracy of the result is limited by the noncausality of radioactive decay. The decay of any specific atom is purely a random event, but it occurs with an isotope-specific probability [reflected in the decay constant, Eq. (5.3)]. The results of a repeated decay experiment exhibit a frequency distribution which can be described by the mean $\bar{N}$ and its standard deviation $\sigma_n$ (Sect. 4.2; Fig. 4.2). The value of $\sigma_n$ is obtained from the square root of the total number of observed events $N$ and $\sigma_n$ from the counting rate $n$ and the counting time $t$ as follows:

$$\sigma_n = \sqrt{N}/t = \sqrt{n/t} \tag{5.9}$$

It can be seen that by lengthening the measurement time, the value of $\sigma_n$ and hence the width of the confidence interval $n \pm \sigma_n$ (Sect. 4.2), can be decreased. This has economic and technical limits in practice. The latter are set by the long-term stability of the detector and the electronics (Currie 1968, 1972).

The lower limit of detection is determined by the background counting rate $n_0$ and the efficiency $\eta$ of the detector: $n_0$ is the counting rate for a sample containing no radioactivity; it is caused by radioactive impurities in the materials of the detector itself and within the shielding. The efficiency $\eta$ of a detector is given by the ratio of the number of registered counts to the number of decay events.

Determination of the minimum activity a sample must have so that it can be distinguished from the background activity is complicated (e.g., Currie 1972). In general, the lower the background and the greater the efficiency of the detector, the lower the minimum detectable activity. A measure of this is the factor of merit $g$

(e.g., Polach 1987a):

$$g = \frac{\eta n_{st}}{\sqrt{n_0}} \tag{5.10}$$

where $n_{st}$ and $n_0$ are the counting rates of a standard and the background, respectively.

Thus, according to Eq. (5.10), the factor of merit $g$ can be increased by enlarging the volume of the detector and/or decreasing the background. As the background counting rate is often proportional to the size of the detector, there are physical limits on the construction of a counter. Individual solutions to the problem are manifold (e.g., Oeschger and Wahlen 1975). Low-level techniques still have prospects for improvement (Polach 1987a).

A number of measures are used to lower the background:

– *Shielding of the detector* with lead (the best is old lead, i.e., lead that contains no $^{210}$Pb) or iron containing no $^{60}$Co, and boron-containing paraffin for neutron absorption. Measurement in a deep underground laboratory is also a very effective means of lowering the background (Oeschger and Wahlen 1975).
– *Construction of the detector with low-activity materials,* free of natural radioactive impurities (e.g., quartz or electrolytic copper, Geyh 1967).
– *Pulse-height discrimination* to limit the measurement to pulses that have amplitudes corresponding to the energy range of the radiation to be measured (Fig. 5.5).

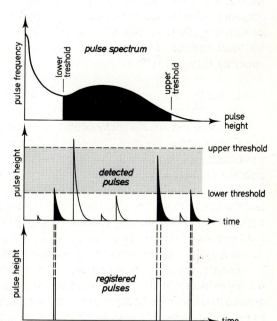

**Fig. 5.5.** Principle of pulse-height discrimination: Pulses which have an amplitude greater than a lower threshold or smaller than a given upper threshold (*shaded peaks*) are registered, the others are ignored

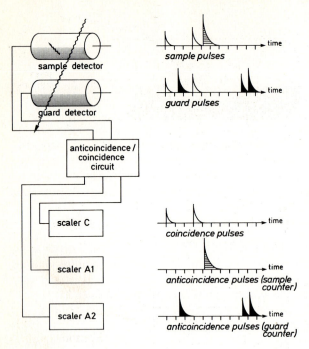

**Fig. 5.6.** Principle of an anticoincidence/coincidence counting system: Pulses detected simultaneously by the sample and guard detectors are registered as coincidence pulses, the others as anticoincidence pulses. The anticoincidence pulses of the sample counter, apart from a small correction, are proportional to the specific sample activity

– *Pulse-shape discrimination:* Some detectors, e.g., proportional counters (Sect. 5.2.2.1), produce pulses whose rise times depend on the type and energy of the radiation. Only the pulses within the range of amplitudes and rise times of interest are measured.

– *Anticoincidence and ( β-β or γ-γ ) coincidence measurements:* These are made with an inner detector containing the sample and a surrounding guard detector (Sects. 5.2.2.1 or 5.2.2.2). This arrangement makes it possible to distinguish the sample radiation from other radiation, e.g., hard cosmic radiation, chiefly muons, which pass through even thick lead layers and produce simultaneous, i.e., coincident, pulses in the inner detector and the outer detector. These are distinguished electronically from the pulses measured only by the sample detector, i.e., in anticoincidence (Fig. 5.6). The Oeschger-type counter is an especially effective realization of this technique (Geyh 1967).

### 5.2.2.1  Gas-Filled Proportional and Geiger–Müller Counters

Owing to their simple construction, proportional counters are still used for measuring $\alpha$, $\beta$, and $\gamma$ radiation down to 5 keV (Tsoulfanidis 1983). They consist of a gas-filled, sealed metal tube with an electrically insulated wire mounted along the axis of the tube. The gas in the tube is a noble gas, carbon dioxide, a hydrocarbon, or a mixture of these at a pressure of 1–30 bar. Often the gas is the sample itself.

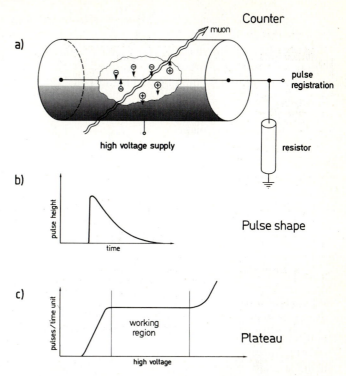

**Fig. 5.7 a–c.** Principle of proportional and Geiger–Müller counters: **a** The electrons and ions produced in the counter gas by ionizing radiation are strongly accelerated by the electric field between the anode wire and the cathode tube. Collisions with the gas molecules in the tube produce avalanches of electrons and ions **b**, which are registered as electric pulses **c**. The graph shows the plateau region of the working voltage

If an appropriately high voltage of several kilovolts is applied between the wire (the anode) and the tube (the cathode), ionizing radiation will produce an avalanche of ions and electrons, which is registered as an electrical pulse (Fig. 5.7). Within the plateau region, the counting rate is rather constant and independent of the voltage. Because the amplitude of the pulses produced in a proportional counter is proportional to the energy of the incident radiation, the background counting rate $n_0$ can be kept low by employing pulse-height discrimination (Fig. 5.5). A maximum energy resolution of 20% is attained, the minimum recovery time is about 20 $\mu$s.

Geiger–Müller counters have a similar construction to that of proportional counters, but the counting gas always contains a mixture of a noble gas and a gaseous hydrocarbon. Application of Geiger–Müller counters to geochronological problems is very limited as they only allow determination of total activity. The reason for this is that the amplitudes of the pulses are independent of the energy of the radiation. Pulse-height discrimination cannot be used to reduce the background counting rate, which is larger than that of proportional counters. However, no additional electronic amplification is necessary. The recovery time is about 100 $\mu$s.

To maximize the detection efficiency, the counting gas is generally prepared from the sample to be dated. For example, carbon dioxide ($CO_2$), methane ($CH_4$), ethane ($C_2H_6$), or acetylene ($C_2H_2$) is used for determinations of radiocarbon and tritium. The activity of argon-39 is measured by mixing it with methane under high pressure in very small counters (Lehmann and Loosli 1984).

### 5.2.2.2  Scintillation Counters

For the detection of radiation, scintillation counters use transparent phosphors which emit photons when they interact with ionizing radiation. The number of photons is proportional to the absorbed energy. Two photomultiplier tubes arranged on opposite sides of the sample produce coincident signals from the flashes of light, which are counted electronically (Fig. 5.8). Each photomultiplier consists of a photocathode, focusing electrodes, and ten or more dynodes to amplify the current produced by the impinging electrons from the photocathode.

The quality of a modern scintillation counter depends primarily on its long-term stability; this is because the thermal noise of the photomultiplier tubes is largely suppressed electronically (e.g., Polach et al. 1984). Only very short ($< 20$ ns) coincident events are counted. Pulse-shape discrimination is used additionally when K-electron emitters are measured. Extremely low background counting rates and very high sensitivity can be attained by making the measurements in a laboratory deep underground (Oeschger and Wahlen 1975).

Solid scintillation counters utilize thallium-activated sodium iodide (NaI(Tl)), cesium iodide crystals, or plastic scintillators in various shapes. The $\gamma$-sensitivity of NaI (Tl) is 30% better than that of CsI and significantly better than that of plastic scintillators. In contrast, the time resolution of plastic scintillators is the best of the three (only a few nanoseconds), but the energy resolution is only about 8–10% since 300 keV is needed to produce an electron pair, compared to 30 keV for NaI (Tl) crystals.

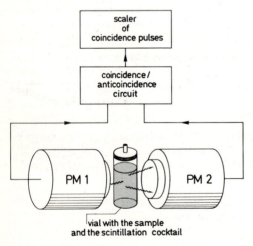

**Fig. 5.8.** Principle of a liquid scintillation counter (LSC): Photons are produced in the scintillation "cocktail" by radiation. These photons are detected by two photomultipliers (*PM*) arranged on opposite sides of the sample. The use of two photomultipliers makes it possible to discriminate the coincident pulses from the sample radiation from the non-coincident pulses caused by thermal noise. The coincident signals are fed to a summing amplifier and 1–3 lower and upper limit pulse-height analyzers and counters. Modern LSC counters also have high-voltage stabilization, pulse-height ratio discrimination, variable coincidence bias, external cosmic flux detectors, and a pulse-shape analyzer coupled with a microcomputer

Liquid scintillation counters (LSC) use a mixture of a scintillation "cocktail" and the sample, which has been dissolved or suspended in an appropriate solvent, e.g., water, toluene, benzene, or alcohol. For example, the sample is converted to benzene for the measurement of radiocarbon. The scintillation cocktail consists of a scintillator (e.g., PPO) and an optical frequency converter (e.g., POPOP). The latter absorbs the photons produced by the scintillator and emits light with a frequency that corresponds to the maximum sensitivity of the photocathode. The sample is changed automatically in commercial liquid scintillation counters, permitting the measurement of large numbers of samples with a minimum of handling. Water for tritium measurement, for example, is mixed with "Instagel" to produce the scintillation cocktail. Electrolytic enrichment is common for low tritium levels (IAEA 1981a).

The advantages and disadvantages of modern liquid scintillation counters relative to proportional counters (Sect. 5.2.2.1) are about even. A technical revolution was started in the 1980s when electronic optimization and anticoincidence guards coupled with multi-channel analysis drastically improved LSC instruments for radiocarbon analysis (Polach 1987b). This improvement, however, is obtained at considerable expense. A commercially available instrument (the Quantulus low-level LSC) permits the dating of a radiocarbon sample in 15 ml of benzene with a precision of 0.2% up to ages of 64 ka (Sect. 6.2.1).

### 5.2.2.3 Semiconductor Detectors

Semiconductor detectors are solid-state "ionization chambers" (Tsoulfanidis 1983). They consist of a coaxial or layer arrangement of p-type germanium or silicon crystals in which lithium has been deposited on the surface (doping) to compensate for impurities and lattice defects (n-type). Another type (Fig. 5.9) consists of high-purity germanium crystals free of doping (intrinsic Ge detectors). Ionizing radiation that penetrates a semiconductor produces electrons and positive holes in the "intrinsic" zone between the p- and n-type layers; a potential field across

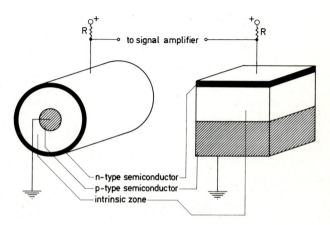

**Fig. 5.9.** Principle of a semi-conductor detector: Ionizing radiation passing through the intrinsic zone between the layers of p- and n-type semiconductors produces electrons and positive holes, which move in an electric field existing across the detector. The resulting pulses of electric current are then counted. Coaxial and planar detectors are shown as examples (e.g., Tsoulfanidis 1983)

to signal amplifier

n-type semiconductor
p-type semiconductor
intrinsic zone

the detector causes these electrons and positive holes to move, yielding a pulse of electric current, which is then counted.

Semiconductor detectors produce extremely short pulses (nanoseconds) and require only about 3–4 keV to produce an ion pair. They thus provide superior resolution for X-ray and $\gamma$-spectrometric measurements. The measurable energy range stretches from several keV to several MeV with up to 100% detection efficiency (Oeschger and Wahlen 1975).

Silicon semiconductor detectors are used for $\alpha$- and $\beta$-spectrometry. They operate at ambient temperatures. Silicon surface-barrier detectors are preferred for $\alpha$-spectrometry (e.g., for U, Th, Ra, and Po analysis).

Germanium lithium-drifted (Ge/Li) semiconductor detectors are ideal for measuring $\gamma$-emitters and low-energy $\beta$-emitters. The detection efficiency for $\gamma$-radiation reaches 100% with $\pi$-detectors. If the detector is not made of extremely pure germanium, it must always be kept below $-150°C$ to stabilize the lithium dopant. Detectors of very pure germanium need to be brought to this temperature only during the measurement.

To avoid radiation losses due to absorption, solid samples (in very thin layers) are placed on the detector in a vacuum chamber. Owing to their low volume, the background of semiconductor detectors is so small that except for extremely low-level measurements, lead or mercury shielding is not necessary. The pulses are registered with a multi-channel analyzer.

### 5.2.3   Measurement of Stable and Long-Lived Isotopes: Atom Counting Methods

A general description of the ion-beam technologies has been given by Bird et al. (1983).

### 5.2.3.1   Mass Spectrometry (MS)

The mass spectrometer is one of the most important instruments of the geochronological researcher for measuring isotope ratios very precisely (McDowell 1963; Milne 1971; Duckworth et al. 1986). The mass spectrometer is also used to determine element concentrations indirectly by isotope dilution analysis (Sect. 5.2.4.1). Mass spectrometric determination of isotope ratios is the major step in nearly all of the dating methods described in Sect. 6.1, with the exception of the chemical lead (Sect. 6.1.12) and lead/alpha (Sect. 6.1.13) methods, which are now obsolete. U/Th dating (Sect. 6.3) is also profiting from the use of mass spectrometry (Edwards et al. 1986/87). The precision of $\pm 0.05\%$ required for $^{87}Sr/^{86}Sr$ is usually easily obtained. Even better precision has been obtained for special isotope geochemical studies. The measurement of isotope ratios of $10^{-6}$ requires the detection of $10^3–10^6$ atoms.

The main components of a mass spectrometer are the ion source, the ion separation system (magnet and analyzer tube), and the detector (ion collector).

The sample is ionized in the ion source. In the simplest ion source, a purified gaseous sample is ionized by an electron beam emitted from an incandescent filament. Solids are placed on a filament (e.g., Ta, W, Pt, or Rh) and thoroughly degassed before being ionized thermally. The positive ions are accelerated by an electric field ($\sim 10\,kV$), focused to a beam, and deflected through a curved path by a homogeneous magnetic field. The radius of the curve is dependent on the strength of the magnetic field and the energy, charge, and mass of the ionized particle. After acceleration by the electric field, all particles with the same charge have the same energy, hence the lighter nuclides will be deflected by the magnetic field more than the heavier ones. Thus, the magnetic field separates the ion beam into beams of ions of the same mass. An ion collector is placed behind a slit so that it detects only one of these beams. By varying the strength of the electric field in which the ions are accelerated or the magnetic field in which they are deflected, the entire ion spectrum can be registered (Fig. 5.10). In the simplest construction, the collector is a Faraday cylinder that is grounded via a high-resistivity resistor. The current produced in the collector is amplified and measured. If extremely low sample quantities are to be measured, the ions are counted with a photomultiplier.

Mass spectrometers with double or multi collectors measure the isotopic abundances of two or more (up to five) nuclides simultaneously. Fluctuations in the ion beam are compensated for and the precision of the measured ratio is significantly increased.

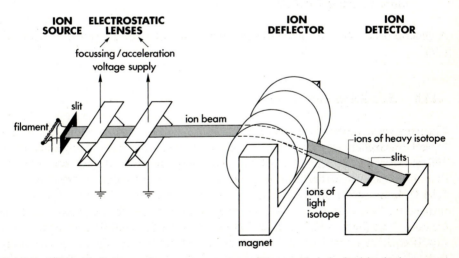

**Fig. 5.10.** Schematic diagram of a mass spectrometer: The sample is ionized in the ion source (different for solid and gaseous samples). The positive particles are focused into a beam and accelerated by an electric field to the same energy. They are deflected by a magnetic field into curved paths whose radius decreases with decreasing mass and increasing charge of the particles. The abundances of the ions in the separated ion beams can be measured by varying the strength of the electric or magnetic field rather than changing the position of the ion collector. The desired isotope ratio is obtained from the ratio current produced in the collector by the appropriate beams

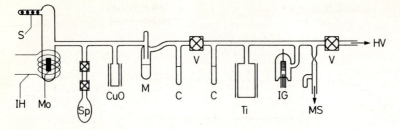

**Fig. 5.11.** Scheme of argon extraction and purification (after Dalrymple and Lanphere 1969). *S* sample; *Mo* molybdenum crucible; *IH* induction heater; *SP* spike, *CuO* CuO furnace; *M* molecular sieve; *C* charcoal fingers; *V* valves; *Ti* Ti furnace; *IG* ionization gauge; *MS* mass spectrometer; *HV* high vaccum

This is all done in an ultra-high vacuum ($10^{-7}$–$10^{-9}$ torr) to avoid collisions of the ions with gas molecules. Multiply-charged ions and molecular fragments make separation of isobar ions difficult or impossible.

Gas samples must be purified before being introduced into the mass spectrometer. Because they are inert, the noble gases are commonly purified chemically. A purification line for the K-Ar method is shown in Fig. 5.11 as an example. A second isotope ratio of the element of interest is usually measured so that a correction can be made for the content of atmospheric gases. Owing to a possible shift in the isotopic composition during the measurement due to mass fractionation, extrapolation must be made to the time the gas inlet was opened.

### 5.2.3.2 Accelerator Mass Spectrometry (AMS)

The development of the accelerator mass spectrometer technique began in the late 1970s for the measurement of ratios of the long-lived radiocarbon and the stable carbon isotopes (Bennet et al. 1977; Muller 1977; Nelson et al. 1977). In the meantime, it has become a successful procedure for atom counting of other long-lived cosmogenic radionuclides (Sect. 6.2), such as $^{10}$Be, $^{26}$Al, $^{32}$Si, $^{36}$Cl, $^{41}$Ca, and $^{129}$I. It could gradually replace decay counting techniques (Polach 1987a). The decisive advantage of the AMS technique is that atom ratios up to $10^{-17}$ can be measured with sample sizes on the order of one milligram or less. Decay counting requires sample sizes of several grams or more. Spectacular successes have been achieved with the analysis of $^{14}$C and $^{10}$Be (Wölfli et al. 1984; Gove et al. 1987) and $^{36}$Cl (e.g., Bentley et al. 1986a, b; Phillips et al. 1986b).

Tandem accelerators are used for the AMS method. The principle is essentially that of an ultrasensitive mass filter (Fig. 5.12) in which the particle flux at the detector has been reduced to a rate that allows the particles to be characterized by their total energy and their rate of energy loss when stopped by matter. High energies are necessary for the characterization of the particles and to allow stripping to multiply-charged ions (eliminating molecules). Prerequisite for AMS is that the element to be analyzed forms stable negative ions.

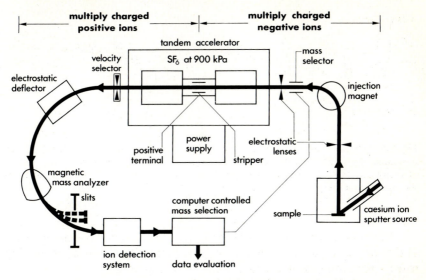

**Fig. 5.12.** Schematic diagram of an accelerator mass spectrometer: Negative ions are produced from the sample in a *cesium sputter source*. These anions are accelerated and focused by *electrostatic lenses*. They then pass through an *"injection" magnet* into the *tandem accelerator*, where they gain energies of several MeV. A stripper removes electrons from the ions and the positive ions thus produced are repelled from the positive terminal of the tandem accelerator, refocused, separated by deflection in electrostatic and magnetic fields, and counted in an $E$ ion detector and a $dE/E$ detector

    The solid sample $(1-10\,\text{mg})$ is bombarded with cesium ions in a sputter source to produce a large number of negative ions $(1-10\,\mu\text{A})$, which are then given an initial acceleration (to $100-400\,\text{keV}$) and double-focused into a beam. An "injection" magnet selects ions with a particular mass/charge (m/e) ratio and directs them to the positive terminal of the accelerator (which is in a high-pressure tank containing an insulating gas). After the negative ions have gained an energy of several MeV or more (typically two to several tens of MeV), they are passed through a foil or gas to remove electrons (electron stripping), producing multiply-charged positive ions. This stripping also breaks up most of the disturbing molecules and molecular fragments in the beam (Coulomb explosion). The positive ions produced by the stripping are repulsed by the positive terminal and further accelerated, doubling their energy. After refocusing, the ions are deflected by electrostatic and magnetic fields to remove unwanted molecular fragments and charge states. The beam then passes through a slit and a velocity filter, ending at the detector, which can be a time-of-flight detector, a gas ionization detector, or a silicon surface-barrier detector. These detectors measure both total energy and the rate of energy loss ($dE/E$ counter). Isobar ions (Table 6.3) can be distinguished with such detectors. The AMS system reduces disturbing ions by a factor of $10^{11}-10^{13}$! Three-dimensional plots of the isotopic composition can be obtained (Fig. 5.13).

    Because isotopic ratios are measured for age determinations, AMS is used for at least two isotopes separately. Since the current of the ion beam can be stabilized over periods as long as milliseconds only with difficulty, the mass-dependent,

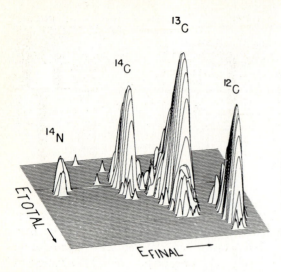

**Fig. 5.13.** Three-dimensional representation of the carbon isotope spectrum (measured by the Rochester laboratory); $E_{total}$ is the total ion energy and $E_{final}$ is a function of the ion range and the energy loss rate $dE/dx$; the vertical axis is the logarithm of the number of counts (Mahaney 1984)

focusing and separation field potentials are switched back and forth at millisecond intervals by a fast-beam-pulsing system (Suter et al. 1984). The more abundant isotope is measured for a shorter time than the less abundant isotope. In the Zurich AMS laboratory, for instance, radiocarbon is counted for ten periods of 50 s each, interrupted 6000 times for $20\,\mu s$ to count $^{12}C$ plus $200\,\mu s$ to count $^{13}C$. This procedure makes it possible to attain a reproducibility for $^{14}C$ in modern carbon, for example, of down to 0.3% within 2 h, which corresponds to that of precise conventional decay counting methods. The superiority of the AMS technique is illustrated, for instance, by the fact that as little as 0.5 mg of carbon, for example, are needed to obtain in about an hour, data with the same precision as otherwise obtained with decay counting techniques using samples of several grams and requiring more than 30 h. This is the case even though only up to about 2% of the $^{14}C$ atoms are detected. But the price of the accelerator unit for the AMS technique, as well as its maintenance, is so high that an analysis is still more expensive than that done by conventional decay techniques.

### 5.2.3.3 Resonance-Ionization Spectrometry (RIS)

Resonance-ionization spectrometry is an ultrasensitive, very selective, and highly efficient technique for element and isotope analysis. It utilizes narrow-band lasers tuned to a specific wavelength to selectively excite the element or isotope to be analyzed from the ground state to a resonance state without interference from much more abundant elements or isotopes. A second, tunable dye laser is used to ionize the atoms by resonant excitation (Fig. 5.14), which are counted with a proportional counter (Sect. 5.2.2.1) or by mass spectrometer. Detection of as little as just as few atoms is possible (Fassett et al. 1985).

A so-called atom buncher is used to enhance the sensitivity for counting individual atoms (Hurst et al. 1982, 1985; Chen et al. 1984). It is coupled with a

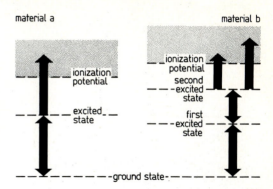

**Fig. 5.14.** Resonance ionization process: A turned laser is used to raise an atom to an excited state. A second dye laser is then used to ionize the excited atom (*a*). Some substances require a third (or even further) dye laser for ionization owing to intermediate excited states (*b*). The ions produced in the last step are counted with a proportional counter

quadrupole mass spectrometer. The individual atoms to be counted are first condensed on a Be-Cu target cooled with liquid helium. These atoms are then released by a laser after a defined time delay so that the RIS laser can ionize all of the atoms when the detector is ready. The probability for the ionization of krypton-81, for example is enhanced dramatically (Lehmann et al. 1985; Thonnard et al. 1987).

## 5.2.4  Other Analytical Techniques

A comprehensive description of the various analytical techniques most frequently applied in geochronology is given together with references by Tite (1981).

### 5.2.4.1  Isotope Dilution Analysis (ID)

Isotope dilution analysis (Webster 1960; Faure 1977) is used for extremely precise ($\leqslant \pm 1\%$) quantitative determination of very small quantities (submicrogram) of individual elements in solid, liquid, or gaseous substances. To determine the unknown quantity $N_p$ of the element (in terms of number of atoms) in the sample, a known quantity $N_s$ of that element is added to the sample; this added amount is called a spike or tracer. This spike contains the element with an artificial, precisely known isotopic ratio $R_s$ of the isotopes 1 and 2, different from the natural composition $R_p$. For solid samples, the spike is added during or after digestion of the sample in acid. When the sample and spike are mixed to a homogeneous composition ($N_p + N_s$), a new isotopic ratio $R_m$ is obtained, which is measured with a mass spectrometer. The maximum sensitivity of the mass spectrometer or radiometer is obtained when the two isotopes 1 and 2 used for quantitative determination have similar abundances. The ratio of sample to spike is selected so that this is the case. This is illustrated for Rb/Sr dating (Sect. 6.1.3) in Fig. 5.15.

For an element with two natural isotopes, the unknown quantity $N_p$ of the element in the sample is obtained using

$$N_p = N_s \frac{(R_m - R_s)(1 + R_p)}{(R_p - R_m)(1 + R_s)} \tag{5.11}$$

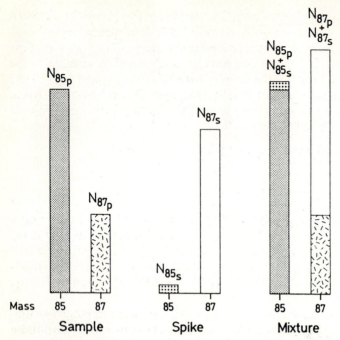

**Fig. 5.15.** Isotope dilution analysis (using the element rubidium as example): A known quantity $N_s$ (in terms of number of atoms) of a spike of known isotope composition (in the case of rubidium: $N_{85s}$ and $N_{87s}$; 85 and 87 = atom mass numbers; $s$ = spike) is added to a sample of unknown quantity $N_p$ containing the isotopes 85 and 87 ($N_{85P}$ and $N_{87P}$; $p$ = sample). The mixture has to contain approximately the same quantity of both isotopes to obtain the maximum precision of the mass spectrometer

where

$R_m$ = the isotopic ratio $N_1/N_2$ in the mixture,

$R_p$ = natural isotopic ratio $N_1/N_2$ in the sample (to be determined),

$R_s$ = the isotopic ratio $N_1/N_2$ in the spike (known), and

$N_s$ = quantity of the element in the spike (in terms of number of atoms).

When $R_p$, $R_s$, and $N_s$ are known, it is sufficient to measure the isotopic ratio of the mixture $R_m$ in order to determine the quantity of the element in the original sample. If the element has more than two isotopes, there are systems of equations that can be solved to correct for any mass fractionation that may occur during the mass-spectrometric measurement.

The main advantages of isotope dilution analysis can be summarized as follows:

1) The chemical treatment need not be quantitative: After weighing the spike and sample and mixing them, it is only necessary to determine the isotopic ratio of any part of the mixture, which can be done extremely precisely on a mass spectrometer.

2) The analysis is free of any influence from other elements: Interfering isotopes of other elements are removed chemically before the mass-spectrometric measurement.

3) It has extraordinarily high sensitivity: The amount of sample can be increased with decreasing concentration of the element being determined.
4) It is possible to measure more than two natural isotopes of one element at the same time: The natural isotopic composition is measured in addition to the concentration of the element in the sample. If double spikes are used (e.g., Rb and Sr or Pb and U), it is not necessary to split the sample. This lessens the analytical error since (a) fewer weighings are necessary and (b) both elements are measured on the same aliquot of the sample.

The main disadvantage is that the method is very time-consuming.


### 5.2.4.2   Neutron Activation Analysis (NAA)

Neutron activation analysis (Fleming 1976; Henderson 1984) is an extremely sensitive, but very complicated multi-element method that can be used for many purposes. For most elements, concentrations between 1 ppm and 100% can be very precisely measured ($\pm 2$–5%) with very low detection limits. The most appropriate elements for the method are the halogens, the REE, the platinum group, uranium, and thorium.

The sample (50–100 mg) is placed together with a standard in plastic, quartz-glass, or aluminum containers in a nuclear reactor to be activated by irradiation with slow (thermal) or fast neutrons. This activation is dependent on the total neutron flux $\phi$, the concentration of the target nuclei, and their effective cross sections $\sigma_1^*$. The radionuclides are produced mainly by $(n, \gamma)$ reactions and after irradiation of the sample, the $\gamma$-radiation is analyzed with a gamma-spectrometer. The concentration $c$ is obtained from the counting rates $n$ of the sample and the standard:

$$\frac{c_{\text{sample}}}{c_{\text{standard}}} = \frac{n_{\text{sample}}}{n_{\text{standard}}} \tag{5.12}$$

The determination of the absolute concentration is more difficult than the determination of relative concentrations because the effective cross-sections are usually not known exactly.

Undesired, short-lived isotopes produced together with the desired ones by the irradiation of the sample are eliminated by waiting several weeks before the gamma-radiation is measured. Depending on the objective of the analysis and the type of sample, the sample is analyzed as a whole (rock and mineral samples) or the elements to be analyzed are separated chemically (radiochemical activation analysis). In the latter case, a stable tracer isotope (spike) is added before the irradiation step so that quantitative separation is not necessary.

Each element in the separated fraction has a characteristic gamma-radiation, which is measured with a solid-state counter (Sect. 5.2.2.3). This is done preferentially with a high-resolution Ge(Li) detector, whose pulses are sorted according to their amplitude by a multi-channel analyzer. A $\gamma$-spectrum is obtained from the cumulative signals. Earlier, Geiger–Müller counters (Sect. 5.2.2.1) or NaI (Tl) scintillation counters (Sect. 5.2.2.2) in $\beta/\gamma$ or $\gamma/\gamma$ coincide arrays were used.

The maximum precision obtained by neutron activation analysis is about $\pm 1\%$. The detection limit is as low as $< 10^{-6} \mu g$ for some elements and may be improved by increasing the neutron flux. For very low concentrations, it is preferable to determine absolute concentrations from the neutron flux and the effective cross-section.

Neutron activation analysis is more suitable for quantitative analysis than for the qualitative analysis of specific elements. Activation by thermal neutrons provides the best results with the REE (rare-earth elements), Ta, Sc, Re, Pt, Hf, Cd, and As. Lighter elements (e.g., Li, Be, B, F, N, O, Si, P, Cl, Fe, Cr, and Ga) require irradiation with fast neutrons.

### 5.2.4.3 Flame Photometry, Atomic Absorption Spectrometry (AA) and Inductive Coupled Plasma Analysis (ICP)

Flame photometry (sometimes called optical emission spectrometry) and atomic absorption spectrometry (Dalrymple and Lanphere 1969; Price 1972; Thompson and Reynolds 1978; Welz 1983) are closely related analytical methods based on the thermal production of free atoms of metallic elements in a dissolved sample by a gas flame. The basis of flame photometry – an optical emission spectrum – is just the opposite of that of AA analysis. In emission spectrometry, photons are emitted when an atom in an excited state returns to its ground state; atomic absorption analysis is based on the production of an excited state by the absorption of a photon. Advantages of both methods are the speed with which they can be conducted and that chemical separation of the individual elements is not necessary. In geochronology, flame photometry is often used to determine potassium for the K/Ar (Sect. 6.1.1) and K/Ca (Sect. 6.1.2) methods, sometimes for determining rubidium for the Rb/Sr method (Sect. 6.1.3).

A very exact determination of element concentrations $\geq 0.02\%$ can be made with a *flame photometer*. It is based on the fact that when heated to a sufficiently high temperature, each element emits a line spectrum with characteristic wavelengths. This radiation, whose wavelengths for most elements are within the range of visible light, is produced by the transition of electrons from higher to lower energy levels. Its intensity is proportional to the concentration of the element. Most elements require a very high excitation temperature (normally a gas-oxygen flame); the alkali metals, for which a gas-air flame is sufficient, are an exception. Thirty to forty elements, mainly metals, can be analyzed within a concentration range of 100 ppm to 10%.

The mineral or whole-rock sample (5–200 mg) is digested in a hydrofluoric–sulfuric acid mixture (sometimes also with the addition of nitric acid). The solution is then sprayed into a gas flame where it is atomized. The emitted radiation is observed through a filter or prism. The intensity of an especially strong line or line series of the spectrum of each element is measured. The concentration of the element is obtained by comparison of the intensity of the unknown sample with the intensity of a standard solution. Potassium is usually measured with an internal lithium standard for calibration. For samples with more than 0.5% $K_2O$, the precision of the method is better than 1%.

Besides the determination of the main components of a sample, *AA spectrometry* is especially suitable for quantitative analysis of traces of about 40 metallic elements with a precision of at least 2%. The limit of detection is in the range of 0.1–1 ppm. If light containing the spectrum of a specific element is passed through a gas mixture containing free, unexcited atoms of that element, the characteristic wavelengths of this element will be partially absorbed, producing the line spectrum of that element, which for most elements is in the visible and ultraviolet part of the electromagnetic spectrum. Since the intensity of absorption is directly dependent on the number of atoms present that are capable of absorption, there is a linear relationship between the extinction of the sample and the concentration of the element. Of course, a calibration curve has to be prepared first.

The most important parts of an AA spectrometer are a light source, an absorption part (e.g., an acetylene flame in which the sample is atomized), a monochromator (normally a grid monochromator), a detector (usually a photomultiplier), an amplifier, and a recorder. The atomization of the sample can also be done without a flame in a heated graphite tube; for many elements, this significantly improves the detection limit. The chemical preparation of the sample (10–100 mg) is similar to that for flame photometry. The method is of only secondary significance for geochronological analysis (e.g., determination of potassium for the K/Ar method), but it is of importance as a rapid, routine method for preselection of samples.

*Inductive coupled plasma analysis* (Thompson and Walsh 1983) is a special form of emission spectrometry that uses very complicated equipment. The atomization and ionization of the sample is done by transformation of the sample aerosol into a plasma. This plasma is produced by inductive heating of a gas (usually argon, sometimes nitrogen) in the coil of a high-frequency wave generator. The ionization temperature is about 8000 K.

The main advantage of ICP analysis is the low detection limit, especially for elements that are difficult to atomize and thus cannot be measured by AA spectrometry at very low concentrations [e.g., rare-earth elements (REE), alkaline earth elements, boron, silicon, uranium, and tantalum]. This is because elements with a high affinity for oxygen tend to form oxide or hydroxide radicals in the AA flame, which do not dissociate further. This does not occur in the ICP plasma.

Further advantages relative to AA analysis are the possibility of determining several elements simultaneously and the lower susceptibility to chemical disturbances. On the other hand, the reproducibility of the ICP method is, in general, not as good as AA analysis.

### 5.2.4.4  Ion-Microprobe (IPM) and Laser Microprobe Mass Analysis (LAMMA)

The *ion-microprobe* is a very sophisticated analytical instrument that is slowly finding increasing application in the geosciences (e.g. the SHRIMP – sensitive high mass-resolution ion microprobe at the University of Canberra). The probe is used for the isotopic analysis (ppm to ppb) of very small samples (diameter 1–20 $\mu$m), e.g., the center or margin of crystalline grains in rocks. The first age determinations using the U/Th/Pb method (Sect. 6.1.9) were carried out on individual zircon grains (Compston et al. 1983); ages can also be obtained using monazite, sphene, apatite, and feldspar.

The sample is placed in a focused beam of high-energy primary ions. Some of the atoms ejected during erosion or sputtering of a very small area on the surface of the sample are ionized. These secondary ions are introduced into a mass spectrometer (Sect. 5.2.3.1) by a high-potential field ($\sim 10 \, kV$). Very high mass resolution is required since the sample is not chemically pretreated, i.e., possible disturbing elements or compounds are not eliminated. IMP allows quantitative analysis of major and trace elements, depth profiling, and in situ isotope analysis.

The *laser microprobe mass analysis* was developed for dating lunar samples by the $^{39}Ar/^{40}Ar$ method (Sect. 6.1.1.2). Individual mineral grains down to a weight of 0.2 $\mu$g have been measured with this probe. It has also been tested (Maluski and Schaeffer 1982) and successfully applied to terrestrial samples (van Bogaard et al. 1987).

A polished section (ca. 1–2 mm thick, 4–6 mm in diameter) is irradiated with neutrons in a nuclear reactor to induce the following reaction: $^{39}K \, (n, p) \, ^{39}Ar$. It is then placed in a vacuum chamber built into a microscope and attached to a laser (which can be focused to within a few $\mu$m) and a mass spectrometer. The laser beam is focused on the potassium-bearing minerals in the sample to free argon in them, which is then measured mass spectrometrically.

### 5.2.4.5  X-Ray Fluorescence Analysis (XRF)

X-ray fluorescence analysis (Schroll 1975; Johnson and Maxwell 1981; Williams 1987) is one of the most important analytical methods of geochemistry for elements with atomic numbers greater than 22. The main application for geochronology is the routine preliminary examination of samples to select the most suitable ones for further testing or to determine the optimum amount of spike. The comparatively large analytical error for trace elements (3–15%) generally excludes direct application for geochronological purposes. In some laboratories, however, techniques have been developed to determine elements and element ratios, e.g., Rb/Sr (Sect. 6.1.3), with sufficient precision for age determination.

X-ray fluorescence analysis is a purely physical method. It is based on the fact that when an element is excited by high-energy radiation (e.g., X-rays) it irradiates light (fluorescence) with a wavelength characteristic for that element. This emission spectrum consists of a continuous spectrum of X-rays (bremsstrahlung) and a few spectral lines. The intensity of the fluorescence depends on various parameters but chiefly on the concentration of the element in the sample. The spectral lines of the fluorescence are produced by the relaxation of the excited electrons in the inner shells. For the heavier elements (atomic number $Z = 50$–60) only the spectral lines of the L series are suitable for the analysis: for the lighter elements, only those of the K series.

An XRF apparatus consists of a radiation source (e.g., an X-ray tube), collimator, crystal, detector, amplifier, pulse-height discriminator, and recording unit. Both wavelength-dispersive and energy-dispersive systems are available.

Primary X-rays are produced when a substance (chemical element) is placed in a beam of electrons in a high vacuum. X-ray tubes are constructed with a Rh, Cr, W, Mo, Ag, or Au anode.

Rock or mineral samples for analysis are ground very fine (grain size $\leqslant 40\,\mu$m) in an agate mill and either mixed with a binding agent (e.g., paraffin or phenylformaldehyde resin) and pressed or melted with a flux, e.g., lithium tetraborate ($Li_2B_4O_7$), to a glass in the form of a tablet. The details of the method vary from laboratory to laboratory.

The element concentrations are determined by comparing the values obtained for the sample with the values obtained on the same apparatus for international standard samples. Corrections must be made, especially for mass absorption and interelement excitation.

# 6 Radiometric Dating Methods

## 6.1 Parent/Daughter Isotope Ratios as a Geochronometer

Except for Quaternary times, the major dating methods for geology and petrology are based on parent/daughter isotope ratios. The most important of these methods for dating igneous and metamorphic rocks and their minerals are the K/Ar (Sect. 6.1.1), Rb/Sr (Sect. 6.1.3), U/Th/Pb (Sect. 6.1.9), and Sm/Nd (Sect. 6.1.6) methods. For example, only with these methods can time relationships be determined for the Precambrian. This is also true for younger basement rocks, for instance, for deciphering the development of an orogen, especially where there are no associated sediments that can be dated by other means.

The methods discussed in this chapter are based on radioactive decay [see Sect. 5.1, Eqs. (5.1 to 5.6)]: For each parent atom that decays, a stable daughter isotope is formed, either directly or as the end-product of a decay series. The probability of the decay of a specific radionuclide within a specific time is given by the decay constant $\lambda$.

The radiometric "clock" based on this physical law can be compared to an hourglass: For each grain of sand that disappears from the upper (parent) reservoir, a grain is added to the lower (daughter) reservoir. If we stop the hourglass after some time (corresponding to the analysis of a sample in the laboratory), then we can determine how long the clock had been running from the ratio of the sand in the two reservoirs:

1) It must be known how many grains of sand drop from the one reservoir into the other per unit of time.
2) When the hourglass was started the lower (daughter) reservoir must have been empty (or the initial amount must be known exactly).
3) The hourglass must not have any holes, i.e., no sand may be added to it from a source other than the upper reservoir and no sand that falls from the upper reservoir may be lost.

Applied to radiometric dating, condition 1) corresponds to the decay constant $\lambda$ and conditions 2) and 3) are boundary conditions that are of great importance for isotopic age determination. Condition 3) corresponds to the closed system requirement; data from the radiometric methods discussed in this chapter could not be interpreted if this condition is not fulfilled. A significant difference between an hourglass and a radiometric "clock" must not be overlooked: Whereas the number of

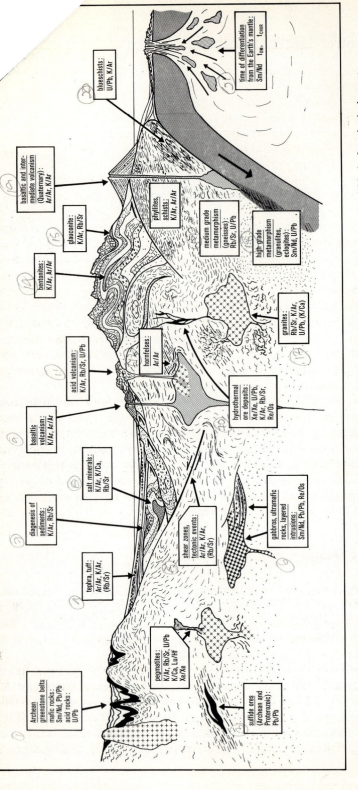

**Fig. 6.1.** Scheme illustrating the applicability of the different methods for dating hard rocks in terms of rock type and kind of geological event

sand grains that fall from one reservoir to the other per unit time is a function only of grain size and diameter of the neck of the hourglass and not the number of grains present, the radioactivity of an isotope is a function of the number of parent nuclides present, because the decay constant is an expression of the probability that any particular nuclide will decay. If a large number of parent nuclides are present, many of them will decay within a certain period of time; if there are only a few radioactive nuclides present, fewer will decay within the same time period. The ratio of decay events per unit time is a linear function of time for an hourglass and an exponential function for the radiometric clock [Eqs. (5.1 to 5.5)].

The half-lives of the various radioactive isotopes that are suitable for dating purposes vary greatly, from several years (e.g., $^{210}$Pb) to many billion years (e.g., $^{187}$Re). This makes geochronological studies possible over the entire range of possible ages. The methods discussed in this section however, are all based on *long-lived* radionuclides with half-lives between about 0.7 Ga ($^{235}$U) and > 40 Ga ($^{187}$Re). To be suitable for dating purposes, the half-life must, on the one hand, be short enough to have produced a measureable amount of daughter isotope since time zero of the sample being studied, on the other hand, it must be long enough that a measureable amount of parent isotope is still present. This means that the methods of this section are not suitable for young geological units such as unconsolidated sediments, but as a rule can be used only for rocks and minerals older than the Quaternary (except for the K/Ar and Ar/Ar methods, which reach into the Quaternary period, for which they yield important information). For this reason, we refer to these methods as the "hard-rock" methods.

The spectrum of applications for which the methods of this section can be used in the field of geology is extremely wide. The different types of rocks that may yield a date for a particular geological event are shown in Fig. 6.1 together with the applicable methods. The list can never be viewed as complete because the range of application of the individual methods is being continually extended. Despite great progress during the last several decades, there are still frequently occurring rock types that cannot yet be dated or for which the age can only be estimated, e.g., ore deposits and many sediments.

## 6.1.1  Potassium/Argon ($^{40}$K/$^{40}$Ar) Method***

We divide this method into two subchapters: (1) the conventional K/Ar method and (2) the $^{39}$Ar/$^{40}$Ar method. Although the latter was originally only a modern analytical variation of the former, its use of the incremental heating technique allows it to be applied to problems for which the conventional method is not suitable, thus qualifying it to be considered an independent method.

Potassium is the eighth most abundant element in the Earth's crust and forms numerous minerals. It has three natural isotopes: masses 39 (93.2581%), 40 (0.01167%), and 41 (6.7302%) (Garner et al. 1976). Potassium-40 is radioactive with branched disintegration (Fig. 6.2): by $\beta^-$-emission (E = 1.32 MeV) to $^{40}$Ca (88.8%), by $\beta^+$ emission (0.001%), and by two kinds of electron capture (11% and 0.16%) to $^{40}$Ar. These currently accepted isotope abundances, as well as the decay

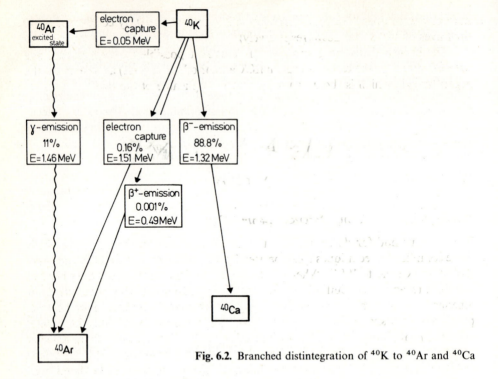

**Fig. 6.2.** Branched distintegration of $^{40}$K to $^{40}$Ar and $^{40}$Ca

**Table 6.1.** Decay constants for $^{40}$K and isotopic
abundances of potassium and argon

Decay constants ($10^{-10}a^{-1}$)

| | | |
|---|---|---|
| $\lambda_\beta$ | | 4.962 |
| $\lambda_{EC}$ | | 0.581 |
| $\lambda_{total}$ $(=\lambda_\beta + \lambda_{EC})$ | | 5.543 |

Isotopic abundances

| | (%) | | (% atm.) |
|---|---|---|---|
| $^{39}$K | 93.2581 | $^{36}$Ar | 0.337 |
| $^{40}$K | 0.01167 | $^{38}$Ar | 0.063 |
| $^{41}$K | 6.7302 | $^{40}$Ar | 99.600 |

constants, were recommended by the Subcommission for Geochronology of the
IUGS in 1976 for the calculation of K/Ar ages (Steiger and Jäger 1977).

   After helium, argon is the second most abundant noble gas in the rocks and
minerals of the Earth ($10^{-6}$ to $10^{-10}$ cm$^3$ $^{36}$Ar/g at STP, Alexander 1978). It is the
third most abundant component of the atmosphere (0.934 vol %). It has three
natural stable isotopes: masses 36 (0.337% of atmospheric argon), 38 (0.063%), and
40 (99.600%). The isotope ratio $^{40}$Ar/$^{36}$Ar$_{atm}$ = 295.5 (Nier 1950) is used by
convention to correct for atmospheric argon in age calculations (Steiger and Jäger
1977). In addition to these three isotopes, small amounts of the naturally

radioactive isotopes $^{37}$Ar and $^{39}$Ar (Sect. 6.2.7) are produced by cosmic radiation (half-lives of 35.0 d and 269a, respectively).

The branched distintegration of $^{40}$K makes it possible to carry out an age determination via the calcium branch (K/Ca method, Sect. 6.1.2) as well as via the argon branch, which is the much more widely applicable of the two.

## 6.1.1.1   Conventional Potassium/Argon ($^{40}$K/$^{40}$Ar) Method***

(Von Weizsäcker 1937; Aldrich and Nier 1948)

### Dating Range, Precision, Materials, Sample Size

Standard method for determining conventional ages greater than 3–5 Ma. With more complicated techniques it can be used for ages down to several thousand years (Gillot and Cornette 1986). Whenever possible, several samples from the object of interest are analyzed. Suitable for dating are potassium-bearing minerals from igneous, metamorphic or sedimentary rocks: feldspars (sanidine, anorthoclase, plagioclase), feldspathoids (leucite, nepheline), mica (biotite, muscovite, phlogopite, lepidolite), amphibole (hornblende), and infrequently also pyroxene from igneous rocks (Schaeffer and Zähringer 1966; Dalrymple and Lanphere 1969; Hunziker 1979). Glauconite, illite, and clay minerals from sediments can also be dated (e.g., Odin and Dodson 1982). A precision as good as 2–4% (2$\sigma$) is usually attainable for the age determination. Whether the dates are to be interpreted as the age of crystallization, cooling, sedimentation, or diagenesis must be decided from case to case. Particularly problematic is the interpretation of the K/Ar dates of salt minerals (sylvite, carnallite, cainite, langbeinite, polyhalite) (e.g., Wardlaw 1968; Lippolt 1977). Phenocryst minerals (sanidine, biotite, hornblende) and authigenic minerals (feldspars) in tuffs and bentonites can yield important stratigraphic marks (e.g., Hellmann and Lippolt 1981). A K/Ar dating of the clay minerals or other K-bearing minerals (e.g., celadonite) in ore veins and silicification zones can made in an attempt to determine the formation age of hydrothermal ore deposits (Bonhomme et al. 1983; Staudigel et al. 1986). Attempts have been made to date fault zones using phyllitic mineral fractions from them (Kralik and Riedmüller 1985).

Whole-rock dating can be done successfully with young, very fresh, mafic volcanic rocks, particularly basalts (e.g., Pankhurst and Smellie 1983). Volcanic glasses and tectites can also be dated (e.g., Hall et al. 1984); however, the results may not be very reliable in every case. The K/Ar method is also a standard method for dating meteorites and lunar samples (e.g., Pepin et al. 1972; Alexander et al. 1977; Sect. 6.5); however, for these samples it is being increasingly replaced by the $^{39}$Ar/$^{40}$Ar method (Sect. 6.1.1.2).

The dating of fossil groundwater with an age of $5 \times 10^4$ a or more has developed as a specialized application of the K/Ar method (Gerling et al. 1967; Seifert 1978).

The rock samples must be fresh and unaltered. The amount of sample necessary varies greatly depending on the age of the sample, the mineral being dated and its potassium content, as well as its proportion in the rock. A hand specimen may be sufficient in one case, 50 kg may be necessary in another case.

## Basic Concept

Using the concentrations of $^{40}K$ and radiogenic $^{40}Ar^*$ (in $cm^3$ Ar(STP)/g sample or mol/g), the radiometric age of the sample can be calculated as follows:

$$^{40}Ar^* = \frac{\lambda_{EC}}{\lambda_{total}} \, ^{40}K \, (e^{\lambda_{total}t} - 1), \tag{6.1}$$

which leads to

$$t = \frac{1}{\lambda_{total}} \ln \left( \frac{\lambda_{total}}{\lambda_{EC}} \frac{^{40}Ar^*}{^{40}K} + 1 \right) \tag{6.2}$$

When the new values of the constants are placed in the equation, the following formula is obtained:

$$t = 1.804 \times 10^9 \ln \left( 9.540 \frac{^{40}Ar^*}{^{40}K} + 1 \right) \tag{6.3}$$

For ages of less than about $10^7$ a, the logarithmic factor can be replaced by $9.540(^{40}Ar^*/^{40}K)$ and $t = 1.72 \times 10^{10}(^{40}Ar^*/^{40}K)$. The resulting error for an age of $4 \times 10^7$ a is only about 1%.

Because fractionation of the potassium isotopes in geological processes can be generally excluded, it is only necessary to know the present $^{40}K$ content of a sample to solve the age equation. The age can thus be calculated directly from the total potassium content and the values in Table 6.1.

The K-Ar ages calculated from the age equation (6.2) are model ages based on the following assumptions:

1) When the radiometric clock is started, there is no $^{40}Ar^*$ in the sample or only negligible amounts.
2) The rock or mineral has formed a closed system since time $t_0$.
3) Compared with the age to be determined, the closing of the system occurred quickly.

Damon (1970) proposed that instead of the "ideal" K-Ar clock, which takes only radioactive decay into consideration, a differential equation for a "real" K-Ar clock be used in which diffusion of $^{40}Ar$ into and out of the crystal is taken into consideration.

Closure temperatures: The decay of $^{40}K$ continually produces radiogenic $^{40}Ar^*$. What is special about the K-Ar method is that the daughter nuclide is a noble gas, which is normally not incorporated into minerals and is not bound in the mineral in which it is formed. In spite of its gaseous character, it is retained and accumulated by most crystal lattices at relatively low temperatures, due to its large atomic radius

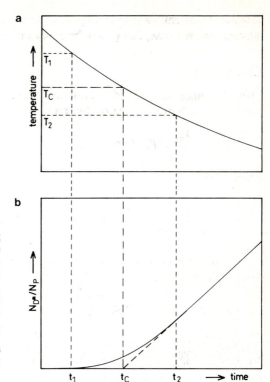

**Fig. 6.3 a, b.** Principle of closure temperatures (after Dodson 1979): **a** cooling curve of a given system; **b** increase in the ratio of radiogenic daughter to radioactive parent isotope $(N_D{}^*/N_P)$ as a function of time. Between times $t_1$ and $t_2$ (corresponding to temperatures $T_1$ and $T_2$), there is a transition range in which part of the daughter nuclide is lost, i.e., only incomplete accumulation of $N_D{}^*$ occurs. The closure temperature $T_c$ is defined as the temperature of the system at time $t_C$

(1.9 Å). But the system is very sensitive to thermal influences: loss by diffusion occurs at elevated temperatures; in melts, accumulation of argon is no longer possible. A graph illustrating the accumulation of argon is shown in Fig. 6.3. Below a critical temperature, $T_2$, all the argon produced is retained by the crystal; at high temperatures, none. In between there is a transition range in which partial accumulation occurs. The critical "blocking" or "closure" temperature, $T_c$, differs not only from mineral to mineral, but also depends on the rate of cooling (Harrison 1981) and fluid activity. The theoretical aspects have been discussed in detail by Dodson (1973, 1979). $T_c$-values determined for the K-Ar system are given in Table 6.2.

Analogous to the Rb/Sr method (Sect. 6.1.3), the data obtained using the K/Ar method can be presented in an isochron plot (e.g., Hayatsu and Carmichael 1970; Shafiqullah and Damon 1974). In this graph the $^{40}Ar/^{36}Ar$ ratios of the samples are plotted versus the $^{40}K/^{36}Ar$ ratios. Cogenetic samples yield a regression line whose slope is proportional to their age; the y-intercept gives the initial argon isotope ratio. The isochron graph, however, is not widely used in K/Ar geochronology because its use is more complicated than, for example, in the Rb-Sr system. Apparent isochrons can appear as lines between radiogenic and atmospheric argon. K-Ar isochrons are used primarily for dating glauconites and diamonds and in cosmochronology. Musset and McCormack (1978) have proposed the use of three-dimensional correlation diagrams $(^{40}Ar-^{36}Ar-K_2O)$ for interpreting K/Ar data.

**Table 6.2.** Range of closure temperatures for the
K-Ar system

| | |
|---|---|
| Hornblende | $500 - 700°C$ |
| Phlogopite | $400 - 470°C$ |
| Muscovite | $350 \pm 50°C$ |
| Phengite | $350 \pm 50°C$ |
| Biotite | $350 - 400°C$ |
| Feldspars | ca $230°C$ |

A special variation of the K/Ar method uses accumulations of radiogenic noble gases for dating groundwater older than about $7 \times 10^4$ a (e.g., Gerling et al. 1967; Tetzlaff et al. 1973; Seifert 1978). If the rate of argon accumulation can be determined for a given reservoir and the $^{40}Ar^*$ concentration is high enough to be measured mass spectrometrically, the age $t$ of the water can be calculated as follows:

$$t = k(^{40}Ar^*/^{40}Ar_{atm}), \tag{6.4}$$

where $K =$ (theoretical soluble $Ar_{atm}$)/(Ar accumulation rate).

Analogous ages can be calculated from $^4He^*/^{40}Ar_{atm}$, $^4He^*/^{20}Ne_{atm}$, and $^4He^*/r$ (where r is the helium accumulation rate). Corrections must be made for the solubility of the noble gases in water and for loss due to diffusion.

### Sample Treatment and Measurement Techniques

The preparation of mineral and whole rock samples is done according to the scheme given in Sect. 5.2.1.1 (Fig. 5.3). The samples (also whole-rock and glass samples) may not be pulverized (loss of argon), but are broken up, sieved, and analyzed in fractions with a narrow grain-size range, as much as possible $> 100 \mu m$. Loss of argon occurs with some minerals (e.g., feldspar) if the sample is ground to a grain size of less than $100 \mu m$ (e.g., Honda et al. 1982). At no time may the samples come in contact with HCl (interference of the masses 36 and 38 in the mass spectrometric determination of the argon).

To solve the age equation, the present $^{40}K$ and $^{40}Ar^*$ concentrations must be determined. For $^{40}K$, only an analysis of total potassium is needed (see Basic Concept). This can be done by wet chemistry, AA (Sect. 5.2.4.3), or XRF (Sect. 5.2.4.5); with especially great presicion by flame photometry (Sect. 5.2.4.3), isotope dilution analysis (Sect. 5.2.4.1), or neutron activation analysis (Sect. 5.2.4.2). For very low concentrations ($< 0.1\%$), the best precision by far ($\pm 2\%$) is obtained by isotope dilution analysis (using a $^{41}K$ spike). But, in general, flame photometry is preferable owing to its greater simplicity; its accuracy with potassium concentrations greater than $1\%$ is comparable with that of isotope dilution analysis (standard deviation of $< 1\%$). In neutron activation analysis, $^{41}K$ is converted to $^{42}K$, which decays by beta and gamma emission with a half-life of $12.36 h$ and is determined radiometrically. An exact knowledge of the irradiation parameters is not necessary if a monitor mineral with a known potassium concentration is irradiated at the same time. The beta or gamma activities are measured with semi-conductor detectors (Sect. 5.2.2.3) and multi-channel analyzers.

The $^{40}$Ar* concentration is usually determined by isotope dilution analysis. The sample is heated in an ultra-high vacuum and melted at temperatures between 1300 and 2000°C; the argon is extracted together with other gases. The special furnaces for this purpose normally consist of a water-cooled quartz-glass cylinder containing a molybdenum crucible heated inductively. For more detail, see Dalrymple and Lanphere (1969), Hunziker (1979) and Roth and Poty (1989).

The extracted argon, together with the other gases, is led into a high-vacuum apparatus over hot copper (II) oxide, which oxidizes hydrogen and carbon monoxide. The water and carbon dioxide that are produced are frozen out in a cold trap or removed using a molecular sieve; the reactive gases are absorbed on a titanium screen (getter); helium can be removed from the remaining gas by fractional distillation. An $^{38}$Ar spike is added for the isotope dilution analysis; this is obtainable with a purity of 99.9997%. Corrections must be made for atmospheric and possibly other extraneous argon. For this purpose a value of 295.5 is used for the $^{40}$Ar/$^{36}$Ar ratio of atmospheric argon; any $^{40}$Ar contributing to a higher ratio than this is considered to be radiogenic. The error thus introduced is insignificant for ages greater than about 300 Ma since in this age range more than 90% of the argon is usually radiogenic. But the proportion of atmospheric argon in very young samples (in unfavorable cases up to 98%) is a problem which can affect the accuracy of the age determination. The amount of radiogenic argon is usually given in cm$^3$ $^{40}$Ar*STP/g sample or in mol/g sample. The amount of sample needed for the argon determination depends on the age of the sample and its potassium concentration.

Techniques have been developed for precise dating of very young and K-rich samples, as well as basalt (from about 5 Ma down to several thousand years). These techniques permit mass spectrometric determination of the radiogenic argon concentration without addition of a spike, i.e., avoiding the small error introduced by isotope dilution. A detailed discussion with examples has been published by Cassignol and Gollit (1982) and Gillot and Cornette (1986). Very old samples can be dated with minute amounts of pure mineral extracts.

Alternatively, the radiogenic argon concentration can also be determined volumetrically or by neutron activation analysis. However, the volumetric method cannot be used for young samples and it must always be accompanied by a mass spectrometric analysis of the argon isotope ratios. Neither method attains the precision of isotope dilution analysis.


## Scope and Potential, Limitations, Representative Examples

*Scope and Potential.* The conventional K/Ar age determination method has developed in the some 40 years since its first application to one of the most widely used routine methods of geochronology of hard rock samples. The comparatively "simple" measurement techniques and the wide range of application of this dating method are equally responsible for this development. The main advantages are:

1) the wide distribution of potassium minerals, especially in magmatic and metamorphic rocks;
2) the fact that non-radiogenic argon generally occurs only in negligible amounts in minerals, i.e., aside from atmospheric contamination, which can be corrected for, the origin of any argon in potassium minerals can normally be only the decay of $^{40}K$;
3) the comparative ease and accuracy of measurement of very small amounts of argon, due to its inert character; and
4) the half-life of $^{40}K$, which is of such a convenient size (1.25 Ga) that measureable amounts of radiogenic argon can also be present in geologically very young samples, making it possible to date samples from all periods of the Earth's history.

Today, K/Ar dating is possible down to ages of a few thousand years (Gillot 1985; Jäger et al. 1985). Highly involved analytical techniques are required, but the method fills the gaps between the periods covered by the uranium disequilibrium methods (Sect. 6.3) and the radiocarbon (Sect. 6.2.1) and $^{10}Be$ (Sect. 6.2.3) dating methods. K/Ar dating has therefore become a useful tool for Quaternary geology as well as for archeology.

*Igneous and metamorphic rocks:* The main application of the K/Ar method is the dating of igneous and metamorphic rocks. Whole-rock samples are often used for the determination of the age of slate and phyllite. But especially in polymetamorphic areas, mica fractions are clearly preferable. Recrystallized potassium-rich white mica, formed from illite already under anchizonal conditions, provides reliable dates for low-grade metamorphism, even for grain-size fractions $< 2\,\mu m$ (e.g., Hunziker 1979; Kralik et al. 1987). With low-grade metamorphism, however, care must be taken that no detrital mica is present.

The radioactive clock of a mineral begins to "tick" when its temperature falls below its closure temperature. Thus, the K/Ar clocks of the various minerals in a rock unit that is cooling very slowly start at different times (cooling age), making it possible to reconstruct the cooling process (e.g., end of metamorphism, tectonic uplift). However, it must be mentioned that the concept of closure temperatures is not undisputed. Some mineral ages for samples from the Alps have been also interpreted as crystallization ages (e.g., Hänny et al. 1975). If during progressive or retrograde metamorphism the crystallization of a datable mineral occurs below its critical blocking temperature and if this temperature is not reached during the further course of metamorphism, the K-Ar dates correspond to the metamorphic crystallization age. Crystallization ages are also provided by concordant mineral age values of very rapidly solidified igneous rocks.

The requirement that only the freshest material be used is generally valid for the K-Ar age determination. The mica minerals (biotites) form the only special case. As a result of the layered lattice of these minerals, potassium and argon are lost in the same proportions during alteration (base-exchange theory, Kulp and Engels 1963); this means that even micas that have been somewhat altered (e.g., slightly chloritized biotite) may still yield reliable K/Ar ages.

*Basalts:* An important application of the K/Ar method is the dating of young mafic volcanic rocks, e.g., basalts and dolerites (whole-rock samples), which could otherwise not be dated using the Rb/Sr or U/Pb methods (e.g., Kaneoka et al. 1982; Seidemann et al. 1984). Synsedimentary volcanic rocks thus permit a temporal classification of stratigraphic sequences. Whole-rock samples must be fine-grained to dense because from coarse-grained samples it would not be possible to obtain sample aliquots that are sufficiently homogeneous for separate analysis of potassium and argon (Engels and Ingamells 1970). Basaltic glass, in contrast to acid glasses, has a very poor argon retentivity and is therefore unsuitable for K/Ar dating. It is very important to examine carefully the interstitial phases to determine whether they are glassy or crystalline since, as a rule, it is here where the highest potassium concentrations are (Mankinen and Dalrymple 1972). Basaltic whole-rock samples should be very fresh because alteration and devitrification affect not only the geochemistry of the alkali (potassium) elements but also the argon retentivity. One criterion for this is the water content, which for volcanites should not exceed 1% (Kaneoka 1972). In the case of dikes, samples from the interior of the dike generally yield the best age values (Hunziker 1979).

*Pyroclastic rocks:* For pyroclastic rocks and acid volcanites (e.g., tuff, bentonite), dating is usually possible only via phenocryst minerals (e.g., mica, sanidine, hornblende), in some cases also via authigenic minerals (Hellmann and Lippolt 1981). Although several recent studies have used volcanic glass from pyroclastic rocks for K/Ar dating (e.g., Hall et al. 1984), they should be interpreted with caution because even siliceous volcanic glass is subject to low-temperature alteration involving a high degree of alkali mobility, which naturally affects the K/Ar dating method (Cerling et al. 1985).

*Sediments:* The K/Ar method is finding increasing use for the direct dating of sediments: (1) detrital minerals may provide information on the source area; (2) the sedimentation or diagenesis age can be obtained by analysis of clay minerals (e.g., Bonhomme 1982). For the latter purpose, pure 1-Md micas and minerals of the glauconite-montmorillonite family are the most suitable. Pure montmorillonite, however, is unsuitable for K/Ar dating due to its expandable interlayers. Criteria for the selection of suitable glauconites have been published by Odin (1982b). Glauconites with $K_2O$ contents below 7% may be used only with reservation. According to Odin, the ideal sediment for a K/Ar age determination contains $> 10\%$ glauconitic minerals, if possible in the $> 200$-$\mu$m fraction; the sediment should contain $< 10\%$ carbonate and in the clay mineral fraction, $< 10\%$ kaolinite. The samples must be absolutely fresh; the influence of groundwater has a negative effect.

*Evaporite minerals:* Although dating of potassium salts was the first application of the K/Ar method (e.g., Aldrich and Nier 1948; Smits and Gentner 1950), it is today relatively unimportant (e.g., Pilot and Rösler 1967; Wardlaw 1968; Brookins et al. 1985). The reason for this is the special properties of salt minerals, the first of which is the poor retentivity of most of them. Hygroscopic salts are particularly difficult to work with. Another unfavorable property is that salts are almost always metamorphic or even polymetamorphic rocks (recrystallization of some already below 100°C). But this may represent a future application for the dating of salts: the determination of the age of very low-temperature metamorphic events.

K/Ar age determinations have been made on sylvite, carnallite, cainite, langbeinite, and polyhalite. Owing to their potassium content, the minerals rinneite, glaserite, schoenite, leonite, and syngenite are also of interest. A comprehensive review has been published by Lippolt (1977).

*Ore deposits and tectonic events:* Clay minerals and micas (e.g., illite and celadonite) whose genesis is unambiguously epithermal are a possibility for dating hydrothermal deposits (e.g., Bonhomme et al. 1983; Halliday and Mitchell 1983; Staudigel et al. 1986) with the K/Ar method. Adularia or muscovite formed hydrothermally in a silicification zone, as well as phyllitic fractions ($< 1.5 \mu$m) from fault zones, can also be used for dating tectonic events (e.g., Kralik and Riedmüller 1985). In each case, very careful examination of the mineralogical and geochemical conditions is necessary, as well as the separation of a pure mineral fraction. The age of formation of porphyry copper deposits can in some cases be determined via syngenetic minerals (Walther et al. 1981).

*Extraterrestrial materials:* The K/Ar method is often used for dating lunar samples and meteorites (e.g., Eberhardt et al. 1971; Pepin et al. 1972; Alexander et al. 1977; Sect. 6.5) but in the last decade it has been largely replaced by the $^{39}$Ar/$^{40}$Ar method (Sect. 6.1.1.2).

**Limitations.** In most cases, K/Ar ages correspond to the stratigraphic classification of the sample and to dates obtained using other radiometric methods. This shows that the basic assumptions are realistic. There are, however, exceptions in which these conditions are not fulfilled. For example, metamorphosed samples may yield a distorted age value if the radiogenic argon that had previously accumulated in the material was not completely driven out before recooling. In this case, a mixed age without geological significance is obtained (Fig. 6.4). Such partial argon loss is possible because diffusion is both temperature *and* time dependent. A review of the literature on diffusion in silicate minerals and glasses has been published by Freer (1981). Combined K-Ca-Ar analyses on authigenic sanidine suggest that diffusion of argon is several orders of magnitude faster at low temperatures than extrapolation from high-temperature data would indicate (Marshall et al. 1986).

Loss of $^{40}$Ar leads to a reduction of the radiometric age. In general, loss of argon can occur as a result of any of several factors, such as too little retentivity of the mineral with respect to argon, thermal and dynamic metamorphism, recrystallization, melting, or mechanical deformation. Shock, interestingly enough, does not seem to disturb the K/Ar system, as experiments with plagioclase feldspars have shown (Jessberger and Ostertag 1982).

A second source of error in K/Ar dating is the possible occurrence of extraneous argon. A distinction is made between excess $^{40}$Ar (i.e., $^{40}$Ar that is incorporated into the rock or mineral by some process other than the in situ decay of $^{40}$K, e.g., diffusion) and inherited $^{40}$Ar (i.e., produced by in situ decay of $^{40}$K *before* the dated

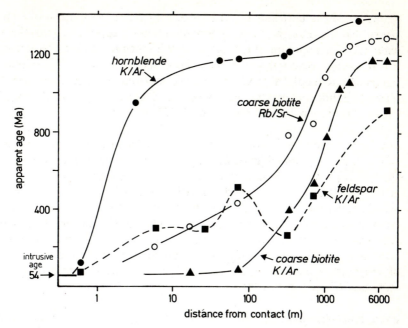

**Fig. 6.4.** Increase in mineral ages of Precambrian rocks from the Front Range in Colorado as a function of distance from the contact with a Tertiary intrusion (Eldora stock, 54 Ma) (after Hart 1964). The differences in the paths of the apparent age values for the different minerals result from differences in the closure temperatures of the minerals (cf. Fig. 6.3)

event). Inherited $^{40}$Ar can, for example, be $^{40}$Ar* that was not completely driven off at the time of metamorphism. Other possibilities are the inclusion of older, argon-bearing fragments in a melt or incomplete degassing of a magma. Excess and inherited argon cannot always be distinguished. Excess argon has been demonstrated in the minerals beryl, cordierite, tourmaline, hornblende, pyroxene, biotite, phlogopite, feldspar, sodalite, diamond, and olivine, in fluid inclusions (e.g., quartz and fluorite), and in basaltic rocks. The use of the isochron method for such samples is promising, as shown by Zashu et al. (1986) for diamonds from Zaire.

Problems with excess argon often occur in the dating of high-pressure metamorphic rocks. Control measurements on cogenetic minerals that contain no potassium should be made in this case. Under high-pressure conditions, argon is also taken up by quartz, for example. Whole-rock dates of such rocks usually cannot be interpreted.

To recognize loss of argon or the presence of extraneous argon, Harper (1970) proposed the use of a plot of $^{40}$Ar versus $^{40}$K. However, interpretation of K/Ar dates for samples with a complex history can be extremely difficult. Several episodic argon losses over long periods of time, possibly associated with the infiltration of extraneous argon, make reconstruction impossible.

***Representative Examples.*** The effects of contact metamorphism on the K/Ar ages of minerals have been studied by Hart (1964) on minerals in Precambrian rocks from the Front Range in Colorado. The country rocks are high-grade gneisses, schists, amphibolites, and pegmatites; they are intruded by numerous Tertiary magmatic rocks. The minerals in the Precambrian rocks along a profile 6 km long beginning at the contact of a quartz monzonitic intrusive body (the Eldora stock) were analyzed using various geochronological methods. The apparent K/Ar ages for the hornblende, biotite, and potassium feldspar are shown in Fig. 6.4, together with the Rb/Sr dates for the biotite. The difference in the behavior of the different minerals with respect to thermal overprinting, caused by different argon retentivities, can be seen very clearly.

An example of the application of K/Ar whole-rock dating to largescale geodynamic problems is a study of the K/Ar geochronology of the South Shetland Islands off the tip of the Antarctic Peninsula (Pankhurst and Smellie 1983). About seventy K/Ar whole-rock ages obtained for low-potassium tholeiitic and andesitic volcanic and intrusive rocks provide evidence for continual volcanic activity from

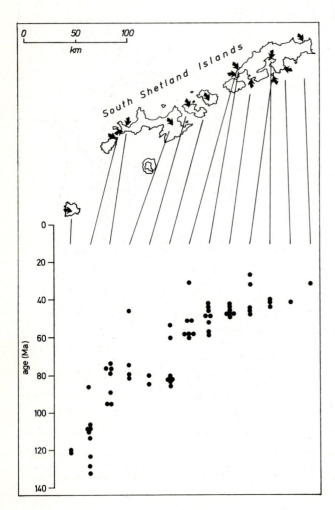

Fig. 6.5. Sketch map of the South Shetland Islands show sampling locations and a gr of the K/Ar dates obtained for tholeiitic and andesitic rocks from these areas. The *dots* reflect the northeastwards drift of the main phase of volcanic activity (after Pankhurst and Smellie 1983)

130 Ma to 30 Ma (Jurassic/Cretaceous to Oligocene/Miocene) and recent activity with more alkalic magma character about 2 million years ago. The volcanic activity during the main phase moved steadily from southwest to northeast (Fig. 6.5). The K/Ar dates thus provide important evidence for the plate tectonics of this area.

   A detailed study on the effects of regional uplift, deformation, and meteoric/hydrothermal metamorphism on the K/Ar dates for biotite has been carried out for the southern part of the Idaho Batholith (Atlanta Lobe) (Criss et al. 1982). The apparent K/Ar ages for this area are generally younger than the true

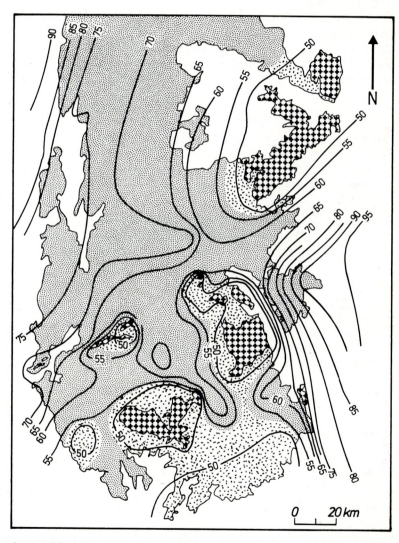

**Fig. 6.6.** Sketch map of the Atlanta Lobe of the Idaho Batholith showing K/Ar age isolines (in Ma) for biotite (simplified after Criss et al. 1982). Eocene intrusions are indicated by +; *stippling* indicates where the biotites lost virtually all their argon during the Eocene meteoric/hydrothermal event

intrusion ages of the biotite-bearing host rocks (probably $> 100$ Ma). These ages vary between 95 Ma at the western and eastern edges of the batholith and 37 Ma in the central part, with a gap between 53 and 46 Ma. The regional distribution of the K/Ar biotite ages is shown in Fig. 6.6. In the central part of the region, which has lower apparent ages, there are epizonal Eocene plutons were K/Ar ages of 44 $\pm$ 6 Ma. This value may be considered to be very close to the crystallization age of the plutons. These Eocene plutons were apparently responsible for extensive meteoric/hydrothermal alteration of the Mesozoic country rocks. This resulted in partial chloritization of the biotite, connected with a lowering of its $K_2O$, deuterium, and $^{18}O$ contents and also its K/Ar age. In addition to this lowering of the K/Ar age by meteoric/hydrothermal processes, Criss et al. also demonstrated the influence of regional uplift on K/Ar biotite ages. A uniform rate of uplift of 0.10–0.14 mm/a for the period from the Late Cretaceous to the Early Tertiary was calculated from these dates.

### Non-Chronological Applications

The isotope geochemistry of argon plays a major role in research dealing with cosmic noble gases (e.g., Heymann 1977) and with the development of the Earth's atmosphere (e.g., Fanale 1971; Ozima 1975; Cadogan 1977; Fisher 1978; Sarda et al. 1985). Alternative models discussed for the latter are a continual degassing of the Earth or a two-step degassing history in which a large amount of gas was lost during the first phase. The very high $^{40}Ar/^{36}Ar$ ratio found during the last several years in rocks derived from the mantle (up to 25,000 in MORB) make the latter appear more probable.

Sarda et al. (1985) have recently shown that by careful selection of the samples (e.g., unweathered glass) and appropriate analytical techniques (e.g., stepwise heating, see Sect. 6.1.1.2 on the $^{39}Ar/^{40}Ar$ method), argon isotopes can be used as geochemical tracers. They found a negative correlation between the $^{40}Ar/^{36}Ar$ and $^{87}Sr/^{86}Sr$ ratios in MORB glasses, which can be interpreted as a result of the mixing of magma from two different mantle reservoirs.

### 6.1.1.2   Argon/Argon ($^{39}Ar/^{40}Ar$) Method**

(Sigurgeirsson 1962)

### Dating Range, Precision, Materials, Sample Size

In principle, this is a modification of the conventional K/Ar method (Sect. 6.1.1.1) whih has important advantages but because it is analytically so complicated to carry out, it has been used by only a few laboratories. The method is used for dating minerals from igneous or metamorphic rocks that contain potassium but not too much calcium, e.g., mica (biotite, muscovite, phlogopite), feldspar (sanidine, K-feldspar, plagioclase), alunite, and amphibole (Merrihue and Turner 1966; Dalrymple and Lanphere 1971, 1974; Dallmeyer 1979; Albarède 1982; McDougall and Harrison 1988). Glauconite, as well as illite and clay minerals from sedimentary rocks, are only sometimes suitable (Halliday 1978); more promising are the first datings of potassium-bearing sulfides (e.g., rasvumite and bartonite) and pyrite (Czamanske et al. 1978; York et al. 1982). Whole-rock samples of shale, phyllite, and hornfels (Reynolds and Muecke 1978), as well as volcanic rock (Hall and York 1984), can also be dated, but the results are sometimes difficult to interpret. Sedimentary sequences can be dated only by using syngenetic volcanic rocks or phenocrysts of them. The time of eruption of Quaternary volcanic rocks can be determined using a laser dating technique on single grains (e.g., van den Bogaard et al. 1987). Information on the thermal history of sedimentary basins can be obtained by the analysis of detrital feldspar (Harrison and Bé 1983). The possibility of using the $^{39}Ar/^{40}Ar$ method to date low-temperature tectonic events (e.g., mylonitization) also seems to be promising (Baksi 1982; Lancelot et al. 1983).

The argon measurements are interpreted via "degassing" spectra. The method can be used to determine ages ranging from those of meteorites (4.5 Ga) down to a few ten thousand years. The ability to recognize extraneous argon and argon losses is of special advantage for very young specimens.

The $^{39}Ar/^{40}Ar$ method has been used with great success for dating lunar samples and meteorites (e.g., Turner 1971, 1977; Sutter et al. 1971; Stettler and Albarède 1978; Bogard et al. 1979; Rajan et al. 1979; Wang et al. 1980; Niemeyer 1983; Sect. 6.5). Silicate enriched, lithic components (clasts) are usually used instead of pure minerals.

The accuracy of the $^{39}Ar/^{40}Ar$ plateau ages is often better than that of conventional K/Ar ages ($\leqslant 1\%$). Prerequisite for this is extremely fresh material.

### Basic Concept

To solve Eq. (6.2) for the conventional $^{40}K/^{40}Ar$ method, separate determinations of the potassium and radiogenic argon concentrations are necessary (cf. Sect. 6.1.1.1). For the $^{39}Ar/^{40}Ar$ method, the sample is irradiated in a nuclear reactor with fast neutrons ($E > 1$ MeV), inducing the reaction

$$^{39}K(n, p)^{39}Ar.$$

Argon-39 is instable and decays to $^{39}K$ by $\beta^-$ emission with a half-life of 269 a. The $^{39}Ar$ thus produced can be analyzed without any problems (Sect. 6.2.8). The potassium in the irradiated sample is represented by $^{39}Ar$ according the following formula (Mitchell 1968):

$$(^{39}Ar_K) = (^{39}K) \cdot t_{ir} \int_{E=0}^{E=max} \phi(E) \cdot \sigma_n(E) \cdot dE \tag{6.5}$$

where $\phi(E)$ (in neutrons/s/cm$^2$) and $\sigma_n(E)$ (in cm$^2$) are irradiation parameters, $t_{ir}$ is the length of time the sample was irradiated, and $E$ is the energy of the neutrons.

Division of Eq. (6.4) by Eq. (6.5) yields

$$\left(\frac{^{40}Ar^*}{^{39}Ar_K}\right) = \frac{\lambda_{EC}}{\lambda_{total}}\left(\frac{^{40}K}{^{39}K}\right)\frac{1}{t_{ir}}\frac{(e^{\lambda_{total}t}-1)}{\displaystyle\int_{E=0}^{E=max}\phi(E)\sigma_n(E)dE} \tag{6.6}$$

At the present time, the irradiation parameters cannot be determined with sufficient precision. The errors associated with this factor are greatly reduced when a monitor mineral with a known K/Ar age $t_{monitor}$ is irradiated and analyzed together with the unknown sample. For each irradiation, a $J$ value can be calculated as follows:

$$J = \frac{(e^{\lambda_{total}\cdot t_{monitor}}-1)}{(^{40}Ar^*/^{39}Ar_K)_{monitor}} \tag{6.7}$$

This value, together with the $^{40}Ar^*/^{39}Ar_K$ ratio measured by mass spectrometry (Sect. 5.2.3.1), can be used to determine the age of the sample:

$$t = \frac{1}{\lambda_{total}}\ln\left\{J\cdot\left(\frac{^{40}Ar^*}{^{39}Ar_K}\right)+1\right\} \tag{6.8}$$

The fundamental, simple form of the $^{39}Ar/^{40}Ar$ method as represented by this equation is seldom used. One recently developed application is the laser dating of single mineral grains.

In the usual form of the Ar/Ar method, the sample is degassed stepwise (incremental heating) (e.g., Lanphere and Dalrymple 1971). In this way, several determinations are made on the same sample yielding a spectrum of apparent ages, whereby each age value is associated with a specific heating temperature. These apparent ages are plotted versus the accumulative percent $^{39}Ar$ released (Fig. 6.7). Using this technique, secondary disturbances of the K-Ar system can be recognized: Geochronologically significant events cause a plateau in the degassing spectrum. The plateau should cover at least three degassing steps in which at least 50% of the degassed $^{39}Ar$ was obtained and whose age values lie with in the $2\sigma$ range (Berger and York 1981; Albarède 1982). The mean weighted according to the analytical errors of the individual steps is taken as the plateau age.

The degassing spectra often yield lower apparent ages for the first degassing steps than for the plateau age (Fig. 6.7). This may be attributed to a loss of $^{40}Ar^*$ by diffusion from the potassium positions that are the least retentive, e.g., lattice defects, microfissures, crystal surfaces, and interstitial positions.

Analogous to the isochron plots of the conventional K/Ar method (Sect. 6.1.1.1), an isochron age determined by plotting $^{40}Ar/^{36}Ar$ versus $^{39}Ar/^{36}Ar$ can also be used for a $^{39}Ar/^{40}Ar$ dating (e.g., Roddick 1978). In this case, a determination of the atmospheric argon component is not necessary. An advantage of incremental heating is that an isochron can be determined from a single sample (Fig. 6.8). The isochrons are calculated using the least-squares regression method (Williamson 1968; York 1969; Minster et al. 1979).

### Sample Treatment and Measurement Techniques

Separation and preparation of minerals is done according to the scheme given in Sect. 5.2.1.1 (Fig. 5.3). The samples may not come in contact with HCl (interference in the mass spectrometric measurement of the argon isotopes). Whole-rock samples are not pulverized, but are broken up, sieved, and analyzed in fractions with a narrow grain-size range (e.g., 60–85 mesh) in the same way as the minerals. For very small grain sizes ($< 10$–$30\,\mu m$), $^{39}Ar$ losses due to recoil effects may be a serious problem.

The samples are sealed in quartz tubes and irradiated, together with a monitor sample, with fast neutrons (ca. $10^{18}/cm^2$) in a nuclear reactor. After a period of time to allow the short-lived radionuclides to decay (at least 2–3 weeks), the sample can be analyzed. It is heated inductively in a molybdenum crucible in an ultra-high vacuum to drive off the argon. The technique of incremental heating requires a great deal of analytical experience.

At each temperature step of the degassing procedure, the sample is heated for an exactly defined period of time (e.g., 1 h). The extracted gas from each stage of heating is purified separately (cf. Sect. 6.1.1.1), after which the isotopic composition of the argon is determined with a mass spectrometer (Sect. 5.2.3.1). An age is calculated for each stage. In addition to corrections for atmospheric argon, corrections must be made for argon isotopes produced during the neutron irradiation from isotopes other than $^{39}K$ (primarily calcium) (Brereton 1970).

Using a special laser fusion technique (Sect. 5.2.4.4), reliable $^{39}Ar/^{40}Ar$ dates can also be obtained for individual mineral grains or phenocrysts of 1 mg or less (York et al. 1981; Maluski and Schaeffer 1982, van den Bogaard et al. 1987), even for Quaternary ages. Prerequisite for this is a homogeneous potassium and argon distribution in the crystal. When such conditions are present, a simple, flat age spectra is obtained, indicating that single-step fusion ages can yield the correct eruption age. This is the case for uncontaminated Quaternary sanidine samples (e.g., Lippolt et al. 1986).

Several different standard samples have been introduced internationally for evaluating the analytical results (Roddick 1983). For more details on $^{39}Ar/^{40}Ar$ measurement techniques, refer to McDougall and Harrison (1988).

### Scope and Potential, Limitations, Representative Examples

*Scope and Potential.* Since, in principle, the $^{39}Ar/^{40}Ar$ method is only a further development of the conventional K/Ar method, its applications are generally the same as those discussed for the K/Ar method (Sect. 6.1.1.1). This is especially the case for closure temperatures and the geological interpretation of the mineral ages (cf. Sect. 4.1.2). To answer the question as to whether a $^{39}Ar/^{40}Ar$ mineral age represents a cooling age or the age of primary formation, additional information is generally necessary, e.g., how near the surface the intrusion ascended or the maximum temperature during metamorphism.

The main advantages of the $^{39}Ar/^{40}Ar$ method with respect to the conventional K/Ar method are as follows:

1) Potassium and argon are measured simultaneously on the same sample aliquot, thus eliminating errors due to inhomogeneity.
2) The technique of stepwise degassing permits the distribution of the argon isotopes in the sample to be checked so that possible argon loss or extraneous argon can be recognized. Differences in the $^{40}Ar^*/^{39}Ar$ ratios for the individual degassing steps indicate disturbance of the K-Ar system, which would mean that the conditions for the calculation of a conventional K/Ar age are not present.
3) The calculation of "plateau" ages, which provides a more accurate age, uninfluenced by loss of argon at the rim of the mineral grain.

The $^{39}Ar/^{40}Ar$ method has proven to be a powerful tool for resolving small age differences between minerals and recognizing disturbances of the K-Ar system. The precision of the method is determined chiefly by the uncertainty in the conventional K/Ar age obtained for the monitor sample. The ability to measure potassium concentrations as low as 10 ppm makes it possible to obtain Ar/Ar ages even for minerals that contain little potassium, e.g., plagioclase, garnet, kyanite, and quartz.

*Dating of minerals:* Sanidine, muscovite, and phlogopite are excellently suited for dating with the $^{39}Ar/^{40}Ar$ method. Biotites analyzed by the $^{39}Ar/^{40}Ar$ method must be very pure and unweathered, in contrast to the conventional K/Ar method, for which a certain amount of chloritization has no effect on the age obtained. Good results are also obtained with hornblende, although sometimes with degassing spectra that are rather difficult to interpret. Terrestrial plagioclase often yields U-shaped spectra, which yields only inadequate age information (e.g., Maluski 1978). In lunar samples, crystallization ages are obtained that are as precise as those obtained with the Rb/Sr method (about 1%; Turner 1977). Shock effects, which can be present in lunar rocks, apprently have little or no affect on the $^{39}Ar/^{40}Ar$ age of plagioclase, as is shown by experiments up to 52.5 GPa on labradorites from anorthosite occurrences in Minnesota and Labrador (Jessberger and Ostertag 1982).

Analogous to the conventional K/Ar method, the analysis of hornblendes opens up the possibility of obtaining radiometric ages for basic igneous rocks. Since the closure temperatures for hornblendes are very high (about 620–750°C according to Berger and York 1981, see also Sect. 6.1.1.1), the hornblende retention age for rocks that were not subjected later to high-grade metamorphism should be very close to the formation age (e.g., Harrison and McDougall 1980).

*Volcanic rocks:* A major application of the $^{39}Ar/^{40}Ar$ method is the dating of young basaltic rocks. The analytical precision of the plateau technique makes it possible to determine ages into the Quaternary ($10^4$–$10^5$ a) with an error of only a few percent (e.g., Hall and York 1984; Fuhrmann and Lippolt 1985, 1986). The samples, however, must be extremely fresh. Specimens containing glasses seem to be unsuitable.

The $^{39}Ar/^{40}Ar$ method can also be used to determine the formation age of tephra and bentonites via the analysis of primary volcanic sanidine phenocrysts or authigenic potassium feldspar. Its superiority over the conventional K/Ar method is especially convenient for calibration of the geological time scale using syngenetic tuffs or bentonites (e.g., Hellmann and Lippolt 1981). Tephra sometimes forms isochronous markers that are widespread over vast areas, providing an important key for unraveling stratigraphic and volcanic events, especially in the Quaternary. The $^{39}$Ar/$^{40}$Ar laser dating technique seems to have a very high potental for solving tephrochronological problems (York et al. 1981; van den Bogaard et al. 1987). This is because it requires no more than a single mineral grain (sanidine or phlogopite). Unaltered phenocrysts are often found not only in unweathered, but even in highly altered tuff and bentonite. Because the laser fusion technique can be used on a single grain, it is easy to detect contaminating, older material (Lo Bello et al. 1987).

*Slates and hornfels:* Whole-rock samples of slates and hornfels can yield precise plateau ages for very fine-grained material (Reynolds and Muecke 1978). These ages can be interpreted as the dates of contact metamorphic events in the case of hornfels and the date of regional metamorphism in the case of slates. Detrital feldspar in such metasediments, however, can cause considerable distortion of the degassing spectra.

*Ores and tectonic events:* The application of the $^{39}$Ar/$^{40}$Ar method to potassium-bearing sulfide ores (e.g., rasvumite, bartonite, and pyrite) (Czamanske et al. 1978) is in the development stage at present. It remains to be seen whether a usable possibility for dating ore deposits results. Further applications for the $^{39}$Ar/$^{40}$Ar method include the reconstruction of the thermal history of sedimentary basins (Harrison and Bé 1983), the indirect dating of young volcanic rocks via incompletely degassed xenoliths (Gillespie et al. 1982), and the determination of the time in which low-grade tectonic structures were formed (Lancelot et al. 1983).

**Limitations.** The main disadvantages of the method are the complicated analytical procedures and the long period of time between irradiation of the sample and the mass spectrometric measurement (most of the induced radioactivity of the sample lasts about half a year). Moreover, the irradiation parameters cannot be measured with adequate precision; not even the measurement of a monitor sample is sufficient to completely remove this uncertainty. Neutron flux gradients during the activation step in the nuclear reactor can be determined by measuring several monitor samples, but this means considerably more analytical work.

A limit inherent in the method is caused by loss or redistribution of $^{39}$Ar as a result of recoil effects during irradiation of the sample. The $^{39}$Ar produced by the nuclear reaction $^{39}$K$(p, n)^{39}$Ar recoils with a maximum energy of about 300 keV and a distance of about 0.1 $\mu$m in silicates (Mitchell 1968). Thus, $^{39}$Ar can leave the sample and be lost during irradiation if the material is very fine-grained ($< 4\,\mu$m) (Alexander et al. 1977). Illite and other clay minerals (Halliday 1978), as well as glauconite are, therefore, not suitable for dating with the $^{39}$Ar/$^{40}$Ar method. Well crystallized glauconites were shown to loose 17–29% of their $^{39}$Ar during irradiation (Foland et al. 1984). In contrast, radiogenic $^{40}$Ar* was retained quantitatively.

Recoil of argon-39 is also a problem for the interpretation of $^{39}$Ar/$^{40}$Ar spectra obtained from biotites. Even if the biotite under the microscope appears to be homogeneous and unaltered, TEM studies show that they may contain substantial amounts of submicroscopic alteration products, e.g., chlorite (Hess et al. 1987). Redistribution of $^{39}$Ar between biotite and intercalated alteration products resulting from the recoil of the $^{39}$Ar on formation during irradiation is assumed, which in combination with other effects during the stepwise degassing leads to a distortion of the degassing spectra. The individual steps of such spectra yield ages that cannot be interpreted geologically, although the mean $^{39}$Ar/$^{40}$Ar age (which corresponds to the conventional K/Ar age) gives a geologically meaningful value (Rittmann 1984).

Whole-rock dating of geologically old igneous rocks is not recommended, especially for acid rocks due to the migration of the alkali elements that often occurs. However, reliable $^{39}$Ar/$^{40}$Ar plateau ages of 3.1–3.5 Ga have been reported by Martinez et al. (1984) for komatiite from the Barberton Mountains, South Africa.

*Representative Examples.* A degassing spectrum for a hornblende from Jebel Shayi Gabbro in Saudi Arabia is shown in Fig. 6.7 (after Fleck et al. 1976). The plateau age derived from four degassing steps is 615.3 ± 8.3 Ma, which correlates with both an orogenesis of the Arabian Peninsula and the Pan-African event. An age of 616.4 ± 4.2 Ma is obtained from the isochron graph (Fig. 6.8) for the same degassing steps.

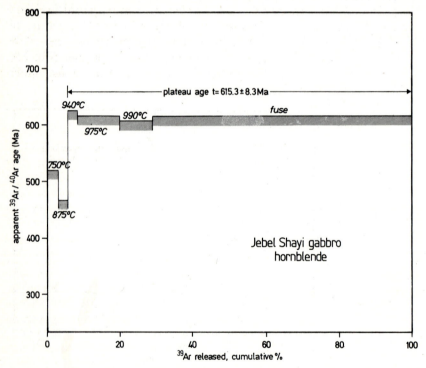

**Fig. 6.7.** An example of a degassing curve for the $^{39}$Ar/$^{40}$Ar method using a hornblende from Jebel Shayi Gabbro in Saudi Arabia (Fleck et al. 1976). The apparent $^{39}$Ar/$^{40}$Ar ages for the individual degassing steps are plotted versus the cumulative percentage of the degassed $^{39}$Ar. The steps used for calculating a "plateau" age are marked

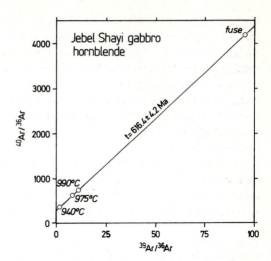

**Fig. 6.8.** Isochron graph of the plateau steps in Fig. 6.7

$^{39}$Ar/$^{40}$Ar mineral ages for several metamorphosed basic intrusive rocks within a high-grade metamorphic area of the Grenville Province near Haliburton, Ontario, have been used for studying the geothermometric history of the area (Berger and York 1981). The closure temperatures for hornblende, biotite, plagioclase and potassium feldspar were calculated according the theory of Dodson (1979) from the $^{39}$Ar/$^{40}$Ar degassing spectra (Sect. 6.1.1.1). These values, together with the age values obtained from them, were used to construct a cooling curve for the area (Fig. 6.9) that is in good agreement with the petrological evidence. Thus, a maximum

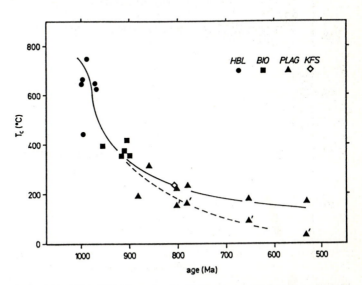

**Fig. 6.9.** Cooling curve derived from $^{39}$Ar/$^{40}$Ar dates for the Haliburton mafic intrusions and the local Haliburton Highlands (after Berger and York 1981). The dates suggest high cooling rates between 5 and 10°C/Ma for times older than about 900 Ma, and substantially lower cooling rates for younger ages (0.3–0.5°C/Ma)

temperature of $> 700°C$ (hornblende values) was reached more than 1 billion years ago, after which the temperature sank to about 200°C, depending on the different uplift and cooling rates, 500–700 million years ago. Most of the cooling curve between about 900 and 500 Ma indicates a slow cooling history for the study area.

Individual K-feldspar and phlogopite crystals from late Pleistocene tephra deposits in the eastern part of the Eifel volcanic field in Germany have been dated by the $^{39}Ar/^{40}Ar$ laser fusion technique (van den, Bogaard et al. 1987). Their results indicate that the Hüttenberg tephra from the Wehr volcano erupted $213,000 \pm 4000$ years ago and the Rieder volcanic complex erupted over a period lasting at least from 470,000 BP to 410,000 BP. These dates are in good agreement with other recent radiometric age studies of this region (e.g., Fuhrmann and Lippolt 1986). Sanidine with apparent ages greater than the eruption age was detected in abundance, particularly in the Rieder phonalite lapilli.

### Non-Chronological Applications

See Sect. 6.1.1.1, conventional K/Ar method.

### 6.1.2   Potassium/Calcium ($^{40}K/^{40}Ca$) Method*

(Ahrens 1951; Backus 1955)

### Dating Range, Precision, Materials, Sample Size

This method is occasionally used for dating minerals that have a very high potassium concentration and practically no calcium ($K/Ca > 50$), such as lepidolite, muscovite, biotite, potassium feldspar, and salt minerals (e.g., Sylvite and langbeinite), for ages of more than about 60 Ma (Coleman 1971; Wilhelm and Ackermann 1972; Heumann et al. 1979; Ovchinnikova et al. 1980). The amount of sample necessary is 30 mg–20 g, depending on the technique used. Recent advances in mass spectrometry have extended the applicability of the method to granitic rocks and their constituent minerals (biotite, potassium feldspar, plagioclase) and to sedimentary rocks (Marshall and DePaolo 1982; DePaolo et al. 1983).

### Basic Concept

Potassium has three natural isotopes: masses 39 (93.2581%), 40 (0.01167%), and 41 (6.7302%). Potassium-40 is radioactive with dual decay: by $\beta^-$ emission ($E = 1.32$ MeV) to $^{40}Ca$ (ca. 89%) and by electron capture (EC) and subsequent $\gamma$-emission ($E = 1.46\%$) to $^{40}Ar$ (ca. 11%) (see Fig. 6.2). They decay constants were established by conventional means (Steiger and Jäger 1977): $\lambda_\beta = 4.962 \times 10^{-10} a^{-1}$ and $\lambda_{EC} = 0.581 \times 10^{-10} a^{-1}$. Thus, $\lambda_{total}$ ($= \lambda_\beta + \lambda_{EC}) = 5.543 \ 10^{-10} a^{-1}$. The branching ratio $\xi_\beta$ ($= \lambda_\beta/\lambda_{total}) = 0.8952$.

Six isotopes of calcium occur in nature: masses 40 (96.9821%), 42 (0.6421%), 43 (0.1334%), 44 (2.0567%), 46 (0.0031%), and 48 (0.1824%) (Russell et al. 1978).

Radioactive decay of $^{40}K$ leads in time to an enrichment of $^{40}Ca$ relative to any of the other calcium isotopes (i) according to the following basic equation:

$$\left(\frac{^{40}Ca}{^{i}Ca}\right) = \left(\frac{^{40}Ca}{^{i}Ca}\right)_0 + \left(\frac{^{40}K}{^{i}Ca}\right)\frac{\lambda_\beta}{\lambda_{total}}(e^{\lambda_{total}t} - 1) \tag{6.9}$$

Analogous to the Rb/Sr method (Sect. 6.1.3, see Fig. 6.12), this enrichment process can be illustrated in a K/Ca isochron graph, e.g., $^{40}Ca/^{42}Ca$ versus $^{40}K/^{42}Ca$. In such a plot points for samples of the same age all lie along a straight line (isochron) whose slope is a function of the time $t$ since the last homogenization of the calcium isotopes (e.g., $[^{40}Ca/^{42}Ca]_0$). Starting with the age equation [derived from Eq. (6.9)]:

$$t = \frac{1}{\lambda_{total}}\ln\left[\left(\frac{^{40}Ca^*}{^{40}K}\right)\frac{\lambda_{total}}{\lambda_\beta} + 1\right] \tag{6.10}$$

the following equation can be derived for the isochron method:

$$t = \frac{1}{\lambda_{total}}\ln\left(\frac{(^{40}Ca/^{42}Ca) - (^{40}Ca/^{42}Ca)_0}{(^{40}K/^{42}Ca)(\lambda_\beta/\lambda_{total})} + 1\right) \tag{6.11}$$

A plot of $^{40}Ca/^{44}Ca$ versus $^{40}K/^{44}Ca$ can also be used to obtain an isochron age (e.g., Ovchinnikova et al. 1980).

Due to the short half-life of $^{40}K$, e.g., relative to that of $^{87}Rb$, the function for the enrichment of $^{40}Ca$ with respect to time is strongly curved.

### Sample Treatment and Measurement Techniques

The preparation of mineral and whole-rock samples is done according to the scheme given in Fig. 5.3. A grain-size fraction between 100 and 250 $\mu m$ is selected for the mineral samples. The samples are digested in a mixture of HF and $HClO_4$. Calcium and potassium are quantitatively separated by ion exchange chromatography. The potassium is normally determined by flame photometry (Sect. 5.2.4.3) or mass spectrometry (Sect. 5.2.3.1). If a lithium-rich mineral (e.g., lepidolite) is being analyzed, the flame photometer must be calibrated with something other than an internal Li standard. A $^{41}K$ spike is used in the mass-spectrometric isotope dilution analysis (Sect. 5.2.4.1; Heumann et al. 1979).

The most important condition for the success of K/Ca dating is a very precise measurement of the calcium isotope ratio. This is especially true for the isochron method. Russell et al. (1978) have developed a high-precision mass spectrometric technique for measuring Ca isotope ratios. Spikes of $^{43}Ca$ or $^{42}Ca$ are used for the isotope dilution.

### Scope and Potential, Limitations, Representative Examples

**Scope and Potential.** The K/Ca method can, in principle, be used in the same way as the Rb/Sr method (Sect. 6.1.3): The age of a sample can be determined if the initial calcium isotope ratio ($^{40}Ca/^{42}Ca$ or $^{40}Ca/^{44}Ca$) is known. If the sample has a very high K/Ca ratio, any error in the initial Ca isotope ratio will have little effect on the age determined. If a number of cogenetic samples (minerals or whole-rock samples) are analyzed using the isochron method, the initial Ca isotope ratio and the age can be determined simultaneously.

Crystallization ages can be determined using the K/Ca method on minerals that have very high K concentrations and practically no Ca, e.g., pegmatite minerals like lepidolite, muscovite, and potassium feldspar (Coleman 1971; Heumann et al. 1979; Ovchinnikova et al. 1980), and salt minerals, e.g., sylvite, langbeinite, and carnallite (Wilhelm and Ackermann 1972; Heumann et al. 1979; Baadsgaard 1987).

Very little is known as yet about the behavior of the K-Ca system in metamorphic processes. The K-Ca system seems to be more stable than the K-Ar or even the Rb-Sr systems. Ovchinnikova et al. (1980) have reported closed system behavior in Archean lepidolites from the Soviet Union (K/Ca age: $3.0 \pm 0.3$ Ga), whereas the K-Ar and Rb-Sr systems in these minerals were open during later geological events. Baadsgaard (1987) has reported similar behavior for salt minerals (halite, sylvite, and carnallite) in potash horizons in the Devonian Prairie Evaporite in Saskatchewan.

Progress in the mass spectrometry of Ca has made it possible to also determine the age of normal granitic minerals and whole-rock samples, but this application is still limited due to the very specialized high-sensitivity mass spectrometry. This possibility for dating is of particular interest for acid igneous and metamorphic rocks containing little zircon.

In addition, it is also possible to date diagenetic minerals (e.g., glauconite, feldspar, and salts) in marine sediments. This possibility assumes that these minerals at the time of their formation had the same Ca isotope ratios as sea water, which is similar to that of the Earth's mantle ($^{40}Ca/^{42}Ca = 151.016$ or $^{40}Ca/^{44}Ca = 47.119$; Marshall et al. 1986). Since this may be viewed as a constant, the initial Ca ratio of all diagenetic minerals in marine sediments is considered to be known.

*Limitations.* The big disadvantage of the K/Ca method is that not only potassium but also calcium is one of the major elements of the Earth's crust. The isotope of interest, $^{40}Ca$, makes up $> 97\%$ of this "common" calcium. This makes an exact determination of the amount of radiogenic $^{40}Ca$ extremely difficult (danger of contamination!). In addition, mass spectrometric analysis of Ca is not easy owing to fractionation effects and a high ionization potential. These disadvantages limited the method until recently to the dating of very old (Precambrian) minerals with a very high K concentration and practically no Ca.

*Representative Examples.* Marshall and DePaolo (1982) introduced the isochron technique for K/Ca age determination. Analyzing biotite, potassium feldspar, plagioclase, and a whole-rock sample, they obtained a K/Ca age of $1041 \pm 32$ Ma for granite from Pikes Peak in Colorado (Fig. 6.10). This result is in good agreement with the corresponding Rb/Sr isochron age of the Pikes Peak batholith ($1008 \pm 13$ Ma, Barker et al. 1976). They determined an initial $^{40}Ca/^{42}Ca$ ratio of $151.024 \pm 0.016$.

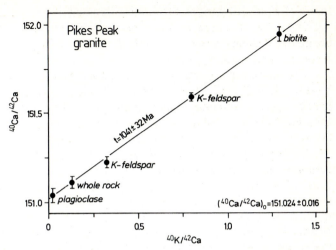

**Fig. 6.10.** K/Ca isochron plot for Pikes Peak granite (after Marshall and DePaolo 1982). The corresponding Rb/Sr isochron age is $1008 \pm 13$ Ma (Barker et al. 1976)

The first ages of authigenic minerals were determined by DePaolo et al. (1983) and Marshall et al. (1986). In both studies, authigenic sanidine from sedimentary rocks were used.

### Non-Chronological Applications

The K-Ca system is the only isotopic system that provides direct information about the geochemical behavior of some of the major elements involved in magmatic processes. This was recognized as early as 1932 by Holmes, whose proposal for using the K-Ca system to clarify the genesis of granitic rocks represents the conceptual beginning of isotope geochemistry. But due to technical difficulties, there is little data available on calcium isotope ratios in terrestrial rocks.

As a result of the large relative differences in the masses of the calcium isotopes (masses 40 to 48), isotopic fractionation occurs during geological processes. According to Russell et al. (1978) isotopic fractionation of calcium during geological processes reaches a maximum of $\Delta = 2.5‰$. However, when the isotope ratios are normalized relative to a constant ratio that does not involve a radiogenic or radioactive isotope (e.g., $^{42}Ca/^{44}Ca = 0.31221$), fractionation resulting from differences in mass is removed from consideration. When this is done, it is found that the isotopic composition of calcium in meteorites, lunar samples, and material from the Earth's mantle is extraordinarily constant (average $^{40}Ca/^{42}Ca = 151.016 \pm 11$; Marshall and DePaolo 1982). In potassium-rich crustal matter, $^{40}Ca/^{42}Ca$ ratios up to 151.9 can be expected due to radiogenic $^{40}Ca$.

### 6.1.3   Rubidium/Strontium ($^{87}$Rb/$^{87}$Sr) Method***

(Hahn and Walling 1938)


*Dating Range, Precision, Materials, Sample Size*

This is a relatively time-consuming standard method for determining (conventional) ages, normally for those greater than 10 Ma. Both mineral and whole-rock samples are used. Magmatic, metamorphic and (with some limitations) sedimentary rocks or minerals can be dated. The interpretation of the dates as cooling, intrusion, metamorphism, or diagenesis (early or late) age must be clarified from case to case (e.g., Faure 1977, 1986; Jäger 1979). Minerals with a high Rb/Sr ratio are required for a pure mineral age (e.g., muscovite, biotite, K-feldspar, phengite, adularia, phlogo-pite, lepidolite, leucite). Minerals that are cogenetic with these but which have low Rb/Sr ratios (such as plagioclase, apatite, epidote, garnet, ilmenite, hornblende, and pyroxene) are also analyzed for mineral isochrons. Authigenic clay minerals (e.g., smectite, illite, and glauconite) isolated from sediments can also be dated (Clauer 1979, 1982; Montag and Seidemann 1981), as well as rubidium-rich salt minerals, such as carnallite, rinneite, and langbeinite (e.g., Lippolt and Raczek 1979a, b; Brookins et al. 1985).

Ore deposits and mineralizations may be dated with the Rb/Sr method via syngenetic minerals like muscovite, biotite, or adularia (e.g., Richards et al. 1982; Walther et al. 1982). Indirect dating of an ore deposit is also possible using calcite (Ruiz et al. 1984) or clay minerals. Alterations, as well as tectonic events, also seem to be datable if this resulted in the formation of suitable new minerals (e.g., Reymer 1982; Wickman et al. 1983; Piasecki 1985).

When isochrons are used to date whole-rock samples, at least four samples should be used with Rb/Sr ratios that are as far apart as possible. The selection should be done after having determined a large number of Rb/Sr ratios using a routine method (e.g., XRF, Sect. 5.2.4.5; AA, Sect. 5.2.4.3). The Rb/Sr method is little suited for basic and ultrabasic samples with low Rb/Sr ratios ($<$ ca. 0.3). The method is one of the principal ones used for dating meteorites and lunar samples (Sect. 6.5; see e.g., Papanastassiou and Wasserburg 1971, 1973; Faure and Powell 1972; Tatsumoto et al. 1976; Nyquist 1977; Minster and Allègre 1979, 1981).

Depending on the strontium content, ca. 20 mg (apatite) to 300 mg (muscovite) of a mineral is needed for the analysis. Aliquots of 100–500 mg are taken from whole-rock samples of 25–40 kg. The precision of the method is 1–5%.


*Basic Concept*

The alkali metal rubidium does not form any minerals of its own, but is found in potassium minerals substituted for potassium. It has two natural isotopes: masses 85 (72.1654%) and 87 (27.8346%) (Catanzaro et al. 1969). Rubidium-87 is radio-active, decaying by $\beta^-$ emission ($E_{max} = 275$ keV) into strontium-87 with a half-life of $\tau = 48.813$ Ga, corresponding to a decay constant of $\lambda = 1.42 \times 10^{-11}$ a$^{-1}$. For calculating Rb/Sr ages, this decay constant and a $^{85}$Rb/$^{87}$Rb ratio = 2.59265 were recommended in 1976 by the Subcommission for Geochronology of the IUGS (Steiger and Jäger 1977). The values $1.47 \times 10^{11}$ $a^{-1}$ and 1.39 were used earlier. Changes in the $^{85}Rb/^{87}Rb$ ratio resulting from radioactive decay of $^{87}Rb$ can be neglected in most cases due to the large half-life.

Strontium forms a few minerals of its own, but also substitutes for calcium and, more rarely, potassium in their minerals. It has four natural isotopes with the masses 88 (82.58%), 87 (7.00%), 86 (9.86%), and 84 (0.56%). The following ratios are used by IUGS convention: $^{84}Sr/^{86}Sr = 0.056584$; $^{86}Sr/^{88}Sr = 0.1194$ (Steiger and Jäger 1977). The latter value was determined by Nier in 1938; it is used for correcting mass spectrometer measurements of strontium isotopes for fractionation effects. The strontium isotopic composition of rocks is changed by the formation of radiogenic $^{87}Sr$ by the decay of $^{87}Rb$.

The ratio of $^{87}Rb$ and radiogenic $^{87}Sr*$ (in terms of number of atoms per unit weight of mineral or rock) is used to calculate age as follows:

$$t = \frac{1}{\lambda} \ln \left[ \left( \frac{^{87}Sr*}{^{87}Rb} \right) + 1 \right] \tag{6.12}$$

This equation does not take primary strontium into consideration. With only a few exceptions, almost all minerals and rocks contain primary non-radiogenic strontium with differing isotopic compositions and thus also differing concentrations

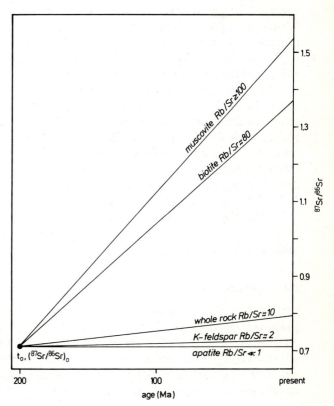

**Fig. 6.11.** A Compston-Jeffery diagram showing a single-stage case for the evolution of the strontium isotope composition. At time $t_0$, the rock, and thus also the individual minerals, was homogenized with respect to the $^{87}Sr/^{86}Sr$ ratio. Thereafter, the $^{87}Sr/^{86}Sr$ ratio of each component of the rock evolves along a line that has a slope of $\lambda(^{87}Rb/^{86}Sr)$

of non-radiogenic $[^{87}Sr]_0$. Since $^{87}Sr* = [^{87}Sr]_{total} - [^{87}Sr]_0$, knowledge of the primary concentration of $^{87}Sr$ would be necessary to solve Eq. (6.12). This problem can be solved if it is assumed the geological event to be dated caused a homogenization of the strontium isotopes within the rock sample. This means that at time zero ($t_0$) all of the minerals of the rock had the same strontium isotope ratio (Fig. 6.11). But the Rb/Sr ratio varies from mineral to mineral: Thus, the time-integrated growth of the $^{87}Sr/^{86}Sr$ ratio caused by in-situ decay of $^{87}Rb$ will be different in the different minerals of the rock. This is illustrated in a Compston-Jeffery diagram (1959) shown in Fig. 6.11; the $^{87}Sr/^{86}Sr$ ratio is plotted versus time $t$.

By tranformation of Eq. (6.12),

$$^{87}Sr = (^{87}Sr)_0 + {}^{87}Rb(e^{\lambda \cdot t} - 1) \tag{6.13}$$

If Eq. (6.13) is divided by $^{86}Sr$, which is stable and not involved in radioactive decay, then the following equation for a straight line is obtained:

$$\frac{^{87}Sr}{^{86}Sr} = \left(\frac{^{87}Sr}{^{86}Sr}\right)_0 + \frac{^{87}Rb}{^{86}Sr}(e^{\lambda \cdot t} - 1) \tag{6.14}$$

Since $t \ll \tau$ for all terrestrial samples, the term $(e^{\lambda t} - 1)$ can be simplified to $\lambda t$. The slope of the individual growth lines then is equal to $\lambda[^{87}Rb/^{86}Sr]$ and can be calculated from the Rb/Sr ratios of the individual rock components. Since the present $^{87}Sr/^{86}Sr$ and Rb/Sr ratios can be measured, the time since the last isotopic homogenization, as well as the initial strontium isotope ratio, can be determined via the intersection of the growth lines of at least two congenetic minerals (Fig. 6.11).

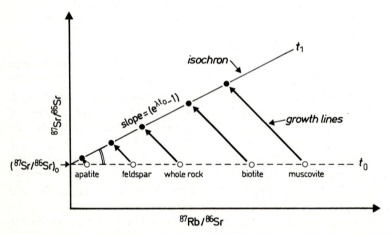

**Fig. 6.12.** An isochron diagram (Nicolaysen 1961) showing a single-stage case for the evolution of the $^{87}Sr/^{86}Sr$ and Rb/Sr ratios: the minerals and the whole-rock sample have different Rb/Sr ratios. At time $t_0$ (= the last isotopic homogenization) these ratios all lie along a horizontal line (i.e. with the same $^{87}Sr/^{86}Sr$ ratio). With increasing age of the rock, the sample points move along growth lines with a slope of $-1$. At any time $t_1$ they all lie along a straight line (called an isochron) whose slope is a function of the time since $t_0$ and whose y-intercept is the initial strontium isotope ratio

The form of the Rb/Sr method generally used today is the calculation of isochrons and their presentation in Rb-Sr isochron diagrams. This method is based on the work of Nicolaysen (1961) and Allsopp (1961). The $^{87}Sr/^{86}Sr$ ratios of the samples are plotted versus the $^{87}Rb/^{86}Sr$ ratios (see Fig. 6.12). Equation (6.14) is the basis for this plot, the slope being here ($e^{\lambda t} - 1$), in contrast to the Compston-Jeffery model. At time zero of the system ($t_0$, i.e., the time of the last strontium isotope homogenization), all of the cogenetic minerals and thus also the whole-rock sample lie on a horizontal line and have the same $^{87}Sr/^{86}Sr$ ratio ( = initial strontium isotope ratio) but, depending on the mineral, different $^{87}Rb/^{86}Sr$ ratios. Since a $^{87}Sr$ atom is formed from each $^{87}Rb$ that decays, the points for the different cogenetic minerals move on parallel lines with a slope of $-1$. But within any one period of time, more $^{87}Rb$ decays in the rubidium-rich minerals than in minerals containing little rubidium. Thus, the points for cogenetic samples at any given time $t$ in the parent-daughter isotope correlation all lie on a straight line (i.e., isochron). The slope of this line is a function of the time since the last isotopic homogenization. The greater the slope of the isochron, the older the rock or group of cogenetic rocks. The intercept of these isochrons with the $y$-axis yields the initial strontium isotope ratio $(^{87}Sr/^{86}Sr)_0$ of the system.

An age determined in this way from several cogenetic samples is called an isochron age. A number of mathematical procedures based on the principle of the least squares are available for calculating the isochrons and their standard deviations (McIntyre et al. 1966; Brooks et al. 1968; Williamson 1968; York 1966, 1969). The most used method is that of York; a FORTRAN computer program utilizing this method is included as an appendix in a book by Faure (1977). A model age can be obtained from a single mineral or rock sample (with a high Rb/Sr ratio) if a geologically meaningful initial isotope ratio can be used. The accuracy with which the age can be determined depends primarily on the $^{87}Rb/^{86}Sr$ ratio.

### Sample Treatment and Measurement Techniques

For the dating of minerals, the separation and preparation of the samples is done by the standard methods described in Sect. 5.2.1.1. The preparation of particularly pure mineral samples ( > 99.5%) is absolutely necessary, especially for young samples. Special methods are necessary for the separation of clay minerals, including sedimentation and centrifugation (Clauer 1979). For isochron ages of whole-rock samples, routine analytical methods (e.g., XRF, Sect. 5.2.4.5, or AA, Sect. 5.2.4.3) are used to select samples with the widest possible variation in their Rb/Sr ratios. The samples must be very fresh, since the Rb-Sr system is very sensitive to weathering. The weight of the whole-rock sample should be 25–40 kg for acid rocks; smaller samples are also suitable for basic rocks.

The $^{87}Sr/^{86}Sr$ ratio and the rubidium and strontium concentrations (or the Rb/Sr ratio) in the samples must be determined for the age equation. The determinations of rubidium and strontium can be done by XRF (Pankhurst and O'Nions 1973) or flame photometry (Sect. 5.2.4.3), rubidium can also be determined by neutron activation analysis. Much greater accuracy, however, is obtained by isotope dilution analysis (Sect. 5.2.4.1), which for strontium can be done together with the mass spectrometric measurement (Sect. 5.2.3.1) of the $^{87}Sr/^{86}Sr$ ratio. For this purpose, the carefully prepared aliquots (ca. 100–300 mg) are digested with super-pure acids (e.g., a $HNO_3$–HF mixture) in teflon or platinum crucibles. If the samples are treated in glass vessels, contamination with normal strontium can occur, especially for mica minerals, which leads to incorrect dates for young samples. To determine abundances by isotopic dilution analysis, $^{84}Sr$ and /or $^{87}Rb$ spikes are added separately to the solution (or, better, a highly enriched $^{87}Rb$-$^{84}Sr$ double spike). The amount of spike is chosen on the basis of the normal chemical analysis of the sample.

Rubidium-87 and Strontium-87 have the same atomic mass and thus cannot be distinguished mass spectrometrically. Thus, to avoid errors in the measurement, the two elements must be quantitatively separated chemically before the analysis. This can be done in various ways. One method that has proven itself is the chromatographic separation of rubidium and strontium with ion exchange resins. Other possibilities for rubidium separation are adsorption by $Zr_3PO_4$ or precipitation as the perchlorate. The isotopic compositions of the fractions are then measured mass spectrometrically. It is usual to use constant statistical errors (a priori errors) for the isotope ratios when calculating the age (e.g., 0.05% for $^{87}Sr/^{86}Sr$ and 1.2% for $^{87}Rb/^{86}Sr$), even when the individual measurements have a smaller analytical error.

Special analytical techniques for determining Rb/Sr dates on microscopic samples have been developed for studying extraterrestrial material (Papanastassiou and Wasserburg 1981; Sect. 6.5).

### Scope and Potential, Limitations, Representative Examples

**Scope and Potential.**  Although the first dating with the Rb/Sr method was reported in 1943 (Hahn et al.), the method did not become widely used until after 1950 when the modern, solid-source mass spectrometer was introduced (Sect. 5.2.3.1). Today, it is one of the most important standard methods of geochronology.

The significance of the Rb/Sr method results from the fact that rubidium and strontium, due to their geochemical relationship to potassium and calcium, respectively, very closely mimic the behavior of these two elements, which are major elements in magmatic processes. Rubidium and strontium, therefore, are very important trace elements, which can convey important information, particularly for the petrology of granites. While strontium in granitic systems substitutes chiefly in early-phase minerals like apatite or plagioclase, rubidium is incompatible and therefore becomes enriched in residual melts. This leads to a large variability in the Rb/Sr ratio during differentiation and therefore provides ideal conditions for the isochron method. This is particularly true for granitic rocks, the dating of which has been the main application of the Rb/Sr method. In addition to the dating of igneous rocks, the determination of the time of recrystallization of metamorphic rocks is an important application of the Rb/Sr method.

*Cooling ages and closure temperatures:* In accordance with the concepts for determining isotopic ages given in Sect. 4.1.2, the Rb/Sr method has the following prerequisites:

- The mineral or rock being dated was formed rapidly compared with its age, i.e., by a single, short event and not the sum of processes taking a long time.
- The minerals and rocks formed a closed system for the entire time after they were formed. No rubidium or strontium was added or removed by circulating solutions or metamorphic overprinting.

Even if these conditions are fulfilled, the ages obtained from different cogenetic minerals need not be concordant with each other or with the age obtained with a whole-rock isochron. The dates can be discordant without necessarily being wrong. This is due to differences in the "closure" temperatures (see Sect. 4.1.2) of the different minerals. The following values are suggested for the Rb-Sr system (Jäger 1979): muscovite $500 \pm 50°C$; phengite $500 \pm 50°C$; biotite $300 \pm 50°C$. Higher closure temperatures (above $400°C$) have also been proposed for biotite (e.g., Del Moro et al. 1982). Only when the temperature falls below these critical values does isotopic exchange with the immediate surroundings of the mineral grain cease and the radiometric clock begin to run. Conversely, the minerals "open" again if they are heated above the critical temperature, i.e., a homogenization of the isotopes occurs again. Extensive theoretical discussions on closure temperatures have been published by Dodson (1973, 1979). In any case, the problem of closure temperatures seems to be closely connected with the presence or absence of fluids.

In most cases, minerals yield only cooling ages. This is basically true also for igneous rocks, since the closure temperatures of the minerals are distinctly below the solidus of magmatic melts. Whether the various minerals defining an internal isochron will yield concordant or discordant closure temperatures (and therefore the age values) depends on whether closure in the minerals occurred simultaneously (e.g., Cavazzini 1988). A model for the determination of the closure temperature of internal isochrons has been proposed by Ganguly and Ruiz (1986). Only for igneous intrusions that cooled very rapidly can the mineral age be interpreted as the age of the intrusion. In this case, the ages of the various cogenetic minerals must be concordant. For volcanic rocks, it is thus always the time of eruption that is dated. For metamorphic rocks, the time of (metamorphic) mineral formation can be dated if the mineral was formed at temperatures below its critical "opening" temperature (Dodson 1976) and this temperature was not exceeded during the remaining course of metamorphism.

*Igneous rocks and metamorphic events:* When subjected to thermal stress, the Rb-Sr system in whole-rock samples is considerably more stable than in minerals. For example, whole-rock isochrons for igneous rocks usually give the time of intrusion. This is also true for igneous rocks subjected to later metamorphism, since although isotopic exchange of strontium occurs between the different minerals when fluid phases are present during metamorphism, the bulk isotopic composition is not, as a rule, disturbed if the whole-rock sample is large enough (this is the reason for the large sample sizes of 25–40 kg).

The rubidium-strontium isotopic evolution in an igneous rock is shown in Fig. 6.13 for a two-stage case. If during a metamorphic event (time $t_1$) the rock

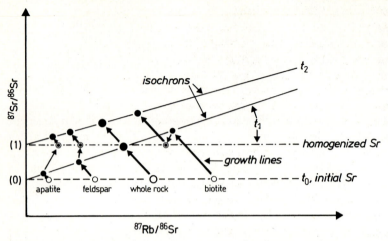

**Fig. 6.13.** An isochron diagram (Nicolaysen 1961) showing a two-stage case for the evolution of the $^{87}Sr/^{86}Sr$ and $^{87}Rb/^{86}Sr$ ratios. At time $t_0$, i.e., the formation of the rock and its minerals, all of the mineral phases have the same initial isotopic ratio $(^{87}Sr/^{86}Sr)_0$. The decay of $^{87}Rb$ causes the $^{87}Sr/^{86}Sr$ and $^{87}Rb/^{86}Sr$ ratios to evolve along straight lines with a slope of $-1$. At time $t_1$, when isotopic homogenization occurred, the points lie along an isochron with a slope depending on the time interval $t_1 - t_0$. This homogenization results in new initial values, from which new growth lines develop, producing a new isochron

formed at time $t_0$ is heated above the closure temperatures given above, then the strontium isotopic composition is homogenized, partially or completely, depending on the temperature and duration of the event. Complete homogenization is usually attained within the amphibolite facies. However, homogenization of the strontium isotopes occurs only within a small volume, the whole-rock sample remains a closed, undisturbed system. The $^{87}Sr/^{86}Sr$ ratios of all the minerals in the rock are changed to the average ratio of the whole-rock sample (assuming complete homogenization). Thereafter, the points for the mineral components move on new growth lines (Nicolaysen 1961).

Metamorphism thus leads to a loss of the age information stored in the minerals. For this reason, minerals from metaigneous rocks cannot be used to date their magmatic origin, but only the last metamorphic event. This, however, is not a disadvantage, because the whole-rock samples remain uninfluenced. As can be seen in Fig. 6.13, the $^{87}Sr/^{86}Sr$ ratio of the metamorphically homogenized strontium (time $t_1$) is higher than the original initial isotope ratio. Metamorphosed cogenetic whole-rock samples with differing primary Rb/Sr ratios (e.g., different differentiation products of a magma), therefore, have different strontium isotope ratios from sample to sample. This is illustrated in Fig. 6.14. An isochron through these sample points yields the primary magmatic age even with later metamorphic overprinting. In this way, a two-stage history can be reconstructed from a metamorphically overprinted igneous rock: whole-rock dating yields the intrusion age, mineral dating yields the time of metamorphism.

Although it does not fit the conventional model for Rb/Sr isochron dating, resetting of Rb/Sr whole-rock isochrons by high-grade metamorphism (granulite facies) has been reported (e.g., Burwash et al. 1985).

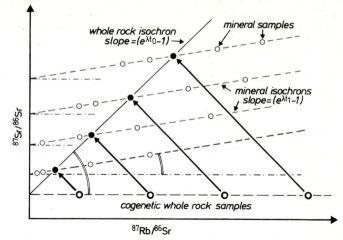

**Fig. 6.14.** Whole-rock and mineral isochrons for samples with a two-event history (two-stage case). Several cogenetic rock samples with differing Rb/Sr ratios are shown. A metamorphic event affects only the isotopic compositions of the minerals but not those of the whole-rock samples. The whole-rock dates, therefore, lie along an isochron whose slope is a function of the time since time $t_0$, the mineral dates lie along concordant isochrons whose slopes are a function of time $t_1$

*Ore deposits, mineralization, and tectonic events:* Various attempts have been made to date alteration of granitoid rocks (e.g., Wickman et al. 1983; Turpin 1985). The Rb/Sr method has been successfully applied for the dating of ore deposits or mineralizations when syngenetic minerals, e.g., muscovite, biotite, and adularia, are present (e.g., Walther et al. 1981; Richards et al. 1982; Mensing and Faure 1983). Indirect determination of the approximate formation age of hydrothermal mineralizations by the analysis of minerals with a low Rb/Sr ratio (e.g., calcite and fluorite) has been proposed by Ruiz et al. (1984). Prerequisite for this is a high Rb/Sr ratio in the country rock. This application is based on the assumption that the isotopic composition of the strontium in the mineral at the time of mineralization was the same as that of the country rock. The present difference between the strontium isotope ratios of the country rock and the mineralized rock is then a function of both time and the Rb/Sr ratio in the country rock.

Very low-grade metamorphic resetting of the strontium isotope radiometric clock of magmatic rocks and minerals at temperatures below 250°C has been reported by André and Deutsch (1985) for Ashgillian ignimbrites from the Brabant Massif in Belgium. This resetting appears to be due to tectonic stress along a fault zone. Thus, Rb/Sr dating makes it possible to determine the time of tectonic movements. This possibility is also provided by minerals formed in fault or ductile shear zones (e.g., Reymer 1982; Kralik and Riedmüller 1985; Piasecki 1985). Fluid transport seems to be the major process for resetting the Rb-Sr system in this type of environment (Hickman and Glassley 1984).

*Stratigraphic sequences:* The dating of pyroclastic rocks such as tuffs and bentonites is possible, as a rule, only via their phenocryst minerals (e.g., biotite, muscovite, sanidine). This is a proven procedure for assigning radiometric ages to stratigraphic sequences (and therefore very useful for time-scale problems), but this is done more often with the K/Ar method (Sect. 6.1.1.1) than with the Rb/Sr method (Baadsgaard and Lerbekmo 1982).

*Sediments:* The direct dating of sediments by the Rb/Sr method is, in principle, possible by the analysis of authigenic minerals. Clay minerals (e.g., Bonhomme et al. 1966; Clauer 1979, 1982) are considered to be very promising. Reliable results depend on (a) the presence of fine-grained argillaceous sediments (silt, sandstone, and slightly metamorphosed rocks are unsuitable), (b) the occurrence of early diagenesis and (c) the use of both the Rb/Sr and K/Ar methods (Bonhomme 1982). But this requires a good knowledge of the genesis of the different clay minerals in the rock because only those minerals can be used for dating that were formed or transformed during sedimentation or diagenesis (e.g., Clauer 1976; Morton 1985). Glauconite, illite, smectite, palygorskite, and mixed-layer minerals are suitable (e.g., Harris 1982). Preliminary attempts at dating using cherts and zeolites (phillipsite) have also been made. A review and discussion of sediment dating using the Rb/Sr method has been published by Clauer (1982).

*Evaporite minerals:* Rb/Sr dating using evaporite minerals has not yet been very successful. Minerals like carnallite, langbeinite, rinneite, kainite, leonite, and sylvite have favorable Rb/Sr ratios (e.g., Lippolt and Raczek 1979a, b; Brookins et al. 1985; Baadsgaard 1987). Some of the results, however, can be interpreted only with difficulty: In general, they do not give sedimentation ages, but metamorphism ages since most salt rocks are polymetamorphic rocks. Because metamorphic recrystallization of evaporite minerals begins below 100°C, weak and episodic thermal events can also influence their Rb-Sr system. Thus, Rb/Sr dating of salts could perhaps be used to date such low-temperature metamorphic events; prerequisite for this would be a better understanding of the isotopic behavior of evaporite minerals.

*Cosmochronology:* The Rb/Sr method is a major method in cosmochronology for dating meteorites and lunar material (Sect. 6.5) (e.g., Papanastassiou and Wasserburg 1971, 1973; Birck et al. 1975; Tatsumoto et al. 1976; Nyquist 1977; Minster and Allègre 1979, 1981). Stony meteorites, as well as silicate inclusions in iron meteorites can be dated. Not only whole-rock samples but also mineral phases (e.g., feldspar, pyroxene, olivine, and ilmenite) are used. Most stony meteorites yield a Rb/Sr whole-rock age of $4.6 \pm 0.1$ Ga, which is interpreted as the crystallization age of the parent body of the meteorites. It has also been shown that the Rb/Sr clock in meteorites and lunar rocks can be reset by shock effects as well as metamorphic events (Minster et al. 1979). Reviews have been written by Wetherill (1971), Faure and Powell (1972), and Nyquist (1977), numerous papers have appeared in *Geochimica et Cosmochimica Acta* and its supplement volumes *Proceedings of Lunar Science Conferences.*

**Limitations.** The application of the Rb/Sr method is limited by the long half-life of [87]Rb, which permits the dating of rocks younger than 10 Ma only in rare cases. In Upper Cenozoic rocks, only minerals with very high Rb/Sr ratios (e.g., biotite, muscovite, sanidine, and leucite) can be dated. An earlier disadvantage of the method was the inexact knowledge of the half-life (values varied by as much as $6\%$ from one another); this is no longer a problem since the decay constant $\lambda$ was set by convention to $1.42 \times 10^{-11}\,a^{-1}$ (Steiger and Jäger 1977). However, when older age data are compared, attention must be paid to possible differences in the $\lambda$-values.

The prerequisite for isochron dating that a cogenetic rock system be isotopically homogeneous at time $t_0$ has been called increasingly in question in recent years, even for igneous rocks. Some granites formed from crustal material by anatectic melting have yielded only poorly defined isochrons. In some cases it has been shown that the scatter is not caused by secondary post-magmatic disturbances, but by incomplete homogenization of the anatectic melt (e.g., Oberkirch biotite granite in the Black Forest, FRG).

The Rb-Sr whole-rock system is very sensitive to post-magmatic alteration and weathering processes (e.g., Compston et al. 1982; Schleicher et al. 1983). Reductions in the measured age of up to $20\%$ have been reported for weathered granite and volcanic rocks (Fullagar and Ragland 1975). Radiogenic $^{87}$Sr is easily leached from minerals, such as biotite, and just as easily substituted in calcium minerals, such as plagioclase and apatite. The prerequisite of a closed system is often not met at whole-rock sample size when this happens. A lower measured age may also result from a gain of rubidium. In any case, care must be taken to use only very fresh sample material.

Basic rocks, in contrast to granitic ones, usually have Rb/Sr ratios that are too small to be suitable for the Rb/Sr whole-rock dating method. However, advances in mass spectrometric analysis have opened up further possibilities for these cases during the last several years. A special aspect is presented by the dating of acid igneous rocks with a high Rb/Sr ratio, especially acid volcanic rocks (e.g., Gale et al. 1979; McKerrow et al. 1980). The Rb-Sr system in these rocks is often disturbed in such a way that the linearity of the sample points is retained in the isochron graph, thus producing apparent isochrons with reduced age values ("rotated isochrons", Schleicher et al. 1983). This may also occur for rocks that macroscopically appear unaltered. Without an age determination with another method for comparison, it is often not possible to recognize such an isochron age as false.

Since the temperature can rise and/or fall over long periods of time during metamorphism, the dating of metamorphic events using minerals is often only an approximation. Better possibilities are offered by whole-rock dating of small samples (thin slabs) of about hand specimen size or smaller, for example from different layers of a migmatite (e.g., Krogh and Davis 1973; Hofmann and Grauert 1973; Hofmann 1979). However, even at this small scale, geological events may not produce complete isotopic homogenization in all cases, but may only smooth out local $^{87}$Sr/$^{86}$Sr gradients. The use of such disequilibrium profiles for dating has been proposed by Bachmann and Grauert (1986).

The dating of sediments using whole-rock isochrons usually yields unsatisfactory results because the analytical values usually have too large a scatter (e.g., Clauer 1982). The reason for this is the occurrence of different detrital components (Hofmann et al. 1974; Clauer 1976). Cordani et al. (1978) have shown, however, that sediments (especially claystones) are sometimes deposited in a sufficiently homogeneous state to allow whole-rock dating with enough accuracy to obtain stratigraphically meaningful ages. Whether this is the rule is controversial because in only a few cases has whole-rock dating been successful for determining the sedimentation or diagenesis age of a sedimentary sequence (e.g., Moorbath 1969; Gebauer and Grünenfelder 1974).

***Representative Examples.*** Rb/Sr ages determined for metamorphosed Mesozoic granitic rocks from southeastern China (Jahn et al. 1976) are an example of a well-defined two-stage evolution of the Rb/Sr isotope system. Two major thermal events are obtained from the whole-rock and mineral isochrons: $165 \pm 13$ Ma and 90 to 120 Ma, respectively. These events were connected with periods of rapid spreading of the Mesozoic Pacific Ocean floor and include two phases of the Yenshan orogeny.

The internal isochron for an adamellite obtained from plagioclase, K-feldspar, chlorite + biotite, and a whole-rock sample is shown in Fig. 6.15a. Various internal isochrons for the study area in southeastern China are shown in Fig. 6.15b together with the isochron for the whole-rock samples from which the analyzed minerals were taken. The initial strontium ratios obtained from these internal isochrons vary systematically with the Rb/Sr ratios of their whole-rock samples, showing the influence of a more recent metamorphic homogenization of the strontium isotope compositions (compare with Fig. 6.14).

An example of a Quaternary Rb/Sr age determination is the dating of 300,000-year old tuff samples from the Alban Hills in Italy by Radicati di Brozolo et al. (1981) using internal isochrons for leucite, biotite, and pyroxene. In this case the very high K and Rb concentrations in the magmas of the Roman Comagmatic region made it

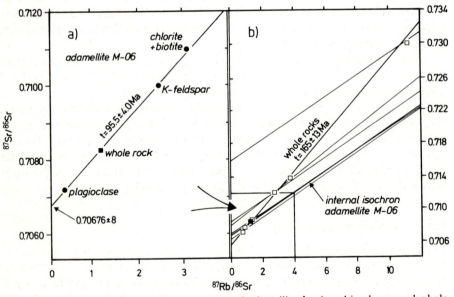

**Fig. 6.15. a** Mineral isochron for a metamorphosed adamellite; **b** mineral isochrons and whole-rock isochron for several granitic rocks from southeastern China (Jahn et al. 1976)

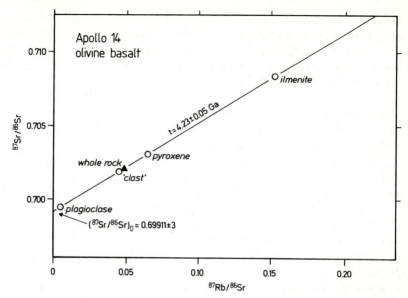

**Fig. 6.16.** An internal Rb/Sr isochron for olivine basaltic clast (sample 14305,122 from Apollo-14; modified after Taylor et al. 1983). ("*clast*" = a fragment probably contaminated with matrix)

possible to determine Rb/Sr ages; these high concentrations result in high $^{87}$Sr* concentrations despite the very young age of the magmas. This is particularly true for leucite. The isochron ages obtained for three tuff samples range from $380 \pm 20$ ka to $330 \pm 20$ ka and are in good agreement with the $^{39}$Ar/$^{40}$Ar ages of the leucites. The data for each tuff sample yielded a well-defined, uniform initial $^{87}$Sr/$^{86}$Sr ratio. Different tuffs, however, showed small differences in their initial $^{87}$Sr/$^{86}$Sr ratios, thus indicating different magma sources or assimilation of different materials during the ascent and extrusion of the magmas.

A Rb/Sr age determination on an olivine basaltic clast from breccia sample 14305 taken from the Fra Mauro Formation during Apollo-14 (Taylor et al. 1983) is shown in Fig. 6.16. A well-defined isochron corresponding to an age value of $4.23 \pm 0.05$ Ga was obtained from four mineral samples (ilmenite, pyroxene, plagioclase, and "clast") and a whole-rock aliquot. The initial value for the sample is very low ($0.69911 \pm 0.00003$). These results suggest that mare-type basaltic volcanism occurred in the Fra Mauro area at least 4.2 billion years ago. This disproves the thesis that there was a hiatus between the plutonism of the lunar highlands ($> 4.4$ Ga) and mare-type volcanism ($< 3.95$ Ga) (Taylor 1982).

### Non-Chronological Applications

The isotopic composition of strontium is one of the most important parameters available to modern geochemistry for investigating the origin of terrestrial material. No more than a brief survey of the major applications of the geochemistry of strontium isotopes can be given within the scope of this book; the reader is directed to the extensive literature in the field for further information (e.g., Faure and Powell

1972; O'Nions et al. 1979; Allègre et al. 1979; DePaolo and Wasserburg 1979; Jahn et al. 1980; Faure 1986; Allègre 1987).

*Initial isotope ratio:* As described in the "Basic Concept" section above, the isochron technique provides not only the age of cogenetic rocks (whole-rock samples), but also their initial isotope ratio. The "initial" strontium isotope ratio is representative of the strontium isotope ratio of an igneous rock at the time of its formation, i.e., the isotopic composition of the melt from which it crystallized. Its value is dependent on the previous history of the strontium, particularly on the Rb/Sr ratio of the system in which the strontium was incorporated. A limiting factor is that only a "single-stage model" can be used, in contrast to the U-Th-Pb system (Sect. 6.1.9). This means that for a given case, it is not possible to directly determine the number of systems or the time in which strontium was involved.

No significant (i.e., measureable) amount of fractionation of strontium isotopes is known to occur in any geochemical process. Thus, a melt has the average isotope composition of the precursor material. If fractionation did occur, it would be automatically eliminated by normalization of the raw data using the Nier value (1938) of 0.1194 for $^{86}Sr/^{88}Sr$.

*Crustal and mantle rocks:* One of the main pieces of information provided by strontium isotope ratios is whether an igneous rock is derived from a melt from the Earth's upper mantle or from the crust itself. This is due to the fact that the isotopic evolution of strontium has been different in these two major units of the Earth. Crustal rocks are enriched in rubidium relative to the mantle and thus often have significantly higher Rb/Sr ratios. Thus the enrichment of radiogenic $^{87}Sr$ from the decay of $^{87}Rb$ is considerably higher in the Earth's crust than in the mantle.

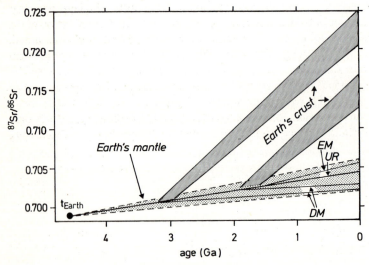

**Fig. 6.17.** A representation of the development of strontium isotopic evolution within the Earth's mantle and crust. *DM* = depleted mantle; *EM* = enriched mantle; *UR* = uniform reservoir = bulk Earth

Consequently, the time-dependent $^{87}Sr/^{86}Sr$ ratio is significantly higher. This difference in evolution is shown in Fig. 6.17.

The $^{87}Sr/^{86}Sr$ ratios for the present upper mantle range between about 0.702 and 0.706. However, enriched parts of the mantle have been reported (e.g., Vollmer et al. 1984), probably due to a large-scale recycling of lithospheric material (e.g., Hofmann and White 1982). $^{87}Sr/^{86}Sr$ ratios for continental crustal rocks are distinctly higher ( > ca. 0.708); in epizonal, acid granites with high Rb/Sr ratios, values as high as 1.5 have been measured. This large variation in crustal rocks is due to differences in their residence times in crustal subsystems that have different Rb/Sr ratios and which were formed or modified in different, sometimes several, orogenies.

Very low initial strontium isotope ratios of less than about 0.706 in igneous rocks thus indicate derivation of the magma from the upper mantle, or at least participation of mantle material. This is true, in general, for basaltic rocks, but in some cases also for granitic rocks (e.g., the Salisbury Pluton in North Carolina, for which $Sr_i = 0.7032$; Fullagar et al. 1971). The strontium isotope geochemistry of carbonatites also provides evidence for their origin from the Earth's mantle (e.g., Faure and Powell 1972; Bell et al. 1982).

In contrast, high initial strontium ratios in granites generally demonstrate their origin from crustal material by anatectic melting. In basic rocks for which proof of mantle origin is provided by other petrogenetic evidence, elevated initial $^{87}Sr/^{86}Sr$ ratios indicate contamination of the magma by crustal components, e.g., in the case of some alkaline volcanic rocks (Bell and Powell 1970; Downes 1984; Schleicher et al. 1990) and continental basaltic rocks, or they may show that crustal material was directly involved in the production of the magma (e.g., andesites).

*Sr-Nd correlation diagrams:* The interpretation of strontium isotope ratios in combination with those of neodymium (see Sect. 6.1.6) has proven to be very useful, particularly for the geochemical characterization of the mantle. The heterogeneity, as well as the metasomatism, of the mantle, the recognition and determination of the duration of its differentiation, and the interaction between mantle and crust are all problems to which the isotope geochemistry of both strontium and neodymium can be applied. The information provided by the individual systems can be increased by applying both methods. The data is usually plotted in a graph of $\varepsilon_{Nd}$ versus $\varepsilon_{Sr}$. The $\varepsilon_{Sr}$ value is defined as the relative deviation (in parts per $10^{-4}$) from the mean strontium composition of a uniform reservoir ($Sr_{UR}$) of the primitive mantle (which corresponds to the bulk earth composition) at any time $t$ according to the following equation:

$$\varepsilon_{Sr_t} = \frac{(^{87}Sr/^{86}Sr)_{sample_t} - (^{87}Sr/^{86}Sr)_{UR_t}}{(^{87}Sr/^{86}Sr)_{UR_t}} \, 10^4, \tag{6.15a}$$

where

$$\left(\frac{^{87}Sr}{^{86}Sr}\right)_{UR_t} = \left(\frac{^{87}Sr}{^{86}Sr}\right)_{UR_0} + \left(\frac{^{87}Rb}{^{86}Sr}\right)_{UR} (e^{\lambda t} - 1) \tag{6.15b}$$

The values of 0.7045 and 0.0827 for $^{87}Sr/^{86}Sr_{UR}(0)$ and $^{87}Rb/^{86}Sr_{UR}$, respectively, were proposed by De Paolo and Wasserburg (1976).

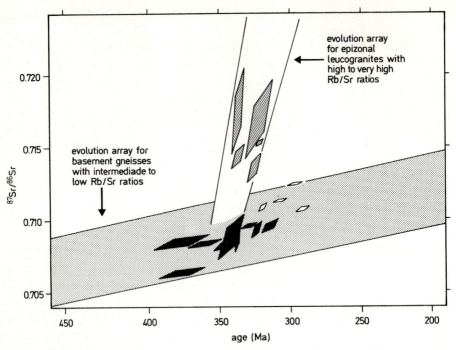

**Fig. 6.18.** Evolution of the strontium isotopic composition during the Variscan in basement rocks in the Black Forest (FRG): The *rhomboids* give the 2σ error of the radiometric ages and of the corresponding initial strontium isotope ratios.

   *black* granodioritic to normal granitic, parauthochthonous and intrusive rocks with normal Rb/Sr ratios; *gray* epizonal leucocratic intrusive granites with very high Rb/Sr ratios; *white* Late Variscan, mainly rhyolitic volcanites ("quartz porphyries") and granite porphyry

*History of specific rock units or geologic regions:*  In addition to a rough, but in most cases unambiguous classification of most igneous rocks as being derived from crustal or mantle material, strontium isotope geochemistry can provide very precise information about the history of a specific rock unit or geological region. For example, the strontium isotope evolution in the basement rocks of the Black Forest area (FRG), which is part of the Moldanubian of Central Europe, is shown in Fig. 6.18 (compilation of values from various authors in the literature, converted and supplemented from a paper by Brewer and Lippolt 1974). The pre-Variscan rocks of this area consist of a sequence of various types of gneiss and gneissic granite subjected to anatectic metamorphism 490 million years ago (Hofmann and Köhler 1973). The evolution of the strontium isotopic composition of this unit is shown by the shaded area. During the Variscan orogeny, the gneissic basement was intruded many times by granitic magmas. The times of these intrusions are indicated by the error parallelograms in Fig. 6.18 The strontium isotope composition indicates that the origin of most of the granites can be explained by direct anatectic melting of the gneissic basement rock. The granites are mostly parauthochthonous to intrusive granodiorites to normal granites. At the climax of the Variscan orogeny, large-scale magma systems formed in which very high Rb/Sr ratios were created by

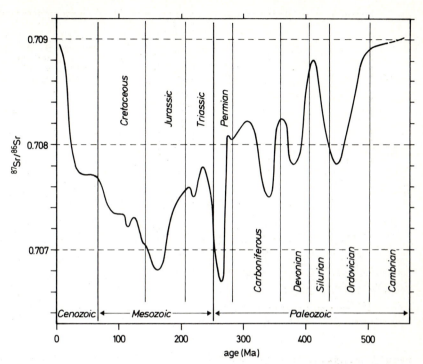

**Fig. 6.19.** Fluctuations in the strontium isotope ratio in marine carbonates during the Phanerozoic (after Faure 1986). Low ratios are thought to be due to times of high volcanic activity, high ratios indicate high influx rates of continental matter (i.e. high erosion rates)

fractionation (Emmermann 1977). These magmas were the source of very leucocratic, epizonal granite intrusions with high initial strontium isotope ratios (shaded error parallelograms). The youngest Variscan igneous rocks in this area are rhyolitic volcanites. The strontium isotope ratios clearly show that these rocks are not relics of the Variscan granite series, but of anatectic remelting of the basement (Lippolt et al. 1983).

*Ocean water and marine sediments:*  An increasingly important area of research is the isotope geochemistry of strontium in ocean water and marine sediments. The isotopic composition of strontium in sea water appears to be homogeneous worldwide with a present value of 0.7090 for $^{87}Sr/^{86}Sr$. This value is influenced by three major reservoirs with different strontium isotope compositions:

1) continental crustal rocks with high ratios (mean: ca. 0.720),
2) young oceanic volcanic rocks (mean: ca. 0.704), and
3) marine sediments with intermediate ratios.

Chaudhuri and Clauer (1986) have suggested a fourth reservoir: continental groundwater in contact with sea water. Because differing proportions of these reservoirs entered the oceans during the Earth's history, the strontium isotope ratio in ocean water has varied with time, as shown in Fig. 6.19 (after Faure 1986). For the periods in which the ratio is greatly different from the mean it is possible to roughly date marine sediments using only the $^{87}Sr/^{86}Sr$ dates for their authigenic carbonate components. The reader is referred to the literature for further information (e.g., Tremba et al. 1975; Veizer and Compston 1976; Faure 1982, 1986; Burke et al. 1982).

## 6.1.4  Lanthanum/Cerium ($^{138}La/^{138}Ce$) Method

(Tanaka and Masuda 1982)

### Dating Range, Precision, Materials, Sample Size

This is a recently developed dating method that has, as yet, been applied in only a few cases. Mineral phases from both basic rocks (e.g., pyroxene and plagioclase) and acid rocks like gneiss and pegmatite (e.g., allanite, apatite, and sphene) can be dated via isochron plots. Owing to the large half-life of $^{138}La$, the La/Ce method is suitable only for very old (Archean) rocks. A sample size of 0.5–1 g is needed.

### Basic Concept

Lanthanum has two natural isotopes with masses of 138 (0.089%) and 139 (99.911%). Lanthanum-138 is radioactive with branched decay: $\beta^-$ emission to $^{138}Ce$, which emits 789 keV $\gamma$-rays, or electron capture (EC) to $^{138}Ba$, which emits 1436 keV $\gamma$-rays. The values published for the long half-lives of $^{138}La$ have a wide scatter: Values for $\tau_\beta$ vary from $2.2 \times 10^{11}$ to $4.7 \times 10^{11}$ a. Tanaka and Masuda (1982) and Shimizu et al. (1986) used a value of $2.69 \pm 0.24 \times 10^{11}$ a, corresponding to a decay constant $\lambda_\beta$ of $2.58 \times 10^{-12}$ a$^{-1}$. The branching ratio (EC/$\beta$) is $0.51 \pm 0.01$ and the combined half-life ($\tau_{tot}$) is $1.03 \pm 0.02 \times 10^{11}$ a, corresponding to $\lambda_{tot} = 6.73 \times 10^{-12}$ a$^{-1}$ (Sato and Hirose 1981).

Cerium has four natural stable isotopes: masses 136 (0.19%), 138 (0.25%), 140 (88.48%), and 142 (11.08%).

The age determination is derived as follows:

$$t = \frac{1}{\lambda_{total}} \ln\left[ \left(\frac{^{138}Ce^*}{^{138}La}\right) \frac{\lambda_{total}}{\lambda_\beta} + 1 \right] \tag{6.16}$$

The isochron plot of $^{138}Ce/^{142}Ce$ versus $^{138}La/^{142}Ce$ is based on the following equation, analogous to the isochron plots made for the Rb/Sr (sect. 6.1.3) and Sm/Nd (sect. 6.1.6) methods:

$$\left(\frac{^{138}Ce}{^{142}Ce}\right) = \left(\frac{^{138}Ce}{^{142}Ce}\right)_0 + \frac{\lambda_\beta}{\lambda_{total}} \left(\frac{^{138}La}{^{142}Ce}\right)(e^{\lambda_{total} \cdot t} - 1), \tag{6.17}$$

where $(^{138}Ce/^{142}Ce)_0$ is the initial cerium isotope ratio.

## Sample Treatment and Measurement Techniques

The mineral and whole-rock samples are prepared according to the scheme given in sect. 5.2.1.1 (Fig. 5.3). A sample size of 0.5–1 g (pure mineral fraction or ground whole-rock) is usually needed. The samples are digested in HF and $HClO_4$, then dissolved in HCl and divided into two aliquots. A composite REE spike is added to one of the aliquots. Lanthanum and cerium in the two aliquots are then separated on ion-exchange columns and measured mass spectrometrically (sect. 5.2.3.1) (Tanaka and Masuda 1982).

## Scope and Potential, Limits, Representative Examples

**Scope and Potential.** Only a few La/Ce age determinations have been published as yet—from one laboratory. A promising advantage of the method appears to be that it can be used not only for basic rocks but also for acidic rocks.

**Limitations.** The La/Ce method may develop into a new geochronometer, although the uncertainty in the La/Ce ages obtained so far are sizable, owing to the large uncertainties in the half-life of $^{138}$La, as well in the analytical technique. The long half-life of $^{138}$La restricts the La/Ce method to very old (Archean) rocks.

**Representative Examples.** The first application of the method was on a gabbro from the upper zone of the Bushveld complex in South Africa, for which an internal isochron using the minerals pyroxene and plagioclase and a whole-rock aliquot yielded and age of 2390 ± 480 Ma. This age agrees within the limits of error with the Sm/Nd isochron age of 2050 ± 90 Ma obtained from an aliquot of the same sample solutions. An age determination of pegmatite from Mustikkamaki, Finland, yielded an age (allanite-apatite-sphene isochron) of 2099 ± 64 Ma (Shimizu et al. 1986). A La/Ce isochron age of 3074 ± 436 Ma was obtained by the same authors for an Amitsoq gneiss from Greenland.

## Non-Chronological Applications

Because both La and Ce belong to the rare-earth group, as is the case for the Sm-Nd system, the combined application of both systems is a promising new tool for isotope geochemistry.

## 6.1.5 Lanthanum/Barium ($^{138}$La/$^{138}$Ba) Method

(Nakai, Shimizu and Masuda 1986)

## Dating Range, Precision, Materials, Sample Size

The application of this recently developed dating method is restricted to minerals rich in rare-earth elements (e.g., allanite, monazite, and epidote) and which come from very old geological units (Archean). For minerals with a very high REE content (e.g., allanite), a single grain (down to a mm across) is sufficient. Both mineral isochron ages and mineral model ages can be calculated.

### Basic Concept

Lanthanum has two natural isotopes with masses of 138 (0.089%) and 139 (99.911%). Lanthanum-138 is radioactive with branched decay: $\beta^-$ emission to $^{138}$Ce or electron capture (EC) to $^{138}$Ba. The values published for the long half-lives of $^{138}$La have a wide scatter. Nakai et al. (1986) used a value of $4.44 \pm 0.15 \times 10^{-12}$ $a^{-1}$ for the electron-capture decay constant ($\lambda_{EC}$), corresponding to a half-life $\tau_{EC}$ of $1.56 \pm 0.05 \times 10^{11}$ a. The combined decay constant ($\lambda_{tot}$) is $6.73 \pm 0.13 \times 10^{-12}$ $a^{-1}$, corresponding to a half-life ($\tau_{tot}$) of $1.03 \pm 0.02 \times 10^{11}$ a (Sato and Hirose 1981).

Barium has seven natural isotopes: masses 130 (0.106%), 132 (0.101%), 134 (2.417%), 135 (6.592%), 136 (7.854%), 137 (11.23%) and 138 (71.70%). The isotopic composition of barium is assumed to be effectively constant. This means that the initial barium isotope ratio ($^{138}$Ba/$^{137}$Ba)$_0$ can also be assumed to be invariable for common rocks: $6.38969 \pm 0.00012$. This makes it possible to determine model ages from only one sample. An isochron age is calculated analogously to Eq. (6.17) using a plot of $^{138}$Ba/$^{137}$Ba versus $^{138}$La/$^{137}$Ba. The isochron lines are fitted to the points using the method of York (1969).

### Sample Treatment and Measurement Techniques

The mineral samples are prepared according to the scheme given in Sect. 5.2.1.1 (Fig. 5.3). The samples are digested and dissolved in acid. The barium and lanthanum concentrations are determined by isotope dilution analysis (Sect. 5.2.4.1). The isotopic compositions are measured mass spectrometrically (Sect. 5.2.3.1) (Nakai et al. 1986).

### Scope and Potential, Limits, Representative Example

**Scope and Potential.** Although the La/Ba geochronometer is suitable for only a limited number of minerals, it may achieve some significance for studies of the evolution of the continental crust, especially in the Precambrian. A particular advantage is that model ages can be determined using a single specimen, or even a single mineral grain. The fact that epidote, a common product of low-grade metamorphism, seems to be datable with this method, opens up the opportunity to date such events in ancient shield areas.

**Limitations.** Owing to the extremely long half-life for the decay of $^{138}$La by electron capture, the La/Ba method is suitable only for very old (Archean) rocks via their minerals. For Precambrian rocks, only in REE-rich minerals, such as allanite and monazite, which have La/Ba ratios greater than about 300, is the accumulation of radiogeneic $^{138}$Ba sufficiently large to cause significant changes in the isotopic composition of barium (Nakai et al. 1986).

*Representative Example.* The first dating reported (Nakai et al. 1986) was of a Amitsoq gneiss (western Greenland), for which an age of $2408 \pm 24$ Ma was obtained. The isochron (allanite, epidote, whole rock) yields an initial $^{138}Ba/^{137}Ba$ ratio of $6.38968 \pm 0.00028$, which is in agreement with the ratios obtained for reagent-grade $BaCl_2$ (Merck), JB-1 standard basalt, and BCR-1 standard basalt.

### 6.1.6 Samarium/Neodymium ($^{147}Sm/^{143}Nd$) Method**

(Wahl 1941, Lugmair 1974)

### Dating Range, Precision, Materials, Sample Size

Developed in the last decades, this is an analytically highly involved method for dating old, especially basic igneous rocks ($>$ca. 50 Ma). High-grade metamorphism in amphibolite to granulite facies can also be dated, as well as "crustal residence ages" of continental areas. Whole-rock samples as well as minerals (e.g., garnet, hornblende, pyroxene, plagioclase, ilmenite, and apatite) are used. Whole-rock and mineral model ages can be determined as well as isochron ages (e.g., Hamilton et al. 1977; McCulloch and Wasserburg 1978; De Paolo et al. 1982; Pettingill et al. 1984; De Paolo 1988). It may be possible to date hydrothermal ore deposits directly by analysis of primary pitchblende using the Sm/Nd method (Fryer and Taylor 1984).

The samarium/neodymium method is especially suitable for determining the age of basic and ultrabasic igneous rocks (e.g., basalt, comatiite, anorthosite, peridotite, andesite, diorite, and monzonite), for which the Rb/Sr method (Sect. 6.1.3) cannot be used. The original crystallization age of metamorphosed igneous rocks (e.g., basic metavolcanites from greenstone belts) can also be determined using the Sm/Nd isochron technique. In contrast, model ages give the time of differentiation and separation of a magma reservoir from the Earth's mantle.

Due to the great resistance of the Sm-Nd system to secondary alteration, the samples do not necessarily need to be extremely fresh. Aliquots of 10–500 mg are used for the analysis of whole-rock samples.

The Sm/Nd method has been used with great success for the dating of the solidification age of lunar samples and of chondrites and achondrites (Sect. 6.5) (e.g., Lugmair 1974; Nakamura et al. 1976; Papanastassiou et al. 1977; Jacobsen and Wasserburg 1980, 1984).

### Basic Concept

Samarium and neodymium are rare-earth elements (REE). Samarium has seven natural isotopes: masses 144 (3.1%), 147 (15.0%), 148 (11.3%), 149 (13.8%), 150 (7.4%), 152 (26.7%), and 154 (22.7%). Neodymium also has seven natural isotopes: masses 142 (27.13%), 143 (12.18%), 144 (23.80%), 145 (8.30%), 146 (17.19%), 148 (5.76%), and 150 (5.64%). $^{147}Sm$ and $^{148}Sm$ decay by $\alpha$-emission to $^{143}Nd$ and $^{144}Nd$, respectively. The half-life of $^{148}Sm$ is too long ($7 \times 10^{15}$ a) to be geochronologically relevant; however, the decay of $^{147}Sm$ (half-life $\tau = 1.06 \times 10^{11}$ a, $\lambda = 6.539 \times 10^{-12}$ a$^{-1}$;

Lugmair and Marti 1978) can be used for dating. $^{144}$Nd is also radioactive and decays by $\alpha$-emission to $^{140}$Ce; but due to the extremely long half-life of $2.1 \times 10^{-15}$ a, this decay can be neglected. The following isotopic ratios (Lugmair and Marti 1978) are normally used for calculating the age values: $^{148}$Nd/$^{144}$Nd $= 0.241572$ and $^{150}$Sm/$^{149}$Sm $= 0.53406$.

Analogous to the Rb/Sr method (Sect. 6.1.3), isochrons can be constructed in a plot of $^{143}$Nd/$^{144}$Nd versus $^{147}$Sm/$^{144}$Nd (see Fig. 6.12). This is based on the assumption that at the time of crystallization, cogenetic whole-rock and mineral samples with different $^{147}$Sm/$^{144}$Nd ratios have the same initial $^{143}$Nd/$^{144}$Nd ratio, which lies between about 0.506 and 0.516, depending on the age of the sample. Cogenetic samples lie at all times on a straight line (isochron) whose slope is a function of the age of the sample and whose y-intercept gives the initial $^{143}$Nd/$^{144}$Nd ratio. The general age equation for the Sm-Nd system is analogous to that for the Rb-Sr system:

$$t = \frac{1}{\lambda} \ln \left[ \left( \frac{^{143}\text{Nd}^*}{^{147}\text{Sm}} \right) + 1 \right] \tag{6.18}$$

The following isochron equation can be derived from Eq. (6.18):

$$\left( \frac{^{143}\text{Nd}}{^{144}\text{Nd}} \right) = \left( \frac{^{143}\text{Nd}}{^{144}\text{Nd}} \right)_0 + \left( \frac{^{147}\text{Sm}}{^{144}\text{Nd}} \right) (e^{\lambda t} - 1), \tag{6.19}$$

where $(e^{\lambda t} - 1)$ is the slope. Several mathematical procedures are available for calculating the regression lines (isochrons) (McIntyre et al. 1966; Brooks et al. 1968; York 1966, 1969).

The calculation of Sm/Nd model ages is based on the fact that a uniform evolution line for the $^{143}$Nd/$^{144}$Nd ratio can be defined for the source regions of both continental crustal rocks and oceanic basalts (DePaolo and Wasserburg 1976b). This evolution line, which is that of the Earth's mantle, is based on a Sm/Nd ratio that is identical with that of a common chondritic reservoir (CHUR = chondritic uniform reservoir). The present $(^{143}\text{Nd}/^{144}\text{Nd})_{\text{CHUR}}$ ratio $R_{\text{CHUR}}$ is given by the following equation:

$$R_{\text{CHUR}} = R_{\text{CHUR}_t} + \left( \frac{^{147}\text{Sm}}{^{144}\text{Nd}} \right)_{\text{CHUR}} (e^{\lambda \cdot t} - 1) \tag{6.20}$$

where $R_{\text{CHUR}} = 0.512638$ (Wasserburg et al. 1981) and $(^{147}\text{Sm}/^{144}\text{Nd})_{\text{CHUR}} = 0.1967$ (Jacobsen and Wasserburg 1980).

$R_{\text{CHUR}_t}$ is the $^{143}$Nd/$^{144}$Nd ratio of this reservoir (i.e., of its evolution line) at any time $t$ (Fig. 6.20). If a magma is formed and separated from the CHUR reservoir at time $t$, then the initial $^{143}$Nd/$^{144}$Nd ratio $R_0$ is equal to $R_{\text{CHUR}_t}$. However, the processes responsible for magma formation in the mantle cause chemical fractionation of samarium and neodymium relative to the source region. If the fractionation factor $f_{\text{Sm/Nd}}$ is defined as follows:

$$f_{\text{Sm/Nd}} = \frac{(\text{Sm/Nd})_{\text{sample}}}{(\text{Sm/Nd})_{\text{CHUR}}} \tag{6.21}$$

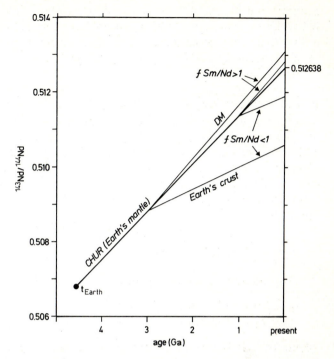

**Fig. 6.20.** Graph showing the evolution of the neodymium isotope ratio since the condensation of the Earth at time $t_{Earth}$. The rate of increase in the $^{143}Nd/^{144}Nd$ ratio of a reservoir or rock unit is dependent on the Sm/Nd ratio in that reservoir or rock unit; for the Earth as a whole, this rate is assumed to be chondritic ($CHUR$ = chondritic uniform reservoir). Evolution lines are shown for two reservoirs separated from CHUR. A crustal reservoir separated from CHUR with a Sm/Nd ratio differing from the CHUR value by a factor of $f_{Sm/Nd} < 1$ will have a present-day $^{143}Nd/^{144}Nd$ ratio lower than CHUR, whereas a mantle reservoir (DM), depleted by the extraction of LIL elements and LREE by a crust-creating event, will result in a higher present-day Nd isotope ratio

Then $f_{Sm/Nd} = 1$ for the CHUR evolution line. For most crustal rocks $f_{Sm/Nd} < 1$, i.e., neodymium is enriched relative to samarium in the crust during fractionation and separation from the mantle. The generation of continental crust from a mantle reservoir, however, causes depletion of LIL elements and LREE in this reservoir. Depleted mantle material ($f_{Sm/Nd} > 1$, DM in Fig. 6.20) has an evolution line different from CHUR (e.g., DePaolo 1981a). When model ages are calculated, the parameters for depleted mantle should be used rather than those for CHUR evolution: $^{143}Nd/^{144}Nd_{DM_0} = 0.513114$, $^{147}Sm/^{144}Nd_{DM} = 0.222$ (Michard et al. 1985).

The model assumes that the Sm/Nd ratio is only insignificantly altered by subsequent geological processes, such as metamorphism and/or erosion and sedimentation, i.e., that since the time of mantle differentiation the Sm/Nd system has been a closed one. A straight line can then be calculated from the present $^{143}Nd/^{144}Nd$ and $^{147}Sm/^{144}Nd$ ratios, whose intercept with the DM (or CHUR) evolution line gives a model age for the time of differentiation from the Earth's mantle:

$$t_{DM} = \frac{1}{\lambda} \ln\left(\frac{(^{143}Nd/^{144}Nd)_{DM} - (^{143}Nd/^{144}Nd)_{sample}}{(^{147}Sm/^{144}Nd)_{DM} - (^{147}Sm/^{144}Nd)_{sample}} + 1\right) \qquad (6.22)$$

Model ages for the differentiation of the crust (so-called crustal residence ages) can thus be determined from representative samples of continental crust (e.g., shales) (O'Nions et al. 1983; Allègre and Rousseau 1984).

To characterize the source regions in terms of isotope geochemistry, it is customary to give the deviation from the CHUR evolution line at time $t$ as $\varepsilon_{Nd}$ [see Eq. (6.23)]. A procedure for calculating $\varepsilon_{Nd}$ values very precisely from Sm/Nd isochrons has been published by Fletcher and Rosman (1982).

### Sample Treatment and Measurement Techniques

The extraction and preparation of mineral samples are described in Sect. 5.2.1.1 (Fig. 5.3). The mineral must be very pure, often hand-picked. Pulverized whole-rock samples do not need to be a chemical aliquot of larger sample, as is the case for the Rb/Sr method (Sect. 6.1.3).

Mass spectrometric isotope dilution analysis (Sects. 5.2.3.1 and 5.2.4.1) is the only possibility for determining the neodymium isotope ratios and element concentrations with the required precision. Due to mass interference between samarium and neodymium, a quantitative chemical separation of the two elements is necessary before the mass spectrometric measurements. Various procedures have been described for this separation (e.g., Nakamura et al. 1976; O'Nions et al. 1976; Jahn et al. 1980). Between 10 and 500 mg of sample are dissolved with acid (e.g., HF–$HNO_3$) in a teflon container. Neodymium and samarium spikes or a double spike is added to an aliquot of the dissolved sample for the isotope dilution analysis. The REE are separated from the other elements by ion exchange; various techniques can be used to separate samarium from neodymium, e.g., electroplating, teflon powder, or $\alpha$-hydroxyisobutyric acid. The sample extracts are then analyzed using a solid-source mass spectrometer.

The neodymium isotope ratios are normally determined using the unspiked part of the digestion solution; the samarium and neodymium concentrations are determined using the spiked fractions. However, if a $^{150}Nd$ spike is used, isotopic composition and concentration of neodymium can be determined at the same time. High precision is required of the mass spectrometric measurements because the natural variations in the isotope ratios are extremely small. Corrections for mass fractionation must be made. International standards (e.g., the LaJolla standard; Lugmair and Carlson 1978) are available for standardization.

### Scope and Potential, Limitations, Representative Examples

**Scope and Potential.** This method has had perhaps the most rapid development of all those discussed in Sect. 6.1. This is true not only for the development of the method and its applications, but also in terms of the number of laboratories that have begun or are beginning to use it. The method opens up a wide spectrum of possibilities for dating and for isotope geochemistry studies.

*Mineral dating:* The total range of Sm/Nd ratios in rocks and minerals is much smaller than, for example, that for Rb/Sr ratios. The chondritic Sm/Nd ratio is 0.302; the ratio for terrestrial rocks varies between about 0.1 for leucocratic granite and about 0.4 for ultramafic rocks. Most rock-forming minerals have similar ratios; an exception is garnet, whose REE distribution pattern shows a very high Sm/Nd ratio (up to 2). Garnet is thus the best material for Sm/Nd mineral dating. Ratios as favorable as those in garnet are sometimes found in pitchblende, which can be used for direct dating of hydrothermal ore deposits (Fryer and Taylor 1984).

*Basic igneous rocks:* One of the most important advantages of the Sm/Nd method is its application to basic and ultrabasic rocks which could not otherwise be dated with either the Rb/Sr method (because the Rb/Sr ratio is too low, see Sect. 6.1.3) or the U/Pb method (because they contain little or no zircon, see Sect. 6.1.9). Simplified, it can be said that the more basic the rock, the lower the LREE concentration and the higher the $^{147}Sm/^{144}Nd$ ratio, thus providing ideal conditions for isochron dating. Layered intrusions are especially promising for age dating owing to the close genetic relationship of rocks with very different Sm/Nd ratios.

The REE patterns of basic volcanic rocks are considerably less influenced by weathering or metamorphism than those of the alkali and alkaline earth elements (e.g., Herrman et al. 1974; Smewing and Potts 1976). This is especially true for the Sm/Nd ratio. This is the reason for a second significant advantage of the Sm/Nd method: The crystallization age of basic magma can be determined from whole-rock samples in spite of secondary events such as metamorphism. This makes it possible to date Archean basic volcanic rocks (Hamilton et al. 1977, 1978; McCulloch and Compston 1981; Jahn et al. 1982).

*High-grade metamorphic rocks:* The fact that the Sm-Nd system is considerably less sensitive to metamorphism than the Rb-Sr system means that the two systems used together can possibly be used to date even high-grade metamorphic more exactly events. Little is known as yet about the closure temperatures of the Sm/Nd system in the various rock-forming minerals. They are assumed to lie between 500 and 700°C (Cliff et al. 1983). This makes it possible to date the cooling of high-grade metamorphism in granulite facies with Sm/Nd mineral isochrons. The fact that this is done with minerals that play a major role in geothermometry and geobarometry (e.g., garnet) may lead to new innovations in the reconstruction of metamorphic histories.

*Model ages:* Sm/Nd model ages ($t_{DM}$ or $t_{CHUR}$ ages; McCulloch and Wasserburg 1978) provide an estimate of the time that a rock unit has had a different Sm/Nd ratio from that of the Earth's mantle. Prerequisite for the model is that secondary geological processes, such as weathering, spilitization, or metamorphism, do not, or only negligibly influence the Sm-Nd system after differentiation of the primary melt from the Earth's mantle.

*Crustal residence ages:* Determination of so-called "crustal residence ages" ($t_{CR}$) is based on Sm/Nd model ages applied to samples of continental crust. The $t_{CR}$ age gives the time up to which the $^{143}Nd/^{144}Nd$ ratio was identical with that of the mantle reservoir from which the original crustal rock differentiated. Since the Earth's crust was apparently formed from parts of the mantle depleted in LREE (Allègre and Ben Othman 1980; DePaolo 1981b), the Sm/Nd parameter for depleted mantle (DM) (e.g., DePaolo 1981a) should be used rather than that of the CHUR growth line. $t_{DM}$-Ages should be sufficiently accurate approximations of crustal residence ages even if the crustal segment has been eroded and deposited as sediment (O'Nions et al. 1983). Representative crustal samples could be composite samples obtained by mixing numerous individual samples (e.g., gneisses, shales, and igneous rocks) from extensive crustal areas (McCulloch and Wasserburg 1978) or also samples of sediment for which a very large source area can be assumed. Very fine-grained clastic sediments (e.g., shale) especially fulfill this condition with respect to the Sm-Nd system (e.g., Haskin et al. 1966; Taylor and McLennan 1981); with such samples, the growth and evolution of continents can be traced through geological periods of time (e.g., O'Nions et al. 1983; Allègre and Rousseau 1984).

*Extraterrestrial samples:* The first application of the Sm/Nd method was the dating of lunar samples and meteorites (basaltic achondrite) (e.g., Lugmair 1974; Naka-mura et al. 1976; Jacobsen and Wasserburg 1980; Sect. 6.5). The method has proved to be extremely valuable in this field. Mainly the solidification age is determined with this method.

**Limitations.** The small variations in the Sm/Nd ratio mentioned above, together with the long half-life ($1.06 \times 10^{11}$ a), are responsible for the very small differences in $^{143}Nd/^{144}Nd$ ratios. This is the reason for the most significant disadvantage of the method: owing to limits deriving from the attainable analytical precision, only old rocks (older than about 50 Ma) can normally be dated with sufficient precision. Moreover, it has the disadvantage that the laboratory procedures are very involved. In addition, the behavior of the Sm-Nd system in geological processes is still little understood.

Basaltic magmas formed in the Earth's mantle with Sm/Nd ratios near the CHUR value usually yield only poorly defined isochrons owing to the small scatter in the isochron plots. Moreover, *whole-rock isochron dating* does not seem to be entirely free of disturbances. Post-emplacement disturbances of the Sm-Nd system have, for example, been demonstrated for two anorthosite samples from the Blue Ridge Mts. of central Virginia (Pettingill et al. 1984). Sometimes whole-rock isochrons yield age values that are obviously too high, compared with the U/Pb zircon ages (Sect. 6.1.9) and the stratigraphic assignment (e.g., Cattell et al. 1984; Chauvel et al. 1985). This probably results from mixing of material from isotopically different mantle sources. The fact that the Sm/Nd ratios vary little makes such isochrons highly sensitive to variations in the initial $^{143}Nd/^{144}Nd$ ratio.

Since little is known as yet – in comparison with the K-Ar (Sect. 6.1.1), Rb-Sr (Sect. 6.1.3), or U-Pb (Sect. 6.1.9) systems, for example – about the Sm-Nd system, there is a certain amount of uncertainty in the model ages. Moreover, the model ages are dependent on the mantle evolution line used (e.g., CHUR or DM). The fact that the Earth's mantle has been shown to be heterogeneous with regions depleted as well as enriched in LREE is a further uncertainty. This especially has an effect on all samples whose Sm/Nd ratio varies little from the chondritic value (0.302), owing to the poor analytical definition of the intercept with the growth line for the mantle.

***Representative Examples.*** DePaolo and Wasserburg (1979b) obtained an age of $2701 \pm 8$ Ma from an internal Sm/Nd isochron using plagioclase, clinopyroxene, orthopyroxene, and a whole-rock sample from a gabbro in the Stillwater layered mafic intrusive complex (Fig. 6.21a). The initial $^{143}Nd/^{144}Nd$ ratio was 0.508248 $\pm$ 0.000012. Whole-rock samples from the Stillwater complex ranging from anorthosite to pyroxenite were found to lie on the same isochron (within experimental error), indicating they all have the same age and initial ratio (Fig. 6.21b). In contrast, the Rb-Sr data have a large scatter, indicating resetting of the Rb-Sr system. This example clearly shows the high resistence of the Sm-Nd system to metamorphic resetting.

By combining whole-rock Sm/Nd data with Sm/Nd mineral isochrons, Cliff et al. (1983) were able to assign an age of early Laxfordian to the granulite facies metamorphism in the Lewisian Basement of the Outer Hebrides of Scotland. A minimum age of $1860 \pm 50$ Ma was obtained from garnet-pyroxene-amphibole-plagioclase internal isochrons for two anorthosite samples; this was interpreted as

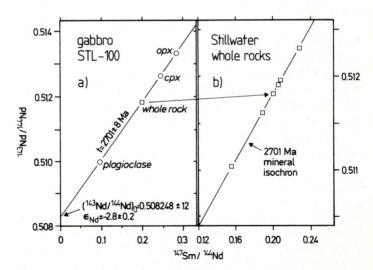

**Fig. 6.21 a, b.** Sm-Nd isochron diagrams for rocks and minerals from the Stillwater layered mafic complex (redrawn after DePaolo and Wasserburg 1979b): **a** internal mineral isochron for a gabbro sample; **b** data for six whole-rock samples and the 2701-Ma isochron from (**a**). All the whole-rock data fit the isochron and therefore confirm that the mineral isochron age represents the crystallization age

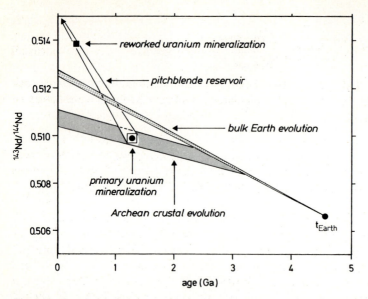

**Fig. 6.22.** Graph showing the neodymium isotope evolution of the Collins Bay uranium deposit in northern Saskatchewan (after Fryer and Taylor 1984). The data suggest Proterozoic primary mineralization of uranium and a Phanerozoic event which reworked the pitchblende reservoir resulting in the present uranium mineralization

the cooling age of a high-pressure, high-temperature metamorphic event (maximum temperature ca. 825°C, Wood 1975). A maximum age of $2180 \pm 60$ Ma results from the whole-rock isochrons of the anorthosites.

As an example for a very young Sm/Nd isochron age, Zindler et al. (1983) have dated the Ronda ultrabasic complex (Spain), obtaining an isochron age for whole-rock, garnet, clinpyroxene, and plagioclase of $21.5 \pm 1.9$ Ma.

Fryer and Taylor (1984) determined a mineral isochron age of $1281 \pm 80$ Ma on primary pitchblende from the Collins Bay unconformity-type uranium deposit (northern Saskatchewan). The initial $^{143}Nd/^{144}Nd$ ratio of 0.50988 indicates it is derived from Archean crustal material (see Fig. 6.22). Partial remobilization of the uranium and the REE at about 350 Ma is indicated by some of the samples. Both age values are in good agreement with known events in that area.

### Non-Chronological Applications

Neodymium isotope studies have become a major aspect of isotope geochemistry since the introduction of this method. The reason for this is the special geochemical properties of the REE: They are very immobile, and a significant change in the Sm/Nd ratio due to fractionation is hardly possible owing to the similarity of the geochemical behavior of these two elements.

Neodymium isotope ratios are usually represented in the form of $\varepsilon_{Nd}$ values, which are defined as the relative deviation (in parts per $10^4$) from a chondritic $^{143}Nd/^{144}Nd$ ratio at a given time $t$ [in analogy to Eq. (6.15)] (DePaolo and

Wasserburg 1976a)

$$\varepsilon_{Nd_t} = \frac{(^{143}Nd/^{144}Nd)_{sample_t} - (^{143}Nd/^{144}Nd)_{CHUR_t}}{(^{143}Nd/^{144}Nd)_{CHUR_t}} \cdot 10^4, \qquad (6.23)$$

where $(^{143}Nd/^{144}Nd)_{sample_t}$ is the initial Nd ratio derived from a whole-rock isochron and $(^{143}Nd/^{144}Nd)_{CHUR_t}$ is calculated from Eq. (6.20) for the time t given by the whole-rock isochron.

Due to the high stability of the Sm-Nd system with respect to secondary geological events, it is possible to demonstrate very old heterogeneities in the Earth's mantle using initial $\varepsilon_{Nd}$ values determined from the isochrons of Archean igneous rocks (e.g., Depaolo and Wasserburg 1979b; McCulloch and Compston 1981).

*Nd-Sr correlation diagrams:* The combination of neodymium isotope data with strontium isotope data has proved to be especially effective, above all for discriminating between rocks that are derived from the Earth's mantle and those that were formed by anatectic melting of the Earth's crust or rocks that have been contaminated with crustal material. In the plot of $\varepsilon_{Nd}$ versus $^{87}Sr/^{86}Sr$ shown in Fig. 6.23, points for rocks with a mantle isotope composition lie within the shaded area (called the mantle array). Crustal rocks, in contrast, usually yield negative $\varepsilon_{Nd}$ values and often very high $^{87}Sr/^{86}Sr$ ratios.

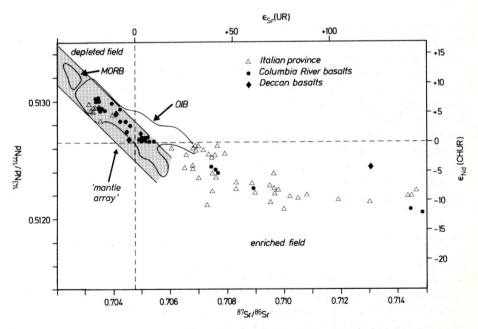

**Fig. 6.23.** Nd-Sr isotopic correlation diagram for basalts from mid-ocean ridges (MORB), oceanic islands (OIB), and selected volcanic rocks from continental environments (namely the Italian Province) (data from various authors); the *shaded area* marks the so-called *"mantle array"*, which refers to the isotopic compositions thought to represent depleted upper mantle

The pattern of mantle samples in the array is explained by the occurrence of regional depletion or enrichment of LIL elements and LREE in the mantle at some time(s) in the geological past. In addition to the fact that the Nd and Sr isotope systems correlate, the "mantle array" of recent oceanic basalts demonstrates the depletion of a significant part of the upper mantle, not only relative to the crust, but also to the primordial Earth. This appears to be due to fractionation during the separation of material from the mantle to form continental crust.

Fields for basaltic rock from mid-ocean ridges (MORB), oceanic islands (OIB), and selected continental environments (e.g., the Italian province) are shown in Fig. 6.23. MORB and OIB originated from sources made up of residual solids remaining after extraction of partial melts of undifferentiated ("primitive") mantle. In contrast, the Nd-Sr isotopic pattern of volcanic rocks in the Italian province is explained in two ways: (1) magma generation by partial melting or by assimilation of crustal rocks (e.g., Taylor and Turi 1976) or (2) magma generation from heterogeneous mantle sources with high Rb/Sr ratios but low Sm/Nd ratios similar to crustal material ("subcontinental enriched mantle"; e.g., Hawkesworth and Vollmer 1979).

As an example of a combined Nd-Sr isotopic study, data for basalts from the Columbia River area and northern California (Carlson et al. 1981) are shown in Fig. 6.24. The data shows considerable isotopic variability, suggesting that mixing occurred after separation of the primary magmas from their mantle sources. The line is a calculated hyperbolic mixing line between the most depleted of the basalts and a metasedimentary xenolith from one of the other basalts.

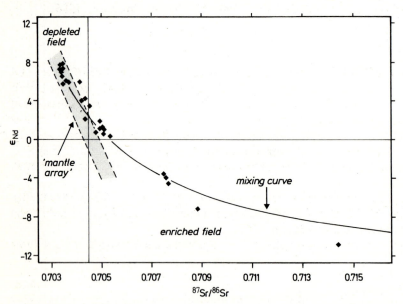

**Fig. 6.24.** Variation in Nd and Sr isotopic composition of basalts from the Columbia River area and northern California (after Carlson et al. 1981). The *hyperbolic line* is a calculated mixing curve between a basaltic end member and a metasedimentary crustal xenolith

For more information on the geochemistry of neodymium isotopes, the reader is referred to the literature (e.g., DePaolo and Wasserburg 1979a; Allègre and Ben Othman 1980; Basu and Tatsumoto 1980; DePaolo 1981b; McCulloch and Chappel 1982; Jacobsen et al. 1984; Patchett and Bridgwater 1984; Allègre 1987; DePaolo 1988).

## 6.1.7  Lutetium/Hafnium ($^{176}$Lu/$^{176}$Hf) Method*

(Herr, Merz, Eberhardt and Signer 1958)

### Dating Range, Precision, Materials, Sample Size

This is an occasionally used method for determining formation ages (more than about 500 Ma) of minerals (e.g., gadolinite, xenotime, priorite) that have very high concentrations of rare-earth elements (REE) and very low concentrations of hafnium (Boudin and Deutsch 1970; Owen 1974). Advances in mass spectrometry have made it possible to apply the method to meteorites and lunar samples (Patchett and Tatsumoto 1980a, 1981; Sect. 6.5), as well as whole-rock systems (Pettingill and Patchett 1981). To determine the initial hafnium isotope ratios, zircon, baddeleyite, and eudialyte are especially suitable. Between 0.5 and 1.5 g of sample are normally necessary. For terrestrial whole-rock samples, very large quantities (30–40 kg) of starting material are desirable so that the samples are homogeneous with respect to the Hf-bearing accessory minerals. Hafnium model ages can be calculated from Hf isotope ratios in zircon (Patchett et al. 1981; Kinny and Compston 1986).

### Basic Concept

Two isotopes of lutetium occur in nature: masses 175 (97.4%) and 176 (2.6%). $^{176}$Lu is radioactive with branched decay: by $\beta^-$ decay ($E_{max} = 0.43$ MeV) to $^{176}$Hf (ca. 97%) and by electron capture (EC) to $^{176}$Yb (ca. 3%). Only the $\beta^-$ decay is of importance for age determinations. The $\beta^-$-decay constant $\lambda$ for $^{176}$Lu was determined by Patchett and Tatsumoto (1980a) from the slope of a eucrite isochron to be $1.94 \pm 0.07 \times 10^{-11}$ a$^{-1}$ (which corresponds to $\tau = 3.57 \pm 0.14 \times 10^{10}$ a). This decay constant has recently been confirmed ($1.93 \pm 0.03 \times 10^{-11}$ a$^{-1}$; Sguigna et al. 1982).

Hafnium has six natural isotopes: masses 174 (0.16%), 176 (5.2%), 177 (18.6%), 178 (27.1%), 179 (13.74%), and 180 (35.2%). Hafnium-174 is radioactive and decays with the emission of $\alpha$-particles to $^{170}$Yb.

Analogous to the Rb/Sr method (Sect. 6.1.3), the age is calculated as follows:

$$t = \frac{1}{\lambda} \ln \left[ \left( \frac{^{176}\text{Hf*}}{^{176}\text{Lu}} \right) + 1 \right] \tag{6.24}$$

Because all of the analyzed minerals or whole-rock samples contain primary hafnium, their initial hafnium isotope ratios must be known. The age is obtained by solving the following isochron equation:

$$\left(\frac{^{176}\text{Hf}}{^{177}\text{Hf}}\right) = \left(\frac{^{176}\text{Hf}}{^{177}\text{Hf}}\right)_0 + \left(\frac{^{176}\text{Lu}}{^{177}\text{Hf}}\right)(e^{\lambda t} - 1) \tag{6.25}$$

To determine an isochron age, at least two cogenetic mineral or whole-rock samples must be measured. If a geologically acceptable initial hafnium isotope ratio is assumed, a model age can also be calculated for individual samples. Zircon is especially suitable for determining initial hafnium isotope ratios due to its high hafnium content (up to 1%) (Patchett et al. 1981).

## Sample Preparation and Measurement

The extraction and preparation of mineral samples or the processing of whole-rock samples is done according to the scheme given in Sect. 5.2.1.1 (Fig. 5.3). Whole-rock samples are very difficult to homogenize with respect to lutetium and hafnium because these elements (in contrast to Rb and Sr for example) are bound almost exclusively in accessory minerals. Grinding and dividing as much sample as possible must, therefore, be done very carefully.

For Lu/Hf dating, the concentrations of lutetium and hafnium, as well as the $^{176}\text{Hf}/^{177}\text{Hf}$ ratio, must be measured. Two alternative methods are used to determine the lutetium and hafnium concentrations: mass spectrometric isotope dilution analysis (Sects. 5.2.3.1 and 5.2.4.1) and neutron activation analysis (Sect. 5.2.4.2). Lutetium can also be determined by flame photometry in concentrations above 10 ppm (Sect. 5.2.4.3).

Mass spectrometric analysis of hafnium is quite easy for zircon (Hf $\geqslant$ 1%), but very difficult for whole-rock samples. For the mass spectrometric determination of the hafnium isotope ratio, as well as the lutetium and hafnium concentrations, the samples are digested in acid in a teflon-lined autoclave. The chemical preparation must be carried out especially carefully since lutetium and hafnium are contained only in mineral components that are very difficult to dissolve (e.g., zircon). Hafnium and lutetium are separated from each other and from the REE and zirconium by ion exchange. A quantitative separation of lutetium and hafnium is very important because both elements, as well as ytterbium, have isotopes with the same mass, namely 176. A high-precision method that can be used routinely for the chemical and mass spectrometric determination has been described by Patchett and Tatsumoto (1980b). The main problem of the mass spectrometric measurement is the very high ionization potential of hafnium. The ratio of recorded ions to atoms loaded onto the filament is about 1 to 30000 (Patchett 1983). Nevertheless, the $^{176}\text{Hf}/^{177}\text{Hf}$ ratio can be determined with a precision of 0.01–0.03%.

Neutron activation analysis is a sensitive method for determining the concentrations of hafnium and lutetium. The metastable $^{180}\text{Hf}$ is produced in the reactor from $^{179}\text{Hf}$ by a $(n,\gamma)$ reaction ($E_\gamma = 0.44$ MeV, $\tau = 5.5$ h) and $^{176}\text{Lu}$ reacts in the same way to $^{177}\text{Lu}$, which then decays by $\beta$ emission ($E_{\text{max}} = 0.5$ MeV, $\tau = 6.75$ d). The advantage of neutron activation analysis is that the complicated chemical preparation of the samples is unnecessary.

An ion microprobe (Sect. 5.2.4.4) can be used to measure the $^{176}Hf/^{177}Hf$ ratio in a single zircon crystal with a precision of better than 1‰ (Kinny and Compston 1986).

### Scope and Potential, Limitations, Representative Examples

***Scope and Potential.*** *Whole-rock and mineral dating:* In whole-rock samples with a complex history, the Lu-Hf system behaves, according to present knowledge, analogously to the Rb-Sr system: The crystallization age of a rock can be dated even when later low- to medium-grade metamorphic events occurred, quasi looking through these events, using whole-rock isochrons. Whereas lutetium occurs mostly in apatite, sphene, and also allanite, hafnium occurs primarily in zircon (up to 1 wt%). Hafnium is enriched in acid rocks more than lutetium during magmatic differentiation: with increasing concentration of the two elements, the Lu/Hf ratio decreases.

Since basic and ultrabasic rocks have a higher Lu/Hf ratio than acidic rocks, it should be possible to date basic rocks that cannot be dated with the Rb/Sr method (Sect. 6.1.3). But such rocks always have a very low hafnium concentration, which complicates the analysis.

*Zircon:* As can be seen in Fig. 6.25, zircon exhibits a very special composition: it contains nearly no lutetium but has a very high hafnium content. For this reason, it

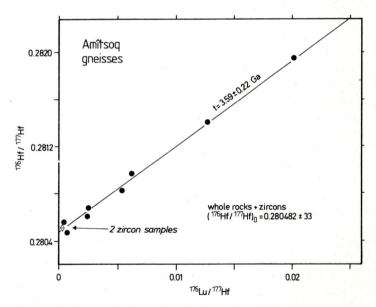

**Fig. 6.25.** Lu/Hf whole-rock isochron age for Amitsoq gneisses in western Greenland (after Pettingill and Patchett 1981). Values for zircon from two gneisses lie at the initial Hf ratio of the whole-rock line despite high-grade metamorphism of the Amitsoq Gneiss at 2.9 Ga, demonstrating the very high stability of the Hf isotope system in zircon

is regarded as particularly suitable for providing information about the initial hafnium isotope ratios. Owing to the very high retention of hafnium by zircon, this is also true even if the host rock has been subjected to high-grade metamorphism. Owing to the ability of zircon to preserve its initial $^{176}Hf/^{177}Hf$ ratio despite the influence of subsequent geological processes on other isotopic systems, a special kind of hafnium model age results: This model age is calculated from the deviation of the hafnium ratios in the zircon of the sample from that of the bulk of the Earth or the mantle evolution line.

*Extraterrestrial samples:* Meteorite and lunar samples (Sect. 6.5), as well as minerals from them, can be dated only with difficulty with the Lu/Hf method because 0.5–1 g of material is necessary, which for extraterrestrial samples is a very large amount. In addition, chondrites have very low hafnium concentrations (0.1–0.2 ppm).

**Limitations.** Due to the considerably greater amount of analytical work involved with the Lu/Hf method, it is used considerably less than the Rb/Sr, U/Pb, and Sm/Nd methods. Minerals with a very high Lu/Hf ratio, such as gadolinite or priorite, are too rare to be of general importance for dating purposes. Age determinations on rock-forming minerals, e.g., apatite and garnet, have not yet been attempted. In contrast to the Rb-Sr system, terrestrial crustal rocks have only very low Lu/Hf ratios (ca. 0.2 or smaller), which makes Lu/Hf dating of these rocks very difficult.

**Representative Examples.** Pettingill and Patchett (1981) determined a Lu/Hf isochron age of $3.59 \pm 0.22$ Ga and an initial $^{176}Hf/^{177}Hf$ ratio of $0.280482 \pm 33$ for whole-rock samples of Amitsoq Gneiss (Greenland) (Fig. 6.25). This age is in agreement with the Rb/Sr whole-rock date and the U/Pb zircon date. This date is either the time of the intrusion of the initial magmatic precursors of the orthogenetic Amitsoq gneisses or that of a very high-grade metamorphic event occurring immediately thereafter. According to the initial hafnium isotope ratio, the precursor material of the Amitsoq Gneiss separated from the mantle at about 3.65 Ga.

Two of the analyzed zircons lie on the whole-rock isochrons (Fig. 6.25) and thus show no change or overprinting of the hafnium isotope ratio, despite granulite-facies metamorphism at 2.8–3.0 Ga.

### Non-Chronological Applications

Research on the geochemistry of the isotopes of hafnium began in the late 1970s and early 1980s (Patchett and Tatsumoto 1980a; Patchett et al. 1981, 1984; Patchett 1983a, b). This research indicates that the isotopic evolution of hafnium started at 4.55 Ga with a $^{176}Hf/^{177}Hf$ ratio of $0.27978 \pm 9$ (initial eucrite value) which has increased by radioactive decay of $^{176}Lu$ to a present value of 0.28286 for the Earth as a whole. This increase corresponds to a chondritic Lu/Hf ratio of 0.240 (Patchett and Tatsumoto 1981). A Lu/Hf ratio of 0.243 was reported for the Orgueil Cl chondrite by Beer et al. (1984).

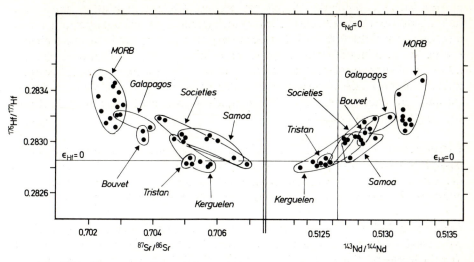

**Fig. 6.26.** $^{176}$Hf/$^{177}$Hf versus $^{143}$Nd/$^{144}$Nd and $^{87}$Sr/$^{86}$Sr in oceanic basalts (after Patchett 1983b): The lines $\varepsilon_{Hf} = 0$ and $\varepsilon_{Nd} = 0$ correspond to present-day undifferentiated (chondritic) compositions

Since hafnium enters the melt more rapidly than lutetium during magmatic process, differentiation during formation of the Earth's crust led to a lower Lu/Hf ratio in continental crustal material and higher ratio in the residual mantle than in the chondritic mantle. This explains the low initial hafnium isotope ratios in rocks of continental crust and the relatively high values for numerous rocks from oceanic regions (e.g., MORB). The $^{176}$Hf/$^{177}$Hf ratios are shown in Fig. 6.26, together with the $^{143}$Nd/$^{144}$Nd (Sect. 6.1.6) and $^{87}$Sr/$^{86}$Sr ratios of several oceanic basalts (from Patchett 1983a), illustrating the opposite behavior of the Sr and Hf systems and the similar behavior of the Hf and Nd systems. The deviations from the present-day chondritic hafnium isotope composition ($\varepsilon_{Hf} = 0$) can be calculated analogously to Eq. (6.15), using $^{176}$Hf/$^{177}$Hf in place of $^{87}$Sr/$^{86}$Sr and a value of 0.28286 for $^{176}$Hf/$^{177}$Hf$_{chondrite}$ (Patchett and Tatsumoto 1981).

An example for the study of extraterrestrial Lu-Hf isotopic evolution is provided by Unruh et al. (1984), who used Lu-Hf and Sm-Nd data to investigate the genesis of lunar mare basalt.

### 6.1.8  Rhenium/Osmium ($^{187}$Re/$^{187}$Os) Method*

(Herr and Merz 1955)

### *Dating Range, Precision, Materials, Sample Size*

Occasionally used method for dating meteorites (especially iron meteorites) (Herr et al. 1967; Luck et al. 1980; Sect. 6.5) and rhenium-rich sulfide minerals that contain little or no osmium (especially molybdenite) and are older than 200 Ma (Herr et al. 1967; Esenov et al. 1970; Luck and Allègre 1982). This method has also been recently applied to ultrabasic magmatic rocks (komatiite, Luck and Arndt 1985). For molybdenite, a sample of 200 mg to 1 g is needed, depending on its age and rhenium content.

### Basic Concept

Of the rhenium in nature, 37.4% is $^{185}$Re and 62.6% is $^{187}$Re. Rhenium-187 decays by $\beta^-$ emission ($E_{max} = 8$ keV) to $^{187}$Os ($\lambda_{Re} = 1.64 \pm 0.05 \pm 10^{-11}$ a$^{-1}$, corresponding to $\tau = 42.3 \pm 1.3$ Ga; Lindner et al. 1989).

Osmium has seven stable isotopes: masses 184 (0.02%), 186 (1.58%), 187 (1.6%), 188 (13.3%), 189 (16.1%), 190 (26.4%), and 192 (41.0%).

The equation for Re/Os age determination is based on radiogenic $^{187}$Os*:

$$t = \frac{1}{\lambda} \ln\left[\left(\frac{^{187}\text{Os}^*}{^{187}\text{Re}}\right) + 1\right] \tag{6.26}$$

Analogous to the Rb/Sr method (Sect. 6.1.3), a plot of $^{187}$Os/$^{186}$Os versus $^{187}$Re/$^{186}$Os may yield an isochron whose slope is a measure of the radiometric age of the sample. The intersection with the $y$-axis corresponds to the initial $^{187}$Os/$^{186}$Os ratio:

$$\left(\frac{^{187}\text{Os}}{^{186}\text{Os}}\right) = \left(\frac{^{187}\text{Os}}{^{186}\text{Os}}\right)_0 + \left(\frac{^{187}\text{Re}}{^{186}\text{Os}}\right)(e^{\lambda \cdot t} - 1) \tag{6.27}$$

### Sample Treatment and Measurement Techniques

The separation of the ore minerals is normally done according to the scheme given in Sect. 5.2.1.1 (Fig. 5.3). Mechanical separation techniques, however, do not seem to be suitable for molybdenite. This mineral should be handpicked from the ore specimens. The rhenium content is determined by neutron activation analysis (Herr et al. 1967; Sect. 5.2.4.3) or by isotope dilution (Luck and Allègre 1982; Sect. 5.2.4.1). The osmium concentration and isotope composition can be determined by ion-sputtering mass spectrometry, inductively coupled plasma mass spectrometry (ICP–MS; Sect. 5.2.4.3), accelerator mass spectrometry (AMS; Sect. 5.2.3.2), or resonance ionization mass spectrometry (RIMS; Sect. 5.2.3.3). The detection limits for all of these techniques are better than 1 ppb for both elements.

For the mass spectrometric isotope dilution analysis (Sects. 5.2.3.1 and 5.2.4.1) of osmium and/or rhenium, the sample is chemically digested and the two elements quantitatively separated from one another, since the mass of interest, 187, occurs in both elements. $^{190}$Os and $^{185}$Re spikes are used. The analysis is done separately for each element. An "ion-sputtering" mass spectrometer (ion microprobe) was developed specially for this analysis (Luck et al. 1980); this instrument provides a precision of 1% on nanogram amounts of Os. The accelerator mass spectrometer (Sect. 5.2.3.2) is a very sensitive tool for analyzing osmium, for which a detection limit as low as 0.01 ppb has been reported by Fehn et al. (1986b).

A special technique has been recently developed for isotopic analysis of only nanogram quantities of osmium using inductively coupled plasma mass spectrometry (Russ and Bazan 1986). An oxidizing agent is used to obtain $OsO_4$ from the sample; $Os^+$ is produced from $OsO_4$ in the plasma and the isotope ratios are measured with a quadrupole mass spectrometer.

### Scope and Potential, Limitations, Representative Examples

*Scope and Potential.* Owing to the low decay constant of $^{187}Re$, the Re-Os system can be used as a cosmochronometer (Sect. 6.5). Rhenium and osmium are both siderophilic and chalcophilic, becoming enriched in both metallic and sulfide phases. Thus, in principle, the Re/Os method makes it possible to date terrestrial sulfide deposits, as well as iron meteorites or metallic phases of other meteorites.

The most important terrestrial mineral for the Re/Os method is molybdenite ($MoS_2$). This ore mineral is unusual in that it contains no "common" osmium (Luck and Allègre 1982) and is also highly enriched in rhenium. This opens the possibility to date samples simply by determining rhenium and osmium concentrations by chemical methods.

*Limitations.* This method has not yet been used routinely because it requires a great deal of time and effort. In contrast to other systems (e.g., Rb-Sr and Sm-Nd), there is still little data available on the isotopic behavior of the Re/Os system in magmatic and metamorphic processes. Re/Os data for basic igneous rocks appear to correlate with MgO content. Osmium seems to be "compatible" with olivine, whereas rhenium is "incompatible" and is concentrated in the melt.

*Representative Examples.* Luck et al. (1980) determined a Re/Os age of 4.55 $\pm$ 0.03 Ga for five iron meteorites and one ordinary chondrite. They obtained a value of 0.805 $\pm$ 0.011 for the initial $^{187}Os/^{186}Os$ ratio of the solar system. These figures yield an age of between 13.3 and 22.4 Ga for our galaxy.

Luck and Arndt (1985) have published a Re/Os isochron for samples from two Archean komatiite flows near Alexo in the Abitibi belt in Ontario. Within the limits of error, the age (2.72 $\pm$ 0.15 Ga) agrees with the Sm/Nd (Sect. 6.1.6), U/Pb (Sect. 6.1.9), and Pb/Pb (Sect. 6.1.10) zircon ages, and the initial $^{187}Os/^{186}Os$ ratio (0.91) is close to the value for mantle material at 2.8 Ga inferred from the work of Allègre and Luck (1980). Magmatic sulfides in the komatiite flow, which have osmium isotope compositions that lie well to the left of the isochron, appear to have lost rhenium during recent leaching.

### Non-Chronological Applications

Allègre and Luck (1980) showed on the basis of Os-rich samples (osmiridium) that the $^{187}Os/^{186}Os$ ratio in terrestrial material is a function of $^{187}Re$ radioactivity. In the Earth's mantle this ratio ranges from 0.805 to 1.04 (Fig. 6.27); in continental crust it varies greatly and ratios as high as 26 occur (for example, in recent granites).

A further possibility for applying the Re-Os system is as a tracer for detecting other conditions than those detected using Sr, Pb, or Nd. Tracer analysis with this

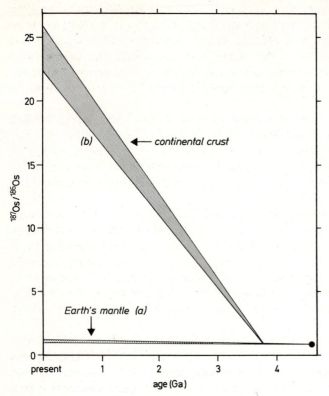

**Fig. 6.27.** Theoretical $^{187}Os/^{186}Os$ growth lines for mantle material; **a** $^{187}Re/^{186}Os = \sim 3.2$, and continental crustal material; **b** $^{187}Re/^{186}Os = \sim 400$ (after Allègre and Luck 1980)

system could be used to recognize various processes in the development of crustal material (e.g., Fehn et al. 1986b). The Re-Os system has received further interest owing to its use as a means of detecting material of possible extraterrestrial origin at the Cretaceous/Tertiary boundary (Luck and Turekian 1983).

### 6.1.9  Uranium/Thorium/Lead Methods*** ($^{238}U/^{206}Pb$, $^{235}U/^{207}Pb$, and $^{232}Th/^{208}Pb$ Methods)

(Boltwood 1907; Holmes 1911)

#### Dating Range, Precision, Materials, Sample Size

A set of time-intensive routine methods for determining radiometric ages of uranium- and thorium-bearing minerals in igneous and metamorphic rocks (zircon, monazite, xenotime, sphene, more rarely apatite, allanite, uraninite, pitchblende, pyrochlore, perovskite, brannerite, coffinite, davidite, baddelleyite, secondary uranium minerals, rutile, and thorite) down to less than 1 Ma (Wetherill 1956, 1963; Tilton 1960; Silver and Deutsch 1963; Wasserburg 1963; Stern et al. 1966; Catanzaro 1968; Doe 1970; Goldich and Mudrey 1972; Allègre et al. 1974; Gebauer and Grünenfelder 1979; Aleinikoff 1983; Wendt and Carl 1985). The most suitable mineral is zircon. Monazite is especially suitable for dating younger rocks, due to its high uranium content. The most reliable ages are obtained from concordant zircon and monazite.

Ages are sometimes also determined using whole-rock samples (e.g., Rosholt and Bartel 1969; Kramers and Smith 1983). The age of hydrothermal mineralizations can be determined via uranium ores or uranium-bearing opal (Ludwig et al. 1980, 1982). High-pressure, low-temperature metamorphism in metabasites can be dated via minerals like sphene, apatite, lawsonite, glaucophane, garnet, and hornblende (Mattinson 1986).

The analytical data is usually evaluated using isochron and/or concordia diagrams. Using the latter for samples from which lead was lost in a single event, both the primary crystallization age and the secondary age of the geological event causing the lead loss can be calculated. However, problems do sometimes arise in the interpretation of the age values (e.g., crystallization, metamorphism, tectonic uplift and stress, metamictization). Using special analytical techniques, the standard deviation of the dates can be lowered even for precambrian rocks to $\pm 1$–2 Ma (Krogh 1982a, b).

The samples should be as fresh as possible; however, reliable zircon U/Pb ages have also been reported from weathered rocks. The sample size (10–80 kg) depends on the content of datable minerals. Normally, several (4–5) cogenetic mineral fractions must be analyzed for each age determination. The amount needed for each analysis (pure fractions) is 1–10 mg, but sub-milligram quantities and even single crystals of zircon, monazite, or thorite have recently become analyzable with new, highly involved analytical techniques. Using an ion microprobe (Sect. 5.2.4.4), several analyses can be made on single crystals of Precambrian zircon (e.g., Williams et al. 1984).

Attempts to date samples of the groundmass of kimberlites, as well as whole-rock samples of carbonatic sediments with the U/Pb method have yielded mostly unsatisfying results. In contrast, detrital zircon from sediments and metasediments provides an indication of the source area of the sedimentary material.

The uranium/thorium/lead methods have also been used with great success on lunar samples and meteorites (e.g., Tatsumoto 1970; Gale 1972; Nunes et al. 1973; Tera and Wasserburg 1974; Chen and Tilton 1976; Manhes et al. 1984; see also Sect. 6.5).

### Basic Concept

Uranium has three natural isotopes with the masses 238 (99.275%), 235 (0.720%), and 234 (0.005%). All three are instable. Uranium-238 and Uranium-235 are the initial nuclides of natural decay series, $^{234}$U is an intermediate member of the uranium-238 decay series. The nonradioactive end-products of the two series are the stable isotopes $^{206}$Pb and $^{207}$Pb. The isotopic ratio $^{238}$U/$^{235}$U is 137.88 (Shields 1960).

The initial nuclide of a third natural decay series is thorium-232, the stable end-product is $^{208}$Pb (see Fig. 6.64).

Lead has four natural isotopes: 208, 207, 206, and 204. Only $^{204}$Pb is entirely nonradiogenic. The other isotopes usually consist of a radiogenic and a nonradiogenic (common lead) component. Thus, the current concentrations of the parent isotopes and those of the lead daughter isotopes accumulated since radioactive equilibrium was established in a closed system lead to three independent possibilities for dating:

$$t = \frac{1}{\lambda_{238}} \ln\left[\left(\frac{^{206}\text{Pb}^*}{^{238}\text{U}}\right) + 1\right], \tag{6.28}$$

$$t = \frac{1}{\lambda_{235}} \ln\left[\left(\frac{^{207}\text{Pb}^*}{^{235}\text{U}}\right) + 1\right], \tag{6.29}$$

and

$$t = \frac{1}{\lambda_{232}} \ln\left[\left(\frac{^{208}\text{Pb}^*}{^{232}\text{Th}}\right) + 1\right], \tag{6.30}$$

where
$^i$Pb* represents the number of atoms of the respective radiogenic lead isotopes and $^{238}$U, $^{235}$U, and $^{232}$Th represent the number of atoms of the respective isotopes.

In addition, a $^{207}$Pb/$^{206}$Pb age can be determined by division of Eq. (6.29) by Eq. (6.28). This method is described in Sect. 6.1.11.

When the following conditions are fulfilled, the three independent ages [Eqs. (6.28–6.30)] are concordant and give the age of the mineral or rock:

1) The three decay constants are known with sufficient precision.
2) The initial lead isotope ratios are known or the initial lead concentration can be neglected.
3) The minerals or rocks have formed a closed system since their formation.

The values of Jaffey et al. (1971) and Le Roux and Glendenin (1963) are recommended by the IUGS for the decay constants of uranium and thorium:

$$\lambda_{238} = 1.55125 \times 10^{-10}\,\text{a}^{-1} \qquad \tau = 4.468\,\text{Ga}$$
$$\lambda_{235} = 9.8485 \ \times 10^{-10}\,\text{a}^{-1} \qquad \tau = 0.704\,\text{Ga}$$
$$\lambda_{232} = 4.9475 \ \times 10^{-11}\,\text{a}^{-1} \qquad \tau = 14.01\,\text{Ga}$$

At time zero there is normally a small proportion of so-called "common lead" present, for which the isotopic composition must be known so that it can be corrected for. The lead isotope composition for the "common lead" correction can be taken from the two-stage lead evolution curve of Stacey and Kramers (1975), or values obtained from cogenetic galena or feldspar can be used (see Sect. 6.1.10 on the common lead method).

*Concordia diagrams:* The ages obtained with the above equations are almost always discordant. The following sequence is usually obtained:

$$t(^{207}Pb/^{206}Pb) > t(^{207}Pb/^{235}U) > t(^{206}Pb/^{238}U) > t(^{208}Pb/^{232}Th)$$

This implies lower model ages than the true age of the rock or mineral. The reason for this discordancy can be secondary loss of lead or gain in uranium and thorium. If the lead loss occurs in a single geological event (e.g., metamorphism or weathering) at time $t_1$, both the primary crystallization age $t_0$ and the secondary age $t_1$ can be determined using a "concordia" diagram (Wetherill 1956, 1963): $^{206}Pb*/^{238}U$ versus $^{207}Pb*/^{235}U$, which corresponds to $(e^{\lambda_{238}t} - 1)$ vs. $(e^{\lambda_{235}t} - 1)$, see Eqs. (6.32) and (6.33). This is possible because isotopes of the same chemical element (i.e., lead) are the end products of all three decay series.

The "concordia curve" (Fig. 6.28) is the curve on which all of the points lie that have the same (i.e., concordant) $^{207}Pb/^{235}U$ and $^{206}Pb/^{238}U$ ages. Its curvature results from the greatly different half-lives of $^{238}U$ and $^{235}U$. If a loss of lead or an overgrowth of the zircons occurs at time $t_1$, then the points for samples with concordant ages are displaced along a straight line between $t_0$ and the zero point of the diagram (i.e., the zero point of the concordia curve, Fig. 6.28a) corresponding to the size of the lead loss. With progressing time, $t_0$ and $t_1$ shift in the same direction by the same amount along the concordia curve and with them the so-called discordia, on which the sample points lie. Their positions are described by the following equation:

$$R = (1 - \Delta)(e^{\lambda_i \cdot t_0} - 1) + \Delta \cdot (e^{\lambda_i \cdot t_1} - 1) \tag{6.31}$$

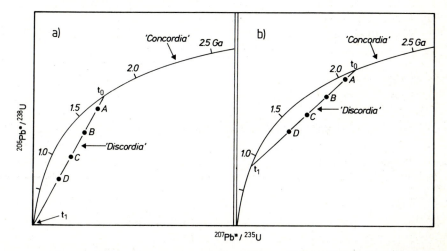

**Fig. 28 a, b.** The principle of U/Pb evolution in a $^{206}Pb*/^{238}U$ vs. $^{207}Pb*/^{235}U$ diagram (concordia diagram): The concordia curve is defined by the locus of all the points which have the same (i.e., concordant) $^{206}Pb/^{238}U$ and $^{207}Pb/^{235}U$ ages. The U/Pb isotope ratios of undisturbed samples formed at time $t_0$ change with time along this curve. **a** If loss of lead occurs at time $t_1$, then the sample points plot along a line between $t_0$ and the point of origin. **b** With progressing time, points $t_0$ and $t_1$ both shift along the concordia curve; the plot of the points of the samples remains linear. These points form a *straight line* (discordia) which intersects the concordia at $t_0$ (primary age) and $t_1$ (secondary age)

where $R = {}^{207}Pb/{}^{235}U$ or ${}^{206}Pb/{}^{238}U$, $i = 235$ or $238$, respectively,
and $\Delta = $ the relative lead loss at time $t_1$.

For at least three cogenetic samples with differing lead losses, the discordia can be calculated as a regression line (Fig. 6.28b), yielding $t_0$ (so-called primary age) and $t_1$ (so-called secondary age) from the intersections of the discordia with the concordia curve. The required cogenetic sample points with differing lead isotopic composition are obtained by dividing, for example, the zircon from a whole-rock sample into several fractions on the basis of magnetic behavior, grain size, crystal-shape, color, or degree of metamictization. If the various minerals of a rock are first separated and then the zircon inclusions are analyzed, different degrees of lead loss are also sometimes observed (Aleinikoff 1983). Several regression methods are available for calculating the discordia lines and their intercepts with the concordia (York 1966, 1969; Ludwig 1980, 1983; Davis 1982). A disadvantage of the usual concordia diagram is the strong correlation between the errors of the variables $x(= {}^{207}Pb*/{}^{235}U)$ and $y(= {}^{206}Pb*/{}^{238}U)$ because $x$ is calculated from the ${}^{207}Pb*/{}^{206}Pb*$ ratio and the $y$ value. This correlation is taken into account in the mathematical procedure developed by Ludwig (1980). The ${}^{207}Pb/{}^{235}U$ ages have a larger error than the ${}^{206}Pb/{}^{238}U$ ages because the ${}^{235}U$ concentration is smaller and the ${}^{207}Pb$ concentration is influenced by "common lead" more than that of ${}^{206}Pb$.

Analogous concordia diagrams can be made by plotting ${}^{208}Pb*/{}^{232}Th$ versus ${}^{207}Pb*/{}^{235}U$ or ${}^{206}Pb*/{}^{238}U$. The interpretation of the data is, however, usually difficult because U and Th may respond differently to disturbances and ${}^{206}Pb$ and ${}^{208}Pb$ may be situated at different sites in a rock or crystal. A further method of representation is the so-called Tera-Wasserburg concordia diagram, in which ${}^{207}Pb*/{}^{206}Pb*$ is plotted versus ${}^{238}U/{}^{206}Pb*$ (Fig. 6.29; Tera and Wasserburg 1972, 1974). This plot avoids the strong correlation of errors associated with the normal

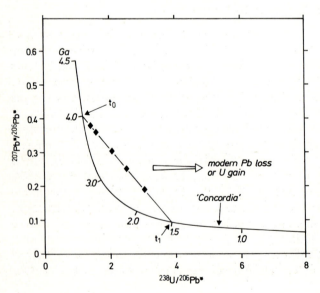

**Fig. 6.29.** Principle of the U/Pb evolution in a Tera-Wasserburg diagram (Tera and Wasserburg 1972, 1974). The concordia here is defined as the locus of all the points that have concordant ${}^{238}U/{}^{206}Pb$ and ${}^{207}Pb/{}^{206}Pb$ ages (Sect 6.1.11). The evolution of the discordia line as a function of $t_0$ and $t_1$ is the same as described for Fig. 6.28. Recent lead losses or recent changes in uranium concentration shift the sample points parallel to the x-axis

concordia method. A further advantage: recent lead losses or uranium input results in a shifting of the points parallel to the x-axis and are thus graphically easier to recognize. Another advantage is the greater curvature of the concordia line, particularly for Paleozoic times. A variation of the Tera-Wasserburg diagram is a plot of $^{206}$Pb*/$^{238}$U versus $^{207}$Pb*/$^{206}$Pb* (Gebauer and Grünenfelder 1978).

*Common lead:* A disadvantage of concordia diagrams is that they can be applied only to the radiogenic lead component of uranium-bearing minerals. Therefore, a correction is necessary for the normally very small, but no negligible amounts of common lead in the accessory minerals (zircon, monazite, etc.). In young samples (less than about 50 Ma), common lead may constitute a significant proportion of the total lead. The isotopic composition of this common lead component is, however, seldom known. Mathematical methods for determining samples with common lead of unknown composition have been proposed by Levchenkov and Shukolyukov (1970) and Wendt (1984). The method of Wendt is based on the calculation of a three-dimensional U-Pb discordia plane where $x = {}^{204}\text{Pb}/^{206}\text{Pb}$, $y = {}^{207}\text{Pb}/^{206}\text{Pb}$, $z = {}^{238}\text{U}/^{206}\text{Pb}$. The advantage is that the measured lead isotope ratios can be used without any common-lead correction to calculate the ages $t_0$ and $t_1$, as well as the isotopic composition of the common lead.

*Isochron diagrams:* Analogously to the Rb/Sr method (Sect. 6.1.3), isochrons can be calculated for cogenetic samples in the graphs of $^{206}$Pb/$^{204}$Pb versus $^{238}$U/$^{204}$Pb, $^{207}$Pb/$^{204}$Pb versus $^{235}$U/$^{204}$Pb, and $^{208}$Pb/$^{204}$Pb versus $^{232}$Th/$^{204}$Pb. They are used for the dating of whole-rock samples, recently also for internal mineral dating. The y-intercepts give the initial isotope ratios of the nonradiogenic (common lead). U/Pb age, however, are not often calculated in this way because the U-Pb system is usually disturbed in whole-rock samples.

### Sample Treatment and Measurement Techniques

The minerals needed for the age determination are separated from up to 80 kg of unweathered material as described in Sect. 5.2.1 (Fig. 5.3). The mineral (zircon) fractions are obtained by sieving, magnetic separation, heavy liquids, and by hand using a stereomicroscope. The minerals must be handpicked to obtain the necessary purity ($> 99.5\%$). Special techniques for the magnetic separation and for the abrasion of crystals as described by Krogh (1982a, b) can help reduce the analytical error in the age determination, sometimes considerably. A cathode luminescence microscope is an important tool for making the internal structure of a mineral visible for interpretaton of the history of the mineral. Individual crystals or crystal fragments (e.g., zircon or thorite) can be analyzed (e.g., Todt and Büsch 1981; Compston et al. 1985/86; Gariepy et al. 1985); normally, 1–10 mg of sample are used per analysis (for a sample age of about 300 Ma and uranium concentrations of 200–500 ppm). Submilligram quantities of zircon have recently become analyzable (e.g., Roddick et al. 1987).

   To determine a U/(Th)/Pb age, the concentrations (expressed as number of atoms) of lead and uranium (or thorium) isotopes of interest must be known. Due to the usually very small amounts of sample (mineral fractions), only mass spectrometry (Sect. 5.2.3.1), which also has the advantage of high precision, is possible for this analysis. For this purpose the mineral fractions are digested in concentrated HF at a temperature of 180–220°C in a Teflon-coated autoclave (e.g., Krogh 1973). The digestion normally takes 4 to 10 days. Aliquots of the solution are used for isotope dilution analysis (Sect. 5.2.4.1) with U, Pb, and Th spikes or with mixed spikes. A $^{235}$U spike is used for uranium, a $^{208}$Pb, $^{205}$Pb or $^{202}$Pb spike is used for lead (the latter two have recently become available; Parrish and Krogh 1987). The separation of U, Th, and Pb is done in most laboratories by ion exchange, sometimes with complexing agents or electrolytic techniques (Barnes et al. 1973). Extremely high-purity chemicals must be used and the laboratory must be kept dust-free since the danger of contamination with common lead is extremely high.

The measurements are carried out on a solid-source mass spectrometer. The so-called silica gel technique (Cameron et al. 1969) has proved to be especially useful; rhenium is usally used for the filament.

A high-resolution ion microprobe (SHRIMP; Sect. 5.2.4.4) can be used to analyze several points on individual, zoned zircon crystals in order to calculate an internal discordia age (e.g., Hinthorne et al. 1979; Compston et al. 1983; Pidgeon et al. 1986). Mainly very old (Early Proterozoic and Archean) samples can be dated successfully with this technique. However, Paleozoic zircon can also be dated down to ages as young as 25 Ma with an ion microprobe via their $^{206}$Pb/$^{238}$U data alone. A standard zircon sample of known age is necessary.

### Scope and Potential, Limitations, Representative Examples

**Scope and Potential.** The U/Th/Pb methods, without a doubt, have great potential for application, comparable to the Rb/Sr (Sect. 6.1.3), K/Ar (Sect. 6.1.1), and Sm/Nd (Sect. 6.1.6) methods. This is especially true for U/Pb dating using a concordia diagram. Originally used for age determinations on uranium- and thorium-rich pegmatite minerals, this set of methods is now applied with great success especially to accessory minerals of metamorphic and magmatic rocks—thanks to advances in mass spectrometry.

The main advantage of the U/Pb method is the fact that the two radioactive uranium isotopes decay with different half-lives to different stable lead isotopes. This means that every mineral that contains uranium has two, completely independent radiometric clocks in it. Secondary disturbances (i.e., loss of lead or uranium input) affect the two decay systems similarly and can, therefore, not only be recognized, but even be used to extend the geochronological information obtained from the analysis by mathematical and graphical treatment of the data.

*Minerals:* The most important mineral for dating with this method is zircon; monazite and sphene are also used successfully. Zircon rarely yields concordant U/Pb ages but well-defined discordia are often obtained. Apatite sometimes also fulfills the requirements for dating (Oosthuyzen and Burger 1973). In addition, a large number of uranium-bearing minerals (e.g., xenotime, allanite, perovskite, uraninite, thorite, pitchblende, pyrochlore, coffinite, brannerite, davidite, bad-deleyite, and rutile) and secondary uranium minerals (e.g., torbernite, autunite, and uranophane) can be used for dating purposes. Garnet, hornblende, and lawsonite can also be analyzed for mineral isochrons.

*Igneous rocks:* Especially granite and granodiorite can be dated very well owing to their relatively high zircon content. The dating of mafic igneous rocks (e.g., diorite and gabbro) or mafic metamorphic rocks (e.g., eclogites and ophiolites) is possible, but require very large samples ( > 50 kg) due to their very low zircon content (e.g., Gebauer et al. 1981; Paquette et al. 1985; Dunning et al. 1986). The analytical points for the zircon in these rocks often plot closer to the concordia curve than those for zircon from granite (Gebauer and Grünenfelder 1979). Different zircons can react very differently to thermal and dynamic secondary overprinting. These differences result in a large number of possibilities for dating and interpretation. The standard model for interpreting discordant data points assumes that a single loss of radiogenic lead occurred in an episodic secondary event after the primary crystallization of the mineral (e.g., zircon) (Wetherill 1956).

     To a certain degree, the interpretation of U/Pb ages depends on the geological age of the sample: For Archean igneous rocks, the primary age (upper intercept with the condordia) can usually be interpreted as the time of the intrusion or extrusion. The secondary age (the lower intercept with the concordia) can be the time of a more recent metamorphic event that was responsible for the loss of lead (compare Figs. 6.31 and 6.32). In many Archean areas the lower intercept gives an age value that cannot be assigned to any known geological event. This secondary age is then viewed as meaningless. It could be due to curvature of the discordia curve caused by continual (Tilton 1960) or temporary diffusion (Wasserburg 1963). This diffusion affects only very old (Archean) rocks, the upper part of the discordia is not affected. Experimentally determined diffusion coefficients (Shestakov 1972) seem to be much too small for such a model, however. Moreover, ion microprobe results with zircons showing concordant Archean ages do not support the assumption of diffusion (e.g., Compston and Kröner 1988).

     For younger (Phanerozoic) igneous rocks derived from the crust, the interpreta-tion of U/Pb age data is often made difficult by a high proportion of radiogenic lead taken up by the magma from their crustal precursor rocks (inherited zircons) or the surrounding rocks during its ascent. In these cases, the secondary age often must be interpreted as the time of intrusion. The upper intersection with the concordia usually then has a relatively large error. It gives an indication of the age of the crustal fragment involved.

*Metamorphic rocks:* For Phanerozoic metasediments (e.g., gneiss and slate), the primary age values for discordant zircons can be interpreted as approximate ages of the corresponding crust or the source area, analogous to the Phanerozoic magmatites mentioned above. The secondary age gives the time of the metamorphic event. In some areas, a relationship has been shown to exist between the degree of metamorphism and the degree of discordance of the zircons (Grauert and Hofmann 1973), probably due to younger overgrowths rather than loss of lead. If the zircon is strongly metamictized, metamorphism temperatures as low as ca. 300°C can lead to considerable loss of lead (Gebauer and Grünenfelder 1976).

In contrast, the primary ages of Archean metasediments are interpreted as the time of mineral formation during high-grade metamorphism. The lower intercept can then be assigned to a more recent metamorphic or other geological event.

Zircon with a highly disordered structural arrangement (domain structure) and elevated concentrations of uranium and other extraneous elements (e.g., REE) tends to recrystallize even at relatively low temperatues (i.e., ca. 300°C) (low-temperature annealing; Pidgeon et al. 1973; Gebauer and Grünenfelder 1976). This process is accompanied by very high to complete loss of lead or exchange of the lead with the environment. Secondary ages in Archean rocks that cannot be interpreted in terms of major geologic events can probably be explained in this way. Zircon annealed (sealed) in this way is often much more stable, even under conditions of high-grade metamorphism (i.e., ca. 600°C).

Analogously, high resistance of the U-Pb system in zircon is reported for Proterozoic granite shocked by impact although the zircons were mechanically disrupted by the impact (Siljan structure in Sweden, Åberg and Bollmark 1985). In contrast, resetting of the U-Pb system by contact metamorphism has been demonstrated (e.g., Hart et al. 1968; Hanson et al. 1971), whereby the U-Pb system in sphene and zircon is distinctly more resistant than the K-Ar system in hornblende and biotite. A gain in uranium and/or thorium can occur, as well as loss of lead.

Monazite is especially suitable for dating young rocks due to its high uranium content. It occurs mainly in felsic igneous and metamorphic rocks. It seems to be very resistant to metamorphic overprinting (Gebauer and Grünenfelder 1979) and, thus, the analytical data for this mineral more often lie along the concordia. However, discordant behavior is also known.

*Tectonics:* The behavior of the U-Pb system in zircon under low-temperature tectonic stress has not yet been fully clarified. If recrystallization occurs during tectonic stress in zircon with a distorted lattice, the U-Pb system is usually opened. On the other hand, even zircon fragments in ultramylonite can remain a closed system at low temperatures, as shown by Lancelot et al. (1983) for samples from a Pan-African low-temperature mylonitic shear zone. Prerequisite for this must be the low temperatures at which the mylonitization occurred. In contrast, the K-Ar system in the shear zones was disturbed: the tectonic event could be dated via $^{39}Ar/^{40}Ar$ age determinations (Sect. 6.1.1.2) on feldspars.

The escape of water through microchannels in metamict zircon may lead to a loss of dissolved radiogenic lead. This process is induced by the release of pressure during crustal uplift, which can then be dated (dilatancy model; Goldich and Mudrey 1972).

*Sediments:* Zircon and monazite from clastic sediments are suitable for obtaining information about the source areas of detrital material (Ledent et al. 1964; Gaudette et al. 1977). Changes in the age information caused by chemical processes during sedimentation, diagenesis, or low-grade metamorphism have not been clearly demonstrated for these minerals.

*Initial disequilibrium:* The treatment of discordant U/Pb ages in concordia plots assumes radioactive equilibrium in both decay series for the entire history of the mineral. This assumption is justified as long as the secondary age is very large compared to the longest half-life within the two decay series ($^{230}$Th and $^{231}$Pa). However, for ages lower than about 50 Ma, it is possible that initial disequilibrium (i.e., at time zero) will cause the age to be too low (for $\Delta[^{230}$Th$/^{238}$U$]_0 < 0$, where $\Delta[^{230}$Th$/^{238}$U$]_0$ is the deviation of the initial activity ratio $[^{230}$Th$/^{238}$U$]_0$ from equilibrium) or too high (for $\Delta[^{230}$Th$/^{238}$U$]_0 > 0$), as demonstrated by Ludwig (1977), Ludwig et al. (1982), and Schärer (1984). Initial disequilibrium can be caused, for example, by fractionation of uranium and thorium between a residual magma and the crystallizing minerals. Therefore, corrections must be made for an excess or deficit of $^{206}$Pb. A "disequilibrium concordia" (Fig. 6.30) that takes initial radioactive disequilibrium into consideration has been derived by Wendt and Carl (1985).

*Isochron dating:* Besides concordia plots, isochron diagrams are also used for U/Pb dating of minerals and whole-rock samples (e.g., Levchenkov et al. 1982; Carl and Dill 1985). Dating of whole-rock samples (e.g., granite) using isochron graphs has shown that, in contrast to the U-Pb system, the Th-Pb system is considerably less disturbed by secondary overprinting (Nkomo and Rosholt 1972). But the geochemical information obtained from this method is more important than the age determination. Age determinations on carbonatic whole-rock samples have been successful in only a few cases (Gerling and Iskanderova 1966; Doe 1970). The $^{206}$Pb$/^{204}$Pb versus $^{238}$U$/^{204}$Pb isochron system used in this case usually

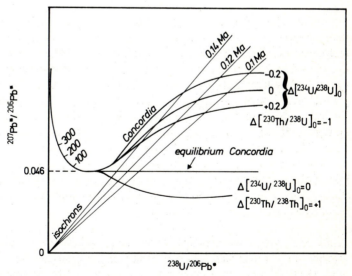

**Fig. 6.30.** A $^{207}$Pb*$/^{206}$Pb* vs. $^{238}$U$/^{206}$Pb* concordia plot for a sample with initial disequilibrium of the daughter isotopes (Wendt and Carl 1985): $\Delta[^{234}$U$/^{238}$U$]_0$ and $\Delta[^{230}$Th$/^{238}$U$]_0$ are the deviations of the initial activity ratios $[^{234}$U$/^{238}$U$]_0$ $[^{230}$Th$/^{238}$U$]_0$ from equilibrium

shows distinct disturbances. For dating kimberlites, attempts have been made using mineral fractions from the matric enriched in perovskite (Kramers and Smith 1983).

The U/Pb isochron method is a promising new approach to the geochronology of high-pressure, low-temperature metamorphism in metabasites (Mattinson 1986). The key mineral is sphene, with its high U/Pb ratio, but other minerals formed by metamorphism in these rocks, like apatite, lawsonite, glaucophane, garnet, and hornblende, can also partition U and Pb in such a way as to provide a range of U/Pb ratios suitable for isochron dating.

*Limitations.* In general, minerals that contain uranium as a major element are significantly more instable with respect to later geological events than uranium-bearing accessory minerals, e.g., zircon or monazite. Whole-rock samples (e.g., sediments or granitic rock) are usually unsuitable for dating purposes owing to the often occurring secondary disturbances of the U/Pb system.

The geological interpretation of the discordant dates in the concordia diagram is often complicated. For example, zircon in polymetamorphic areas (e.g., the Alps) can have been subjected to several episodes of lead loss or crystal growth and, therefore, show very complicated isotopic characteristic. Often these problems are readily solved by ion microprobe analysis (e.g., Black et al. 1986; see also Sect. 5.2.4.4). Another difficulty that may occur is secondary input of radiogenic lead, resulting in Pb/Pb, U/Pb, and Th/Pb ages that are too high (e.g., Williams et al. 1984). Episodic uranium input in connection with geological events has also been established as a possible cause of discordant U/Pb dates for detrital zircon (Grauert et al. 1974). The intercepts of the discordia with the concordia sometimes do not yield geologically meaningful ages, despite linearity of the data; but the upper intercept can usually be interpreted as a minimum age.

Attention must be paid to whether a metamorphic rock is polymetamorphic. A multi-stage history of detrital zircon or monazite can produce a pseudo-linear plot with intercepts between discrete metamorphic events, which are then without geological meaning. The same can be said for the mixing of zircon populations formed at different times in a polymetamorphic rock (Lancelot et al. 1976).

In addition to the question whether a given zircon population or different phases of a single zircon crystal can be considered to be homogeneous and cogenetic, an understanding of the open system behavior of uranium and lead in zircon with respect to pressure, temperature, chemical composition, and secondary defects in the crystal lattice resulting from radiation, hydration, or annealing is extremely important for the interpretation of U-Pb data for zircon (Grauert 1974). Natural crystals of zircon and monazite are multi-phase systems with domains of differing Th/U ratios (Steiger and Wasserburg 1966) and differing trace element concentrations (Köppel and Sommerauer 1974). Amount and manner of inclusion of trace elements in the crystal lattice are decisive for the isotopic behavior of the U-Pb system when subjected to stress and hence determine the degree of lead loss. Alteration zones in metamict zircon generally produce lower U/Pb ages than the unaltered parts (Krogh and Davis 1975). Thus, the arrangement of the analytical data along a discordia can result from the mixing of highly discordant zircon phases with concordant ones (Steiger and Wasserburg 1969); this can occur within a single crystal.

The U/Pb dating method normally cannot be used for very low discordant ages (i.e., < 1 Ma). In these cases, the $^{230}Th/^{234}U$ method (Sect. 6.3.1) seems to be more suitable. However, low U/Pb ages have been reported for ore samples (e.g., Santos and Ludwig 1983) and secondary uranium minerals (Löfvendahl and Åberg 1981; Carl and Dill 1985), but usually with large errors due to the small amounts of radiogenic lead. A prerequisite for dating very young samples is an exact knowledge of the composition of the common lead component.

***Representative Examples.*** Bickford et al. (1981) used the U/Pb method to date granitic and metamorphic rock from the Precambrian basement in Missouri and Kansas. Most of the samples were taken from drill cores. The highest age values (1.610–1.650 Ma) were obtained for the northern part of the study area (cogenetic zircons from 22 rock samples). This unit is intruded by granites with age ranges of 1.450–1.480 Ma and 1.340–1.380 Ma. The concordia diagram of a granite core from Shannon County, Missouri, is shown in Fig. 6.31. The five different zircon fractions A–E all lie on a very well defined discordia, whose upper intercept with the concordia curve gives a crystallization age of 1.473 ± 15 Ma (1σ).

A concordia plot of zircon data for samples of Archean and Early Proterozoic granitoid rocks in the Watersmeet area of northern Michigan is shown in Fig. 6.32 (Sims et al. 1984). The Archean basement consists of tonalitic augen gneiss, unconformably overlain by a biotite gneiss sequence of about the same age. The

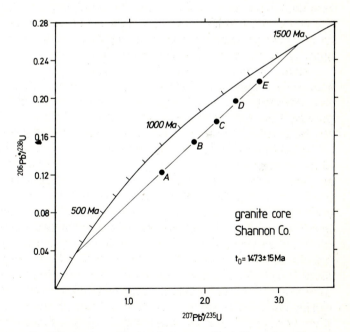

**Fig. 6.31.** Concordia diagram for different zircon fractions from a granitic drill core from Shannon County, Missouri (after Bickford et al. 1981)

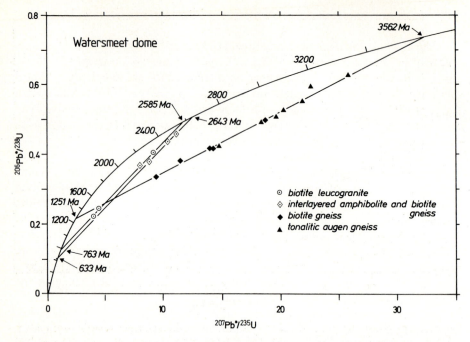

**Fig. 6.32.** Concordia plot of zircon data for samples of granitoid rock in the Watersmeet dome, redrawn after Sims et al. (1984). Discordia lines are shown for augen gneiss and biotite gneiss (upper intercept 3560 Ma), biotite leucogranite (2590 Ma), and intercalated amphibolite and biotite gneiss (2640 Ma)

upper and lower intercepts are about 3560 Ma and 1250 Ma, respectively. A younger sequence consists of interlayered amphibolite and biotite gneiss (about 2640 Ma). All these rocks are intruded by biotite leucogranite with an age of 2590 Ma.

The resistance of zircon to U/Pb resetting in a prograde metamorphic sequence was studied by Peucat et al. (1985). In the Alpe Fjord area of eastern Greenland, Archean detrital zircon in quartzite experienced an episodic lead loss during a major metamorphic event at about 1100 Ma (Grenvillian). But the U-Pb system remained closed during Caledonian metamorphism from the chlorite zone up to and including sillimanite zone conditions.

The studies of Black et al. (1984) on monazite from felsic paragneiss (Napier Complex, Antarctica) show a connection between the color of the monazite and the degree of its lead loss. The cause of both is the submicroscopic structure of monazite, which shows crystalline domains about 1 $\mu$m across, surrounded and held together by non-oriented matrix probably with incomplete atomic orientation. This matrix is considered to have a higher permeability, analogous to zircon.

Carl and Dill (1985) were able to determine the age of secondary uranium mineralizations (uranophane, torbernite, autunite, uranocircite, and saleeite) in basement rocks of northeastern Bavaria using the three-dimensional model of Wendt (1984). A Tera-Wasserburg diagram for torbernite samples from the Grosschloppen uranium occurrence is shown in Fig. 6.33. The upper intercept

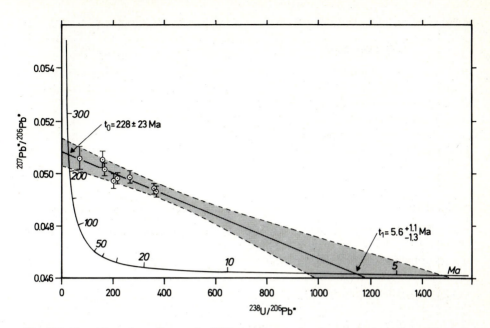

**Fig. 6.33.** Tera-Wasserburg diagram for U/Pb data from torbernite samples from Grosschlopen in northeastern Bavaria (*solid line* discordia; *dashed lines* error hyperbolas) (redrawn after Carl and Dill 1985)

($t_0 = 228 \pm 23$ Ma) in this case is interpreted as the date of primary pitchblende formation, whereas the secondary age ($5.6 \pm^{1.1}_{1.3}$ Ma) is regarded as the age of the secondary uranium mineralization.

### Non-Chronological Applications

See Sect. 6.1.10 (common lead method).

## 6.1.10 Common Lead Method*

(Holmes 1946; Houtermans 1946)

### Dating Range, Precision, Materials, Sample Size

This method is seldom used any more for determining radiometric ages. Its application today is mainly in isotope geochemistry studies of lead, which provide important information about the genesis of rocks and magmas.

The age of mineralization of lead-bearing minerals that contain little or no uranium (e.g., galena, pyrite, nickel sulfides, and feldspar) and the age of formation of igneous rocks (whole-rock samples) can be estimated. The method is dependent to a large degree on the model used and assumes a knowledge of the geochemical parameters of the U-Pb system up to the time of crystallization. For this reason, application of the method is very limited (e.g., Houtermans 1960; Kasanewich 1968; Stacey et al. 1968; Doe 1970; Andrew et al. 1984). This method is closely related to the lead-lead method discussed in Sect. 6.1.11.

An important application of the common-lead method was for calculating the age of the Earth and of meteorites (Murthy and Patterson 1962; Tilton and Steiger 1965; Tilton 1973; Tatsumoto et al. 1973).

### Basic Concept

The expression "common lead" refers to lead that developed in reservoirs whose U/Pb and Th/Pb ratios were never higher than those commonly observed in average crustal or mantle-derived rocks (0.05–1 for U/Pb and 1–10 for Th/Pb). "Common" lead is preserved in minerals whose U/Pb and Th/Pb ratios are so low that the lead isotope ratios have not been changed by the radioactive decay of uranium and thorium since the formation of these minerals. Natural lead has four stable isotopes: $^{208}$Pb, $^{207}$Pb, $^{206}$Pb, and $^{204}$Pb, whose average abundances are 52.4%, 22.1%, 24.1%, and 1.4%, respectively. Only $^{204}$Pb is exclusively nonradiogenic; the other lead isotopes include radiogenic lead originating from the decay of $^{238}$U, $^{235}$U, and $^{232}$Th. Their concentrations (expressed as number of atoms) can be calculated from the following equations, obtained by rearrangement of Eqs. (6.28–6.30):

$$^{206}\text{Pb*} = {}^{238}\text{U} \cdot (e^{\lambda_{238} \cdot t} - 1) \tag{6.32}$$

$$^{207}\text{Pb*} = {}^{235}\text{U} \cdot (e^{\lambda_{235} \cdot t} - 1) \tag{6.33}$$

$$^{208}\text{Pb*} = {}^{232}\text{Th} \cdot (e^{\lambda_{232} \cdot t} - 1) \tag{6.34}$$

The relative abundances of the lead isotopes, therefore, depend on the previous history of the matter from which the minerals and rocks were formed.

*Single-stage models:* If a geological process disturbs a closed U-(Th-)Pb system and the lead is separated from the parent isotopes ($^{238}$U, $^{235}$U, and $^{232}$Th) and included in new minerals containing no uranium or thorium, then the isotopic composition of the lead at the time of separation is preserved in the new minerals. The evolution of the lead isotope composition since time $t$ when U, Th, and Pb formed a closed system for the first time (i.e., the formation of the Earth) can be described with the following equations, derived from Eqs. (6.32–6.34):

$$(^{206}\text{Pb}/^{204}\text{Pb})_t = (^{206}\text{Pb}/^{204}\text{Pb})_0 + \mu(e^{\lambda_{238} t_{\text{Earth}}} - e^{\lambda_{238} t}) \tag{6.35}$$

$$(^{207}\text{Pb}/^{204}\text{Pb})_t = (^{207}\text{Pb}/^{204}\text{Pb})_0 + \frac{\mu}{137.88}(e^{\lambda_{235} t_{\text{Earth}}} - e^{\lambda_{235} t}) \tag{6.36}$$

$$(^{208}\text{Pb}/^{204}\text{Pb})_t = (^{208}\text{Pb}/^{204}\text{Pb})_0 + W \cdot (e^{\lambda_{232} t_{\text{Earth}}} - e^{\lambda_{232} t}) \tag{6.37}$$

where   $(^i\text{Pb}/^{204}\text{Pb})_0$ = the isotopic ratios of the primordial lead,

$t_{\text{Earth}}$ = age of the Earth,

$\mu$ = $^{238}\text{U}/^{204}\text{Pb}$,

137.88 = $^{238}\text{U}/^{235}\text{U}$,

$W$ = $^{232}\text{Th}/^{204}\text{Pb}$,

$t$ = time at which the lead was completely separated and isolated from the radioactive parent elements.

If the isotopic ratios are set to the following values:

$$\alpha = {}^{206}\text{Pb}/^{204}\text{Pb}; \; \beta = {}^{207}\text{Pb}/^{204}\text{Pb}; \; \gamma = {}^{208}\text{Pb}/^{204}\text{Pb},$$

the following equations are obtained:

$$\alpha_t - \alpha_0 = \mu(e^{\lambda_{238}t_{\text{Earth}}} - e^{\lambda_{238}t}) \tag{6.38}$$

$$\beta_t - \beta_0 = \frac{\mu}{137.88}(e^{\lambda_{235}t_{\text{Earth}}} - e^{\lambda_{235}t}) \tag{6.39}$$

$$\gamma_t - \gamma_0 = W(e^{\lambda_{232}t_{\text{Earth}}} - e^{\lambda_{232}t}) \tag{6.40}$$

These equations for the evolution of lead isotope ratios were developed at the same time, independently of each other by Houtermans (1946) and Holmes (1946) and are, therefore, called the Holmes-Houtermans model. The following values are

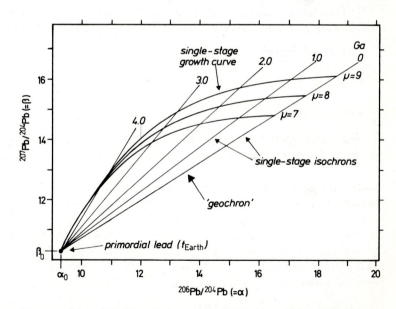

**Fig. 6.34.** Holmes-Houtermans model of the single-stage lead-isotope evolution in an $\beta$-$\alpha$ plot ($^{207}\text{Pb}/^{204}\text{Pb}$ vs. $^{206}\text{Pb}/^{204}\text{Pb}$): Starting from primordial lead with a composition of $(\alpha_0,\beta_0)$ (troilite from Canyon Diablo, Tatsumoto et al. 1973) at time $t_{\text{Earth}}$ (origin of the Earth), the isotopic composition of the lead develops as a function of the U/Pb ratio ($\mu = {}^{238}\text{U}/^{204}\text{Pb}$) along single-stage growth curves. At any time $t$ and independent of $\mu$, the lead in any system lies on an isochron that passes through $(\alpha_0,\beta_0)$

generally used for the calculation:

$$\lambda_{238} = 1.55125 \times 10^{-10} \, a^{-1} \qquad \text{Jaffey et al. (1971)}$$
$$\lambda_{235} = 9.8485 \ \times 10^{-10} \, a^{-1} \qquad \text{Jaffey et al. (1971)}$$
$$\lambda_{232} = 4.9475 \ \times 10^{-10} \, a^{-1} \qquad \text{Le Roux and Glendenin (1963)}$$
$$t_{\text{Earth}} = 4.57 \, \text{Ga} \qquad\qquad\quad \text{Tatsumoto et al. (1973)}$$
$$\alpha_0 \ \ = 9.307 \qquad\qquad\qquad \text{Tatsumoto et al. (1973)}$$
$$\beta_0 \ \ = 10.294 \qquad\qquad\qquad \text{Tatsumoto et al. (1973)}$$
$$\gamma_0 \ \ = 29.476 \qquad\qquad\qquad \text{Tatsumoto et al. (1973)}$$

The values of $\mu$ and $W$ can, in principle, be varied at will. In natural systems they are generally 7–12 or 33–50, respectively. The values selected determine the position of the lead growth curve as shown in Figs. 6.34 and 6.35. These plots show the changes in the lead isotope ratios in closed systems containing uranium and thorium as a function of time. The starting points for primordial lead are based on uranium- and thorium-free mineral phases in meteorites (troilite from the Canyon Diablo meteorite; Tatsumoto et al. 1973). Because Eqs. (6.35 to 6.37) are based on different half-lives, the evolution of the lead ratios is not linear, but occurs along curved lines. A whole family of possible lead growth curves results from the use of different $\mu$ (or $W$) values.

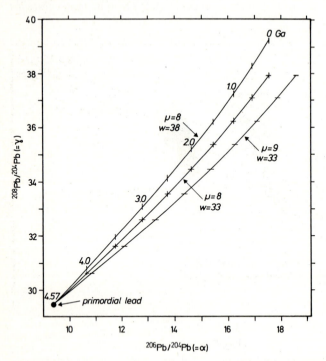

**Fig. 6.35.** Single-stage evolution of lead isotope ratios in a $\gamma$-$\alpha$ plot ($^{208}\text{Pb}/^{204}\text{Pb}$ vs. $^{206}\text{Pb}/^{204}\text{Pb}$). In contrast to the $\beta$-$\alpha$ plot (Fig. 6.34), the isotopic composition of the lead develops not only as a function of the U/Pb ratio ($\mu$), but also as a function of the Th/Pb ratio ($W = {}^{232}\text{Th}/^{204}\text{Pb}$). Three single-stage growth curves are shown for different $\mu$-$W$ values

Division of Eq. (6.39) by Eq. (6.38) yields an equation for a straight line that is independent of $\mu$:

$$\frac{\beta_t - \beta_0}{\alpha_t - \alpha_0} = \frac{1}{137.88} \frac{e^{\lambda_{235}t_{\text{Earth}}} - e^{\lambda_{235}t}}{e^{\lambda_{238}t_{\text{Earth}}} - e^{\lambda_{238}t}} \tag{6.41}$$

This straight line, called an isochron, passes through the point $(\alpha_0, \beta_0)$ (Fig. 6.34). The points for all samples whose lead isotope ratios developed in a closed system between times $t_{\text{Earth}}$ and $t$ lie on this isochron regardless of the value of $\mu$ (single-stage model). An age determination is thus possible, in principle, without a knowledge of $\mu$. Single-stage isochrons are shown in Fig. 6.34 for the age values 1, 2, and 3 Ga; the isochron for an age of $t = 0$ (i.e., today) is called the "geochron" or "zero isochron". Meteorites plot on this isochron, indicating that they have remained closed with respect to uranium, thorium, and lead throughout geologic time.

*Two-stage models:* The isotopic composition of modern lead (e.g., from recent volcanic rocks) does not fit a single-stage evolution model. The lead isotope ratios do not lie on the zero isochron (geochron). That means negative model ages are obtained because the isotopic system is further developed than expected for a single-stage model. In fact, almost all Phanerozoic samples yield negative model ages

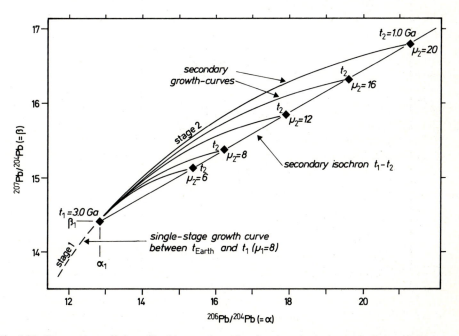

**Fig. 6.36.** Two-stage evolution of lead isotope in an $\beta$-$\alpha$ plot. A geological event at time $t_1$ divides a single-stage lead system into several subsystems with different $\mu$ values. At any subsequent time $t_2$, the lead in all of the subsystems yields points that lie on a straight line (secondary isochron $t_1 - t_2$) that passes through the point $(\alpha_1, \beta_1)$. The time $t_1$ can be extrapolated from this line

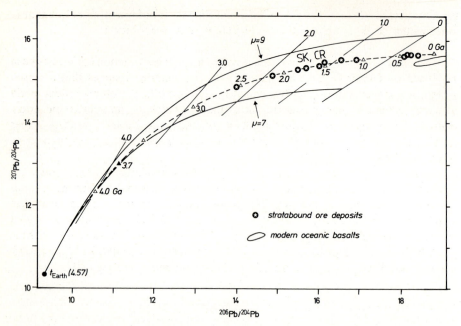

**Fig. 6.37.** An β-α diagram for lead from stratabound deposits (after Köppel and Grünenfelder 1979); *solid lines* growth curves and isochrons from the Holmes-Houtermans model; *dashed line* growth curve after Stacey and Kramers (1975) and Cumming and Richards (1975)

(formerly called "anomalous lead" ages, see Fig. 6.37). This has led to the development of models with two or more stages.

A graph of two-stage evolution is shown in Fig. 6.36. A secondary growth curve results from a change in the values of $\mu$ and $W$ at time $t_1$ (e.g., by separation of crust from mantle material). This line extends into the range of negative ages of the Holmes-Houtermans model. The lead isotope ratios that develop during this second evolution stage – beginning at point $(\alpha_1, \beta_1)$ in any lead system with any value of $\mu$ – lie along a straight line called a secondary isochron, which passes through point $(\alpha_1, \beta_1)$ and, analogous to Eq. 6.41, is described by

$$\frac{\beta_2 - \beta_1}{\alpha_2 - \alpha_1} = \frac{1}{137.88} \cdot \frac{e^{\lambda_{235}t_1} - e^{\lambda_{235}t_2}}{e^{\lambda_{238}t_1} - e^{\lambda_{238}t_2}} \tag{6.42}$$

This equation gives the time $t_1$, e.g., the differentiation which led to different subsystems. Alternatively, if $t_1$ is known, time $t_2$ at which the mineralization of the lead-bearing, uranium- and thorium-free minerals occurred can be calculated.

A two-stage model with differentiation of the Earth into crust and mantle at about 3.7 Ga has been proposed by Stacey and Kramers (1975) (Fig. 6.37). This model is often used in the U/Pb method (Sect. 6.1.9) to correct for common lead. In the first stage of this model, the lead develops with uniform $\mu$ and $W$ values ($\mu_1 = 7.19$, $W_1 = 32.21$) after the first formation of a closed system at 4.57 Ga. Differentiation (separation of the Earth's crust from the mantle) at 3.7 Ga led to an increase in the U/Pb and Th/Pb ratios ($\mu_2 = 9.74$, $W_2 = 37.19$) and, thus, to a greater slope of the growth curve in this second stage of lead evolution.

*Multi-stage models:* A model in which $\mu$ and $W$ are a linear function of time has been proposed by Cumming and Richards (1975). This model assumes that a closed system was never attained within which a radiogenic lead isotope system could develop; instead exchange occurred continuously between the mantle and crust. Both the Cumming-Richards model and that of Stacey and Kramers yield age values for stratabound sulfide deposits that are in good agreement with ages obtained by other methods.

As intensive geochemical studies during the last several decades have shown, the Earth's mantle cannot be viewed as a homogeneous reservoir in terms of its trace element and isotopic composition. This is the case for uranium, thorium, as well as lead. And it is even more true for the Earth's crust. A model of the evolution of lead must take into consideration subsystems with differing $\mu$ and $W$ values, as well as the effects of their interactions, for example in orogenic belts (e.g., Sinha and Tilton 1973; Church and Tatsumoto 1975; Sun et al. 1975; Doe and Zartman 1979).

### Sample Treatment and Measurement Techniques

For lead-bearing minerals that contain no uranium or thorium, only the lead isotope composition needs to be determined. If whole-rock samples are to be investigated in terms of their common lead composition, the uranium, thorium, and lead concentrations may need to be determined in order to correct for in-situ decay of uranium and thorium, using the (often tenuous) assumption of a closed system since the formation of the rock.

The preparation of the samples is described in Sect. 5.2.1.1. It is best to break off ore minerals (e.g., galena) directly from the sample and handpick pure mineral fractions under a stereomicroscope. For galena, a single crystal weighing only a few milligrams is sufficient. For the analysis of whole-rock samples, as well as feldspar, 100–200 mg are needed.

The samples are digested (or only leached) in ultrapure acid; lead, uranium, or thorium is separated by ion exchange. Another method of separation is the electrolytic deposition of the lead. The determination of common lead ratios requires dust-free laboratory conditions and high-purity chemical reagents. Several methods are described in the literature for the chemical preparation of the samples (e.g., Barnes et al. 1973; Sun and Hanson 1975; Ludwig and Silver 1977; Weis 1981).

The lead isotope ratios are measured mass spectrometrically (Sect. 5.2.3.1). The so-called silica-gel technique (Cameron et al. 1969) has proven to be especially useful; rhenium is usually used for the filament. Since the only nonradiogenic isotope of lead is $^{204}$Pb, the mass fractionation that occurs in the mass spectrometric measurement cannot be corrected directly, but only by comparison with a standard (Catanzaro 1967). For geologically young samples (i.e., younger than Tertiary), an age correction is not necessary. For geologically older samples, the uranium, thorium, and lead concentrations of whole-rock or feldspar samples must be determined by mass spectrometric isotope dilution analysis (Sect. 5.2.4.1).

For galena, the lead isotope composition can also be determined with an ion microprobe (e.g., Hart et al. 1981; Deloule et al. 1986; see also Sect. 5.2.4.4).

### Scope and Potential, Limitations, Representative Examples

**Scope and Potential.** Common lead dating is the reverse of nearly all other dating methods. Normally in a radiometric age determination, the time is measured from the event to be dated (e.g., rock formation, cooling, or metamorphism) to the present; this time is defined as the age of the sample. In contrast, in the common-lead method the time is measured from the origin of the Earth (time $t_{Earth}$) to the formation of a mineral (time $t$).

Potassium feldspar and galena lead the list of minerals that contain common lead, i.e., that contain lead but practically no uranium or thorium. Other sulfides (e.g., pyrite and nickel sulfides), feldspars, mica, and melilite are also suitable. Thus, the dating of ore occurrences is the main application of the common lead method. Besides the age information, no other method gives such a detailed picture of the geochemical evolution of magma systems or mineralizations. The common-lead method can also be used for young (Cenozoic) whole-rock samples (especially volcanic rocks), for which corrections must sometimes be applied.

*Mixing lines:* Secondary isochrons can also be interpreted as mixing lines between different reservoirs containing different common lead compositions. The galena within a mining district can even yield the age of the basement through which the mineralizing fluids passed (e.g., Stacey and Hedlund 1983).

Another kind of mixing line was proposed by Andrew et al. (1984) for lead isotope data for galena from southeastern British Columbia: the mixing of lead belonging to different growth curves.

*Age of the Earth:* The age of the Earth can be calculated using Eqs. (6.38–6.41) and "single-stage" lead with an exactly known age of formation (e.g., Holmes 1949; Ostic et al. 1963; Tilton and Steiger 1965). Difficulties in obtaining such ages arise, on the one hand, from the fact that the age of formation of geologically old samples is not often known with the necessary precision, on the other hand, due to the virtual non-existence of "single-stage" lead in young samples whose age is exactly known. An estimate of the age of the Earth is possible, however, using the lead isotope composition of geologically young basalts (e.g., Ulrych 1967). An additional alternative is the dating of meteorites (e.g., Murthy and Patterson 1962; Tatsumoto et al. 1973; Tilton 1973). Another method for interpreting lead isotope data, and a determination of the Earth's age based on it, has been proposed by Manhes et al. (1979, 1980).

**Limitations.** Since several assumptions must be made for this model, the results of the common lead method are more dependent on the given parameters than those of other methods. For this reason the common lead method is seldom used purely as an age determination method.

Single-stage models, e.g. the Holmes-Houtermans model or the one of Collins, Russel, and Farquhar (1953), are not useable for terrestrial samples according to present understanding. Most lead samples cannot be described by such models. Ages based on these models, therefore, have any meaning only in connection with other dating methods.

***Representative Examples.*** Lead samples from a number of stratabound sulfide ore deposits derived at least partially from the Earth's mantle all lie on a common growth curve in both $^{207}Pb/^{204}Pb$ versus $^{206}Pb/^{204}Pb$ and $^{208}Pb/^{204}Pb$ versus $^{206}Pb/^{204}Pb$ plots and, therefore, appeared to early researchers to indicate development of the Earth's mantle from a homogeneous source with a uniform $\mu$ and $W$ ("conformable lead") (Stanton and Russell 1959). Some of the model ages obtained using Eqs. (6.38 to 6.41) are in good agreement with the stratigraphic ages (e.g., Kasanewich 1968). In contrast, almost all Phanerozoic samples yield negative model ages. The points for lead samples for which single-stage evolution was assumed are shown in Fig. 6.37 (modified from Köppel and Grünenfelder 1979), together with the growth curves and isochrons (Holmes-Houtermans model) derived from the values for primordial lead redetermined on troilite from Canyon Diablo (Tatsumoto et al. 1973). The agreement with the stratigraphic ages is poor: For older galena, the Pb model ages are too young and negative ages are obtained for all Phanerozoic samples. The dashed line in the figure is the two-stage growth curve from the model of Stacey and Kramers (1975), which gives a significantly better fit to the data points.

Numerous multi-stage models that satisfy plate-tectonic theory have been developed for "anomalous" multi-stage lead. The model of Doe and Zartman (1979), which simulates the behavior of lead in many different geological environments, is an important example of these models. According to these authors, the upper mantle and the lower and upper continental crust are three separate reservoirs, which exchange material during orogenic events. Growth curves resulting from this model are shown in Fig. 6.41. Other models are discussed by Allègre (1969), Gale and Mussett (1973), Sinha and Tilton (1973), Oversby (1974), Godwin and Sinclair (1982), and Tilton (1983). These models can be used to obtain model ages for galena, sulfide, and feldspar (e.g., Cumming et al. 1982; Thorpe et al. 1986). Schlax and Oldenburg (1984) used a linear inference theory to calculate upper and lower limits for lead isotope growth curves. In this way they obtained model ages with uncertainties of less than 100 Ma for stratiform deposits in all parts of the world.

The lead isotope geochemistry of igneous rocks and ore deposits in southwestern New Mexico were studied by Stacey and Hedlund (1983). In a plot of $\beta$ versus $\alpha$, all of the points lie on a well-defined secondary isochron (Fig. 6.38), which is in good agreement with U/Pb dates on zircon, which place the formation of continental crust in this area between 1.45 and 1.75 Ma. The data plot along the growth curve of Doe and Zartman (1979) for the average orogenic evolution, which possibly indicates an island arc situation. In another study, lead isotopes in nickel sulfides from several deposits in the Thompson Belt in Manitoba were investigated by Cumming et al. (1982). Their dates appear to correspond to events at four distinctly different times: the time of emplacement of the sulfide-bearing ultramafic

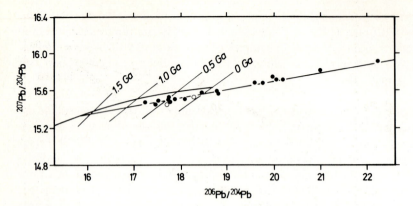

**Fig. 6.38.** Lead isotope data for igneous rocks (*open circles*) and ore deposits (*black dots*) from southwest New Mexico (after Stacey and Hedlund 1983). The growth curve is from the model of Stacey and Kramers (1975)

magma ($2320 \pm 20$ Ma), early folding ($2015 \pm 15$ Ma), retrograde overprinting during the Hudsonian orogeny ($1620 \pm 25$ Ma), and a thermal event associated with the emplacement of a dike swarm ($1125 \pm 60$ Ma).

The interpretation of secondary isochrons as mixing lines is explained by the example shown in Fig. 6.39 for the Neogene (about 55–60 Ma) igneous rocks from the Isle of Skye (Moorbath and Welke 1969). Ultrabasic, basic, intermediate, and acid rocks were analyzed. When the data is interpreted by a conventional single-stage model, all of these rocks contain isotopically anomalous lead. The points for all of the samples plot along a $^{207}Pb/^{204}Pb$ versus $^{206}Pb/^{204}Pb$ line that cuts a conventional growth curve ($\mu = 8.92$) at $t_1 = 3100$ Ma and $t_2 = 60$ Ma. The lead in each of the rocks, therefore, is regarded as a mixture of ancient crustal lead, derived

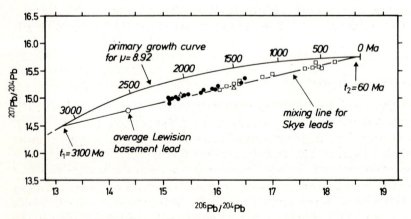

**Fig. 6.39.** Plot of $^{207}Pb/^{204}Pb$ vs. $^{206}Pb/^{204}Pb$ for igneous rocks from the Isle of Skye, Scotland (after Moorbath and Welke 1969). *Black dots* acid rocks; *open squares* basic rocks; *open triangles* ferrodiorite. The data strongly suggest a mixture of ancient crustal lead derived from the underlying Lewisian basement rocks with relatively young, upper mantle lead

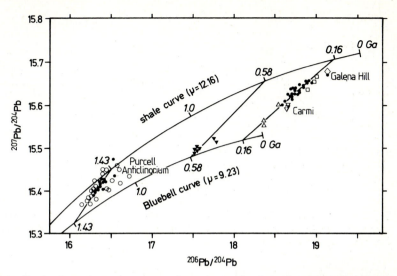

**Fig. 6.40.** Mixing line isochrons between two growth curves resulting from the mixing of lead from two distinct reservoirs, e.g., upper and lower crust or upper crust and mantle. This example is for lead isotope data for galena from southeastern British Columbia (after Andrew et al. 1984)

from the underlying Lewisian basement rocks, with relatively young upper mantle lead.

Andrew et al. (1984) interpreted the lead isotope data for galena from British Columbia in terms of the mixing of lead from two sources whose lead evolution follows different growth curves (Fig. 6.40). Mixing-line isochrons drawn between these two curves at ages corresponding to the best estimates for the mineralization age closely fit the lead isotope data.

Doe and Stacey (1974) used "conformable" galena and modern lead to obtain a single-stage age of the Earth of 4.43 Ga. A similar value (4.47 ± 0.05 Ga) was obtained by Gancarz et al. (1975) from feldspar taken from the Amitsoq Gneiss in western Greenland.

### Non-Chronological Applications

The genetic information provided by the isotope geochemistry of "common lead" is of far greater importance than its application as a dating method. The main aspects have already been discussed in the previous section. Using the isotope geochemistry of common lead, the evolution of a system as a function of time can be included in the geochemical characterization of a geological unit. The isotope geochemistry of lead can provide information especially for the genetic classification of lead-bearing deposits. This is because even a small amount of crustal material mixed with mantle material in an ore will heavily bias the resulting lead composition towards a crustal value, which normally has a $^{207}Pb/^{204}Pb$ ratio greater than that of the mantle. This bias is due to the large difference between the lead contents of crustal and mantle rocks ($> 10$ ppm in the crust, $< 1$ ppm in the mantle).

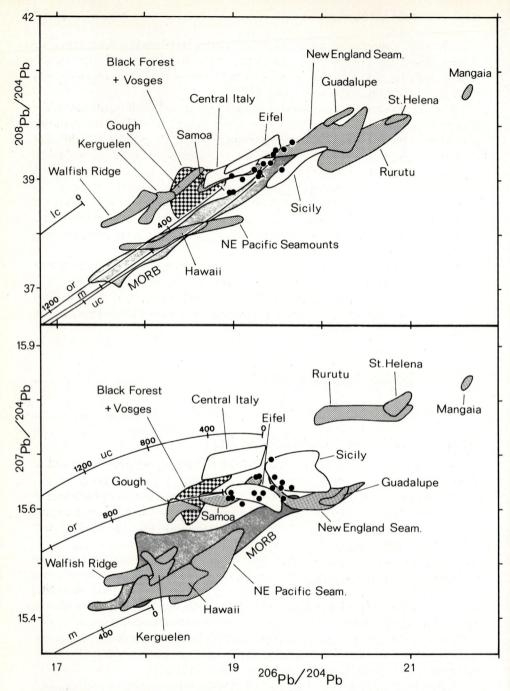

**Fig. 6.41.** Lead isotope data for alkaline volcanic rocks and carbonatites from the Kaiserstuhl area of the Upper Rhine Valley (*black dots*) (Schleicher et al. 1990). For comparison, data for other occurrences of alkaline volcanic rocks in Europe are shown. Also shown are fields for intra-oceanic islands (OIB), MORB, and for Hercynian granitic rocks from Central Europe. Growth curves after the model of Doe and Zartman (1979). *uc* upper crust; *m* mantle; *or* orogeny; *lc* lower crust)

The term "plumbotectonics" (Doe and Zartman 1979) was introduced for models describing the dynamic interaction between the crustal and mantle reservoirs during the chemical evolution of the Earth. A mathematical model was developed by Zartman and Doe (1980, 1981).

The lead isotope geochemistry of volcanic rocks and carbonatites from the Kaiserstuhl alkaline complex in the Upper Rhine Valley, West Germany, is shown in plots of $^{207}Pb/^{204}Pb$ and $^{208}Pb/^{204}Pb$ versus $^{206}Pb/^{204}Pb$ in Fig. 6.41 (Schleicher et al. 1990). The data are compared with the lead isotope compositions of alkaline volcanic rocks in other occurrences in Europe, as well as to MORB, to OIB (ocean island basalts), and to Hercynian granitic rocks of the upper crust, and ore deposits in Western Europe. The data are consistent with the model of magma formation by partial melting of a "plume"-like upper mantle reservoir, which seems to be true not only for the Kaiserstuhl rocks, but also for other alkaline complexes.

For more information on the geochemistry of lead isotopes, the reader is referred to the literature (e.g., Zartman 1974; Unruh and Tatsumoto 1976; Tatsumoto 1978; Doe and Zartman 1979; Chase 1981; Vitrac et al. 1981; Brevart et al. 1982; Anderson 1982; Tilton 1983; Stacey and Stoeser 1983; Thorpe et al. 1986).

## 6.1.11   Lead/Lead ($^{207}Pb/^{206}Pb$) Method*

(Patterson 1955; Sobotovich 1961)

### Dating Range, Precision, Materials, Sample Size

The lead/lead method is closely related to the U/Pb method (Sect. 6.1.9) and is sometimes combined with it in evaluation graphs. It is equally related to the common lead method (Sect. 6.1.10). Intrusion, extrusion and metamorphism ages of granites, volcanites, and gneisses are determined using whole-rock samples and feldspars (K-feldspar and plagioclase) (Sobotovich et al. 1963a, b; Rosholt and Bartel 1969; Nkomo and Rosholt 1972; Oversby 1975, 1976). Sulfides (e.g., pyrite, chalcopyrite, pyrrhotite, and nickel sulfides) can also be used for dating with the lead/lead method (Gulson 1977; Brevart et al. 1986). The isotope data obtained from all minerals that can be dated with the U/Pb method (e.g., zircon, monazite, xenotime, and sphene) can generally also be used for the lead/lead method (e.g., Todt 1976). Both isochron and model ages can be determined.

It is of advantage that not only fresh rocks, but also samples of deeply weathered rocks can yield reliable Pb/Pb dates. Several cogenetic samples (100–500 mg) are needed for an isochron age determination (at least three).

Age determinations can be made on individual zircon crystals with this method using either an ion microprobe (Sect. 5.2.4.4) or a special evaporation technique (Lovering et al. 1976; Kober 1986).

Owing to the low uranium content of whole-rock, feldspar, and sulfide samples, the lead/lead method can normally be used only for very old rocks (Precambrian). Advances in mass spectrometry (Sect. 5.2.3.1) have made it possible to apply the lead/lead method to very old basic to ultrabasic magmatic rocks (e.g., kimberlites, komatiites (Kramers and Smith 1983; Brevart et al. 1986). This is of significance because the Rb/Sr (Sect. 6.1.3) and U/Pb methods are usually unsuitable for these rocks. Attempts have also been made to date metasedimentary carbonate rocks with this method (Taylor and Moorbath 1986) as well as carbonatites (Andersen and Taylor 1988).

The lead/lead method is important for dating meteorites and lunar samples (e.g., Tatsumoto et al. 1976; Manhes et al. 1984; see also Sect. 6.5). The age of the Earth can also be determined with the Pb/Pb method (Patterson 1956; Manhes et al. 1979).

### Basic Concept

Natural lead contains the following four isotopes (average abundances): $^{204}$Pb (1.4%), $^{206}$Pb (24.1%), $^{207}$Pb (22.1%), and $^{208}$Pb (52.4%). Only $^{204}$Pb has no radiogenic component. The amount of radiogenic (*) lead-206 and lead-207 produced in a given time $t$ can be calculated rom the present concentrations (in number of atoms per unit weight) of $^{238}$U and $^{235}$U, respectively, according to the following equations:

$$^{206}\text{Pb*} = {}^{238}\text{U} \cdot (e^{\lambda_{238} t} - 1) \text{ and} \tag{6.43}$$
$$^{207}\text{Pb*} = {}^{235}\text{U} \cdot (e^{\lambda_{235} t} - 1) \tag{6.44}$$

where $\lambda_{238}$ and $\lambda_{235}$ are the decay constants of $^{238}$U and $^{235}$U, respectively.

These equations are also the basis of the conventional U/Pb method (Sect. 6.1.9). By rearrangement and division of Eq. (6.44) by Eq. (6.43), an equation is obtained that is independent of uranium and which can be solved using only the isotopic composition of the lead in the sample:

$$\frac{\left(\dfrac{^{207}\text{Pb}}{^{204}\text{Pb}}\right) - \left(\dfrac{^{207}\text{Pb}}{^{204}\text{Pb}}\right)_0}{\left(\dfrac{^{206}\text{Pb}}{^{204}\text{Pb}}\right) - \left(\dfrac{^{206}\text{Pb}}{^{204}\text{Pb}}\right)_0} = \left(\frac{^{207}\text{Pb}}{^{206}\text{Pb}}\right)^* = \left(\frac{^{235}\text{U}}{^{238}\text{U}}\right)\frac{e^{\lambda_{235} t} - 1}{e^{\lambda_{238} t} - 1} \tag{6.45}$$

where $(^i\text{Pb}/^{204}\text{Pb})_0$ = the initial lead isotope ratio.

The same isotopic ratio $^{238}\text{U}/^{235}\text{U}$ has been found in all but one of the terrestrial, lunar, and meteorite samples that have been analyzed: 137.88 (Shields 1960). The values of Jaffey et al. (1971) are recommended by the IUGS Subcommission on Geochronology for the decay constants:

$$\lambda_{238} = 1.55125 \times 10^{-10}\,\text{a}^{-1}$$
$$\lambda_{235} = 9.8485 \times 10^{-10}\,\text{a}^{-1}$$

Using Eq. (6.45), the value of $t$ can only be approximated by iteration. Because the half-lives of $^{235}$U and $^{238}$U differ, the radiogenic $(^{207}\text{Pb}/^{206}\text{Pb})^*$ ratio varies as a function of time. The relationship betwen age ($t$) and $(^{207}\text{Pb}/^{206}\text{Pb})^*$ values is shown graphically in Fig. 6.42. Tales for determining ages have been published by Stacey and Stern (1973). The ages determined in this way from single samples are called lead/lead model ages.

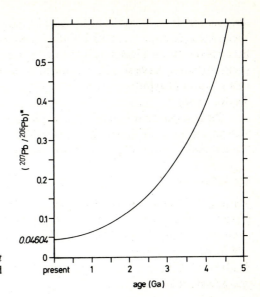

**Fig. 6.42.** Relationship between the age $t$ and the $(^{207}Pb/^{206}Pb)^*$ ratio of a closed uranium-bearing system [Eq. (6.45)]

Isochron ages can be obtained from a graph of $^{207}Pb/^{204}Pb$ versus $^{206}Pb/^{204}Pb$ using the lead isotope ratios determined on several cogenetic samples. Samples of the same age and the same initial $(^{207}Pb/^{204}Pb)_0$ and $^{206}Pb/^{204}Pb)_0$ ratios lie on a straight line with the slope

$$m = \frac{1}{137.88} \frac{e^{\lambda_{235}t} - 1}{e^{\lambda_{238}t} - 1} \qquad (6.46)$$

This slope corresponds to the $(^{207}Pb/^{206}Pb)^*$ ratio in Eq. (6.45) and is a function of the age $t$.

The initial lead isotope ratio cannot always be determined with the lead/lead system; for this purpose, the uranium and thorium concentrations must also be determined (see U/Pb method, Sect. 6.1.9). In many instances, however, the isotopic composition of lead of the feldspar in the rock approximates the initial ratio because the uranium content of feldspar is almost negligible.

## Sample Treatment and Measurement Techniques

Whole-rock samples and minerals are prepared as described in Sect. 5.2.1.1 (Fig. 5.3). Other methods are described by Oversby (1975). To remove lead deposited on the surfaces of the sample, broken samples (fragments ca. 1 cm³) are washed in dilute HCl. To obtain the initial values, the feldspar is leached in acid (HNO₃ and HCl; Ludwig and Silver 1977). Leaching techniques can also be used for whole-rock samples, which are not ground, but only broken into small pieces (about 1 cm³). Depending on the lead concentration, 25–100 mg of sample is needed.

After chemical separation of the lead, a solid-source mass spectrometer (Sect. 5.2.3.1) is used to measure the lead isotope ratios needed for the age determination. Absolute concentrations of uranium and lead are not needed for the lead/lead method. The chemical preparation for the mass spectrometric measurement (e.g., Oversby 1976; Gulson 1977; Manhes et al. 1984) is identical with that for the common lead method (Sect. 6.1.10). After the samples have been digested in very small amounts of ultra-pure acids (e.g., $HF + HNO_3$ or $HF + HClO_4$), the lead is separated by ion exchange techniques. The silica gel technique has proved to be especially useful for the mass spectrometric measurement (Cameron et al. 1969); rhenium is usually used for the filament. Several procedures have been published for the calculation of the regression line for the Pb/Pb isochrons (e.g., York 1969).

A lead/lead age determination is also possible using an ion microprobe (Sect. 5.2.4.4) on indvidual uranium-rich mineral grains (Lovering et al. 1976). A recently proposed technique for dating individual zircon grains is based on the thermal evaporation of radiogenic lead directly in the mass spectrometer (Sunin and Malyshev 1983; Kober 1986).

### Scope and Potential, Limitations, Representative Examples

**Scope and Potential.** The lead/lead method is a promising dating method for two applications: (i) determination of formation ages of crystalline rocks and (ii) dating of meteorites (Sect. 6.5).

*Crystalline rocks and minerals:* Whole-rock Pb/Pb isochrons are not influenced by more recent geological events if these have not led to a complete homogenization of the lead isotopes. This characteritic results from the fact that secondary loss of lead, as a rule, has little influence on the $^{207}Pb/^{20}Pb$ ratio (e.g., Nkomo and Rosholt 1972). This is why Pb/Pb isochron dating may in many cases yield reliable ages for weathered samples from areas where other techniques are inadequate due to the separation of parent and daughter isotopes as a result of weathering (e.g., Gulson et al. 1986). Compared with other methods that may be used for weathered rocks (e.g., zircon U/Pb method, Sect. 6.1.9), the Pb/Pb method is fast and field sampling and sample preparation are simple and inexpensive.

The lead/lead method is quite useful for age determinations using whole-rock and feldspar samples. The U-Pb system is often highly distrubed in such samples, owing to loss of uranium in a surficial environment, making them unsuitable for U/Pb age determination. However, they often yield $^{207}Pb/^{206}Pb$ ages that are concordant with Rb/Sr (Sect. 6.1.3) and Sm/Nd (Sect. 6.1.6) whole-rock isochron ages (e.g., Nkomo and Rosholt 1972; Oversby 1976).

Lead/lead whole-rock ages give the time of the formation of the rock. For igneous rocks that have not been subjected to later metamorphism, this is the time of intrusion or extrusion; for metamorphic rocks, this is the time of the last redistribution and homogenization of the lead isotopes in the rock.

Precambrian metasedimentary carbonate rocks as well as carbonatites sometimes show variations in their lead isotope ratios that are large enough to yield good isochrons using the Pb/Pb method (Taylor and Moorbath 1986; Andersen and Taylor 1988).

*Single-grain dating:* Whole-grain evaporation techniques applied to individual chemically untreated zircon grains like the one developed by Kober (1986) may become a powerful tool for dating zircon, owing to the possibility they offer for distinguishing between different lead components in the same grain. This results from the sequential evaporation of different domains of the grain (e.g., crystalline and metamict parts, older cores, inclusions) that have different lead components. This is made possible by differences in the activation energies of the various parts of the grain.

*Extraterrestrial samples:* The $(^{207}Pb/^{206}Pb)^*$ ratio of radiogenic lead can be measured with a precision of about $0.1\%$. This makes it possible to determine Pb/Pb ages in the billion-year range with an analytical error of only a few million years. The lead/lead method is thus considered to be an extremely sensitive tool for dating lunar samples and meteorites (Sect. 6.5). Its sensitivity comes close to that of the $^{129}I$ method (Sect. 6.2.12), but it has the advantage that it yields an absolute age (Manhes et al. 1984).

**Limitations.** The use of the Pb/Pb isochron method for uranium-rich accessory minerals (e.g., zircon, monazite, and sphene) will probably remain limited because the use of the concordia diagram of the U/Pb method (Sect. 6.1.9) has definite advantages for dating these minerals. For example, geological events that have caused partial reopening of the U-Pb system will normally not be detected in a lead/lead age determination, whereas both the 'primary', as well as the 'secondary' age can be determined with the U/Pb method.

One difficulty is that the uranium content of rocks and feldspars is usually very low ($< 50\,ppm$). This means that only very small amounts of radiogenic lead, especially $^{207}Pb$, accumulate in them. Thus, for analytical reasons, the lead/lead method is best suited for very old ($>$ ca. 1 Ga) and relatively uranium-rich (i.e., generally acid) rocks.

**Representative Examples.** For 3-billion-year-old granite and gneiss from the Ukraine, Sobotovitch et al. (1963a) determined a lead/lead isochron age of 3030 Ma, which is somewhat higher than the K-Ar age for mica from these rocks. An example of very good agreement between Pb/Pb isochron ages and Rb/Sr whole-rock ages is provided by 2950-million-year-old Precambrian gneiss from the Granite Mountains in Wyoming (Nkomo and Rosholt 1972; Peterman et al. 1971). In contrast, Oversby (1976) found evidence for post-emplacement disturbance of the lead isotope system in gneissic granites and granodiorites from Australia (Pilbara Block). The redistribution of the lead ranges in these rocks from little effect to complete equilibrium of the potassium feldspars during later metamorphism.

An example of the application of the lead/lead method to basic and even ultrabasic Archean rocks is an investigation of komatiitic lavas and associated sulfides in three rock complexes by Brévart et al. (1986). The Pb/Pb isochron for rocks from Barberton, South Africa, gives an age of $3.46 \pm 0.07$ Ga, in good agreement with the Sm/Nd age (Sect. 6.1.6). Three distinct komatiitic lava flows in the Munro township in the Abitibi belt in Canada yielded significantly different

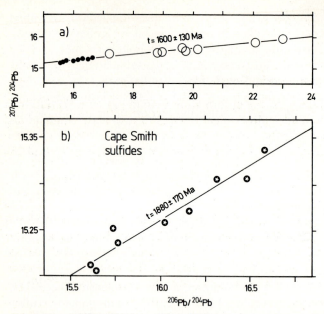

**Fig. 6.43. a** A $^{207}Pb/^{204}Pb$ vs. $^{206}Pb/^{204}Pb$ plot for komatiites (*open symbols*) and sulfides (*dark symbols*) from Cape Smith, Canada (after Brévart et al. 1986): The regression line for whole-rock and sulfide samples yields as age of $1600 \pm 130$ Ma. **b** Corresponding plot for the sulfides alone ($t = 1880 \pm 170$ Ma)

ages: $2.72 \pm 0.02$ Ga, $2.47 \pm 0.13$ Ga, and $2.58 \pm 0.02$ Ga. The first date is interpreted as the emplacement age; the other two dates are interpreted as two later metamorphic events. The age obtained for rocks from Cape Smith in Canada ($1.6 \pm 0.13$ Ga) is lower than those obtained by other methods (Fig. 6.43a), obviously due to a superimposed metamorphic event. In contrast, the isochron for sulfides from the same complex (Fig. 6.43b) gives an age of $1.88 \pm 0.17$ Ga, in agreement with the Sm/Nd method.

An excellent Pb/Pb isochron age of $540 \pm 14$ Ma is reported by Andersen and Taylor (1988) for rocks from the Fen carbonatite complex (Norway), using carbonatites, fenites, and apatite cumilate rocks. The latter show $^{206}Pb/^{204}Pb$ ratios up to 230.

A lead/lead isochron for various meteorites has been published by Chen and Tilton (1976). This isochron includes data for the Allende, Canyon Diablo, Beardsley, Plainview, Murray, Richardton, Modoc, Mezo-Madaras, and Pultusk meteorites. The data yielded a well-defined regression line with a slope of $0.6240 \pm 0.0015$, which corresponds to an age of $4.565 \pm 0.004$ Ga. This value is one of the most precise determinations of the age of a meteoritic parent body (Sect. 6.5).

### Non-Chronological Applications

See Sect. 6.1.10 on the common-lead method.

## 6.1.12   Chemical Lead Method

(Boltwood 1907)

### *Dating Range, Precision, Materials, Sample Size*

This method for determining the age of formation of uranium- and thorium-rich minerals from pegmatites and hydrothermal mineralizations (e.g., uraninite and zircon) has only historical significance any more (e.g., Keevil 1939). It has been replaced by modifications of the lead/alpha method (Sect. 6.1.13) and the lead-210 method (Sect. 6.3.13).

### *Basic Concent*

The lead isotopes $^{206}$Pb and $^{208}$Pb are the stable end-products of the $^{238}$U and $^{232}$Th decay series. Lead-207, as end-product of the $^{235}$U series, can be neglected for this method. The age of minerals that originally contained no lead can be determined from the present ratio of total lead to the uranium or thorium concentration. Because 1 g of $^{238}$U generates $(206/238) \cdot \lambda_{238}$ grams of $^{206}$Pb per year, the age is

$$t = \left(\frac{\text{Pb}}{\text{U}}\right) \cdot \frac{238}{206} \cdot \frac{1}{\lambda_{238}} \quad \text{analogously,} \tag{6.47}$$

$$t = \left(\frac{\text{Pb}}{\text{Th}}\right) \cdot \frac{232}{208} \cdot \frac{1}{\lambda_{232}} \tag{6.48}$$

where Pb, U, and Th are in percent,

$$\lambda_{238} = 1.55125 \times 10^{-10}\,\text{a}^{-1}, \quad \lambda_{232} = 4.9475 \times 10^{-}\,\text{a}^{-1},$$

and the lead is assumed to consist entirely of $^{206}$Pb or $^{208}$Pb, respectively.

If the mineral contains both thorium and uranium, more complicated equations must be used (see Sect. 6.1.13).

### *Sample Treatment and Measurement Techniques*

The uranium or thorium concentration and the lead concentration are determined on very pure mineral samples (for their preparation, see Sect. 5.2.1.1, Fig. 5.3) using standard chemical methods (wet chemistry, AA, XRF, $\gamma$-spectrometry, fluorimetry, etc.). The age is calculated using either Eq. (6.47) or (6.48). Depending on the analytical method used and the concentrations of the elements measured, 0.5–2 g samples are needed.

### *Scope and Potential, Limitations*

The chemical lead method today is of only historical significance. This is due to the numerous souces of error which make it impossible to obtain an exact age with this analytically simple method. For one, the determination of the element concentrations with routine analytical methods is sometime associated with considerable error. A second, particularly disadvantageous source of error is that a basic assumption of the model on which the method is based is not correct, namely that the analytically determined value for lead coincides with the lead produced by the decay of uranium and/or thorium. In fact, 'common' lead, whose origin is not the decay of radioactive nuclides in the mineral being analyzed, must be expected in almost all uranium- and thorium-bearing minerals (expressed by the presence of lead-204).

### 6.1.13    Lead/Alpha Method (Larsen Method)

(Larsen and Keevil 1947)

*Dating Range, Precision, Materials, Sample Size*

This is a rapid and inexpensive method for a rough determination of the minimum age of uranium- and thorium-bearing, originally lead-free minerals (e.g., zircon, xenotime, monazite, and thorite) (Larsen et al. 1952; Gottfried et al. 1959; Grünenfelder and Stern 1960; Hamilton 1965; Delaloye 1979). The age range extends from ca. 10–20 Ma to more than 1.5 Ga.

Several mineral fractions of a rock unit should always be measured in order to obtain a measure of the reliability of the age determination. Between 100 and 200 mg of sample are needed.

*Basic Concept*

The principle of the lead/alpha method is the same as that of the chemical lead method (Sect. 6.1.12). The model on which it is based assumes that the dated mineral contained no lead at the time of crystallization. Radiogenic lead (Pb*) is then produced by the radioactive decay of uranium and/or thorium, which accumulates in the mineral according to the Eqs. (6.32 to 6.34) (Sect. 6.1.10).

If the mineral has formed a closed system with respect to uranium, thorium, and lead since its formation, the ratio of lead concentration (in ppm) to the sum of the $\alpha$-activities of $^{238}U$, $^{235}U$, and $^{232}Th$ ($\alpha$-particles per mg per h) is proportional to the age of the sample. The age can then be calculated as follows:

$$t = f_1 \frac{Pb}{[^{238}U] + [^{235}U] + [^{232}Th]}, \tag{6.49}$$

where $f_1$ is a function of the U/Th atomic ratio:

$$f_1 = \frac{2632 + 624\,(Th/U)}{1 + 0.312\,(Th/U)} \tag{6.50}$$

If only uranium is present, $f_1 = 2632$.
If only thorium is present, $f_1 = 1990$.

If the sample is older than 200 Ma, the decay of the parent isotopes must be taken into consideration using the following correction:

$$t_{cor} = t - 0.5 f_2\, t^2, \tag{6.51}$$

where $f_2$ is also a function of the U/Th ratio: $f_2 = 1.9 \times 10^{-4}$ for the atomic ratio U/Th $= \infty$ and $f_2 = 0.5 \times 10^{-4}$ for U/Th $= 0$. The value of $f_2$ is calculated from the measured U/Th ratio as follows:

$$f_2 = \frac{77.2 + 6.20\,(\text{Th/U})}{4.06 \times 10^{11}\,(1 + 0.312\,(\text{Th/U}))} \tag{6.52}$$

For ages greater than 1.7 Ga, the following correction is used:

$$t_{\text{cor}} = (t - 0.5 f_2 t^2) + 3.4 \times 10^{-9}(t - 0.5 f_2 t^2)^3 \tag{6.53}$$

### Sample Treatment and Measurement Techniques

The minerals to be dated are separated and cleaned using the standard procedures described in Sect. 5.2.1.1 (Fig. 5.3). The lead concentration, which may be in the range of 10–200 ppm, is determined by spectroscopic methods (Waring and Worthing 1953; Rose and Stern 1960) or by any of several other routine analytical procedures, e.g., flame photometry, AA (Sect. 5.2.4.3), XRF (Sect. 5.2.4.5), or ICP (Sect. 5.2.4.3). Depending on the sensitivity of the method selected, as well as on the age and U/Th content of the sample, between 0.1–1 g of sample is needed. The uranium and thorium concentrations (usually 300–3000 ppm and 100–2500 ppm, respectively) are either estimated via the α-activities of their decay products (which can be obtained from the α counting rate; Sect. 5.2.2) or determined by XRF, fluorimetry, autoradiography, or α-spectrometry.

### Scope and Potential, Limitations, Representative Examples

*Scope and Potential.* The advantage of the Larsen method is its simplicity. Time-consuming mass spectrometric isotope-dilution analyses (Sects. 5.2.3.1 and 5.2.4.1), as in the U/Pb method, are not necessary. The lead/alpha method can be used to quickly obtain rough ages for areas that have not been previously studied geochronologically, but it cannot be used for detailed geochronological studies.

*Limitations.* The lead/alpha method assumes that all of the lead in the mineral being studied is radiogenic. This is as a rule not the case: Common, non-radiogenic lead is almost always present, as is demonstrated by the large amount of lead isotope data for zircons, monazites, etc. obtained using the U/Pb method (Sect. 6.1.9). In young minerals, the common lead component can make up a significant proportion of the total lead present and thus make the calculated ages too high. With increasing age the proportion of non-radiogenic lead decreases so that the resulting error becomes increasingly less.

Certain limits are also placed on the method by the fact that relatively large amounts of handpicked, extremely pure sample material are needed. Age determinations on accessory minerals are especially affected by this limitations. Minerals from coarse-grained pegmatoid rocks are, in contrast, suitable.

A further disadvantage is that disturbances of the closed system by recent lead, uranium, or thorium losses or gains cannot be recognized with the Larsen method.

This is, however, often the case, particularly with zircons from older rock units (see Sect. 6.1.9). For this reason, lead/alpha ages for Precambrian rocks can generally be interpreted only as minimum ages (Houtermans 1960).

*Representative Examples.* An extensive comparison of lead/alpha age data with data from the Rb/Sr (Sect. 6.1.3), K/Ar (Sect. 6.1.1), and U/Pb (Sect. 6.1.9) methods, as well as with stratigraphic age classifications, has been made by Gottfried et al. (1959). For rocks younger than Precambrium, they obtained lead/alpha ages (predominantly on zircon) that (on the basis of the values for the decay constants at that time) were in satisfactory agreement with the values obtained with the other methods. Lead/alpha ages for Phanerozoic igneous rocks of known stratigraphic age (data taken from Gottfried et al. 1959) are plotted in Fig. 6.44 versus the 1983 geological time scale (Geological Society of America; Palmer 1983). The uncertainty in the individual lead/alpha ages in the figure is given by the scatter in the mineral ages obtained for the individual rock units; the stratigraphic error is estimated from the data given by Gottfried et al. (1959). It can be seen that the lead/alpha ages

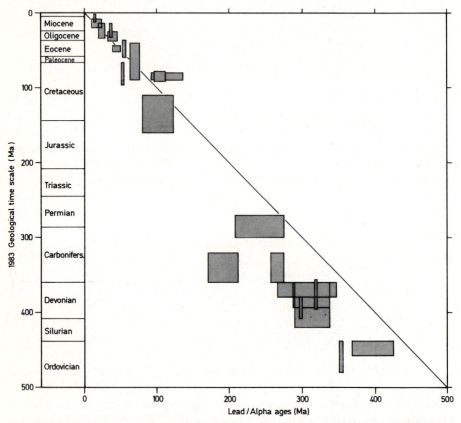

**Fig. 6.44.** Lead/alpha ages of Phanerozoic igneous rocks (data from Gottfried et al. 1959) plotted versus their stratigraphic age (1983 Geological Time Scale, Palmer 1983). The data confirm the unreliability of lead/alpha ages older than Cretaceous

become increasingly lower than the stratigraphic ages with increasing geological age. Lead losses, possibly due to more recent metamorphic events, which is usually the case with Precambrian zircon, evidently influence the results even in the Phanerozoic age range. Lead/alpha age values in this range, analogous to those for the Precambrian, can therefore be viewed only as minimum ages.

## 6.1.14   Krypton/Krypton ($Kr_{sf}/Kr_n$) Method*

(Shukolyukov, Ashkinadze, Kirsten and Jessberger 1976)

### Dating Range, Precision, Materials, Sample Size

This is a seldom used method for determining the krypton retention age of uranium-bearing minerals (e.g., zircon, monazite, xenotime, and britholite) with ages greater than 10 Ma. About 100 mg of sample is needed.

### Basic Concept

The $Kr_{sf}/Kr_n$ method is based on the production of krypton by spontaneous fission and thus differs from the $^{81}Kr$ dating method (Sect. 6.2.10) for meteorites (Marti 1967, 1982; see also Sect. 6.5), which is based on the production of krypton by spallation.

Atmospheric krypton has six stable isotopes: $^{78}Kr$, $^{80}Kr$, $^{82}Kr$, $^{83}Kr$, $^{84}Kr$, and $^{86}Kr$. In uranium-bearing minerals and rocks, radiogenic krypton isotopes $Kr_{sf}$ are produced by spontaneous fission of uranium-238. The concentration of these isotopes is a measure of the age of the mineral or rock according to the following equation:

$$t = \frac{1}{\lambda_{238}} \ln \left( \frac{\lambda_{238}}{\lambda_{sf238}} \frac{(^i kr_{sf})}{(^{238}U)^i \zeta_{sf}} + 1 \right), \tag{6.54}$$

where $\lambda_{238}$   = decay constant for $^{238}U = 1.55125 \times 10^{-10}\, a^{-1}$ (Jaffey et al. 1971),
  $\lambda_{sf238}$ = decay constant for the spontaneous fission of $^{238}U = 8.57 \pm 0.42 \times 10^{-17}\, a^{-1}$ (Thiel and Herr 1976, see also Sect. 6.4.7),
  $^i Kr_{sf}$ = krypton isotope with a mass $i$ produced by spontaneous fission of $^{238}U$,
  $^i \zeta_{sf}$ = fission yield for the isotope $^i Kr$.

To determine the uranium content, the sample is irradiated with thermal neutrons in a nuclear reactor. This produces krypton ($Kr_n$) from the neutron-induced fission of $^{235}U$. Because the $^{238}U/^{235}U$ ratio (137.88; Shields 1960) is believed to be constant in all terrestrial samples (Hamer and Robins 1960), the $Kr_n$ produced is proportional to the $^{238}U$ content. The isotopic spectrum of the $Kr_n$ differs from that of the $Kr_{sf}$. Conversion of Eq. (6.54) produces the following equation, which can be used to calculate the age of the sample:

$$t = \frac{1}{\lambda_{238}} \ln\left(\frac{\phi\sigma_{235}R_i\lambda_{238}{}^i\xi_n}{137.88\lambda_{sf238}{}^i\xi_{sf}} + 1\right) \tag{6.55}$$

where $R_i = {}^iK_{sf}/{}^iKr_n$,

   $\phi$ is the integrated neutron flux of the reactor, and

   $\sigma_{235}$ is the effective cross-section of the induced spontaneous fission of $^{235}U$.

Because it is produced by neutron-induced fission of $^{235}U$ but is not present in the isotopic spectrum of the spontaneous fission of $^{238}U$, the isotope $^{85}Kr$ is used as reference for the determination of the $Kr_n$ concentration. The age is calculated on the basis of krypton-86 (i.e., $i = 86$).

The irradiation parameters are specific for each material and generally cannot be determined very precisely. This difficulty is avoided by irradiating a monitor mineral of known age, $t_{monitor}$, together with the sample. The age of the unknown sample is then obtained using the following equation:

$$t = \frac{1}{\lambda_{238}} \ln\left(\frac{R_i}{(R_i)_{monitor}}(e^{\lambda_{238}t_{monitor}} - 1) + 1\right) \tag{6.56}$$

The use of a monitor mineral eliminates the error due to the inexactness of the decay constant $\lambda_{sf}$.

### Sample Treatment and Measurement Techniques

The minerals are prepared using the standard procedures described in Sect. 5.1.1.1 (Fig. 5.3). The grain size of mineral samples should not be too small ($100-500\,\mu m$) since otherwise some of the $Kr_n$ will be lost due to fission recoil displacement (ca. $15\,\mu m$) in the reactor, which would result in ages that are too high.

After having been irradiated in quartz glass tubes, the samples are heated and degassed stepwise in an ultra-high vacuum analogously to the $^{39}Ar/^{40}Ar$ and $Xe/Xe$ methods (Sects. 6.1.1.2 and 6.1.15.2, respectively). The isotopic composition of the noble gas fraction is measured mass spectrometrically (Sect. 5.2.3.1). A limiting value for $R_i$ is determined from the isotopic spectrum so that the age can be calculated using Eq. (6.55) or (6.56). Correction for atmospheric krypton is possible using the $^{82}Kr$ concentration.

### Scope and Potential, Limitations, Representative Example

*Scope and Potential.* The Kr/Kr method will probably not be widely used because it can be used only for minerals that can be dated more easily with other methods (e.g., U/Pb, Sect. 6.1.9). But an advantage is that no absolute concentrations need to be determined, only the isotopic composition of a single element. Moreover, zircon data for the Botnavatn igneous rock complex in Norway provide evidence that the U-(Kr + Xe) system in zircon is more resistant to metamorphism than the corresponding U-Pb systems (Hebeda et al. 1985). Therefore, in combination with

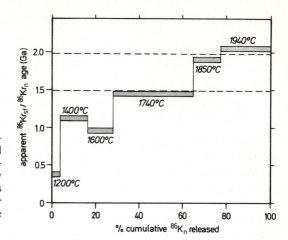

**Fig. 6.45.** Degassing spectrum (apparent $^{86}Kr_{sf}/^{86}Kr_n$ age plotted versus $\%^{86}Kr_n$ released) of a Precambrian monazite (Shukolyukov et al. 1976). *Upper dashed line* gives known age of the mineral; *lower dashed line* gives mean age of all the individual degassing steps

other methods, the Kr/Kr method can in some cases yield additional information.

*Limitations.* A main disadvantage of the Kr/Kr method is that minerals containing little uranium have too little radiogenic krypton to be able to measure it with the necessary precision. Old, uranium-rich minerals (e.g., xenotime) have usually lost part of their noble gas content as a result of metamictization. By stepwise degassing (see Sect. 6.1.1.2), however, gas losses can be recognized and corrected for, if they are not too extreme, by direct correlation of uranium and krypton in the different parts of the sample.

*Representative Example.* A degassing spectrum for a Precambrium monazite from the Ukraine (Shukolyukov et al. 1976) is shown in Fig. 6.45. The ratio of the spontaneous to induced krypton components in the fraction degassed at the highest temperatures is not disturbed by natural $Kr_{sf}$ losses. The apparent age of this fraction is in agreement with the known age of the mineral. Distinct loss of $Kr_{sf}$ is shown by the fractions degassed at lower temperatures; this loss greatly lowers the average age calculated from all of the $Kr_{sf}/Kr_n$ ratios.

### Non-Chronological Applications

Krypton isotope ratios are used in the study of extraterrestrial rocks (e.g., Eugster et al. 1967b; Alaerts et al. 1979; Matsuda et al. 1980; Bogard et al. 1984; see also Sect. 6.5). For example, information on the origin and history of meteorites can be obtained.

Isotopic studies on terrestrial krypton in the atmosphere, as well as in rocks, provide information for reconstruction of the degassing history of the Earth and the development of the atmosphere (e.g., Canalas et al. 1968; Phinney 1972; Podosek et al. 1980; Ozima and Zashu 1983). Less information can be obtained from krypton, however, than from xenon (see Sect. 6.1.15.1).

## 6.1.15  Xenon Methods*

Analogous to the K/Ar methods (Sect. 6.1.1), we divide the U/Xe method into two subchapters:

1) the older $U/Xe_{sf}$ method (Sect. 6.1.15.1), which is nearly obsolete, and
2) the $Xe_{sf}/Xe_n$ method (Sect. 6.1.15.2), which was developed from it.

### 6.1.15.1  Uranium/Xenon (U/Xe$_{sf}$) Method

(Khlopin and Gerling 1947)

### Dating Range, Precision, Materials, Sample Size

A seldom used method for determining xenon retention ages of uranium-bearing minerals and ores (e.g., monazite, zircon, xenotime, euxenite, uraninite, and samarskite) older than 100 Ma, of terrestrial rocks (e.g., granite) older than about 1 Ga, and of meteorites and lunar samples (Butler et al. 1963; Kuroda 1963; Shukolyukov and Mirkina 1963; see also Sect. 6.5). Depending on their age, sample sizes of 10 mg–2 g are needed for uranium-rich minerals, 3–4 g for granites (whole rock).

### Basic Concept

Atmospheric xenon consists of 9 stable isotopes with the atomic masses 124, 126, 128, 129, 130, 131, 132, 134, and 136. Uranium-bearing minerals and rocks contain several other xenon isotopes ($Xe_{sf}$) produced by spontaneous fission of $^{238}U$. The concentration of these isotopes is a measure of age. The equation for a U/Xe age is analogous to that for krypton [Eq. (6.54)], with $^iXe_{sf}$ instead of $^iKr_{sf}$,

$^i\xi_{sf}$ $\models$ relative fission yield for the xenon isotope $i$, and
$\lambda_{sf238}$ = decay constant for the spontaneous fission of $^{238}U$
$\quad = 8.57 \pm 0.42 \times 10^{-17} a^{-1}$ (Thiel and Herr 1976; see also Sect. 6.4.7).

### Sample Treatment and Measurement Techniques

The preparation of the samples is done according to the scheme shown in Sect. 5.2.1.1 (Fig. 5.3). For the determination of xenon, the samples may not be pulverized (grain size 100–500 $\mu m$).

Uranium (in number of $^{238}U$ atoms) is determined by mass spectrometric isotope dilution analysis (Sect. 5.2.4.1) or with an α-spectrometer; xenon is determined with a mass spectrometer (Sect. 5.2.3.1).

### Scope and Potential, Limitations, Representative Example

*Scope and Potential.* Advances in the mass spectrometry of the noble gases are making it possible to also date minerals containing little uranium. The U/Xe$_{sf}$

method, however, has been replaced by an analytical modification, the $Xe_{sf}/Xe_n$ method (Sect. 6.1.15.2).

*Limitations.* Application of this method is limited. Minerals with little uranium (e.g., apatite) contain too little radiogenic xenon; old uranium-rich minerals (e.g., xenotime) often have lost some xenon as a result of radiation damage. Especially typical uranium minerals (e.g., uraninite, pitchblende, broeggerite, brannerite, and betafite) have been shown to be unsuitable for the xenon method owing to their high xenon losses of 30%–70% (Gerling and Shukolyukov 1959). The independent determination of uranium and xenon by different methods increases the error.

*Representative Example.* Five monazite samples from pegmatite in northern Karelia yielded U/Xe ages between 1770 and 2230 Ma (Shukolyukov and Mirkina 1963). The average of $1960 \pm 150$ Ma agrees well with the U/Pb (Sect. 6.1.9) and K/Ar (Sect. 6.1.1) ages for the rock unit ($1950 \pm 50$ Ma).

### Non-Chronological Applications

The isotope geochemistry of xenon (sometimes called "xenology") provides considerable information about the early history of meteorites and the solar system (e.g., Reynolds 1963; Drozd and Podosek 1976; Alaerts et al. 1979; see also Sect. 6.5).

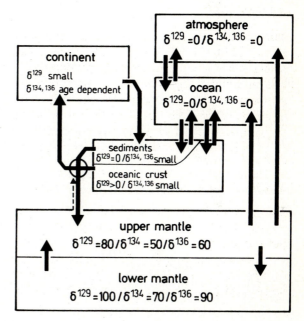

**Fig. 6.46.** Schematic model showing the isotopic compositions of xenon in the various terrestrial reservoirs and the exchange patterns between them (after Staudacher and Allègre 1982)

$$\delta^i = \frac{(^iXe/^{130}Xe)_{sample} - (^iXe/^{130}Xe)_{atm}}{(^iXe/^{130}Xe)_{atm}} \cdot 10^3 \ (‰)$$

The still young study of xenon isotopes in terrestrial rocks has already had some success. Such studies have yielded evidence for natural spontaneous chain reactions at the Oklo mine site (Drozd et al. 1974). Staudacher and Allègre (1982) have shown that deviations in xenon isotope ratio in rocks from the atmospheric ratio can have application in geodynamics. Excess $Xe_{sf}$ has been found in granitoids and crustal rocks; xenon isotope enrichments in mid-oceanic ridge basalts stem from the extinct radioisotopes $^{129}$I and $^{244}$Pu (cf. Sect. 6.5).

The model of Staudacher and Allègre given for xenon in the various terrestrial reservoirs is shown in Fig. 6.46. As can be seen, not only xenon from the fission of $^{238}$U, but also xenon from the fission of $^{244}$Pu and radiogenic xenon from $^{129}$I play a part in the isotopic geochemistry of xenon. For example, comparison of the proportion in terrestrial xenon of $^{129}$Xe produced by the decay of $^{129}$I and $^{136}$Xe produced by fission of $^{244}$Pu (Ozima et al. 1985) suggests that the Earth's interior accreted a few tens of millions of years earlier than the outer parts, from which the atmosphere evolved.

## 6.1.15.2 Xenon/Xenon ($Xe_{sf}/Xe_n$) Method*

(Shukolyukov, Ashkinadse and Komarov 1974a;
Teitsma, Clarke and Allègre 1975)

### Dating Range, Precision, Materials, Sample Size

This is an analytical modification of the U/Xe method (Sect. 6.1.15.1) that has by and large replaced the older form. It is used to determine xenon retention ages greater than about 100 Ma for uranium-bearing minerals (e.g., zircon, monazite, xenotime, and pitchblende), requiring between 1 mg and 1 g, depending on age and uranium content, and for very old rocks ( > ca. 1 Ga) (Shukolyukov et al. 1979; Teitsma and Clarke 1978). With special chemical digestion techniques, the time of formation of zircons can be determined (Kapusta et al. 1983).

The newly developed $^{129}$Xe/$^{136}$Xe$_n$ method can be used to determine the age of uranium-bearing minerals in the range of 5–100 Ma (Meshick et al. 1987).

### Basic Concept

The $Xe_{sf}/Xe_n$ method is based on the same principles as $^{40}$Ar/$^{39}$Ar dating (Sect. 6.1.1.2). Radiogenic xenon isotopes ($Xe_{sf}$) are produced by spontaneous fission of $^{238}$U. The uranium concentration is determined by irradiation of the samples in a reactor with thermal neutrons, whereby xenon is produced by neutron-induced fission ($Xe_n$) from $^{235}$U. Because the $^{235}$U/$^{238}$U ratio is practically constant in all terrestrial samples (Hamer and Robbins 1960), $Xe_n$ is directly proportional to the $^{238}$U content. The isotopic spectrum of the $Xe_n$ differs from that of the $Xe_{sf}$ and the age determination is based on these differences. An equation analogous to

Eq. (6.55) can be used, where

$\lambda_{238}$ = decay constant for $^{238}U$
      = $1.55125 \times 10^{-10} a^{-1}$ (Jaffey et al. 1971),
$\lambda_{sf238}$ = decay constant for the spontaneous fission of $^{238}U$
      = $8.57 \pm 0.42 \times 10^{-17} a^{-1}$ (Thiel and Herr 1976),
$R_i$ = $^iXe_{sf}/^iXe_n$, the xenon isotopes with a mass i produced by the spontaneous
      fission of $^{238}U$ or neutron-induced fission of $^{235}U$, respectively,
$^i\xi_{sf}$ and $^i\xi_n$ = fission yields for the isotope $^iXe$, and $\phi$ and $\sigma_{n235}$ are irradiation
parameters.

If a monitor mineral of known age is irradiated and analyzed together with the sample, it is not necessary to know the irradiation parameters exactly and the error that results from the relatively large uncertainty in $\lambda_{sf238}$ is avoided. It is sufficient if the $Xe_{sf}/Xe_n$ ratios are measured for any isotope i in the sample and the monitor. The age equation is analogous to Eq. (6.56). Usually, the $^{136}Xe$ ($i = 136$) provides the best dating results.

A version of the $Xe_{sf}/Xe_n$ method developed by Meshick et al. (1987), the $^{129}Xe/^{136}Xe_n$ method, is based on the decay of $^{129}I$. This method is also closely related to the $^{129}I$ method (Sect. 6.2.12), but uses $^{129}I$ produced by fission of $^{235}U$ induced by neutrons in nature and, therefore, can be used only for uranium ores (e.g., pitchblende). Owing to the very short half-life of $^{129}I$ ($\tau = 17.2$ Ma), the decay constants for both its production from $^{235}U$ and its own decay must be used. Neutron-induced fission of $^{235}U$ also produces $^{136}Xe$, which can be used as a monitor of the natural neutron flux. The $^{129}Xe/^{136}Xe$ ratio is thus a measure of the age of the uranium ore:

$$\left(\frac{^{129}Xe}{^{136}Xe_n}\right) = \frac{^{129}\xi_n/^{136}\xi_n}{\lambda_{129} - \lambda_{235}}\left(\lambda_{129} - \lambda_{235}\frac{(1 - e^{-\lambda_{129}t})}{(1 - e^{-\lambda_{235}t})}\right), \tag{6.57}$$

where $\lambda_{129} = 4.03 \times 10^{-8} a^{-1}$,
      $\lambda_{235} = 9.8485 \times 10^{-10} a^{-1}$, and
      $^{129}\xi_n$ and $^{136}\xi_n$ = neutron-induced fission yields for $^{129}Xe$ and $^{136}Xe$
      ($= 0.669\%$ and $6.37\%$, respectively).

### Sample Treatment and Measurement Techniques

The preparation of the samples is done according to the scheme given in Sect. 5.2.1.1 (Fig. 5.3). To remove contaminating minerals, zircon samples are leached in $HNO_3$, while the grain size should be between 100 and 500 $\mu m$. If the grain size is smaller, some of the $Xe_n$ escapes already in the reactor as a result of the very large fission recoil displacements of about 15 $\mu m$, leading to ages that are much too high. Normally, neutron absorption by REE and the production of xenon from thorium (e.g., in monazite) must be taken into consideration. To isolate well-crystallized zircon from metamict phases that may have lost part of their radiogenic xenon owing to secondary processes, Kapusta et al. (1983) have proposed a differential dissolution technique.

The xenon isotope ratios are measured with a high-sensitivity gas-source mass spectrometer. Analogously to the $^{40}Ar/^{39}Ar$ method (Sect. 6.1.1.2), the samples are degassed stepwise in five to ten steps at successively higher temperatures. The primary mineral age is taken from the high-temperature plateaus. Corrections for atmospheric xenon are necessary. The $^{129}Xe/^{136}Xe_n$ method does not require stepwise degassing.

### Scope and Potential, Limitations, Representative Examples

*Scope and Potential.* The advantages of the $Xe_{sf}/Xe_n$ method are that

– only the isotope ratios of one element need to be measured and not concentrations. When a monitor mineral is used, the main source of error is the determination of the uranium concentration of the monitor.
– gas losses can be recognized by stepwise degassing. This is due to the direct correlation of xenon and uranium in discrete parts of the mineral grain (see Fig. 6.47).

Very reliable $Xe_{sf}/Xe_n$ age have been reported by Kapusta et al. (1983) for crystalline material isolated from zircon. This suggests that Xe/Xe dating can be a useful tool for determining the time of formation of zircon even when it has been damaged by its own radiation and has been subjected to secondary processes.

In combination with other methods (Rb/Sr, U/Pb, and K/Ar), the Xe/Xe method can provide additional information on the geological history of a rock complex.

The advantage of the $^{129}Xe/^{136}Xe_n$ method that it provides a possibility for determining ages in the range of 5–100 Ma even in the case of partial loss of xenon (Shukolyukov and Meshick 1987). If any xenon is lost by diffusion, $^{129}Xe$ and $^{136}Xe$ are lost in equal proportions and, hence, the $^{129}Xe/^{136}Xe_n$ ratio is not disturbed.

*Limitations.* Up to now, the Xe/Xe method has been used only for minerals with elevated uranium concentrations for which more common methods are available (e.g., U-Pb) or which occur in rocks (e.g, pegmatites) that contain more easily datable

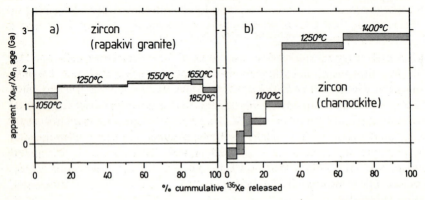

**Fig. 6.47 a, b.** Examples of $Xe_{sf}/Xe_n$ degassing spectra for zircon; **a** after Shukolyukov et al. (1979); **b** after Teitsma and Clarke (1978)

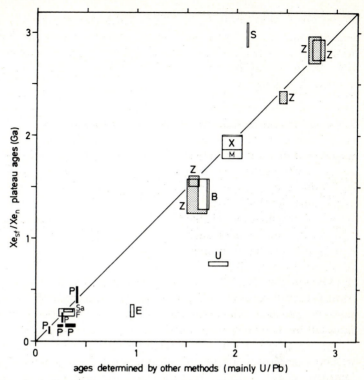

**Fig. 6.48.** Comparison of $Xe_{sf}/Xe_n$ plateau ages with age determinations done using other methods (mostly U/Pb) (data from Shukolyukov et al. 1974b, 1979; Teitsma and Clarke 1978; Shukolyukov and Meshick 1987). The minerals are *B* brithotile; *E* euxenite; *F* fergusonite; *M* monazite; *P* pitchblende; *S* sphene; *Sa* samarskite; *U* uraninite; *X* xenotime; and *Z* zircon

minerals. For this reason, the use of the Xe/Xe method has been limited to comparative studies.

***Representative Examples.*** Degassing spectra for two different zircons are shown in Fig. 6.47: at the left is an undisturbed zircon (Shukolyukov et al. 1979); at the right is a zircon whose margins show heavy xenon losses, possibly caused by geological processes (Teitsma and Clarke 1978). The age of primary mineral formation is obtained from the high-temperature plateau.

In Fig. 6.48, a graphic comparison is shown of $Xe_{sf}/Xe_n$ plateau ages and ages obtained by other methods (primarily U-Pb), sometimes on the same minerals, sometimes on corresponding geological formations. It can be seen that the Xe/Xe method is quite good for zircon, but less so for ores (e.g., pitchblende and uraninite). However, Shukolyukov and Meshick (1987) have reported good agreement between Xe/Xe ages and U/Pb ages obtained for 20 pitchblendes. Moreover, when galenite and molybdenite are present in pitchblende, the $Xe_{sf}/Xe_n$ method yields more reliable ages than the U/Pb method. A whole-rock age has been determined only for a carbonatite (Teitsma and Clarke 1978).

***Non-Chronological Applications.*** See Sect. 6.1.15.1.

## 6.2  Dating with Cosmogenic Radionuclides

A number of nuclear reactions (e.g., neutron capture, spallation, and solar neutrino production in the troposphere, negative muon capture at sea level) of cosmic rays (primary cosmic-ray particles or cascades of secondary low-energy neutrons and protons induced by solar and galactic cosmic radiation and solar-flare cosmic-ray particles) with gas molecules (chiefly nitrogen, oxygen, argon, and krypton) in the stratosphere and troposphere produce many radionuclides (Geiss et al. 1962; Lal 1962; Reyss et al. 1981). Negative muon capture dominates at the Earth's surface (Yogoyama et al. 1977). Negative muon capture, fast muon disintegration, and reactions with neutrons occur in deep underground (Florkowski et al. 1988). Most are produced by $(\alpha, n)$ reactions (Feige et al. 1968) but some are also produced by cosmic particles (Lal 1987b). The long-lived cosmogenic radionuclides can be used for age determinations (Table 6.3). Among the many examples that could be mentioned, radionuclides, such as technetium-97 ($\tau = 2.6\,Ma$) and technetium-98 ($\tau = 4.2\,Ma$), produced in orebodies (e.g., molybdenite) and salt deposits by solar neutrinos could possibly also become useful for geochronological studies (Cowan and Haxton 1982).

Age determinations using cosmogenic radionuclides are made on the basis of a model by Libby (Libby 1946; Anderson et al. 1947; Kamen 1963). This model is based on the following assumptions:

- Cosmogenic isotopes have been produced at constant rates for geological periods of time, periods considerably longer (at least ten times) than the half-lives of the radionuclides. This presupposes a constant intensity of cosmic radiation (e.g., Oeschger et al. 1970) but does not exclude local variation. The production rate of cosmogenic radionuclides is dependent on latitude with a maximum in the temperate zone (Fig. 6.49).
- The cosmogenic radionuclides are retained in constant amounts in the geophysical reservoirs: the biosphere, atmosphere, hydrosphere, and lithosphere, each of which can be considered to be made up of smaller, individual reservoirs.
- The cosmogenic radionuclides have constant mean residence times (MRT) in the various geophysical reservoirs and constant exchange rates between the reservoirs. The MRT are short relative to the half-lives of the cosmogenic radionuclides.

As long as these assumptions are met, the rates of production and decay of a particular cosmogenic radionuclide in a geophysical reservoir are equal.

- Finally, the samples to be dated must have formed closed systems since the beginning of the aging period, i.e., since the geochronological clock was reset to zero (e.g., time of formation).

This means, for example, that steady-state equilibrium exists for gases, e.g., $^{39}Ar$ (Sect. 6.2.8) or gaseous chemical compounds of radiocarbon (Sect. 6.2.1) or tritium (Sects. 6.2.2.1 and 6.2.2.2), and that specific activities are constant with respect to time and place (Lehmann and Loosli 1984). These specific activities are the initial values

**Table 6.3.** Physical dates of cosmogenic radionuclides used for radiometric dating

| Nuclide | $T_{1/2}$ | Age range | Prec % | Units | Prod atom/cm²s | Inv 1000 kg | Act Bq/l | Stable isotope | Isobar isotope | Target | Range of atom ratios |
|---|---|---|---|---|---|---|---|---|---|---|---|
| $^3$H | 12.43 a | 1–100 a | 5 | TU; Bq/l | 0.250 | 0.004 | 1 | $^{1,2}$H | $^3$He | H | $10^{-16}-10^{-21}$ |
| $^{10}$Be | 1.5 Ma | 0.01–15 Ma | 2 | atom*$10^6$/g; AR*$10^9$ | 0.038 | 430 | — | $^9$Be | $^{10}$B | BeO | $10^{-9}-5*10^{-15}$ |
| $^{14}$C | 5730 a | .3–50 ka | 0.2 | pmc | 1.7–2.5 | 75 | 0.01 | $^{12,13}$C | $^{14}$N | C, $CO_2$ | $1.2*10^{-12}-3*10^{-16}$ |
| $^{22}$Na | 2.6 a | 1–30 a | 10 | Bq/l | 0.00005 | — | $4*10^{-5}$ | $^{23}$Na | — | — | — |
| $^{26}$Al | 716 ka | 0.1–5 Ma | 5 | atom*$10^9$/g; AR*$10^{15}$ | 0.00011 | 1,700 | — | $^{27}$Al | $^{26}$Mg | Al | $2*10^{-12}-10^{-15}$ |
| $^{32}$Si | 105–172 a | 100–1000 a | 20 | Bq/kg Si | 0.00016 | 0.001 | $5*10^{-5}$ | $^{28,29}$Si | $^{32}$S | SiH$_3$ | $4*10^{-14}-4*10^{-15}$ |
| $^{36}$Cl | 301 ka | 0.1–3 Ma | 1 | atom*$10^8$/g; AR*$10^{15}$ | 0.0060 | 300— 3000 | $2*10^{-6}$ | $^{35,37}$Cl | $^{36}$S | AgCl | $2*10^{-11}-10^{-16}$ |
| $^{39}$Ar | 269 a | 100–2000 a | 5 | pmar; Bq/l Ar | 0.0056 | 0.022 | $2*10^{-7}$ | $^{40}$Ar | — | — | $9*10^{-16}-10^{-17}$ |
| $^{41}$Ca | 103 ka | 20–400 ka | 20 | AR*$10^{15}$ | — | — | — | $^{40,42}$Ca | $^{41}$K | CaF | $2*10^{-14}-2*10^{-15}$ |
| $^{53}$Mn | 3.7 Ma | 1–10 Ma | 10 | AR | — | — | — | $^{55}$Mn | $^{53}$Cr | MnO | Larger than $3*10^{-10}$ |
| $^{81}$Kr | 213 ka | 0.05–1 Ma | 20 | Bq/l Kr | — | — | $10^{-10}$ | Kr | — | RIS | Smaller than $5*10^{-13}$ |
| $^{85}$Kr | 10.76 a | 1–30 a | 5 | Bq/l Kr | — | — | $10^{-9}$ | Kr | — | — | Smaller than $10^{-11}$ |
| $^{129}$I | 15.7 Ma | 3–80 Ma | 10 | AR*$10^{14}$ | — | — | $10^{-10}$ | $^{127}$I | $^{129}$Xe | AgI | $3*10^{-13}-10^{-14}$ |

Abbreviations: $T_{1/2}$, half-life; Prec, precision; Prod, cosmogenic isotope production rate; Inv, global isotope inventory; Act, mean specific activity in modern water; RIS, resonance ionization spectrometry; AR, activity ratio.

Units: Bq, bequerel; TU, tritium unit; pmc, percent modern carbon; pmar, percent modern argon.

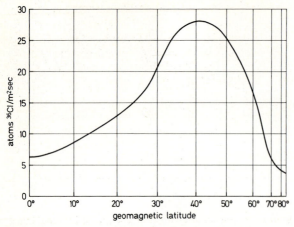

**Fig. 6.49.** Theoretical fall-out distribution of long-lived chlorine-36 as a function of latitude, covering a range of a factor of 5 (after Bentley et al. 1986a)

for age determinations. These initial values can differ depending on the chemical and physical properties of the samples and the processes occurring in the individual geophysical reservoirs or parts of them (reservoir effect, Olsson 1979a). Often, natural exchange and mixing processes (Lal 1962) can be studied by utilizing these differences to obtain information that otherwise cannot be obtained (e.g., Mazor et al. 1986). Gases of cosmogenic radionuclides are present in water and occluded in carbonate deposits and ice.

The cosmogenic radionuclides $^{10}$Be (Sect. 6.2.3), $^{26}$Al (Sect. 6.2.5), and $^{32}$Si (Sect. 6.2.6) in aerosols are washed out of the atmosphere within a few years (a maximum of ten) and deposited on ice (Stauffer 1989), in the soil, or in lacustrine or marine sediments. In the ideal case, they remain there for long periods. If the rates of deposition are constant, the specific activities of these radionuclides will decrease exponentially with increasing depth and age (e.g., Sect. 6.2.3). Since these conditions are rarely fulfilled because the cosmogenic radionuclides formed in solids are irregularly distributed globally in the geological reservoirs and their production rates have fluctuated, dating methods based on two cosmogenic radionuclides with greatly differing half-lives and similar geochemical properties appear to be more reliable than those based on only one radionuclide (Lal 1962) (Sects. 6.2.13 and 6.2.14).

Chlorine-36 (Sect. 6.2.7) is very mobile and highly soluble in water, which is why it has great potential for dating groundwater (e.g., Bentley et al. 1986a, b; Fabryka-Martin et al. 1987). The age of salt crusts (Bagge and Willkomm 1966) and saline soils as well as rates of sedimentation and erosion in arid regions can be determined (Bentley et al. 1986a; Elmore 1986).

The assumptions of the Libby model are fulfilled in nature to a good approximation, but variations in production and exchange rates of cosmogenic radionuclides between the geophysical reservoirs lead to small fluctuations in the initial or saturation concentrations, causing irregular differences between the

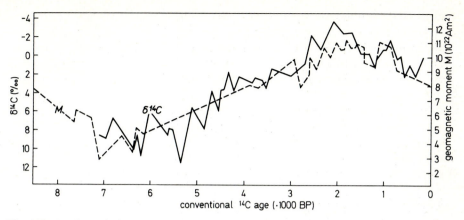

**Fig. 6.50.** Secular variation in the radiocarbon concentration in the atmospheric $CO_2$ and living organic matter and fluctuations in the moment of the Earth's magnetic field (after Bucha 1973)

radiometric time scales and the solar calendar, which makes calibration tables or curves necessary. Factors affecting cosmogenic production are (a) the intensity of the primary cosmic radiation, (b) solar activity, and (c) the geomagnetic field.

As an example, the initial $^{14}$C concentration (Sect. 6.2.1) over the last 10,000 years has been measured with high-precision using tree rings that had been dated dendrochronologically to within a year (Stuiver and Kra 1986). Long-term (Damon and Linick 1986) and short-term (Suess 1986) periods can be recognized in these fluctuations (Fig. 6.50).

The short-term fluctuations in the initial $^{14}$C concentration have periods of several centuries (wiggles; Suess 1986) and are attributed to variations in solar activity (Olsson 1970; Stuiver and Quay 1980a, b).

The long-term trend is apparently sinusoidal with a period of about 11,300 years; it can be attributed to changes in the production of $^{14}$C, with a modulation by the Earth's magnetic field, especially high-latitude components of the non-dipole field (which has a fundamental period of 2400 a; Damon and Linick 1986) (Sect. 7.1). This can also explain the minor changes in $^{10}$Be (Sect. 6.2.3) in a 500-year-long ice profile. In contrast to expectations, the $^{10}$Be concentration was rather constant throughout the profile. Drastic changes in the abyssal circulation of the oceans appeared to be a possible alternative explanation for the variations in the initial $^{14}$C concentration (Keir 1983; Beer et al. 1984). But recent paleo-oceanographic data (Broecker 1987) contradict this possibility. Thus, the discrepancy in the long-term $^{14}$C and $^{10}$Be records indicates complex interactions between the geophysical reservoirs, for which the present models are not yet adequate (Oeschger 1988). At present the long-term trend in radiocarbon concentration can be explained by increased production between 6000 and 8000 BP as a result of reduced geomagnetic shielding against cosmic radiation (Damon and Linick 1986).

It should be possible to recognize long-term changes in the production of cosmogenic isotopes during the Pleistocene with concurrent determination of $^{10}$Be (Sect. 6.2.3) and $^{26}$Al (Sect. 6.2.5) (Lal 1962; Raisbeck et al. 1985). Cold continental ice and pelagic sediments are suitable for such studies because they may be expected to form closed systems with no disturbing chemical reactions to change the initial concentration of cosmogenic isotopes. Erosion rates can be estimated by concurrent $^{14}C$ and $^{10}Be$ analysis (Lal 1987b; Lal et al. 1987). Isotopic analysis of meteorites and cosmic dust (Sect. 6.5) already provides such information (Geiss et al. 1962; Wasson 1963).

The dating methods in this chapter are discussed in order of increasing atomic weight of the cosmogenic isotope. An exception is made for the radiocarbon method (Sect. 6.2.1) because many of the effects which have been studied in detail on it also occur with other cosmogenic radionuclides. Publications describing all absolute dating methods using cosmogenic radionuclides are few (Geyh 1980b; Currie 1982), but the Proceedings of international conferences on the AMS technique are very valuable in this respect (e.g., Wölfli et al. 1984; Gove et al. 1987).

### 6.2.1  Radiocarbon ($^{14}C$) Method***

(Libby 1946; Anderson et al. 1947)

#### *Dating Range, Precision, Substances, Sample Size*

The radiocarbon method is a standard method for determining conventional $^{14}C$ ages (Michels 1973; Currie 1982; Evin 1983; Geyh 1983; Mook and Waterbolk 1983; Gillespie 1984; Gupta and Polach 1985; Taylor 1987b) of organic matter, e.g., wood, charcoal (Simonsen 1983), seeds, leaves, resin, lichen (Olsson et al. 1984), peat, humus, marsh gas, bone (Stafford et al. 1987; Gurfinkel 1987; Stafford et al. 1988), ivory, tissue, horn, hair, mollusc shells, egg shells, secondary carbonate (e.g., travertine and speleothem), soil and sediment, as well as groundwater and ice (10 kg), in the age range of 300 to 50,000 a. Isotope enrichment was used earlier to date very large samples (10–150 g C) up to 70,000 a (Grootes 1978; Stuiver et al. 1978).

Materials that are seldom dated because the results are often unreliable include mortar (Folk and Valastro 1979; van Strydonck et al. 1986) and those that contain only traces of carbon, such as postsherds (De Atley 1980; Johnson et al. 1986; Gabasio et al. 1986), organic matter in rock varnishes (Dorn et al. 1986), and iron objects (Sayre et al. 1982). The determination of the terrestrial age of meteorites and cosmic dust (e.g., Wasson 1963; Brown et al. 1984) is discussed in Sect. 6.5.

The size of the sample is determined by the carbon content, the degree of preservation, the degree of contamination, and the method of $^{14}C$ analysis. Table 6.4 contains values for orientation, which should be confirmed or corrected by the laboratory that is carrying out the analysis.

The now classical decay counting techniques use proportional counters (Sect. 5.2.2.1) or liquid scintillation counters (Sect. 5.2.2.2) and require samples that contain gram quantities of carbon. If miniature counters are used, 10–40 mg C are sufficient (Otlet et al. 1983a). Modern accelerator mass spectrometers (AMS, Sect. 5.2.3.2) need no more than 50–5000 $\mu g$ (Hedges 1981; Mook 1984; Gove et al. 1987).

**Table 6.4.** Carbon content and sample size of common types of samples

| Type of sample | C content % | Sample size Usual | Minimum |
|---|---|---|---|
| Charcoal (dry) (Simonsen 1983) | 50 –90 | 3–   6 g | 50 mg–  1 mg |
| Wood, peat, grain, tissue (dry) | 10 –50 | 6–  50 g | 2–  25 mg |
| Wood, peat (moist) | 2 –10 | 30–  150 g | 10–125 mg |
| Sediment, soil | 0.2– 5 | 50–1500 g | 20 mg–  1 g |
| Bone, teeth (Stafford et al. 1987) | 1  – 5 | 60–  300 g | 20–300 mg |
| Carbonates: coral, ooids, travertine, speleothem | 10 | 30 g | 25 mg |
| Groundwater | $10^{-2}$ | 50–  200 L | 50–200 mL |

The sample sizes given above must be increased as much as tenfold if the sample has a high detrital mineral content, if it is highly contaminated with organic residues (e.g., roots or humic acids), or if it is poorly preserved (e.g., advanced decomposition of peat, weathered shells). In such cases the contamination must be removed or particular fractions must be extracted for analysis, often resulting in the loss of as much as 90% of the sample. Therefore, the following, apparently contradictory rules of thumb should be observed:

*As little material as possible but as much material as necessary!*

*It is better to collect too much material than too little!*

The sample should not be so large that the time span of its formation (sample time width) is as long or longer than the confidence interval of the $^{14}$C age, whose limits are determined by the standard deviation (Sect. 4.2.1). For example, it does not make sense to date a sample to $\pm 40$ a when it took 500 years to form (Mook 1983).

Routine radiocarbon age determinations attain a precision of $\geqslant \pm 40$ a, precision dating of young samples $\geqslant \pm 12$ a. However, the reported standard deviations are frequently too small by a factor of as much as 2 (International Study Group 1982; Stuiver 1982; Stuiver and Kra 1986). Therefore, the use of laboratory-specific error multipliers is recommended. The best precision that can be determined is assumed by conservative researchers to be $\pm 60$–80 a (Stenhouse and Baxter 1983). It is rather questionable whether multiple analysis of the same sample increases the dating precision (e.g., Srdoc et al. 1983). The general relationship between sample size and standard deviation for conventional radiocarbon ages is illustrated in Fig. 6.51.

The age of small pieces of wood (less than 100 rings) which cannot be dated dendrochronologically can, in favorable cases, be dated to within a few years. The $^{14}$C ages of several rings are determined and fitted to the dendrochronological calibration curve by a technique called wiggle matching (Ferguson et al. 1966; Beer et al. 1979; Pearson 1986). Archeological wiggle matching (AWM) is a modification of the wiggle-matching technique and has a potential for widespread application. An apparently linear time scale for $^{14}$C-dated archeological objects must be prepared on the basis of stratigraphic information and quantitative archeological

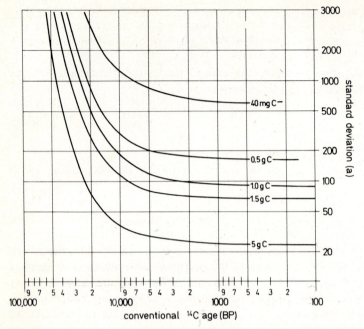

**Fig. 6.51.** An example from the Hannover $^{14}$C Laboratory: Standard deviation as a function of the conventional $^{14}$C age and the amount of carbon in the sample used in conventional counters of various volume and the same counting time of 32 h (Geyh 1983)

assumptions for each problem. The $^{14}$C dates are shifted along the time scale of the calibration curve, similarly to the wiggle-matching procedure. This procedure has been applied to $^{14}$C dates for the early Bronze Age in Turkey; for one site this reduced the apparent span of 350 a reflected by the $^{14}$C data to about 200 a (Weninger 1986).

### Basic Concept

Radiocarbon ($^{14}$C) decays by $\beta^-$-emission ($E_{max} = 158$ keV) with a half-life of 5730 $\pm$ 40 a. For conventional radiocarbon ages, a value of 5568 a is used. Radiocarbon is produced primarily by cosmic radiation from nitrogen:

$$^{14}N(n, p)\ ^{14}C$$

The average specific activity of modern carbon is $13.56 \pm 0.07$ dpm/g C; there is about 75 t of radiocarbon on the Earth. The isotopic ratio of $^{14}$C to $^{12}$C is $1.2 \times 10^{-12}$. The initial $^{14}$C concentration of modern carbon is expressed in a unit of its own: pmc (percent modern carbon = % modern) or ‰ deviation from the standard. Atmospheric $CO_2$ in 1850 was 100 pmc or 0‰ (Stuiver and Polach 1977).

The radiocarbon formed in the upper atmosphere is oxidized to $^{14}CO_2$. The rate is between 1.7 and 2.5 atoms/cm$^2$/s. After mixing with atmospheric $CO_2$ (0.03 vol%), it becomes part of the carbon cycle in the biosphere (e.g., Degens et al. 1984). Assimilated by plants, it enters the food chain and thus becomes part of all living organic matter. Except for slight isotopic fractionation, which can be corrected for [Eq. (6.60)], the specific activity of $^{14}C$ in terrestrial organic matter is the same as in atmospheric carbon dioxide. Subsurface production of radiocarbon is negligible (Zito et al. 1980; Florkowski et al. 1988).

The age of fossil organic matter, which is no longer part of the carbon cycle and has thus lost $^{14}C$ by decay, can be calculated as follows:

$$t = \frac{\tau}{\ln 2} \ln \frac{A_0}{A}, \qquad (6.58)$$

where $A$ is the $^{14}C$ activity of the sample,

$A_0$ is the initial $^{14}C$ activity of a standard or a substance with an age of zero, and

$\tau$ is the conventional half-life.

The net counting rates are used more often than activity values.

$\delta^{13}C$ *correction*: Isotopic fractionation resulting from metabolic processes is responsible for slightly divergent initial $^{14}C$ activities in different kinds of samples. Owing to small differences in the physical properties (e.g., size, atomic weight, reaction rates) of the carbon isotopes, chemical (e.g., precipitation), biological (e.g., assimilation), and physical (e.g., phase transfer) processes lead to characteristic shifts in the isotopic ratios of molecules or radicals (due to isotopic fractionation). Using Eq. (6.59) these shifts can be easily determined for carbon from mass spectrometric measurements (Sect. 5.2.3.1) of the isotopic ratio of the two stable isotopes $^{12}C$ ($\sim 99\%$) and $^{13}C$. The shift in the $^{13}C/^{12}C$ ratio is expressed as $\delta^{13}C$:

$$\delta^{13}C = \frac{R_{sample} - R_{standard}}{R_{standard}} \cdot 1000\text{‰}, \qquad (6.59)$$

where $R_{sample}$ is the $^{13}C/^{12}C$ ratio of the sample and $R_{standard}$ is provided by the PDB standard (Craig 1957) or another carbonate standard (e.g., Solnhofen limestone) calibrated with respect to it. Belemnite from the Peedee Formation in South Carolina was introduced as the PDB standard and by definition had a $\delta^{13}C$ value of 0‰. Secondary standards calibrated to this primary standard have been used since the exhaustion of the Peedee Belemite.

The isotopic fractionation of $^{12}C$ and $^{14}C$ is proportional to that of $^{12}C$ and $^{13}C$, the former being 2.3 times greater than the latter (Saliege and Fontes 1984). Thus, the $\delta^{13}C$ value is used to correct $^{14}C$ activities [Eq. (6.60)] so that all values are referred to the same initial $^{14}C$ activity (Stuiver and Polach 1977). According to international convention, a fractionation factor of 2 is used instead of 2.3. In addition, all conventional $^{14}C$ ages have the same reference value for $\delta^{13}C$ ($-25\text{‰}$) and must be corrected as follows:

$$A_{corr} = A\left\{1 - \frac{2(\delta^{13}C + 25)}{1000}\right\}, \qquad (6.60)$$

where $A$ and $A_{corr}$ are the measured and corrected $^{14}C$ activities, respectively.

Isotopic fractionation can occur during formation of the material or during chemical processing of the sample. The shifts in $^{14}C$ ages obtained with Eq. (6.60) are usually much less than 80 years for the most often dated samples; these are samples derived from plants which assimilate $CO_2$ by the C3 pathway (Calvin cycle). These plants convert carbon dioxide to 3-phosphoglycerate (a 3-carbon compound), in contrast to other plants (e.g., sugar cane, grasses, corn, and millet), which convert it to 4-carbon dicarboxyl acids (Hatch-Slack cycle). Conventional $^{14}C$ ages of plants that assimilate by the C4 pathway, mostly from semiarid areas, are corrected by about $-200$ a (Stuiver and Polach 1977), not including any other corrections that may be necessary (Tauber 1983).

In addition to being required to correct for isotopic fractionation, $\delta^{13}C$-values provide information about the origin and geochemical history of substances, which for example, is valuable for reservoir corrections. In archeology, $^{13}C$ in collagen (better termed protein extract) is used as a tracer to indicate the diets of prehistoric peoples (van der Merwe 1982; Lewin 1983), although it has become evident that $\delta^{13}C$ values change during diagenesis of collagen (Tuross et al. 1988). Bone apatite does not seem to be as suitable for this purpose as collagen (Schoeninger and DeNiro 1982). In geology, the $\delta^{13}C$-values for the methane in natural gas provide information on the maturity of the source material (e.g., Schoell 1983; Whiticar et al. 1986) and those for petroleum are used as an exploration tool (Stahl and Faber 1984). $\delta^{13}C$-values may also be useful for distinguishing between marine and terrestrial matter (Table 6.5). Moreover, they yield paleoclimatic information (e.g., Stuiver and Braziunas 1987).

*Conventional $^{14}C$ ages:* According to international agreement (convention), conventional $^{14}C$ ages must fulfil certain requirements (Stuiver and Polach 1977) so that

**Table 6.5.** $\delta^{13}C$-Values (‰ PDB) for various substances (Vogel and Ehhalt 1963; Stuiver and Polach 1977)

| Substance | | $\delta^{13}C$ (‰) |
|---|---|---|
| Terrestrial organic matter (wood, charcoal, peat) from humid regions (assimilation by the C3-cycle) | | $-35$ to $-20$ |
| Salt marsh and desert plants, tropical grasses from semiarid and arid regions (assimilation by the C4-cycle) | | $-16$ to $-9$ |
| Carbonic acid in groundwater with no noticeable chemical reaction with the aquifer rocks | | $-20$ to $-10$ |
| Bone and collagen (terrestrial) | | $-26$ to $-18$ |
| NBS oxalic acid standard (by definition), | old batch | $-19$ |
| NOX | new batch | $-17.8$ |
| Carbonic acid in groundwater after migration of $CO_2$ from underground, dissolution of lime or salts | | $-10$ to $+2$ |
| Speleothem and travertine (precipitation of bicarbonate) | | $-10$ to $+10$ |
| Atmospheric $CO_2$ | | $-8$ to $-7$ |
| "Spaghetti stalactites" on concrete structures (precipitation of calcium hydroxide) | | $>-30$ |
| Marine carbonates and organisms | | $-2$ to $+2$ |

they can be compared worldwide:

– The reference year for conventional $^{14}$C ages is AD 1950. This is indicated with the letters bp or BP (for "before present"), e.g., 5000 BP.
– NBS (National Bureau of Standards, Washington, D.C.) oxalic acid is used as standard for time zero. The first batch of NBS oxalic acid has been exhausetd and a new one (NOX) has been introduced. ANU (Australian National University, Canberra) sugar is used as a secondary standard. It is available free of charge.
– The half-life of 5568 a introduced by Libby is used for calculating conventional $^{14}$C ages. The physical half-life of $5730 \pm 40$ a is used for data relevant to geophysics. These dates are about 3% larger than the corresponding conventional $^{14}$C ages.
– Equation (6.60) is used to correct the measured $^{14}$C activities to $-25‰$ ($\delta^{13}$C correction) before conversion to $^{14}$C ages [Eq. (6.58)].

*Calibrated radiocarbon ages:* Despite this international convention, conventional $^{14}$C ages differ from actual ages given in solar years (DeVries effect). This is because Libby's assumption that the initial $^{14}$C concentration has not changed over geological periods of time is not fulfilled. This was recognized by DeVries when he determined radiocarbon ages for wood that had been precisely dated dendro-chronologically (DeVries 1958). Since then, deviations of the $^{14}$C time-scale from the solar time-scale have been determined for the last 10,000 years (Olsson 1970; Stuiver and Kra 1986). Several trends are recognizable:

– The long-term trend with a period of about 11,300 a has been explained by changes in the cosmic ray production of $^{14}$C modulated by the Earth's magnetic dipole moment and a high-latitude component of the non-dipole field, which has a fundamental period of about $2280 \pm 410$ a and a peak amplitude of $7.0 \pm 1.3‰$ (Bucha 1973; Damon and Linick 1986; Fig. 6.50). Recent $^{10}$Be measurements seem to put this explanation in question; in this case, changes in oceanic abyssal water circulation could contribute to these fluctuations (Keir 1983; Beer et al. 1984). But Broecker (1987) contradicts this hypothesis on the basis of $^{14}$C measurements on coeval planktonic and benthic foraminifers. At present, the best explanation has been given by Damon and Linick (1986).
– Medium-term trends show up as differences between the two time-scales by several centuries ("wiggles"; Suess 1986) and appear to correlate with changes in the magnetic properties of the solar wind (solar variability) (Castagnoli and Lal 1980; Neftel et al. 1981). A quantitative measure of these fluctuations in $^{14}$C production of up to 2% is given by the Aa indices and sunspot number (Stuiver and Quay 1980a,b). A frequency analysis of the Holocene atmospheric $^{14}$C record indicates that the sun acts similar to a harmonic oscillator. Each of the medium-term variations in the initial $^{14}$C concentration begins with a rapid increase within a period of 30 years, followed by a decrease to the original value in about double that time (de Jong et al. 1979; Bruns et al. 1980a).
– Short-term fluctuations within a sun-spot cycle, which coincide with solar flares and variations in the magnetic field of the Earth, lead to dating errors of 25 years at a maximum.

**Table 6.6.** Calibrated age (cal BC) intervals after dendrochronological correction of conventional $^{14}C$ dates for samples formed within 1 year, for samples formed over a period of 100 years with a constant growth rate (Mook 1983), and samples formed over a 100-year period with a growth rate that increases according to a quadratic function (Geyh 1983)

| Time span of formation $^{14}C$ age (BP) | 1 year Uniform growth | 100 years Uniform growth | 100 years Increasing growth rate |
|---|---|---|---|
| $4650 \pm 50$ | 3360–3380 cal BC<br>3390–3490 cal BC<br>3505–3515 cal BC | 3360–3490 cal BC | 3375–3515 cal BC |
| $4700 \pm 100$ | 3360–3530 cal BC<br>3550–3630 cal BC | 3360–3610 cal BC | 3375–3630 cal BC |
| $4750 \pm 50$ | 3380–3390 cal BC<br>3410          cal BC<br>3490–3505 cal BC<br>3515–3530 cal BC<br>3550–3630 cal BC | 3490–3610 cal BC | 3515–3630 cal BC |

For routine $^{14}C$ age determinations, internationally accepted calibration curves and tables for the correction of raw $^{14}C$ data are available (Stuiver and Kra 1986). These were prepared on the basis of precise radiocarbon measurements on dendrochronologically dated wook made by an international team of radiocarbon researchers and dendrochronologists: M.G.L. Baillie, B. Becker, T.F. Braziunas, M. Bruns, D.M. Corbett, P.E. Damon, A.F.M. De Jong, C.W. Ferguson, B. Kromer, T.W. Linick, A. Long, W.G. Mook, K.O. Münnich, G.W. Pearson, J.R. Pilcher, F. Qua, P.J. Reimer, M. Rhein, H. Schoch-Fischer, and M. Stuiver. These replace the various inaccurate curves previously used. Calibrated $^{14}C$ ages are given the prefix cal, e.g., 5000 cal BC, 1000 cal AD, or 2950 cal BP.

Different calibration curves and tables are available for different time spans represented by the samples (Mook 1983). Samples formed within one year, for example, almost always yield larger dendrochronologically corrected age intervals than samples formed over a longer time span. Often a sample contains different amounts of material for the different parts of the time span represented by the sample (Table 6.6). For example, the inner, older rings or a branch contain significantly less mass than the outer, younger ones. In such cases, weighted averages must be used.

Dendrochronological corrections cannot be applied to samples that have components of different ages in unknown proportions (e.g., soils) or whose reservoir effect cannot be given exactly (e.g., TDIC in groundwater and travertine).

Within the last 2000 years the differences between conventional $^{14}C$ and actual ages vary within $\pm 200$ years (Fig. 6.52). For the period from 7300 to 2000 years ago, this difference increases to about $-800$ years (Fig. 2.1). This difference increases to about 1100 yr between 8000 and 11,000 BP. The difference for the pre-Holocene to about 35,000 BP could be larger (up to 5000 a) (Stuiver 1978; Vogel 1983). One reason for this may be that the atmospheric $CO_2$ concentration was about 30% less during the last pleniglacial than that of the Holocene (Neftel et al.

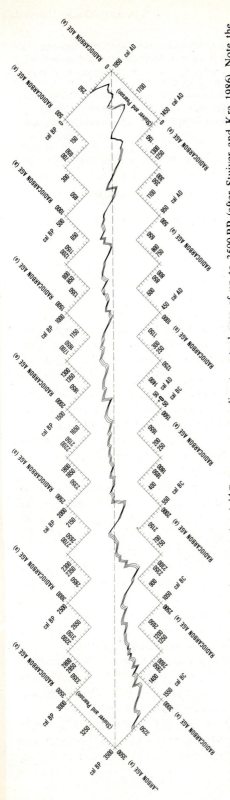

**Fig. 6.52.** High-precision calibration curve for conventional $^{14}C$ ages corresponding to actual ages of up to 3500 BP (after Stuiver and Kra 1986). Note the increasing deviation between the conventional and calibrated $^{14}C$ time-scales with increasing age beginning about 2000 BP

1982), whereas the production rate of cosmogenic radionuclides, according to $^{10}$Be measurements, was about twice as large (Raisbeck et al. 1981).

Age intervals obtained by calibration of conventional radiocarbon dates are larger than the corresponding confidence intervals if they coincide with the nearly horizontal parts of the calibration curves (Fig. 6.52), e.g., the intervals AD 80–120, 130–210, 250–320, 440–530, 690–760, 780–880, 900–980, 1040–1160, 1500–1620, and 1810–1910.

For the first millenium BC (Fig. 6.52), the wiggles are so large that even with precise conventional ages, resolution is usually no better than $\pm 500$ a (Ottaway 1983).

The principle of the calibration of a conventional radiocarbon age using the Calibration Issue of the journal *Radiocarbon* (Stuiver and Kra 1986) is illustrated in Figs. 6.53 and 6.54. Computer programs are usually used for the calibration and to calculate probability distributions.

An important aspect of dendrochronologically corrected $^{14}$C ages is that they can be used to determine the duration of an event, which is not possible with conventional $^{14}$C dates alone.

*Upper dating limit:* The maximum $^{14}$C age (indicated by " > " before the age) that can be obtained from a sample is determined by the detection limit of $^{14}$C activity measurements. This maximum $^{14}$C age is the minimum age of the sample! A probability of 97.7% is usually chosen for the confidence interval of the maximum

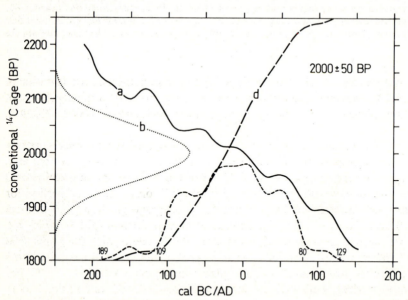

**Fig. 6.53.** Rather ideal example of the calibration of a conventional $^{14}$C date (after Stuiver and Pearson in Stuiver and Kra 1986) conducted with a computer program by van der Pflicht and Mook (1989). Due to rather small undulations in the calibration curve (*a*) the Gaussian probability distribution for the conventional date (*b*) is converted to a rather similar probability distribution (*c*). The ranges of the calibrated date for various probabilities can be read from the integrated probability function (*d*)

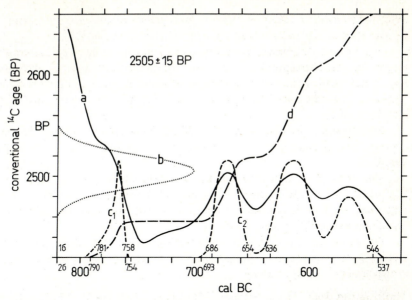

**Fig. 6.54.** Example of the calibration of a conventional $^{14}$C date (after Stuiver and Pearson in Stuiver and Kra 1986) conducted with a computer program by van der Pflicht and Mook (1989). Due to large undulations (wiggles) in the calibration curve (*a*), the Gaussian probability distribution for the conventional date (*b*) is converted to at least two ranges of calibrated age with differing probabilities and a complex probability distribution ($c_1$ and $c_2$). The ranges of the calibrated date for various probabilities can be read from the integrated probability function (*d*). Under other conditions, even more than two ranges may be found. It must be kept in mind that even ranges with a very low probability should not be neglected because they may contain the true age

age ($2\sigma$ criterion, Sect. 4.2.1). This corresponds to a minimum standard deviation of 4000 years. Depending on the duration of the measurement and the quality (factor of merit) of the apparatus used, the upper dating limit ranges from 35,000 to 70,000 a.

*Units:* The $^{14}$C results from geochemical and geophysical studies are frequently given not as conventional $^{14}$C ages, but in terms of $^{14}$C activity – as per mill depletion or enrichment relative to 0.95 times the activity of NBS oxalic acid standard – sometimes with, sometimes without $\delta^{13}$C correction or correction for the sample age. Special symbols are used for geochronological and geochemical studies (Stuiver and Polach 1977): Percent of modern carbon (pM or pmc) is a unit of $^{14}$C activity relative to the corrected (0.95) activity of NBS oxalic acid standard. When isotopic fractionation is neglected, the per mill deviation from this standard is given as d$^{14}$C (for geochronological studies) or $\delta^{14}$C (for geochemical studies), both defined analogously to the $\delta^{13}$C value [Eq. (6.60)]. D$^{14}$C and $\Delta^{14}$C are the corresponding values corrected for isotopic fractionation ($\delta^{13}$C). The two delta values can be calculated with or without age correction (for instance, dendrochronologically dated wood). The $^{14}$C activity of dendrochronologically dated samples is, for example, calculated with age correction to determine the initial $^{14}$C activities of the past.

## Sample Preparation and Measurement Techniques

In the case of organic samples, roots, insects, etc. are removed manually or mechanically, preferentially after microscopic inspection. Humic acids and carbonate are usually removed by chemical procedures. The samples are then burned. If carbonate is to be dated, the sample is treated chemically to remove up to a few mm of the surface, followed by acid treatment to convert the carbonate to carbon dioxide. After purification, the $CO_2$ is measured for $^{14}C$ directly with a proportional counter (Sect. 5.2.2.1) or converted first to methane, acetylene, or ethane. If a liquid scintillation counter (Sect. 5.2.2.2) is used, the samples are converted to benzene via acetylene. The duration of the measurement is normally 2 days, maximum is a week. Many laboratories wait 4–6 weeks after the chemical preparation before starting the activity measurements so that the radon, which would distort the measurements, can decay. It is recommended to date more than one sample for a particular problem (Nydal 1983).

Groundwater samples are dated via the total dissolved inorganic carbon (TDIC) extracted from them in the field. The sample are either degassed and the $CO_2$ precipitated (Linick 1980) or, more commonly, the TDIC is precipitated by the addition of alkaline earth cations in basic solution (IAEA 1983). Occasionally, ion exchange resins are used (Fröhlich et al. 1974).

The samples used to be converted to carbon monoxide for isotopic enrichment in thermal diffusion columns. This process took several weeks.

The main advantage of accelerator mass spectrometry (Sect. 5.2.3.2) is that it makes it possible to date samples with carbon contents in the milligram to microgram range (e.g., Nelson et al. 1986) (Table 6.4). Therefore, the main objective of pretreatment procedures is the extraction of uncontaminated or especially suitable fractions or material not affected by the reservoir effect that occurs only in traces (Lister et al. 1984; Stafford et al. 1987). The pretreated samples are converted directly (Vogel et al. 1984) or via acetylene to graphite. Charcoal or thin carbon deposits are also ideal target materials (Bonani et al. 1984). Further improvements in target preparation are in development (Gove et al. 1987), e.g., measurement of the gas adsorbed on titanium (Bronk and Hedges 1987). A precision of 5‰ has been approached (Kromer et al. 1987). About $\frac{1}{2}$ h is needed per sample. Laser enrichment may make it possible in the future to date samples up to 100 ka. This technique will take advantage of differences in the stability of isotopically lighter and heavier formaldehyde molecules. However, contamination problems limit age determination to about 40,000 a, for samples of 50–100 μg to about 15,000 a (Wölfli 1987).

## Scope and Potential, Limitations, Representative Examples

**Scope and Potential.** The application of the $^{14}C$ method extends to all sectors of geoscientific and archeological research on the late Quaternary. The possibilities for application have been considerably increased by the development of the AMS technique (Sect. 5.2.3.2), which has extended the range of sample sizes down to the microgram range (Gove et al. 1987). The proceedings of the 9th, 10th, 11th, and 12th International Conferences on Radiocarbon Dating (held in Los Angeles in 1976, in

Bern and Heidelberg in 1979, in Seattle in 1982, and in Trondheim in 1985, respectively) contain a large number of case studies. The first of these proceedings were edited by Berger and Suess (1979). The others were edited by Stuiver and Kra and published in issues 22 (1979), 25 (1983), and 28 (1986), respectively, of the journal *Radiocarbon*.

***Limitations.*** *Contamination:* The biggest problem in radiocarbon of old samples is contamination with young allochthonous carbon (Olson 1963). It sometimes causes considerable differences between the apparent conventional $^{14}$C age that is determined and the actual age of the autochthonous, uncontaminated fraction of the sample. Contamination can, for example, result from intrusion of younger materials in the sample (e.g., roots, humic acids) or the introduction of fossil material (e.g., graphite or fossil carbonates, for example, as found in lacustrine sediments). Such contamination can result from man's activities, bioturbation, plant growth, or seepage. Samples with carbon concentrations of less than 5‰ often yield $^{14}$C ages that are many thousand years too high (Olsson 1979b). In such cases, allochthonous carbon obviously predominates; when they have a very low carbon content, such samples can usually be recognized by standard deviations that are many times greater than usual (Fig. 6.51). Contamination can also result from bacterial action (Geyh et al. 1974; Mahaney and Boyer 1986).

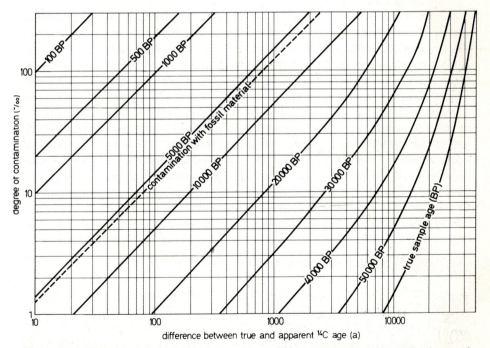

**Fig. 6.55.** Difference between the actual and apparent $^{14}$C ages as a function of the degree of contamination (‰) with organic material of zero, i.e., modern material, and infinite age. The former results in apparent ages that are too small, the latter in ones that are too large (*dashed line*)

Despite careful pretreatment of a sample (removal of visible contamination by hand and chemical separation of allochthonous material, or the extraction of autochthonous fractions), once a sample has been contaminated, it will always give a more or less incorrect $^{14}$C age (e.g., Stafford et al. 1987, 1988). Owing to the multitude of genetically different substances that can be the contaminant, there is no general procedure for decontamination.

The difference $\Delta t$ between apparent and actual age depends on the degree of contamination $k$ (‰) and the $^{14}$C concentration (age) of the admixed (allochthonous) substance $A_k$ (Fig. 6.55). If the contamination consists of fossil carbon, then $A_k = 0$ and the apparent $^{14}$C age $t_f$ is greater than the actual age.

$$\Delta t_f = t_f - t = -\frac{\tau}{\ln 2} \ln\left(1 - \frac{k}{1000}\right) \tag{6.61}$$

$\Delta t_f$ is then independent of the $^{14}$C age of the sample. One percent contamination with fossil material results in an increase of 80 a in the $^{14}$C age.

A sample that has a $^{14}$C concentration A in the autochthonous material and contains some modern carbon ($A_r$) yields an apparent $^{14}$C age ($t_r$) that is smaller than the actual one.

$$\Delta t_r = t_r - t = \frac{\tau}{\ln 2} \ln\left\{1 - \frac{k}{1000}\left(\frac{A_r}{A} - 1\right)\right\} \tag{6.62}$$

One percent contamination with recent carbon yields $\Delta t_r$ values of $-10$, $-200$, and $-7000$ a for samples that are 1,000, 10,000, and 40,000 years old! As can be seen, $\Delta A_r$ can be very large for old, even small, contaminated samples. In extreme cases, Holocene samples cannot be distinguished from Pleistocene samples. Radiocarbon ages of more than about 30,000 a should thus be viewed as minimum ages (Gill 1974).

The presence of contamination can be recognized if different materials from the same location (e.g., wood, peat, or bone) or different autochthonous fractions of the same sample are dated, e.g., cellulose (Olsson 1980a), humic acids, collagen (Taylor 1982; Gurfinkel 1987; Stafford et al. 1988), apatite (Hassan et al. 1977), amino acids (Stafford et al. 1987; Currie et al. 1989), or the outer and inner zones of mollusc shells (sequential decomposition technique).

Contamination caused by improper sample collection, preparation, or storage (Waterbolk 1983) is avoidable (Sect. 3.1) but is not rare (Willkomm 1983). Problems may also arise when very small samples are dated by AMS as it must always be kept in mind that gravitation or the activities of burrowing animals may have altered the stratigraphic level of such pieces.

*Reservoir effect:* Not all datable materials have an initial $^{14}$C concentration that is equal to that of NBS oxalic acid standard or that of preindustrial atmospheric $CO_2$. This kind of material is taken from geophysical reservoirs in which the residence time of the carbon is considerably greater than it is in the atmosphere or the carbon is a natural mixture of material of differing ages (Olsson 1979a, 1983). Phenomena of this kind are summarized under the term "reservoir effect". Conventional $^{14}$C ages of samples taken from such systems appear too old by several hundred or even thousands of years.

The $^{14}C$ ages of marine mollusc shells provide an example: The reservoir correction determined for the Norwegian coast increases from south to north from $-370$ to $-425$ years (Mangerud and Gullikson 1975; Olsson 1980b). A correction of $-510$ years has been determined for Spitsbergen and one of $-750$ years for arctic Canada. In coastal areas with upwelling deep water (Robinson 1981) and the melting of very old ice, as found in the Antarctic, seasonally dependent differences in water ages of up to 2800 years have been determined (Omoto 1983). On many coasts, however, the $\delta^{13}C$ [Eq. (6.60)] and reservoir corrections for the $^{14}C$ ages of mollusc shells approximately cancel each other. But it is more accurate to do each of the corrections separately (Stuiver and Polach 1977) if the necessary data are available to set the time scales for the reservoir correction.

Another example of the reservoir effect is provided by the dates determined for materials formed in southern hemisphere. Their $^{14}C$ ages are larger by 30 years than samples of the same age from the northern hemisphere (Lerman et al. 1970; Stuiver and Kra 1986). The reason for this is a difference in the amount of $CO_2$ exchange between the atmosphere and the ocean, which occurs over a much greater area in the southern hemisphere. For the same reason, the $^{14}C$ concentration in samples from islands is as much as 50‰ smaller than in continental samples (Olsson 1979a, 1983).

The emission of fossil carbon dioxide in volcanically active areas may lower the $^{14}C$ concentration of atmospheric $CO_2$, resulting in $^{14}C$ ages that are much too large and that cannot be recognized on the basis of the $\delta^{13}C$ values (Bruns et al. 1980b; Saupe et al. 1980). The decomposition of limestone by humic acids (Rendzina soil) can also have a considerable effect on the $^{14}C$ concentration in soil $CO_2$ and speleothem (lowering it by as much as 75%!) (Geyh 1970).

The reservoir effect for freshwater systems has been known for three decades under the term "hard-water effect" (Deevey et al. 1954). Due to the participation of fossil soil carbonate in the equilibrium between carbonate and bicarbonate in groundwater, the initial $^{14}C$ concentration in the dissolved inorganic carbon in groundwater is smaller than 100 pmc. Hence, lakes fed by groundwater produce organic materials (e.g., sapropel, shells) that also have an initial $^{14}C$ concentration lower than 100 pmc and which yield conventional $^{14}C$ ages that are up to 1000 years too large. Long-term changes in the size of the reservoir effect of a lake during the Holocene have been demonstrated. The $^{14}C$ dates for the carbonate and organic fractions of lacustrine sediments differ from genetic reasons (Geyh et al. 1971b). One solution to this problem, in order to avoid reservoir corrections, is the application of the AMS technique to terrigenic organic debris extracted from the sediment (Lister et al. 1984). Stromatolites from paleo-lakes in East Africa have also been successfully dated as part of a study of changes of lake levels during the Pleistocene and Holocene (Hillaire-Marcel et al. 1986).

*Suess effect:* The quasi steady-state equilibrium of the natural $^{14}C$ cycle between the atmosphere, biosphere, and the oceans has been disturbed by man since the beginning of industrialization around 1850. Since that time, large amounts of $^{14}C$-free carbon dioxide resulting from the burning of coal and petroleum have been released into the atmosphere. As a consequence, the $^{14}C$ abundance in atmospheric $CO_2$ has decreased by an average of 0.03% annually (Tans et al. 1979; De Jong and Mook 1982). This means, for example, that organic matter formed in 1950 yields apparent $^{14}C$ ages that are about 250 years too high. This observation is known as the "Suess effect" after its discover (Suess 1955); it is also called the industry effect. Radiocarbon data for which a Suess correction was necessary were published in the early 1950's when wood grown between 1850 and 1950 was used as standard instead of NBS oxalic acid.

*Nuclear weapons effect:* Nuclear weapons tests have had the opposite, but more pronounced effect, leading to an increase in the $^{14}C$ concentration in atmospheric $CO_2$ by a factor of up to 2 (Fig. 7.8). The use of this globally distributed tracer for age determinations is discussed in Sect. 7.4. However, it also increases the problem of contamination (e.g., McKay et al. 1986).

**Representative Examples.** There are now so many case studies of the $^{14}C$ method that a selection that would satisfy everyone is no longer possible (Polach 1988). A good source is the Proceedings of the last four international $^{14}C$ conferences.

*Archeology and paleobotany:* Radiocarbon data obtained in archeological and paleobotanic studies have been published in almost all languages. The materials dated most often are wood, charcoal, peat, and mollusc shells. How difficult it is to collect stratigraphically undisturbed samples (Waterbolk 1983) has been described, for example, by Willkomm (1983). In the case of peat, different reservoir corrections are necessary depending on the type of peat. Mollusc shells – used for dating eustatic variations in sea level – may have been transported landwards post-depositionally (MacIntyre et al. 1978).

*Glaciology and paleoclimatology:* Glaciological and paleoclimatic studies use $^{14}C$ age determinations on carbon dioxide occluded in cold ice (Oeschger et al. 1977a; Dansgaard 1981; Andree et al. 1984). These ages are used for correlating the time scales that are obtained from rheological models of glaciers (e.g., Johnsen et al. 1972; Dansgaard et al. 1971; Hammer et al. 1978), as well as to check the validity of these models. Net accumulation and ablation rates of ice sheets can also be determined (Lal et al. 1987).

*Groundwater dating:* The model developed by Münnich (1957, 1968) for ground-water dating is based on carbonate/bicarbonate chemistry. The carbon measured in a $^{14}C$ analysis of groundwater is in the form of $CO_2$, mainly as $HCO_3^-$. Rainwater contains only about 6 mg $CO_2$/L, groundwater contains several orders of magnitude more. This increase is due to the higher concentration of $CO_2$ in soil air (up to 3 vol%), which is produced by the respiration of roots and the decomposition of organic material that has recently died (activity A of $^{14}C = 100$ pmc, $\delta^{13}C = -25‰$). Rainwater infiltrating into the topsoil dissolves this $CO_2$ and this carbonic acid solution then dissolve marine and fossil carbonate in the topsoil (assumed A for

$^{14}C = 0 \, \text{pmc}$, $\delta^{13}C = 0\permil$) with the formation of bicarbonate:

$$CaCO_3 + CO_2 + H_2O \rightleftharpoons Ca^{++}(HCO_3^-)_2 \tag{6.63}$$

The theoretical initial $^{14}C$ concentration after equilibrium between the $CO_2$ and carbonate has been reached (taking into account isotopic fractionation $\varepsilon$ of about $-9\permil$) under closed system conditions is given by

$$A_0 = \frac{C_1 + C_2}{2C_1 + C_2} 100 = \frac{\delta^{13}C_{TDIC}}{\delta^{13}C_{soil} - \varepsilon} 100 \tag{6.64}$$

where $C_1$ and $C_2$ are the molar concentrations of $HCO_3^-$ and $CO_2$, respectively. The relationship between $C_1$ and $C_2$ is expressed by the formula $C_2 = kC_1^3$, where $k$ is a temperature-dependent constant (Wendt et al. 1967). $\delta^{13}C_{TDIC}$ and $\delta^{13}C_{soil}$ are the $\delta^{13}C$ values of the total dissolved inorganic carbon (TDIC) and the soil $CO_2$ (about $-22\permil$), respectively. An improved model has been developed by Pearson and Swarzenki (1974).

According to Eq. (6.64), $A_0$ in newly regenerated groundwater is in the range of 55 to 65 pmc, corresponding to a reservoir correction of $-4500$ to $-3500$ a. $A_0$ increases with increasing TDIC. Reardon and Fritz (1978), Fontes and Garnier (1979), Fontes (1985), and many others have tried to improve the model for the determination of $A_0$ by including isotopic fractionation and mixing. This is difficult to put into practice, however, because the isotopic compositions of the chemical components [Eq. (6.64)] are not accurately known (Tamers 1975) and isotopic equilibrium is usually not attained. In addition, isotopic exchange between the bicarbonate dissolved in pore water and the gaseous $CO_2$ in the unsaturated zone (i.e., an open system) may occur (Wendt et al. 1967; Fontes 1985; McKay et al. 1986), as well as $^{14}C$ accumulation on the surfaces of the carbonate particles in the topsoil during the warm seasons (Geyh 1972a; Salomons and Mook 1976). In addition bacterial metabolism also may change the $\delta^{13}C$ value (Chapelle et al. 1988). These and other effects explain the frequently obtained $A_0$ values between 50 and 85 pmc (Münnich 1968), corresponding to a reservoir correction of about $-1300$ a. In catchment areas where the crystalline basement crops out or quartz sands occur, the initial $^{14}C$ concentration may even be as much as 100 pmc (Vogel 1970; Geyh 1972a), because the carbonate dissolved by the groundwater has been only recently formed from feldspar and atmospheric $CO_2$. Isotopic exchange with the carbonate in the aquifer does not seem to play an important role (Münnich 1968). But concurrent dissolution and precipitation would have a considerable effect (Wigley 1977; Wigley et al. 1978) if it occurs.

For these reasons and owing to other geochemical processes that are difficult to treat (Fontes 1985), empirical and semi-empirical procedures for estimating the initial $^{14}C$ concentration of groundwater have found justification (Geyh 1972a, b; Pearson and Swarzenki 1974; Fontes and Garnier 1979). The reliability of the hydrological interpretation of the results increases if more than one isotope from the same sample is measured (Geyh 1972a, b; Bath et al. 1979; Andrews et al. 1984; Geyh et al. 1984).

Hydrochemical and hydrodynamic processes must be taken into consideration when $^{14}C$ data for groundwater are interpreted (Przewlocki and Yurtsever 1974;

Reardon and Fritz 1978; Wigley et al. 1978; Geyh 1980c; Neretnieks 1981; IAEA 1983; Geyh et al. 1984; Zuber 1986). A prerequisite for reliable interpretation is that water samples of known age composition be analyzed (Egboka 1985). For example, samples containing water from more than one level of the aquifer are usually less suitable than samples derived from only one level.

The wide application of the $^{14}$C method to hydrological problems and the interpretation of the results is covered by the Proceedings of the international conferences on hydrology organized by the International Atomic Agency in Vienna. These are published every 3–4 years. Further suitable sources are the *Guide Book on Isotope Hydrology* (IAEA 1983) and a book in German by Moser and Rauert (1980). Paleohydrogeological aspects have been treated in papers by Andres and Geyh (1970), Bath et al. (1979), and Sonntag et al. (1980).

*Speleothem dating:* The first dating of stalagmites and other speleothem was done using the $^{14}$C method (Franke 1951; Münnich and Vogel 1959; Geyh and Franke 1970). These ages also yield paleoclimatic information (Fig. 6.56). Theoretical aspects of the isotope geochemistry of speleothem have been treated by Hendy (1971). Radiocarbon ages of stalagmites are usually very reliable, neglecting uncertainties of up to 1000 a due to the reservoir effect. However, the samples must be taken along the axis of growth and not radially. Because stalactites grow horizontally as well as vertically, with no definable strata, only very rough ages can be obtained. Isotopic exchange, recrystallization, and later precipitation of modern carbonate in pores and cracks explain apparent $^{14}$C ages of Pleistocene speleothem that are too low (Geyh and Hennig 1986).

*Soil dating:* The problems inherent in the dating of soils have been treated in many papers (Campbell et al. 1967; Geyh et al. 1971a; Scharpenseel and Schiffmann 1977;

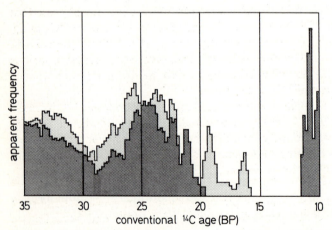

**Fig. 6.56.** Histogram of conventional $^{14}$C dates for stalagmites in European caves north (*dark gray*) and south (*light gray*) of the Alps. The long interruption in speleothem growth reflects the last pleniglacial, which lasted longer in the north than in the south (after Geyh 1970)

Geyh et al. 1983; Matthews 1985). A distinction must be made between radiometric and actual mean ages because the $^{14}$C ages of the different humic acid fractions differ (Scharpenseel 1979). Long-term records of $^{14}$C activity in soils are explained by a process of carbon turnover (Harkness et al. 1986). It is very difficult to date soil carbonate and caliche reliably owing to their complex genesis (Netterberg 1978; Blümel 1982). Therefore, pedogenic soil carbonate in arid regions usually yields apparent $^{14}$C ages that are too large in comparison with ages obtained with other methods (Williams and Polach 1969). This may be due to their having been formed by evaporation of fossil groundwater (Netterberg 1978) and subsequent diagenesis (Blümel 1982). An empirically determined correction factor for the $^{14}$C ages of tufaceous sediments based on lithologic classification seems to be useful (Pazdur 1988).

*Oceanography:* Radiocarbon also functions as a tracer in research on long-term oceanic processes: food chains, currents, hydrodynamic mixing in the oceans (Bien et al. 1960; Lal 1962; Broecker 1981; Stuiver et al. 1983), and formation of marine apatite (Burnett and Kim 1986). In most cases, the organic and carbonate fractions of marine sediments yield different results (Geyh et al. 1974; Geyh 1979). Thus, if pelagic sediments are to be analyzed, especially autochthonous, unsilicified foraminifers should be carefully extracted (Olsson et al. 1968). Errors can be caused by diagenetic recrystallization (Chappell and Polach 1972). Even individual, handpicked foraminifers can now be dated using the AMS technique (Peng and Broecker 1984; Broecker et al. 1984; Broecker 1987). They found that if the influence of bioturbation is taken into account, the deep-sea ventilation rate can be estimated using differences between the $^{14}$C ages of planktonic and benthic foraminifer species that are stratigraphically coeval (Andree et al. 1986). According to research in the South China Sea, changes in the ventilation of the deep sea did not occur during this time and therefore cannot be used to explain $^{14}$C variations. However, the rate of ventilation of the Atlantic and Pacific was slower during the Ice Age of the Pleistocene than during the Holocene (Broecker 1987). Keir (1983) has discussed the influence of the dramatic climate change at the end of the last ice age on $^{14}$C dating. He has also discussed the increase in the dissolution of carbonate (Keir 1984). Oeschger et al. (1975) showed that the mean residence times of carbon in the various geophysical reservoirs can be estimated properly only by taking into account the actual vertical distribution of $^{14}$C in the oceans.

Present-day marine circulation can be traced using anthropogenic $^{14}$C in ocean waters (Stuiver and Oestlund 1980; Oestlund and Stuiver 1980; Roether et al. 1980; Kromer et al. 1987). Corals are also used for this purpose: Their annual growth rings provide, analogously to tree rings, long-term records on fluctuations in the initial $^{14}$C concentration in the marine environment (Druffel 1982).

*Dating of other materials:* Problems are presented by the dating of terrestrial snail shells (Tamers 1970; Evin et al. 1980; Goodfriend and Hood 1983) and the carbonate in antlers, which yield ages that are too high by up to 3000 a. However, because the deviations from the true age are taxa dependent, the variability of such age anomalies can be reduced to $\pm 300$ a by ecologically appropriate species

(Goodfriend 1987a). A similar problem is involved in the dating of mortar, owing to the presence of fossil lime. This is overcome by the use of a $\delta^{13}C/^{14}C$ mixing line (van Strydonck et al. 1986). AMS can be used for dating organic material in shards (0.2–7% C content) (Bonani et al. 1984). However, the organic material may contain components of different ages, influencing the reliability of the $^{14}C$ ages (Arnal and Audrieux 1986; Gabasio et al. 1986).

### Non-Chronological Applications

Current problems to which radiocarbon is being applied include air pollution in metropolitan areas (Currie et al. 1983; Otlet et al. 1983a; Currie et al. 1984) and nuclear power plants (Levin et al. 1980). Sources of methane as a pollutant can be determined with this method, as has been done in New Zealand (Lowe et al. 1987). Causes of cancer have been investigated by Tamers (1979). Druffel and Mok (1983) have studied the rate of formation of gallstones. There are also other applications in mammalian biology (Bada et al. 1987).

Analysis of cosmogenic $^{14}C$ produced in polar ice by spallation of oxygen (probably present as $^{14}CO$) appears to be a possible supplement to conventional methods of determining net accumulation and ablation rates of ice sheets. Complementary tritium measurements are recommended. However, this method is still in discussion (Lal et al. 1987).

### 6.2.2  Tritium ($^3$H) Methods

The classical tritium method (Libby 1953) was based on the environmental tritium concentration in young water, applying radiometric techniques (Sect. 6.2.2.1). More sensitive methods (Sect. 6.2.2.2) have been developed which utilize radiogenic helium-3, produced by the decay of tritium (Tolstikhin and Kamenskiy 1969).

### 6.2.2.1  Classical Tritium ($^3$H) Method*

(Libby 1953; Kaufman and Libby 1954)

### Dating Range, Precision, Substance, Sample Size

This method was developed for dating ice and groundwater (15–2000 mL) with ages of up to 100 years. It has a precision of about $\pm 5$–10% (Moser and Rauert 1980; IAEA 1979, 1983).

### Basic Concept

Tritium ($^3$H) decays by $\beta^-$-emission ($E_{max} = 18.6\,keV$) with a half-life of 12.43 years. It is produced by cosmic-ray-induced spallation of nuclides in the atmosphere, as well as by particle capture reactions of nitrogen and oxygen. The production rate is

0.25 atoms/cm$^2$/s, two-thirds in the stratosphere and one-third in the troposphere. After oxidation to water it enters the hydrological cycle, in which it has a mean residence time of less than 2 years. The natural level of tritium in precipitation is as much as 25 TU at high latitudes, decreasing to about 4 TU in the equatorial zone (Fig. 6.49) (TU = tritium unit, which is defined as 1 tritium atom per $10^{18}$ atoms of hydrogen, which corresponds to 0.118 Bq/L).

According to the classical model by Kaufman and Libby (1954), the age of the precipitation in ice, for example, can be calculated from the decrease in tritium activity with time using Eq. (6.58). But because the water, for example, discharged from a spring is always a hydrodynamic mixture, models are needed to interpret the $^3$H activity of spring water in terms of mean residence time or mean transit time in the rock (Eriksson 1958; Nir 1964; Przewlocki and Yurtsever 1974; Maloszewski and Zuber 1982). For ocean water, tritium data provide information on currents and mixing (Roether et al. 1980).

### Sample Treatment and Measurement Techniques

The amount of sample that is needed depends on the technique used for measuring the tritium activity (IAEA 1979) (Sect. 7.4). Most often, liquid scintillation counters (Sect. 5.2.2.2) are used, for which the tritium is enriched 10- to 100-fold. The simplest method is to electrolyze 1–2 L of water, during which the isotopically heavier tritium-bearing water molecules will become enriched in the remaining water. Other methods of enrichment utilize the thermal diffusion of hydrogen or, in a few cases, gas chromatography. The $^3$H activities of the samples are measured for several hours or days, depending on the precision required (Jouzel and Merlivat 1981) (Fig. 6.57).

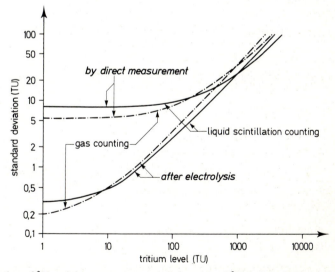

**Fig. 6.57.** Standard deviation of $^3$H activity measurements as a function of $^3$H activity before and after enrichment for different counting systems for the same counting time (after Jouzel and Merlivat 1981)

Proportional counters with guard counters arranged in an anticoincidence circuit (Sect. 5.2.2.1) provide a sensitive method of measurement. With this technique, a tritium activity of 1 TU can be measured using 15-mL samples from which methane, ethane, or propane is prepared for counting. One to two days are needed for the measurement.

Very low concentrations down to 0.005 TU, which are measured, for example, in oceanographic studies using the "regrowth" technique, require the use of a mass spectrometer (Jenkins et al. 1983). The tritium activity is determined via the stable helium-3 produced by the decay of the tritium. The samples (typically ca. 45 mL) are thoroughly degassed and then stored at least half a year in a tightly sealed aluminosilicate container under vacuum. The tritium concentration ($C_{^3H}$) is calculated from the concentration ($C_{^3He}$) of $^3He$ produced during the storage time $\Delta t$ as follows (see also Sect. 6.2.2.2):

$$C_\tau = \frac{C_{^3He} e^{+\lambda \cdot t_s}}{1 - e^{-\lambda \cdot \Delta t}}, \tag{6.65}$$

where $t_s$ is the time elapsed between sampling and the first degassing.

Attempts have recently been made to use the AMS technique (Sect. 5.2.3.2) for tritium analysis (Glagola et al. 1984). If the sensitivity is increased, this method may become important for very small samples of materials that are difficult to obtain.

### Scope and Potential, Limitations

The range of application of the classical tritium method has become rather limited. The natural, cosmogenic $^3H$ activity can no longer be determined in young water because the anthropogenic tritium level in the hydrosphere now exceeds the natural one by one order of magnitude as a result of industrial nuclear activities (Sect. 7.4). Uncontaminated samples with an age of $\leqslant 70$ years occur now only in very deep groundwater and the abyssal parts of the ocean. The $^3H/^3He$ method (Sect. 6.2.2.2) and the tritium injection technique (e.g., Athavale et al. 1983) both provide a way to avoid needing to know the initial tritium activity.

In highly saline groundwater in fractured granite with high uranium, thorium, and lithium contents, a low boron content and a low porosity, tritium levels can reach 0.5 TU, produced by the reaction $^6Li\ (n, \alpha)\ ^3H$ (Andrews and Kay 1982).

### 6.2.2.2  Tritium/Helium-3 ($^3H/^3He$) and Helium-3 ($^3He$) Methods**

(Tolstikhin and Kamenskiy 1969; Kurz et al. 1985)

### Dating Range, Precision, Substances, Sample Size

This method is suitable for dating ice with an age of up to 100 a. The length of time since deep groundwater, hot spring water, and deep ocean water ($> 40$ mL) have had contact with the atmosphere can be estimated in the same time range. Precision is $\pm 2–4\%$ (Moser and Rauert 1980; IAEA 1983; Mamyrin and Tolstikhin 1984).

Cosmogenic helium-3 in surface rocks may allow dating of young basaltic lava flows in the range of the radiocarbon method (Sect. 6.2.1), possibly in an even greater range (Kurz et al. 1985).

### Basic Concept

Tritium decays by $\beta^-$-emission ($E_{max} = 18.6$ keV) to helium-3 with a half-life of 12.43 years. It is produced by cosmic-ray-induced spallation of atmospheric oxygen. After oxidation to water it enters the hydrological cycle. The natural level of tritium in precipitation is as much as 25 TU at high latitudes decreasing to about 4 TU in the equatorial zone (Fig. 6.49) (TU = tritium unit, which is defined as 1 tritium atom per $10^{18}$ atoms of hydrogen, which corresponds to 0.118 Bq/L).

The classical concept of $^3$H dating (Sect. 6.2.2.1) has lost its significance because the initial $^3$H activity is now virtually masked by anthropogenic tritium produced by the atomic bomb tests carried out mainly in the 1960s and contained in emissions from nuclear power plants (Sect. 7.5). This $^3$H activity level differs with time and area and exceeds the natural one by up to four orders of magnitude (Fig. 7.8). Only since the beginning of the 1970's has the specific tritium activity in rain and snow in the northern hemisphere ranged within the narrow limits of 20–40 TU. This level is now maintained primarily by nuclear power plants. In the mid-1980s, the proportion of the tritium activity derived from nuclear power plants became larger than that from the nuclear bomb tests (Eisenbud et al. 1979). Moreover, the natural input of tritium into the various hydrological reservoirs (by precipitation, vapor exchange, and river runoff) has also varied from year to year and place to place. Lower tritium concentrations are measured in coastal areas than inland (continental effect). In addition, mixtures of water from various sources may exist. Moreover, the level is subject to seasonal fluctuation, because summer rains usually contain more tritium than snow. Samples that may be considered to have been closed systems during the past century are found only in cold ice, deep aquifers, and deep ocean water.

It has, therefore, been important to develop a technique to overcome the difficulties of low natural specific $^3$H activity and varying $^3$H input. To determine the effective initial tritium activity, both the $^3$H concentration ($C_{3_H}$) and the concentration of $^3$He produced from the tritium ($C^*_{3_{He}}$) are measured (Tolstikhin and Kamenskiy 1969). The sum is $C_0$ if no gas was lost during the aging of the sample. The age $t$ is calculated as follows:

$$t = \frac{\tau}{\ln 2} \ln \frac{C_{3_H} e^{\lambda t_s} + C^*_{3_{He}} - C^{atm}_{3_{He}}}{C_{3_H} e^{\lambda t_s}} \tag{6.66a}$$

where the term $C^{atm}_{3_{He}}$ is a correction for atmospheric helium-3 ($^3$He/$^4$He$_{atm} \sim 1.4 \times 10^{-6}$), and $t_s$ is the time elapsed between sample collection and the degassing. If the $^3$He concentration is given in cm$^3$(STP)/g H$_2$O and tritium is given in terms of activity ($A_{3_H}$) in TU, then (Weiss and Jenkins 1980):

$$t = \frac{\tau}{\ln 2} \left( 1 + \frac{4.022 \times 10^{14} C^*_{He}}{\rho \cdot A_{3_H}} \right), \tag{6.66b}$$

where $\rho$ is the density (g/mL) of the sample.

Helium-3 is also produced cosmogenically by spallation in surface rocks. Because $^3$He is a stable isotope, long-term processes spanning millions of years can be studied. The production rate is on the order of 140 atoms/g/a at sea level in Hawaii. Helium-3 loss by diffusion in the mineral grains seems to be negligible (Kurz 1986) (Sect. 6.5.5).

### Sample Treatment and Measurement Techniques

To determine the helium-3 content of a sample (Torgersen et al. 1979), a water sample (e.g., 10 mL) is degassed in a vacuum. An especially sensitive mass spectrometer (Sect. 5.2.3.1) is used for a quantitative analysis of the gas for $^3$He. The radiogenic portion of this amount is obtained by subtraction of the atmospheric portion, for which a $^3$He/$^4$He ratio of $1.384 \times 10^{-6}$ is assumed. The continuous flux of $^3$He from the Earth's crust will cause errors that cannot be corrected for (Weiss and Jenkins 1980; Torgersen and Ivey 1985).

Another technique, the "$^3$He-ingrow" technique (Clarke et al. 1976), which has a very low detection limit of about 0.005 TU, is used to determine the tritium activity of 40-mL samples that have been thoroughly degassed and then stored in tightly sealed containers for 6–12 months. The $10^{-16}$ to $10^{-14}$ mL$_{STP}$ of $^3$He produced by the decay of tritium during this time is determined mass spectrometrically after cryogenic adsorption.

The age of the sample is calculated using Eq. (6.66a or b). The precision is 2–4%. The time between sample collection and analysis must be taken into consideration because the short half-life of tritium requires a decay correction of 5% per year.

### Scope and Potential, Limitations, Representative Examples

*Scope and Potential.* Owing to its sensitivity, the $^3$H/$^3$He method is used for dating Antarctic ice, as well as in studies of large-scale mixing of oceanic waters (Jenkins and Rhines 1980), of vertical mixing in limnic systems, and on groundwater storage (Torgersen et al. 1979; Maloszewski and Zuber 1983; Schlosser et al. 1988). Extensive application is prevented by the high price of the mass spectrometer, which is only partly compensated for by the low cost of the analysis in comparison to the conventional tritium decay counting.

The accumulation of cosmogenic $^3$He in surface rocks theoretically makes it possible to determine ages from a few hundred years to many hundred thousand years. Therefore, this method has great potential in geology and geomorphology, as well as archeology. It should be possible to determine erosion rates (Kurz 1986).

*Limitations.* The main problem of the $^3$He/$^3$H method is loss of gas, either previous to sample collection due to natural processes or during sample treatment (Weiss and Jenkins 1980). Even loss of gas through the glass walls of containers during storage has been observed; hence, special precautions are necessary for sample storage. Laboratories using this method will give advice on how to store the samples.

Samples of temperate glacier ice are not suitable (Schotterer et al. 1977) owing to complex diagenetic changes and non-uniform isotopic composition.

Subsurface production of tritium by the reaction $^6Li\,(n,\alpha)\,^3H$ in granitic rocks with a low boron content, low porosity, and high uranium, thorium, and lithium contents may create $^3H$ levels of $> 1\,TU$ in highly saline water (Andrews and Kay 1982; Florkowski et al. 1988).

*Representative Applications.* Oxygen utilization rates in the oceans have been determined by Jenkins (1980). Weiss and Roether (1980) have studied the global tritium budget to obtain a better understanding of the water balance of the oceans taking into consideration the humidity of the air, evapotranspiration, and input from the rivers.

The interpretation of tritium activity in the water of marine and limnic systems requires sophisticated models that take into consideration the complexity of the hydrodynamic processes (e.g., migration, gas exchange at the lake or ocean surface, and vertical diffusion in the epilimnion) over periods of days to decades (Torgersen et al. 1979; Jenkins 1980; Weiss and Jenkins 1980; Maloszewski and Zuber 1983; Mamyrin and Tolstikhin 1984). The best results are obtained with multiple-isotope analysis (Smethie et al. 1986; Salvamoser 1986).

Deep groundwater has been dated by Seifert (1978).

### Non-Chronological Applications

The wide scope of application of this method in environmental studies can be seen in the Proceedings to an IAEA symposium in 1978 (IAEA 1979) (e.g., regional distribution of tritium; biological, limnic, and oceanic systems), although anthropogenic tritium now predominates.

The combined measurement of tritium and helium-3 in lake water in Canada was used to define a helium prospecting index, which appears to be of value for uranium prospecting (Top and Clarke 1981).

## 6.2.3 Beryllium-10 ($^{10}$Be) Method**

(Arnold 1956)

### Dating Range, Precision, Substances, Sample Size

This widely applied method has a potential for chronostratigraphic studies of carbonate-free pelagic sediments with poor biostratigraphic and paleomagnetic records (0.5–1 g) (Amin et al. 1975; Mangini et al. 1984; Sharma and Somayajulu 1984), for dating manganese nodules (50–100 g) (Guichard et al. 1978; Ku et al. 1979a; Krishnaswami et al. 1982; Segl et al. 1984; Mangini et al. 1986), and ice (1–2 kg) (Beer et al. 1984) which have an age of about 100 ka to about 15 Ma. The precision of AMS dating of sequential core samples is 2–5% (Raisbeck et al. 1978; Dansgaard 1981). Soils are little suited (Monaghan et al. 1983; Pavich et al. 1986), but $^{10}$Be produced in terrestrial quartz may make it possible to estimate erosion rates (Nishiizumi et al. 1986). Oil may be datable (Bourles et al. 1984). In addition, weathering, erosional, exposure, and burial histories of rocks can be studied (Klein et al. 1986). Cosmic ray exposure ages of meteorites (10–20 mg) have been estimated (Sarafin et al. 1984).

In combination with $^{26}$Al and $^{36}$Cl analyses, the method is used routinely for determining terrestrial ages of meteorites (Sect. 6.5) (Nishiizumi et al. 1983a; Ehmann and Kohman 1958; Tuniz et al. 1984).

## Basic Concept

Beryllium-10 decays by $\beta^-$-emission ($E_{max} = 555$ keV) with a half-life of 1.51 $\pm$ 0.06 Ma. It is produced by cosmic radiation from nitrogen, oxygen, and carbon (two-thirds in the stratosphere and one-third in the troposphere) ($3.8 \times 10^{-2}$ atoms/cm$^2$/s $\pm$ 25%, Reyss et al. 1981; Monaghan et al. 1985). Very little comes from cosmic dust. The amount produced varies with latitude by as much as factor of 3 (Lal 1987a; Fig. 6.49). Residence time of $^{10}$Be is about 1 year in the stratosphere, several weeks in the troposphere, and shorter than 1500 years in the oceans (Yokoyama et al. 1978; Mangini et al. 1984). Undissolved particles remain in the ocean water for only 0.5–30 years (Kusakabe et al. 1982). Marine deposition of $^{10}$Be correlates strongly with clay sedimentation. All of these residence times are too short for $^{10}$Be and the stable isotope $^9$Be to become homogenized, not even in the oceans.

Beryllium-10, bound to aerosols, finds its way into soils and cold ice, as well as marine and limnic sediments. Almost always associated with iron, it is relatively immobile at pH values over 5 (Pavich et al. 1986).

Due to the relatively short residence time of $^{10}$Be in the atmosphere, changes in the rate of production rapidly influence the input of $^{10}$Be into terrestrial materials. The production rate is determined mainly by the cosmic ray flux and its modulation by heliomagnetic and geomagnetic shielding effects. This explains the good correlation between $^{14}$C and $^{10}$Be changes. Even the 11-year cycle of cosmic radiation is recognizable (Beer et al. 1988a, b). Beryllium-10 fluctuations in precipitation are governed mainly by climatic effects (Lal 1987a), particularly precipitation rate, mixing, and exchange between the stratosphere and troposphere.

Variations in the specific $^{10}$Be activity in precipitation depend on the time-scale taken into consideration: On an annual basis, the initial $^{10}$Be/$^9$Be ratio in ice varies by a factor of 4, but during the past 1 Ma it has varied by only $\pm$ 6% and over the last 7–9 Ma it may viewed as constant (Ku et al. 1982). From 5000 BP to 1000 BP, $^{10}$Be input remained nearly constant, but during the last ice age and the Little Ice Age (Raisbeck et al. 1981; Beer et al. 1984) it was larger by a factor of at least 2 (Fig. 6.58). Such short-term variations correlate with variations in $^{14}$C and the solar activity. For this reason, there is no generally valid, constant initial $^{10}$Be/$^9$Be ratio and therefore it is not generally possible to date individual samples. However, variations in $^{10}$Be seems to be valid globally (Beer et al. 1987) and are controlled mainly by climatic conditions, particularly precipitation. In polar ice, for instance, the $^{10}$Be concentration for the Wisconsin glacial period peake at about three times the Holocene value mainly due to a corresponding decrease in precipitation (Monaghan 1987). Variations in the $^{10}$Be/$^9$Be ratio in marine sediments may be due to

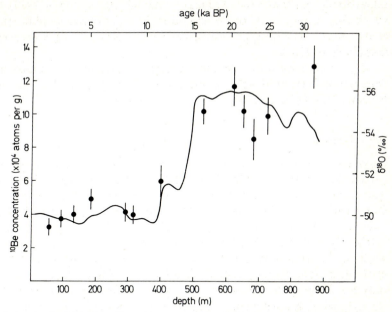

**Fig. 6.58.** Variation in the initial beryllium-10 concentration in the ice core taken at Dome C in Antarctica over the last 30 Ka (*dots*). The ages are derived from $\delta^{18}O$ values (*solid line*) (after Raisbeck et al. 1981)

(1) changes in sedimentation rate, (2) input of melt and river water, (3) chemical and compositional (e.g., clay content) changes in the sediment, (4) variations in geomagnetic field intensity (not yet proven) (Raisbeck et al. 1985), and (5) fluctuations in the intensity of cosmic radiation (Bourles et al. 1988). These causes preclude a general and precise application of $^{10}$Be for dating of marine samples.

Therefore, it is preferable to make several age determinations on a carefully selected series of samples from a core. If the $^{10}$Be concentration decreases exponentially, a constant $^{10}$Be input and a uniform sedimentation rate $r$ (in cm/ka) can be assumed. The age of the sediment at a depth $d$ (in cm) can then be calculated from the $^{10}$Be concentration $C$ (atoms/g dry sediment which has an in situ density $\rho$ in g/cm$^3$) as follows:

$$C = \frac{3.15 \times 10^{10}P}{\rho r} e^{-\lambda t}, \tag{6.67}$$

where $t$ (in years) is the time since the deposition of the sediment (Fig. 6.59) and $P$ is the rate of $^{10}$Be input (atoms/cm$^2$/s), which can be rather precisely estimated from the natural initial specific activity (0.3–10 dpm $^{10}$Be/kg) [Eq. (5.8)].

Beryllium-10 is also produced in continental rock exposed at the Earth's surface by cosmic-ray-induced spallation of silicon and oxygen. The concentration of this radionuclide can be used to determine upper limits for the rate of gradual removal of the overburden by erosion and weathering, i.e., removal of the shielding of rock or ice or the rate of accumulation. This upper limit for the erosion rate corresponds to a

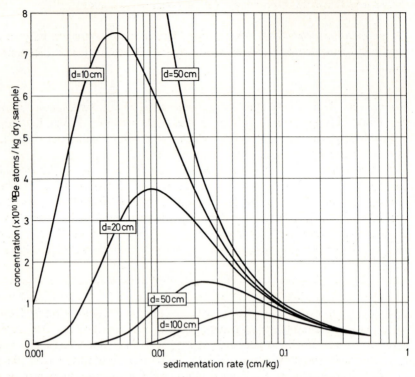

**Fig. 6.59.** Theoretical relationship between $^{10}$Be concentration (*atoms/kg dry sample*) and sedimentation rate $r$ (cm/ka) for various depths (*d*) in the profile assuming a $^{10}$Be input of $4.5 \times 10^{-2}$ atoms/cm$^2$/s and a density of 1.5 g/cm$^3$ (after Amin et al. 1975)

lower limit on the amount of time that the samples have not be shielded (Nishiizumi et al. 1986).

The terrestrial age of chondrites is estimated from a few grams of sample using a saturation concentration (about $18.8 \pm 0.9$ dpm/kg) (Sect. 6.5).

### Sample Treatment and Measurement Techniques

Samples of sediment or manganese nodules (several grams) are gound and digested in 50% HF or leached with 8N HCl in a Teflon crucible (Everst 1964; Brown et al. 1985). Leaching is less time-consuming. Iron is removed with isopropyl ether, manganese is precipitated with NaOH. The beryllium is then extracted with ion exchangers after adding 1 mg of $^{9}$BeSO$_4$ to serve as a carriers to determine the yield.

The target for the AMS measurement is prepared by heating the extracted beryllium hydroxide to yield 0.5–2 mg BeO, which is then mixed with copper and silver. The interfering isobar is $^{10}$B. The detection limit is about $5 \times 10^{-15}$ for $^{10}$Be/$^{9}$Be or $5 \times 10^5$ atoms. About $\frac{1}{2}$ h is needed per sample. The $^{10}$Be concentrations are given in atoms/g (a typical range is $10^3$–$10^9$ atoms/g), $^{10}$Be activity is given in dph/kg.

## Scope and Potential, Limitations, Representative Examples

*Scope and Potential.* The $^{10}$Be method is one of the most rapidly developing methods owing to its importance for dating Quaternary and Neogene marine seidments.

*Limitations.* Due to its short residence time in the atmosphere, cosmogenic beryllium-10 is regularly distributed in the geophysical reservoirs. It is hoped that this advantage will be overcome by making multi-isotope analyses, including $^{26}$Al (Sects. 6.2.5 and 6.2.13) and $^{36}$Cl (Sect. 6.2.14).

The method has its limits for terrestrial samples because beryllium is mobile at pH values below 5. This is a reason why attempts to determine the global beryllium balance have not been successful (e.g., Monaghan et al. 1983). Beryllium-10 analyses of soils yield unreliable dates due to organic chelation and leaching. Instead, rates of erosion, weathering, and leaching of exposed continental rocks (10–20 g samples), as well as rates of accumulation, can be estimated (Monaghan et al. 1983; Nishiizumi et al. 1986).

Beryllium-10 dating of individual samples of marine sediment is difficult because differences in the initial $^{10}$Be concentration of up to one order of magnitude might exist. The processes for this in the oceans are not well understood. However, in areas with high biological productivity, the vertical flux of $^{10}$Be exceeds the production rate. The deficit is compensated by lateral input. These areas are sinks for $^{10}$Be. Such laterally transported $^{10}$Be makes it difficult to set up a $^{10}$Be balance, but makes it possible to study paleoclimatic-oriented erosion, as well as oceanic circulation (Krishnaswami et al. 1982; Mangini et al. 1984).

The concentration of $^{10}$Be in ice is governed not only by the production rate but also by transport and deposition processes. Hence, variations up to a factor of 10 can be observed during a day.

*Representative Examples.* *Paleoclimatology:* Some of the most successful case studies have been $^{10}$Be age determinations combined with U/Th measurements (Sect. 6.3) on manganese nodules (Segl et al. 1984). Four periods of growth were delimited and distinct time markers of paleooceanographic relevance were dated:

1) 12–14.5 Ma, the beginning of the expansion of the Antarctic glaciation;
2) 5.7–6.7 Ma (mean 6.2 Ma), the development of the modern bottom water circulation and the end of the glaciation of the West Antarctic sea;
3) 2.9–3.3 Ma (mean 3.2 Ma), the commencement of glaciation of the northern hemisphere with the formation of the ice masses at the two poles and the closing of the Panama Strait; and
4) 1.1 Ma, the beginning of the Quaternary, corresponding to a large change in the global ice volume.

Beryllium-10 studies on a red clay core (APC-3) from the central North Pacific yielded very similar time markers (Mangini et al. 1984). Beryllium-10 diffusion in manganese nodules that have growth rates of > 2 mm/Ma is negligible (Mangini et al. 1986). The profiles confirm the hypothesis that $^{10}$Be production between 400,000 and 800,000 years ago was about the same as it is today.

*Dating of soils:* Whether the dating of soils is possible is still being researched (Pavich et al. 1986).

### Non-Chronological Applications

Concurrent $^{10}$Be and $^{26}$Al (and $^{10}$Be and $^{36}$Cl) analyses on ice and pelagic sediments (Lal 1962) are promising for obtaining information on long-term fluctuations in the production of cosmogenic radionuclides (Sects. 6.2.13 and 6.2.14) and more exact age determinations (Dansgaard 1981). This kind of analysis may yield estimates of the duration and magnitude of the decrease in magnetic intensity during the Brunhes-Matuyama reversal (Raisbeck et al. 1985).

Analysis of $^{10}$Be in lava (about 10 g) provides an indication of the composition and origin of the source lava and makes it possible to distinguish between sedimentary matter and material derived from subducted lithospheric plates (Brown et al. 1982). Such analyses also hold promise for erosion, weathering, and leaching studies (Valette- Silver et al. 1986). A good description of the state-of-the-art is given by Brown (1987). Combined $^{10}$Be and $^{26}$Al analysis of quartizite samples from elevations of 1000–4000 m in Antarctica, California, and India yielded erosion rates of $1.7 \times 10^{-3}$ to $2 \times 10^{-5}$ cm/a (Nishiizumi et al. 1986); the theory is described in detail by Lal (1987b). Exposure and burial histories of glass from the deserts of Libya have also been studied (Klein et al. 1986).

Beryllium analyses of tektites have yielded information on their genesis (Pal et al. 1982).

Measurement of $^{10}$Be in ice sheets should make it possible to estimate accumulation and ablation rates (Lal et al. 1987).

### 6.2.4  Sodium-22 ($^{22}$Na) Method

(Fleyshman, Kanevskiy, Gritchenko 1975)

### Dating Range, Precision, Substances, Sample Size

Sodium-22 is recommended for dating shallow groundwater and surface water (500–5000 L) with ages of up to about 30 a.

### Basic Concept

Sodium-22 ($\tau \sim 2.6$ a) is a gamma-emitter (E = 1.28 MeV). Besides cosmogenic production, it was also produced by the nuclear weapons tests (Fleishman et al. 1987). After cessation of the tests in 1964, non-steady-state conditions with respect to $^{22}$Na input existed into the 1970's (Sect. 7.4). Since 1980, steady-state activities of cosmogenic $^{22}$Na have been approached in the atmosphere and hydrosphere (production rate at 60°N averages $13 \times 10^{-2}$ Bq/m$^2$/a). An exponential model

(Eriksson 1958) is used to convert $^{22}$Na data for lake and lake discharge water into mean residence times. Well-mixed conditions are a necessary assumption (IAEA 1983).

*Sample Treatment and Measurement Techniques*

Cations in the water are absorbed on an ion-exchange resin. After elution with HCl, the solution is evaporated and the residue calcined. Barium chloride is removed by dissolving it in ethanol. The specific activity of the $^{22}$Na (annihilation gamma-photon, 0.51 MeV) is measured with a low-level coincidence $\gamma$-$\gamma$-spectrometer (NaI(Tl) scintillation counter, Sect. 5.2.2.2). Soft water (precipitation and some lake water) requires a counting time of 20–40 h, hard water can take up to 200 h.

*Scope and Potential, Limitations, Representative Examples*

This method is proposed for hydrological studies in which ages of a century or less are to be measured. It suffers, however, from the very low activity of the $^{22}$Na and, therefore, long counting times. As only one case study has been made, its potential is still unclear.

The mean residence times of the water of Lake Ostrovito and the water discharging from Lake Ladoga (Neva River) in the Soviet Union have been estimated to be 13 ± 1 and 28 ± 2 years, respectively (Fleishman et al. 1987). There was no comparison, however, made with the results of conventional studies.

## 6.2.5   Aluminum-26 ($^{26}$Al) Method*

(Lal 1962; Tanaka, Sakamoto, Takagi, and Tschuchimoto 1964)

*Dating Range, Precision, Materials, Sample Size*

The aluminum-26 method has a wide range of application for dating ice (1–3 kg using the AMS technique), pelagic sediments and manganese nodules (Guichard et al. 1978) with an age of 0.1–5 Ma. Samples of several hundred grams are necessary for conventional decay counting techniques. Samples of only a gram are needed for the AMS technique. Precision depends on the age of the sample and ranges from 5–10% (Amin et al. 1966; Reyss et al. 1976; Raisbeck et al. 1983).

Weathering and erosion rates of continental rocks can be determined (Nishiizumi et al. 1986) and earth-surface processes are studied (Klein et al. 1986).

Cosmic-ray-exposure ages and terrestrial ages of meteorites (Sects. 6.5.6 and 6.5.7, respectively) and lunar samples have been routinely determined up to 2 Ma (Barton et al. 1982).

*Basic Concept*

Aluminum-26 decays with the emission of a positron ($\beta^+$) ($E_{max} = 1170$ keV) with a half-life of 716 ± 32 ka. Two 0.511-MeV gamma photons are emitted (annihilation radition) when it decays. Aluminum-26 is produced only from atmospheric argon by spallation induced by cosmic radiation. The rate of production is strongly dependent on latitude (Fig. 6.49) and averages $1.1 \times 10^{-4}$ atoms/cm$^2$/s (Reyss et al. 1981). Contrary to earlier assumptions, only about 10% of the $^{26}$Al on the Earth has its origin in cosmic dust (Reyss et al. 1976).

The geochemical behavior of $^{26}$Al is probably very similar to that of beryllium-10. Aluminum-26 has an average residence time of about 1 year in the stratosphere and a maximum of 1500 years in dissolved form in the oceans (Yokohama et al. 1978). This time is too short for it to become evenly distributed in the atmosphere and other geophysical reservoirs. As a result, the initial $^{26}$Al activity is dependent on latitude. As is also the case for $^{10}$Be (Sect. 6.2.3), its production has fluctuated considerably in the past. The $^{26}$Al/$^{10}$Be method (Sect. 6.2.13) is an attempt to overcome these difficulties. The ratio of the rate of production of $^{26}$Al to that of $^{10}$Be is nearly constant between $3.8 \pm 0.6 \times 10^{-3}$ (Raisbeck et al. 1983) and $6 \times 10^{-3}$ (Reyss et al. 1981).

Aluminum-26 is found in sediments and soils, bound mainly in clay minerals. The specific activity ranges from 0.1 to 10 dpm/kg dry weight (corresponding to $0.5 \times 10^9 - 50 \times 10^9$ atoms/g).

Aluminum-26 ages cannot be determined for individual samples due to the variations in the initial $^{26}$Al concentration. But it is possible to obtain $^{26}$Al dates for sediment cores. If the $^{26}$Al activity decreases exponentially with depth, uniform sedimentation and constant $^{26}$Al input may be assumed and the ages can be calculated via the sedimentation rates using Eq. (6.67) (Sect. 6.2.3).

Aluminum-26 is also produced by cosmic-ray-induced spallation of $^{28}$S, by negative muon capture by $^{28}$Si in the continental silicate rock exposed at the Earth's surface, and radiogenically from decay chain radionuclides (Sharma and Middleton 1989). The concentration of this radionuclide can be used to determine upper limits for the rate of gradual removal of the overburden by erosion and weathering, i.e., removal of the shielding of rock or ice. This upper limit for the rate corresponds to a lower limit on the amount of time that the samples have not been shielded (Hampel et al. 1975; Nishiizumi et al. 1986).

In stony meteorites, $^{26}$Al is produced mainly by three reactions (Sect. 6.5.7):

$^{28}$Si$(n, p + 2n)\,^{26}$Al
$^{27}$Al$(n, 2n)\quad\,^{26}$Al
$^{26}$Mg$(p, n)\quad\,^{26}$Al

The saturation activity $A_0$ in chondrites is 55–65 dpm/kg Si. The exposure age $t_{exp}$ can be obtained from the $^{26}$Al activity $A$:

$$A = A_0 f (1 - e^{\lambda \cdot t_{exp}}) e^{-\lambda \cdot t_{terr}}, \tag{6.68}$$

where $f$ is the shielding factor and $t_{terr}$ is the terrestrial age.

### Sample Treatment and Measurement Techniques

Since aluminum-26 from the atmosphere is expected to be mainly in clay minerals, repeated leaching with HCl should bring out all aluminum of interest. From this HCl solution, iron(III) hydroxide is precipitated with NaOH. In the filtrate, Al(OH)$_3$

is precipitated with $NH_4OH$—buffering with $NH_4Cl$. The yield of this procedure is estimated to be about 90% (Alder et al. 1967).

The preparation of the samples (about 500 g) for radioactive decay counting techniques begins with repeated leaching of the sample with HCl, followed by precipitation of the aluminum with ammonium hydroxide. The aluminum is often purified further with an ion exchange resin. The precipitate is then converted to alumina by heating.

The decay counting is based on the detection of the 0.51-MeV photon pair produced by the annihilation of the $\beta^+$ as well as of the coincident 1.83-MeV gamma ray emitted by $^{26}Al$. A gamma-gamma coincidence NaI(Tl) spectrometer (Sect. 5.2.2.2) is used for this purpose (Reyss et al. 1976; Barton et al. 1982). The measurements take several months.

New possibilities for the application of the method are opened up by the AMS technique (Sect. 5.2.3.2), for which considerably smaller samples are sufficient, thus making it easier to obtain samples. Only several grams of material are needed to extract 10–20 mg of aluminum for mounting on the target after mixing with silver powder (Raisbeck et al. 1979). The disturbing isobar $^{26}Mg$ does not form negative ions. The detection limit, at present, is about $10^9$ atoms or about $10^{-14}$ for the $^{26}Al/^{27}Al$ ratio. However, the $^{26}Al/^{27}Al$ in recent clay sediments is only $10^{-15}$ or even smaller!

### Scope and Potential, Limitations, Representative Examples

The aluminum-26 method was originally proposed for marine studies (Lal 1962). Owing to the large fluctuations in the $^{26}Al$ input and the resulting uncertainty in the initial $^{26}Al$ activity, the combined $^{26}Al/^{10}Be$ method has a greater chance of success (Sect. 6.2.13). For the last several years, the AMS technique (Sect. 5.2.3.2) has been displacing of conventional decay counting techniques. However, the high $^{27}Al$ content of most materials limits application of the AMS method.

The determination of the $^{26}Al$ cosmic-ray-exposure age of extraterrestrial samples (meteorites and lunar rocks) is a standard procedure (Sect. 6.5.7).

Combined $^{26}Al$ and $^{10}Be$ analysis of pelagic and ice samples (Sect. 6.2.13) is promising for independently dating fluctuations in the production of cosmogenic isotopes during the last million years.

### Non-Chronological Applications

Analogous to the chlorine-36 method (Sect. 6.2.7), erosion studies in arid regions can be carried out (Nishiizumi et al. 1986). Earth-surface processes, e.g., weathering, erosion, exposure, and burial histories of glass from the deserts of Libya, have been studied on the basis of in-situ production of both $^{26}Al$ and $^{10}Be$ (Klein et al. 1986).

### 6.2.6   Silicon-32 ($^{32}$Si) Method*

(Lal, Goldberg and Koide 1960)

#### Dating Range, Precision, Substances, Sample Size

Successful applications of the $^{32}$Si method (Lal et al. 1976) are known for marine suspended-particulates and siliceous sediments ($\sim 300$ g). Its potential for snow and ice has also been investigated (Fig. 6.63). Silicon-32 dating of groundwater is still rather problematic (2–20 m$^3$, Lal et al. 1970). The range of ages that can be determined extends from about 100 to 1500 a; the standard deviations are at least $\pm 150$ a (Lal et al. 1970; Clausen 1973; Somayajulu et al. 1973). A sediment sample size of 100–200 g is sufficient. $^{32}$Si/Si ratios down to at least $4 \times 10^{-15}$ can be determined by AMS (Heinemeier et al. 1987).

#### Basic Concept

The $\beta^-$-emitting isotope silicon-32 ($\tau = 105 \pm 18$ a to $172 \pm 4$ a; $E_{max} = 200$ keV) is produced by cosmic-ray-induced spallation of atmospheric argon: $^{40}$Ar (n,4p5n)$^{32}$Si. The rate of production is strongly dependent on latitude (Fig. 6.49). In Central Europe, the deposition rate is between 1 and 20 mBq/m$^2$/a. The specific activity of precipitation ranges from 2–20 mBq/m$^3$ (Franke et al. 1986). Silicon-32 presumably reaches the Earth's surface as monomeric silicic acid in rain and snow. It is rapidly deposited in ice and in limnic and marine sediments and eventually it finds its way into the groundwater. For closed systems (e.g., ice and pelagic sediments), in which radioactive decay determines the decrease in $^{32}$Si content, the age can be determined using Eq. (6.58), assuming constant initial $^{32}$Si activity.

#### Sample Treatment and Measurement Techniques

The silicate is extracted from the water or ice in the field (Gellerman et al. 1988). Several thousand liters of water are acidified to pH = 3 to remove $CO_2$ and carbonates, 100–200 g of iron chloride are added per cubic meter of water, and the mixture is then brought to pH 8 or 9 with ammonium hydroxide. If the silicon content is very low, a silicon carrier ($Na_2SiO_2$) is added. The precipitated iron hydroxide is filtered after several days and dried. (This step can be accelerated by adding a coagulation agent.) About 300 g silica can be obtained from the filter cake (5–10 kg/m$^3$ water) using hydrochloric acid and sodium hydroxide. Another method uses ion exchangers to extract silicon compounds (Fröhlich et al. 1974).

   The radiochemically purified silicon extract is stored in bottles for 7–12 weeks. During this time, radioactive equilibrium is established between the $^{32}$Si and the $\beta^-$-emitting daughter isotope $^{32}$P ($\tau = 14.3$ d; $E_{max} = 1.710$ MeV). After addition of an inactive phosphorus carrier, the remaining silicon is removed using hydrofluoric acid; the "milked" phosphorus is purified by distillation and ignited to $Mg_2P_2O_7$, which is placed on mylar foil and its activity measured for 1–2 months in a $2\pi$-flow counter (Sect. 5.2.2.1). Phosphorus-32 is measured instead of $^{32}$Si because its higher energy $\beta$-radiation is easier to detect than the weak $\beta$-radiation of $^{32}$Si.

Moreover, a silicon extract would be several orders of magnitude heavier than the phosphorus extract.

The $^{32}$Si activity ($A_{Si}$) is calculated from the $^{32}$P activity ($A_P$) as follows:

$$A_{Si} = \frac{\lambda_P A_P}{\lambda_{Si}(1 - e^{\lambda_P \Delta t})},$$ (6.69)

where $\Delta t$ is the length of time the $^{32}$P activity is measured.

The age is calculated using Eq. (6.58) or extrapolated from a plot of $^{32}$P cross-activity as an expontial function of time (e.g., Lal et al. 1960).

AMS is a more sensitive tool; at present it is capable of detecting $^{32}$Si down to an isotope ratio of about $10^{-15}$. Interference from the disturbing isobar $^{32}$S is avoided by using $SiH_3^-$ ions (Heinemeier et al. 1987).

### Scope and Potential, Limitations, Representative Examples

**Scope and Potential.** The $^{32}$Si method could become a bridge between the dating range of the tritium (Sect 6.2.2.1) and the radiocarbon (Sect. 6.2.1) methods. It finds its main application in glaciological, oceanographic, and fall-out studies, in particular: "aqueous geochemistry, biogeochemical cycles of silicon in the oceans, the chronology of glaciers, and biogenic silica-rich sediments in lacustrine and marine environments" (Lal and Somayajulu 1984).

**Limitations.** Until recently, the main problem with the $^{32}$Si method was the wide range of 105–330 years given for the half-life (Clausen 1973; Kutschera et al. 1980). The newest value is $172 \pm 4$ a (Alburger et al. 1986). The geoscientific implications arising from the large uncertainty in the half-life are discussed by Lal and Somayajulu (1984).

The most important parameter for dating with $^{32}$Si determination of the initial $^{32}$Si activity, which is usually in the range of 0.2–20 mBq/m$^3$ for rainwater and fresh snow, depending on latitude. The range of seasonal variation is very large (an average factor of 2). In addition, there are local differences (Fig. 6.49). Attempts have also been made to date ice with this method (Fig. 6.63) (Clausen 1973; Oeschger and Loosli 1977).

The reasonable results for snow are not necessarily transferable to groundwater (Lal et al. 1970; Andrews et al. 1984) because the geochemistry of silicon in the unsaturated zone, which is not yet understood, has to be taken into consideration. For instance, silicon is soluble to a very slight extent and is taken up by some plants. Any of these phenomena may be a reason why the $^{32}$Si ages of groundwater samples may be as much as an order of magnitude smaller than the corresponding $^{14}$C ages (Sect. 6.2.1). Nevertheless, $^{32}$Si in combination with tritium and radiocarbon seems to be at least a suitable tracer for distinguishing young from old water (Fröhlich et al. 1987).

Silicon-32 produced by the nuclear weapons tests was removed from the atmosphere with a half-life of less than a year and was only a short perturbation. Its impact is negligible and does not invalidate the application of the method in glaciology and hydrology for dating older ice and groundwater.

*Representative Examples.* Silicon-32 and lead-210 analyses (Fig. 6.61), as well as other environmental isotope analyses, on surface and core ice samples from the Nehnar Glacier in the Kashmir Valley yielded a very detailed age distribution of the ice. The average rate of ice movement at different elevations was estimated and found to be in very good agreement with direct measurements at the terminus (Nijampurkar et al. 1982).

Silicon-32 water depth profiles in the Pacific have been interpreted using a one-dimensional advection-diffusion model yielding diffusion coefficients and advection velocities of abyssal water (Somayajulu et al. 1973).

### Non-Chronological Applications

According to Lal (1962), silicon is an ideal tracer for studying the circulation of deep water in the oceans. Of course, complicated models are necessary for interpreting the results (Oeschger et al. 1975; Somayajulu et al. 1987). These models have to take vertical and horizontal mixing into consideration, as well as the effects of biological processes. Mixing near the coasts of water with different silicon concentrations (continental, intermediate, and abyssal waters) are reflected in the specific activity of $^{32}$Si in sponges, which contain 4–80 dpm $^{32}$Si/kg SiO$_2$ (Lal et al. 1976). This makes it possible to study the silicon balance in coastal regions as silicon behaves as a conservative chemical tracer.

The mixing of river and marine water in the estuary of the Gironde River in France has been studied and the excess of stable silicon in the low salinity zone was identified to be of anthropogenic, i.e., industrial origin (Nijampurkar et al. 1983).

### 6.2.7 Chlorine-36 ($^{36}$Cl) Method**

(Davis and Schaeffer 1955)

### Dating Range, Precision, Substances, Sample Size

The chlorine-36 method is suitable for dating very old groundwater (Fontes 1985; Fabryka-Martin et al. 1987), saline soils and sediments, as well as lake water (Bonner et al. 1961; Bagge and Willkomm 1966; Elmore 1986), continental rocks (Phillips et al. 1986a; Leavy et al. 1987), and continental ice (0.5–2 kg, up to ca. 150 ka) (Dansgaard 1981) with ages of up to 0.1–3 Ma. It should also be possible to estimate exposure ages of rocks of up to 500 ka (Tamers et al. 1969; Phillips et al. 1986a), as well as erosion rates in arid regions. The precision of the method is ±1–5%. The AMS technique (Sect. 5.2.3.2) requires only 10 mg Cl, which can almost always be extracted from a few grams of solid matter or a few liters of groundwater (Bentley et al. 1986a).

The $^{36}$Cl produced by the nuclear weapons tests is utilized for dating very young ground water (Bentley et al. 1982) and ice (Elmore et al. 1982) with an age of up to a few decades. However, tritium (Sect. 6.2.2.1) and krypton-85 analyses (Sect. 7.4) are less expensive for this.

An important application of the $^{36}Cl$ method is the determination of the cosmic-ray-exposure age of meteorites (up to 800 Ma) (Sprenkel 1959; Goel and Kohman 1963) (Sect. 6.5.6).

### Basic Concept

Chlorine-38 is a $\beta^-$-emitting radionuclide ($\tau = 301 \pm 4$ ka; $E_{max} = 714$ keV) produced by cosmic-proton-induced spallation of argon (mainly $^{40}Ar$), neutron activation [$^{36}Ar\,(n, p)\,^{36}Cl$], and potassium and muon capture by calcium-40 (60% in the stratosphere and 40% in the troposphere). However, the main source is the reaction of thermal neutrons with $^{36}Ar$ in the atmosphere. The rate of total meteoric production is about 60 atoms/m$^2$/s. The production rate at a latitude of 45° is about five times larger than that at the equator or the poles. There is a distinct continental effect. For example, $^{36}Cl/Cl$ increases along the US coast from $2 \times 10^{-14}$ to about $10^{-12}$. Another important source is the reaction of neutrons with $^{35}Cl$ near the Earth's surface. Both $^{36}Ar$ and $^{35}Cl$ have very large capture cross-sections.

The nuclear weapons tests, most of which were carried out before 1964, produced up to 70,000 atoms/m$^2$/s of $^{36}Cl$ from the $^{35}Cl$ in the water of the oceans (Sect. 7.4).

Cosmogenic $^{36}Cl$ finds its way, presumably in submicron aerosols, via precipitation into groundwater and onto ice within weeks of its formation. Its abundance varies greatly: $^{36}Cl/Cl = 10^{-10}$ to $10^{-16}$. Various processes are responsible for this. Relative $^{36}Cl$ abundance is not changed by evaporation of the water, mineral interactions or secondary mineral formation, but it is influenced by the dissolving of chloride. Therefore, the total chloride should always be determined. Underground neutron densities of about $10^4$ neutrons/kg/a in the rock of the aquifers (Feige et al. 1968) produce $^{36}Cl$ from $^{35}Cl$ and play a role in the age determined for water (initial $^{36}Cl/Cl$ ranges from 5 to $30 \times 10^{-15}$) with an age of 1 Ma or greater. If uranium-rich or chlorine-bearing minerals are present, apparent $^{36}Cl$ groundwater ages that are too low may be obtained (Bentley et al. 1986a). A plot of $^{36}Cl/Cl$ versus $^{36}Cl/L$ yields information on groundwater mixing, evaporation, remobilization of chloride, and radioactive decay and subsurface production of chlorine-36.

The initial $^{36}Cl$ abundance is determined empirically from case to case using groundwater dated with another method (Bentley et al. 1986a). Precision is not particularly important due to the large range that can be dated.

The age of salts, caliche, and phonolitic rocks in lavas, as well as exposed continental rocks (Phillips et al. 1986a; Leavy et al. 1987), can be determined due to the in-situ production of $^{36}Cl$ by secondary neutrons (about $8 \times 10^5$ neutrons/kg/a), which penetrate the soil to a depth of up to 2 m. The $^{36}Cl$ abundance in an exposed chloride-containing sediment increases until a saturation activity of $^{36}Cl$, $A_\infty$, is reached. After 1 Ma, 89% of $A_\infty$ [which can be calculated using Eq. (6.70)] has been attained. The $^{36}Cl$ produced from potassium and calcium must also be taken into consideration. If the rock is exposed and $^{36}Cl$ is produced for a length of time $(t_{exp} - t_d)$ (sediment exposure age), followed by a period $t_d$ (decay age) in which the sediment is covered and only $^{36}$Cl decay occurs, the $^{36}$Cl activity $A$ due to both processes is given by the following equation:

$$A = A_\infty \{1 - e^{-\lambda \cdot (t_{exp} - t_d)}\} e^{-\lambda \cdot t_d} \qquad (6.70)$$

Since there are two unknowns, $t_{exp}$ and $t_d$, one of the two must be estimated on the basis of geological considerations. The saturation concentration $A_\infty$ is a function of the intensity of the cosmic radiation at the sampling site and thus is dependent on latitude and elevation. Erosion in arid regions can be determined using the same method (Fig. 6.60).

The cosmic-ray-exposure age of meteorites can be determined on the basis of $^{36}$Cl (Sect. 6.5.6) together with the cosmogenic radionuclide $^{53}$Mn (Sect. 6.2.10) (Nishiizumi et al. 1983a).

### Sample Treatment and Measurement Techniques

Today, most $^{36}$Cl analyses are done using the AMS technique. About 10 mg chlorine is leached from solid samples with water, precipitated as AgCl with silver oxide and sodium hydroxide, and then placed on the target. The detection limit is $10^{-15}$ for $^{36}$Cl/Cl or $10^6$ atoms $^{36}$Cl. The isobar $^{36}$S interferes with the measurement, but its concentration can be reduced during sample preparation and its influence suppressed during the measurement.

Only a few minutes are needed to prepare silicon tetrachloride from silver chloride and silicon at 450°C for liquid scintillation measurements, which take several weeks (Roman and Airey 1981).

### Scope and Potential, Limitations, Representative Examples

*Scope and Potential.* Because chloride does not chemically interact with its environment (not normally affected by sorption or desorption, dissolution or precipitation of most minerals) and is very soluble in water, it has a very simple geochemistry and is an ideal tracer for hydrological, hydrogeological, and glaciological studies for ages of up to 3 million years, e.g., residence time of groundwater (including paleohydrological studies), determination of flow paths, recognition of mixing of water from different sources, and recognition of sources of salinity as well as of evaporite deposits. And last, but not least, erosion rates can be determined (Fabryka-Martin et al. 1987). The anthropogenic $^{36}$Cl produced by the nuclear weapons tests makes it possible to date ice and groundwater in the unsaturated zone and in unconfined aquifers deposited or regenerated, respectively, during the last three decades (Sect. 7.5) (Bentley et al. 1982; Elmore et al. 1982), much longer than with tritium (Sect. 6.2.2.1). This radionuclide is also used to determine chloride transport in the environment.

*Limitations.* The dating of terrestrial rocks or soils in humid regions is not possible with this method due to the solubility of chloride. Chloride is relatively quickly removed and transported to the terminal lake or playa of the hydrogeological system, but the dating of such lakes and playas themselves is possible (Phillips et al. 1983). The greatest uncertainty in age determinations of saline lake deposits with ages over 70 ka is the estimation of the time when the surface was exposed or covered (Bagge and Willkomm 1966; Tamers et al. 1969; Bentley et al. 1986a). Because there are two unknowns (decay age and exposure age) and only one equation, errors of many hundred thousand years can result for the estimate of either one of them (Fig. 6.60). However, the buildup of $^{36}$Cl in exposed rocks, e.g., young volcanic rocks

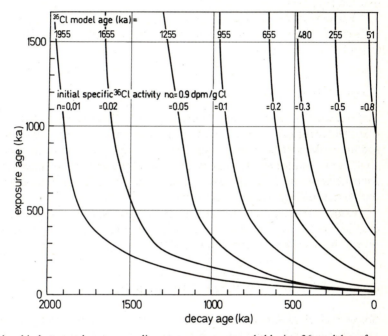

**Fig. 6.60.** Relationship between decay age, sediment exposure age, and chlorine-36 model age for different specific activities of $^{36}$Cl. The estimated maximum possible cosmic-ray-produced saturation activity $n_0$ is equal to 0.9 dpm/g, corresponding to a decay age of zero (Tamers et al. 1969)

**Fig. 6.61.** Calculated build-up of the $^{36}$Cl concentration in a basalt and a caliche suddenly exposed in an outcrop at an elevation of 1500 m. The higher initial $^{36}$C concentration of the caliche is presumed to be due to incorporation of meteoric $^{36}$Cl at the time of formation (after Bentley et al. 1986b)

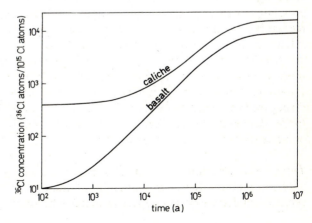

and caliche, acts as a reliable chronometer (Fig. 6.61), as shown by studies in the western United States (Phillips et al. 1986a). Subsurface production of $^{36}Cl$ in the rock matrix [e.g., $^{35}Cl\,(n, \gamma)\,^{36}Cl$, as well as $(n, \alpha)$ reactions; spallation; and to some extent, reaction with negative muons] sometimes makes the interpretation of the $^{36}Cl/Cl$ ratio difficult (Andrews et al. 1986; Florkowski et al. 1988). However, in aquifers that are far from any coast and contain $< 5\,ppm$ U and $< 10\,ppm$ Th, subsurface production does not mask the original $^{36}Cl$ input.

***Representative Examples.*** The most instructive case study is the dating of very old groundwater in the Great Australian Basin. The results are very impressive owing to the very simple geochemistry of the groundwater in this basin. Chlorine-36 ages of up to 1 Ma confirmed the computed ones based on hydrodynamic simulations (Fig. 6.62). The initial isotopic ratio $^{36}Cl/Cl_0$ was estimated to be between 85 and $120 \times 10^{-15}$ (Bentley et al. 1986b).

A detailed $^{36}Cl$ study of the Milk River aquifer in Alberta, Canada, in which the chloride concentrations cover a range of two orders of magnitude, demonstrates that this method can be used to study systems with rather complex hydrodynamics and hydrochemistry. Only one of the proposed hydrodynamic models could explain all of the available data. Chloride infiltration is the best explanation for the elevated chloride concentrations; the groundwater movement in the system is much slower than predicted on the basis of hydrodynamic theory (Phillips et al. 1986b).

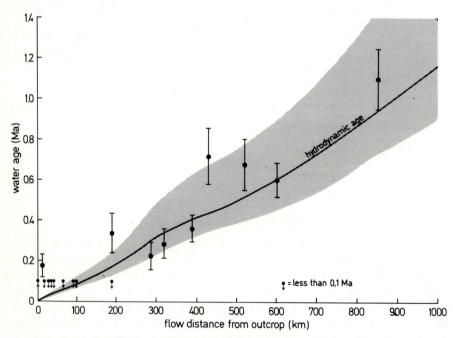

**Fig. 6.62.** Concordant $^{36}Cl$ and hydrodynamic groundwater ages along the southern flow line in the Great Artesian Basin in Australia (after Bentley et al. 1986b)

The dating of halites and other saline sediments in the closed drainage basin of Searles Lake in California yielded ages of up to 1 million years, comparable with ages obtained with other dating methods. In addition, it was concluded that less than 50% of the salts in the terminal drainage were derived from ancient rock sources (Phillips et al. 1983).

### Non-Chronological Applications

The use of $^{36}Cl$ analysis for Quaternary erosion and accumulation studies in arid regions and studies of ancient floods could become a major field of application (Bentley et al. 1986a; Elmore 1986). This is also the case for $^{36}Cl$ analysis of mineral waters in arid regions to determine the source of the salinity (Michelot et al. 1984). Such studies also contribute to our understanding of the dynamics of the chlorine cycle, e.g., a study of the Dead Sea system. Salination by evaporation of meteoric water and leaching of ancient rocks were quantified by $^{36}Cl$ analysis of water from springs, river, and terminal lakes. Ancient rock were shown to be the source of 99% of the stable chloride isotopes (Paul et al. 1986).

Regional evapotranspiration rates can be estimated by comparing the chloride concentration and isotopic composition in groundwater with those in the precipitation from which the groundwater was recharged (Magartiz et al. 1989).

A study of the potential utility of $^{36}Cl$ analysis to differentiate mixing components and to estimate the chloride balance and the mixing of geothermal fluids in hydrothermal systems in the Valles Caldera in New Mexico yielded highly encouraging results (Phillips et al. 1984a). In combination with geothermal fluid mixing, the chloride budget was also determined.

Chlorine-36 produced by the atomic bomb tests has also been utilized to determine velocities of soil water in arid parts of Mexico; rates of 2.5 mm/a were found (Phillips et al. 1984b).

Chlorine-36 input in ice directly reflects variations in $^{36}Cl$ production.

## 6.2.8   Argon-39 ($^{39}Ar$) Method**

(Loosli and Oeschger 1968)

### Dating Range, Precision, Substances, Sample Size

The argon-39 method is used for dating arctic and antarctic ice (up to 4 m$^3$) in the age range of several decades to about 1200 a with a precision of about $\pm 5\%$. Attempts to carry out age determinations on groundwater (ca. 15 m$^3$) have had variable success (Loosli 1983). Great expectations have been recently placed on $^{39}Ar$ analysis of deep ocean water (1.5 m$^3$) within the scope of oceanographic studies (Lehmann and Loosli 1984; Loosli 1988; Smethie et al. 1986). The importance of the $^{39}Ar$ method to determine cosmic-ray-exposure ages for meteorite (several grams) studies (Sprenkel 1959) is discussed in Sect. 6.5.

## Basic Concept

Argon-39 is a $\beta^-$-emitting radionuclide ($E_{max} = 565\,keV$) with a half-life of 269 years. It is formed neutrons produced by cosmic radiation and becomes well mixed in the upper atmosphere by secondary: $^{40}Ar\,(n, 2n)\,^{39}Ar$. Anthropogenic $^{39}Ar$, produced by nuclear weapons tests, makes up less than 5% of the total $^{39}Ar$. The total specific activity in the atmospheric argon is $0.107 \pm 0.04\,dpm/L_{STP}$ Ar, corresponding to $0.03\,dpm/m^3$ water or $0.5\,mBq/m^3$.

Groundwater contains dissolved argon-39 (precipitation contains atmospheric concentrations) and it occurs occluded in ice. Closed systems are assumed in which $^{39}Ar$ activity decreases only due to radioactive decay. The age is calculated using Eq. (6.58).

## Sample Treatment and Measurement Techniques

The gases dissolved in groundwater or occluded in ice ($0.5$–$15\,m^3$) are extracted using a vacuum technique. Already in the field the samples are heated or enough water (ca. 300 L) is sprayed into a vacuum for a period of $0.5$–$15\,h$ to obtain $1$–$2\,L$ of argon. Oxygen is removed by passing the extracted gases through a hot copper furnace ($600°C$). The argon in the remaining gas mixture is separated within $1$–$2$ days by preparative gas chromatography. Quantitative analysis is not necessary since only the ratio of argon isotopes is to be measured. A concurrent $^{85}Kr$ determination (Sect. 7.4) is done to correct for the proportion of atmospheric argon.

The specific $^{39}Ar$ activity is measured for $5$–$30$ days with low-level proportional counters (Sect. 5.2.2.1), $15$–$100\,cm^3$ filled with $0.3$–$2\,L$ of argon at a pressure of up to 30 bar. In the future, resonance ionization spectroscopy (Sect. 5.2.3.3) (Hurst et al. 1982) may replace this complicated procedure since only one liter of water would then be needed.

## Scope and Potential, Limitations, Representative Example

*Scope and Potential.* As a noble gas isotope, $^{39}Ar$ seems to be an ideal tracer for large-scale oceanic mixing and circulation studies (Lal 1962; Loosli 1986; Smethie et al. 1986). Moreover, it is suitable for dating groundwater in the gap between the dating ranges of tritium (Sect. 6.2.2.1) and radiocarbon (Sect. 6.2.1). Chemical reactions that are a disturbing factor for the radiocarbon method are not present (Loosli and Oeschger 1979).

*Limitations.* Ages have been obtained for groundwater that only partly confirm ages determined by other methods. In some cases, $^{39}Ar$ ages deviate by as much as one order of magnitude from the ages calculated on the basis of groundwater hydraulics and those obtained using the radiocarbon method (Loosli 1983; Andrews et al. 1984). Argon-39 cannot be used to date groundwater in granitic rock formations. Production of $^{39}Ar$ by the reaction of neutrons with potassium-39 [$^{39}K\,(n, p)\,^{39}Ar$] (Florkowski et al. 1988), which is always present in rocks, cannot explain these deviations (Loosli 1983). Geyh et al. (1984) propose that these $^{39}Ar$

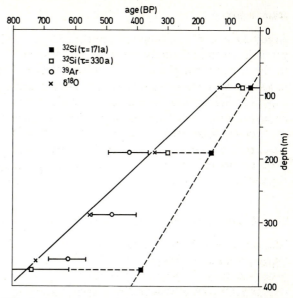

**Fig. 6.63.** Concordant $^{39}$Ar and $\delta^{18}$O dates for the Dye 3 ice core from southern Greenland (after Oeschger and Loosli 1977). The $^{32}$Si dates are in agreement with the argon-39 and oxygen-18 dates only if the earlier used half-life of 330 a was used instead of the presently accepted value of 171 a. It is not known whether the little known chemistry of silicon in ice is responsible for this

anomalies are due to hydrodynamic input of $^{39}$Ar-saturated pore water from the aquitards. The presence of argon-37 ($\tau = 35$ d) in groundwater is an indication of the underground production of $^{39}$Ar and it may become possible to use it to correct for $^{39}$Ar produced underground (Forster, pers. comm.).

*Representative Example.* Dating of arctic and antarctic ice have fulfilled expectations (Fig. 6.63) (Loosli and Oeschger 1979).

### Non-Chronological Applications

Argon-39 analysis of ocean water samples was begun some time ago within the scope of oceanographic studies to provide information on the $CO_2$ uptake capacity of the oceans (Loosli 1988). Because there is no anthropogenic $^{39}$Ar in ocean water, equilibrium exists between the argon-39 in the surface ocean water and that in the atmosphere. Hence, in contrast to $^{14}$C and $^3$H, simple models can be used to interpret the results. Moreover, the specific activity of $^{39}$Ar in ocean water near the surface is an order of magnitude greater than those of the other two isotopes. And last, but not least, the dating range of about one thousand years just covers the range of circulation times in the deep ocean. Multiple-isotope studies involving radionuclides with different half-lives make it possible to investigate short-term and long-term processes (Smethie et al. 1986). Recent episodic erosion rates may be determinable by $^{39}$Ar analysis of terrestrial rocks (Lal et al. 1986).

## 6.2.9   Calcium-41 ($^{41}$Ca) Method

(Raisbeck and Yiou 1979)

### Dating Range, Precision, Substances, Sample Size

It may become possible to use the $^{41}$Ca method to date bones and secondary carbonates with ages ranging from 20 to ~400 ka. This method also has potential for dating groundwater and determining exposure ages of sediments. The measurement can be made only with the AMS technique (Sect. 5.2.3.2), which requires only a fraction of a gram of material (Raisbeck and Yiou 1979). Terrestrial ages of meteorites (Sect. 6.5.7) can also be determined via cosmogenic $^{41}$Ca (Kubik et al. 1986; Henning et al. 1987).

### Basic Concept

The very rare calcium-41 decays mainly by electron capture ($\tau = 103$ ka $\pm 15\%$) with the emission of 3.3 keV X-rays. It is produced at low levels in the atmosphere from the krypton by cosmic radiation and from argon by low-energy alpha-radiation. Most of the $^{41}$Ca is formed by $(n, \gamma)$ reactions in the top meter of the Earth's crust from $^{40}$Ca, which has a high neutron cross-section. The saturation ratio $^{41}$Ca/$^{40}$Ca is $10^{-14}$ to $10^{-15}$ at the Earth's surface, corresponding to an activity of $10^{-3}$ dpm/g Ca, which is much smaller than for other cosmogenic radionuclides.

Calcium-41 has been detected by AMS (Sect. 5.2.3.2) in bone apatite and calcareous sediments. However, the $^{41}$Ca cycle in nature is little understood. Exchange and metabolic processes play a role. If a uniform initial $^{41}$Ca concentration exists, which is rather doubtful, the age could be calculated using Eq. (6.58). More probably, an exposure age is determined because $^{14}$Ca will have also been produced in exposed samples.

Calcium-41 is produced in the metal phase of meteorites by high-energy spallation, which for a meteorite of normal size is insensitive to meteorite size and depth within the meteorite. The first AMS measurement on material from the Bogon iron meteorite yielded a $^{41}$Ca/Ca ratio of $3.8 \times 10^{-12}$, corresponding to a specific $^{41}$Ca activity of 6.9 dpm/kg (Kubik et al. 1986).

### Sample Treatment and Measurement Techniques

A sample containing 5–10 mg of calcium is digested in acid and the calcium extracted with an ion exchanger. The calcium is mounted in the target for the AMS measurement as the metal, oxide, or fluoride mixed with Ag powder. Problems with the interfering isobar $^{41}$K have been solved by using completely stripped $Ca^{20+}$ ions (Steinhof et al. 1987); the limit of the detection of $2 \times 10^{-15}$ achieved so far for $^{41}$Ca/Ca is sufficient for dating terrestrial material. In the case of meteorites, 100–200 mg of metal is sufficient.

*Scope and Potential, Limitations*

*Scope and Potential.* The dating interval of the $^{41}$Ca method is of interest for anthropological studies, as well as for the study of geochemical, biochemical, and hydrological processes involving calcium. However, this method also has potential for application in erosion studies owing to the production of $^{41}$Ca in the topsoil and rock. Calcium-41 determination of terrestrial ages of meteorites will fill the gap between the ranges of the radiocarbon (Sect. 6.2.1.) and the $^{81}$Kr (Sect. 6.2.11) methods.

*Limitations.* It is not yet known whether $^{41}$Ca in calcareous sediments and bones is immobile enough that closed systems can be assumed and isotopic exchange with the calcium in the groundwater can be neglected. In addition, it must be determined whether $^{41}$Ca is more or less uniformly distributed in the various geophysical reservoirs. Another problem may be the subsurface production of $^{41}$Ca in anthropogenic samples. Moreover, the half-life has not yet been confirmed accurately.

## 6.2.10  Manganese-53 ($^{53}$Mn) Method**

(Herpers et al. 1967)

*Dating Range, Precision, Substances, Sample Size*

The manganese-53 method is used to determine cosmic-ray-exposure ages of meteorites (several mg) up to 12 Ma, to establish the occurrence of impact events on the lunar surface in the same time frame, and to estimate the duration and intensity of lunar regolith "gardening" (Herpers et al. 1967; Imamura et al. 1969, 1973; Englert and Herr 1978). Ice (several hundred kilograms) (Bibron et al. 1974) and pelagic sediments (Nishiizumi 1983) with ages of 1–10 Ma have also been dated with a precision of 10–20%. With the application of the AMS technique, sample sizes can be reduced (Korschinek et al. 1987).

*Basic Concept*

Manganese-53 decays by K-electron capture ($\tau = 3.7 \pm 0.6$ Ma). In extraterrestrial matter it is produced mainly from iron by low-energy secondary neutrons and protons of galactic cosmic radiation (GCR) and by solar cosmic ray protons:

$^{54}$Fe $(n, pn)$ $^{53}$Mn; $^{56}$Fe $(n, p3n)$ $^{53}$Mn; $^{56}$Fe $(p, \alpha)$ $^{53}$Mn.

Most of the $^{53}$Mn reaches the Earth in cosmic dust (Bibron et al. 1974). Its specific activity in ice is about 0.8 dpm/$10^6$ kg. Analogously to the $^{10}$Be method (Sect. 6.2.3), deposition rates of pelagic sediments can be calculated with an adaptation of Eq. (6.67).

The cosmic-ray-exposure ages of stony and iron meteorites are determined assuming a constant average $^{53}$Mn production rate in iron by cosmic radiation (Sect. 6.5.6). The saturation activity in meteorites is about $480 \pm 20\%$ dpm/kg Fe.

The terrestrial ages of meteorites (Sect. 6.5.7) are obtained from the decrease in the saturation activity of cosmogenic radionuclides due to radioactive decay.

### Sample Treatment and Measurement Techniques

A sample of a meteorite (0.02–1 g) is digested in nitric acid and hydrofluoric acid; the manganese is then separated by ion exchange techniques. The concentration is determined with a flame photometer (Sect. 5.2.4.3). The $^{53}Mn/^{55}Mn$ ratio of iron meteorites and metallic inclusions in stony meteorites can be determined mass spectrometrically (Sect. 5.2.3.1).

About 3000 kg are needed for dating ice using conventional decay counting techniques (Bibron et al. 1974). Anticoincidence gamma semiconductor spectrometers (Sect. 5.2.2.3) were used previously for measuring the X-rays from the decay of the $^{53}Mn$, a process that took days or even much longer. Neutron activation analysis (Sect. 5.2.4.2) is the most sensitive and most successfully applied method to date. It has been used for several hundred extraterrestrial and terrestrial samples. Iron must be carefully separated from milligram samples. Manganese-53 is converted in a neutron capture reaction to $^{54}Mn$ ($\tau = 312.5$ d), whose 0.835 MeV gamma radiation can be readily measured. The use of AMS makes it possible to reduce sample sizes and to measure $^{53}Mn/Mn$ down to $3 \times 10^{-11}$ (Korschinek et al. 1987).

### Scope and Potential, Limitations

*Scope and Potential.* The manganese-53 method is used routinely to date extraterrestrial samples. The method has been given new impetus by the AMs technique. The use of fully stripped ions has overcome previous difficulties (Korschinek et al. 1987).

*Limitations.* The variability of the production rate of $^{53}Mn$ by cosmic radiation as a function of chemical composition, the size of the extraterrestrial object and the location of the sample in it are the subject of extensive study. Both analysis of meteorites and simulation experiments are used to improve the $^{53}Mn$ method (e.g., Englert et al. 1984).

### Non-Chronological Applications

Manganese-53 analysis of stony and iron spherules from marine sediments has established that they are very probably ablation debris from large stony and iron meteorites, respectively (Nishiizumi 1983).

### 6.2.11  Krypton-81 ($^{81}Kr$) Method*

(Marti 1967; Loosli and Oeschger 1968)

## Dating Range, Precision, Substances, Sample Size

Krypton-81 analysis has proved to be of great value for the determination of cosmic-ray-exposure ages (Marti 1967, 1982; Eugster et al. 1967a) (Sect. 6.5.6) and terrestrial ages (Schultz 1986; Freundel et al. 1986) (Sect. 6.5.7) of meteorite samples (several grams). This method has also been used for dating groundwater and cold ice with isolation ages of 50 ka–1 Ma with a precision of 10% (Lehmann and Loosli 1984). Sample sizes of $\sim 20$ L are needed for analysis by resonance ionization spectroscopy (Sect. 5.2.3.3).

## Basic Concept

Krypton is a trace constituent of air. Two of its isotopes are useful for dating purposes: $^{81}$Kr and $^{85}$Kr (Sect. 7.4). Krypton-81, with an abundance of $10^{-16}$ in air, is produced in chondrites, mainly from Zr, Sr, and Rb. The ratio of $^{81}$Kr to the stable krypton isotopes is used to determine cosmic-ray-exposure ages (Sect. 6.5.6). The terrestrial ages of meteorites (Sect. 6.5.7) are estimated from the exponential decrease in the $^{81}$Kr/$^{83}$Kr ratio resulting from radioactive decay [Eq. (6.58)] after the fall of the meteorite. This decrease must also be taken into account when the true cosmic-ray-exposure age is calculated (Freundel et al. 1986).

Krypton-81 is formed from the stable isotopes krypton-80 (by neutron capture) and kryptron-82, -83, -84, and -86 (by spallation). It decays by K-electron capture with the emission of 13.5-keV X-rays ($\tau = 213 \pm 16$ ka). The activity of atmospheric $^{81}$Kr is about 0.09 dpm/$L_{STP}$Kr, corresponding to an isotopic abundance of $5 \times 10^{-13}$. This means that one liter of "modern" water contains a maximum of 1300 $^{81}$Kr atoms and $2.5 \times 10^{15}$ krypton atoms! The age determination is done analogously to the $^{39}$Ar method (Sect. 6.2.8). Underground production of $^{81}$Kr is negligible (Florkowski et al. 1988).

## Sample Treatment and Measurement Techniques

The relatively high $^{81}$Kr content of meteorites ($\sim 10^{-14}$ cc$_{STP}$/g) makes it possible to measure it mass spectrometrically after at least 1 day of thermal degassing at about 150°C and then at 1700°C. The reference isotope is $^{78}$Kr, which is not cosmogenic.

The krypton is extracted from 0.5–50 L water, analogously to $^{39}$Ar (Sect. 6.2.8), by degassing, separation of the noble gases, and then using preparative gas chromatography to isolate the krypton. Isotopic enrichment (two steps) is necessary before resonance ionization analysis (Sect. 5.2.3.3) can be used to count the $^{81}$Kr atoms. In a liter of water, there is a maximum of 1300 $^{81}$Kr atoms or even fewer (Chen et al. 1984; Lehmann et al. 1985; Thonnard et al. 1987). Contamination with air must be avoided. Such contamination can be checked for by measuring $^{85}$Kr (Sect. 7.4).

## Scope and Potential, Limitations

Determination of cosmic-ray-exposure ages of meteorites via $^{81}$Kr avoids unknown shielding corrections and dependence on chemical composition.

Hydrogeologists request that dating of groundwater with $^{81}$Kr provide isolation dates that are easier to interpret than those obtained by the $^{36}$Cl method (Sect. 6.2.7). Because it is a noble gas, there are no interfering chemical reactions that may have changed the $^{81}$Kr concentration. Subsurface production should be negligible, but contamination with modern atmospheric krypton must be avoided during sampling. There is no interference from isobars.

### 6.2.12  Iodine-129 ($^{129}$I) Method*

(Takagi, Hampel and Kirsten 1974)

*Dating Range, Precision, Substances, Sample Size*

Since the introduction of the AMS technique (Sect. 5.2.3.2), the range of application of the iodine-129 method has expanded rapidly (Elmore et al. 1980). Cosmic-ray-exposure ages of meteorites and lunar rocks (10–20 g) (Sect. 6.5.6) are determined, as well as tellurium ores (several kg) with ages of more than 10 Ma (Takagi et al. 1974; Nishiizumi et al. 1983b) have been dated. Hydrogeological and oceanographic applications are being found (Fehn et al. 1986; Fabryka-Martin et al. 1987). Dating of buried organic matter and its derivatives, e.g., petroleum, (Fehn et al. 1987) as well as marine sediments (several hundred grams), is in development (Fehn et al. 1986a). It is expected that the method can be used for ages of 3–80 Ma (Elmore 1986; Fabryka-Martin et al. 1987); dating precision is about $\pm 10\%$.

*Basic Concept*

Iodine-129 is a $\beta^-$ emitter ($E_{max} = 150$ keV) with a half-life of 15.7 Ma. It is formed in the upper atmosphere by cosmic-ray-induced spallation of $^{129}$Xe. Production varies with latitude and is five times larger at 45° than at the equator or the poles. Underground, it is formed as a product of the spontaneous and neutron-induced fission of $^{238}$U and by cosmic muons (e.g., in tellurium ore):

$$^{130}\text{Te}(\mu, \gamma)\,^{129}\text{I}  \quad \text{and} \quad  ^{128}\text{Te}(\mu, \gamma)\,^{129}\text{I}$$

If the muon flux can be estimated for the layer from which the sample is taken, an $^{129}$I age can be calculated using Eq. (6.70), analogously to the $^{36}$Cl method for terrestrial matter (Sect. 6.2.7). A third source is nuclear weapons tests and nuclear power plants.

In oceans, $^{129}$I has a mean residence time of at least 40,000 a and, therefore, is well-mixed. The natural $^{129}$I/$^{127}$I ratio is $3 \times 10^{-13} - 3 \times 10^{-12}$ and appears to have been rather constant with respect to time and space.

In groundwater, the $^{129}$I/$^{127}$I ratio is determined mainly by the recharge rate, the rate of leaching from the rock of the aquifer, and the amount of in-situ uranium fission. Except for underground production, the $^{129}$I/$^{127}$I ratio is quite constant, to the extent that all processes contributing to it are included in the groundwater dating models (Fabryka-Martin et al. 1985, 1987).

In meteorites and lunar rocks, $^{129}$I is produced mainly by the reaction of low-energy neutrons and protons with tellurium. Cosmic-ray-exposure ages of meteorites up to $10^8$ a are obtained via $^{129}$Xe formed by the radioactive decay of cosmogenic $^{129}$I. Radiogenic $^{129}$Xe (from extinct $^{129}$I) and $^{129}$Xe produced by spallation must be subtracted from this (Sect. 6.5.6). This dating range is of particular interest as it is not covered by any other radionuclide (Marti 1984).

### Sample Treatment and Measurement Techniques

If neutron activation analysis (Sect. 5.2.4.2) is used, the handpicked samples (0.5–2 g) are pulverized and dissolved in acid. After an iodine carrier has been added, lead iodide is precipitated, which is heated to sublime pure iodine. About 300 mg is irradiated with neutrons to convert $^{129}$I to $^{130}$I ($E_\gamma = 1.95$ and 2.34 MeV; $\tau = 12.36$ h). The effective neutron flux is calculated from the activity of the $^{128}$I produced at the same time.

Both gamma lines of the $^{130}$I are distinguished from the interfering gamma cascade of $^{126}$I ($E = 1.46$ MeV) with a highly sensitive NaI(Tl) coincidence spectrometer (Sect. 5.2.2.2). The measurement takes about 12 h per sample.

Sample preparation for the AMS technique (Sect. 5.2.3.2) takes less than an hour. Silver iodide (5–10 mg I) is prepared as target. The measurement is made on $I^{5+}$ accelerated to 30 MeV, which provides for discrimination from $^{129}$Xe. The detection limit at present is $2 \times 10^{-14}$ for $^{129}$I/$^{127}$I or about $10^7$ atoms of $^{129}$I (Elmore et al. 1980; Nishiizumi et al. 1983b).

### Scope and Potential, Limitations

**Scope and Potential.** The first applications of this method have been the study of extraterrestrial samples. The combination of $^{129}$I analysis with the analysis of other short-lived isotopes in extraterrestrial samples provides information on the variability of production rates of cosmogenic isotopes at an early stage in the history of the galaxy. One of the first applications to terrestrial samples was the dating of tellurium ores (Nishiizumi et al. 1983b).

There is potential for dating pelagic sediments and for tracing slow-moving water and brines, detection of young groundwater, and the source of salinity, as well as dating and tracing hydrocarbons (Fehn et al. 1987). Moreover, it should be possible to monitor the mixing of water from different sources (Fabryka-Martin et al. 1987).

**Limitations.** Subsurface production and isotope exchange with the source rock can cause problems of interpretation (Fehn et al. 1986a; Fabryka-Martin et al. 1987; Florkowski et al. 1988). Iodine is strongly sorbed on iron oxides, clay, and all kinds of organic material (Behrens 1982). The release of large quantities of $^{129}$I into the atmosphere by nuclear power plants and reprocessing plants has increased the $^{129}$I/I ratio in the atmosphere to as much as $10^{-4}$ (Fabryka-Martin et al. 1987).

## Non-Chronological Applications

Iodine-129 analysis may become useful for investigations related to long-term nuclear waste storage and to determine the source of brines, oil field brines, crude oil, and mineral deposits (Fabryka-Martin et al. 1985; Fehn et al. 1986a, 1987). Because $^{129}I$ is more soluble in water than $^{238}U$, the ratio of the two isotopes can be used to obtain information on the mobility of the water in the rock (Elmore et al. 1980). The $^{129}I/^{238}U$ ratio in marine sediments yields information on large-scale convective transport, as well as hydrothermal convection. Slow water movement in large basins can also be traced.

### 6.2.13    Aluminium-26/Beryllium-10 ($^{26}Al/^{10}Be$) Method*

(Lal 1962)

### Dating Range, Precision, Substances, Sample Size

This method is used for dating cold ice (several g), marine and lacustrine sediments, coral, fossil deposits of organic matter, oceanic particulate matter, and manganese nodules (several g) in the age range of 0.1–10 Ma (Lal 1962; Yiou et al. 1986). The precision of the method ranges from 5 to 10%.

The method has potential for the study of erosion and weathering of exposed continental rocks and desert glass (Klein et al. 1986). Quartz is extracted from rock samples (10–20 g) (Lal and Arnold 1985; Nishiizumi et al. 1986).

Another application is the determination of shielding-corrected cosmic-ray-exposure ages of meteorites (300–400 mg) (e.g., Graf et al. 1987) (Sect. 6.5.6).

### Basic Concept

Aluminium-26 (Sect. 6.2.5) and beryllium-10 (Sect. 6.2.3) are cosmogenic isotopes with similar geochemical properties, both of which are produced in the upper atmosphere as spallation products of argon. The ratio of their production rates is $3.8 \pm 0.6 \times 10^{-3}$ for terrestrial samples (Raisbeck et al. 1983; Reyss et al. 1981), the ratio of their specific activities is 0.018. In extraterrestrial samples, these ratios may be twice as large.

Both radionuclides have a residence time in the atmosphere of less than 1 year before they reach the Earth's surface in aerosols. Virtually all (99.9%) of these isotopes are deposited in marine sediments. Changes in the production rate may be expected to be mirrored in the activities of these isotopes in the sediments. But the isotopic ratio of the two isotopes should be independent of changes in their production rate.

The effective half-life of the $^{26}Al/^{10}Be$ ratio is 1.30 Ma. Simultaneous measurement of the activities of both isotopes eliminates the uncertainties about the fluctuations in their production and those about the physicochemical and biological conditions that determine the rate of input. The decrease in the atom ratio is consequently only a function of age. From the atom ratios $R_0$ at the sediment surface and at a depth $d$, the age $t$ can be calculated using the following equation:

$$\left(\frac{^{26}Al}{^{10}Be}\right)_d = \left(\frac{^{26}Al}{^{10}Be}\right)_0 e^{-(\lambda_{26} - \lambda_{10})t} \tag{6.71}$$

### Sample Treatment and Measurement Techniques

Preparation of the samples is described in Sects. 6.2.3 and 6.2.5.

### Scope and Potential, Limitations

**Scope and Potential.** The $^{26}Al/^{10}Be$ method has great potential now that the technical conditions for routine use of the AMS analysis (Sect. 5.2.3.2) of $^{26}Al$ and $^{10}Be$ have been obtained (Bourles et al. 1984). Of great interest is the determination of pelagic sedimentation rates and obtaining additional information about changes in the production of cosmogenic isotopes in general. Meterorite studies will also profit from this method (e.g., Graf et al. 1987).

**Limitations.** The question whether diffusion falsifies $^{10}Be$ ages (Mangini et al. 1986) may also be settled by the use of this method since the effect is not the same for both isotopes.

### Non-Chronological Applications

In-situ production of $^{10}Be$ and $^{26}Al$ in exposed terrestrial rocks makes it possible to study erosion, weathering, exposure, and burial processes. The most suitable mineral phase is quartz owing to its low Al content and because it does not absorb rainwater-scavenged cosmogenic $^{10}Be$ (Lal and Arnold 1985; Nishiizumi et al. 1986; Klein et al. 1986).

## 6.2.14  Beryllium-10/Chlorine-36 ($^{10}Be/^{36}Cl$) Method

(Nishiizumi, Arnold, Elmore, Ma, Newman, and Gove 1983a)

### Dating Range, Precision, Substances, Sample Size

The $^{10}Be/^{36}Cl$ method used to be recommended for determining age differences of several tens of ka to many hundred ka between cold ice samples (several g) taken from different places (Nishiizumi et al. 1983a). Precision is supposed to be about $\pm 30$ ka (Dansgaard 1981). The latest results put this method in question (Elmore et al. 1987).

*Basic Concept*

Beryllium-10 (Sect. 6.2.3) and chlorine-36 (Sect. 6.2.7) are cosmogenic spallation products of nitrogen and argon. They rapidly settle out of the atmosphere in aerosols; thus changes in both productions and sedimentation rates are reflected in the input of these isotopes in sediments within a very short time. Therefore, the fundamental prerequisite of a constant input is not fulfilled for either isotope and hence it is not possible to carry out separate age determinations.

This handicap can be overcome by carrying out a double isotope analysis. It is assumed that changes in the accumulation of the snow and the production of cosmogenic isotopes have the same effects on both isotopes so that the decrease in isotope ratio with depth is a function only of age. The effective half-life of the ratio of the two isotopes is 371 ka.

The $^{10}Be/^{36}Cl$ ratios are determined for two samples A and B of different ages. The difference in the ages $\Delta t$ can then be calculated as follows:

$$\Delta t = \frac{1}{\lambda_{36} - \lambda_{10}} \ln \frac{(^{10}Be/^{36}Cl)_A}{(^{10}Be/^{36}Cl)_B} \tag{6.72}$$

*Sample Treatment and Measurement Techniques*

The preparation of the samples and separation of the isotopes is described in Sects. 6.2.3 and 6.2.7. The ice samples are melted, acidified, and chlorine and beryllium carriers are added. Ion exchangers are used for separation. The AMS technique is used to determine the isotope ratios.

*Scope and Potential, Limitations*

The $^{10}Be/^{36}Cl$ method has been applied only once, with only moderate success (Nishiizumi et al. 1983a) because the ice samples were nearly the same age or the isotopic contents were too small. Recent results for an ice core from Camp Century in Greenland show variations in the $^{10}Be/^{36}Cl$ ratio of 4 to 22, precluding the possibility of dating the ice. These variations of up to a factor of 5, however, may reflect paleoclimatic changes (Elmore et al. 1987).

## 6.3 Dating Based on Radioactive Disequilibrium of the Uranium, Thorium, and Protactinium Decay Series: The Uranium/Thorium/Protactinium Methods***

(Boltwood 1907; Pettersson 1937; Piggot and Urry 1939, 1942)

*Dating Range, Precision, Materials, Sample Size*

The uranium/thorium//protactinium methods (Ivanovich and Harmon 1982) are used for dating relatively uranium-rich pelagic sediments, especially handpicked forminifera, alluvial and lacustrine sediments, secondary carbonates (e.g., speleothem and travertine), igneous rocks, especially volcanic rocks, and ferromanganese minerals (all of which require 1–100 g samples). These methods are particularly suitable for fossil coral and manganese nodules (several grams), as well as marine and insular phosphorite, hydrothermal sulfides and oxides with ages up to about 350 ka (maximum 1.5 Ma). Thick, well-preserved mollusc shells, the compact parts of bones (several 10 g) and teeth may also yield reliable ages. Quite promising is the dating of peat and lignite (up to 100 g) in the same dating range (Vogel and Kronfeld 1980). Ice containing appreciable quantities of dust has also been dated successfully (Fireman 1986). With the high-precision measurement of $^{234}U$, $^{238}U$, $^{230}Th$, and $^{232}Th$ the dating range is extended to 500 ka, the precision is improved fourfold ($\pm 0.5\%$), and the sample size is decreased by two to three orders of magnitude (Edwards et al. 1986/87).

The reliability of U/Th ages of carbonaceous marine and terrestrial samples can be assessed on the basis of the following criteria (Thurber et al. 1965):

- The sample must have a uranium concentration of $> 10$ ppb, $> 1$ ppm is better.
- Terrestrial carbonates should have contained no $^{230}Th$ at the time of formation.
- Coral (aragonite, less than 1% calcite), mollusc shells, speleothem, and travertine should be compact, impervious to water, and may show no sign of weathering. They must have formed a closed system (Schwarcz 1980).
- There may be no signs of diagenetic recrystallization, which could have mobilized uranium or subsequent disintegration products (Geyh and Henning 1986). Thus, for example, primary aragonite samples (e.g., mollusc shells or coral) may not contain any calcite.
- The proportion of acid-insoluble residue must be $< 5\%$ and the $^{230}Th/^{232}Th$ activity ratio of terrestrial carbonate should be $> 20$.
- The $^{226}Ra/^{230}Th$ and $^{234}U/^{238}U$ activity ratios of marine samples older than 70 ka should be in the range of $1.0 \pm 0.1$ and $1.14 \pm 0.02$, respectively.
- The radiometric age should be consistent with the stratigraphic data.
- Dates obtained using different methods, e.g., $^{230}Th/^{234}U$ (Sect. 6.3.1), $^{231}Pa/^{235}U$ (Sect. 6.3.2), $^{230}Th$-excess (Sect. 6.3.5), $^{231}Pa$-excess (Sect. 6.3.6), U/He (Sect. 6.3.14), and $^{14}C$ (Sect. 6.2.1), should agree.

If even one of these criteria is not fulfilled, the results cannot be expected to be reliable.

### Basic Concept

The dating methods based on radioactive disequilibrium utilize the time-dependence of geochemical disturbances of the radioactive equilibrium between parent and daughter isotopes of the natural radioactive decay series of $^{238}U$, $^{235}U$, and $^{232}Th$, whose end members are stable lead isotopes (Fig. 6.64). In closed systems, e.g., in crystals or unweathered rocks with ages greater than 400–500 ka, the activities of all members of a decay series are the same. This means that for every element of the series, just as many atoms decay as are formed. The exceptions are the first members (parents) of the series, whose half-lives are so long that the

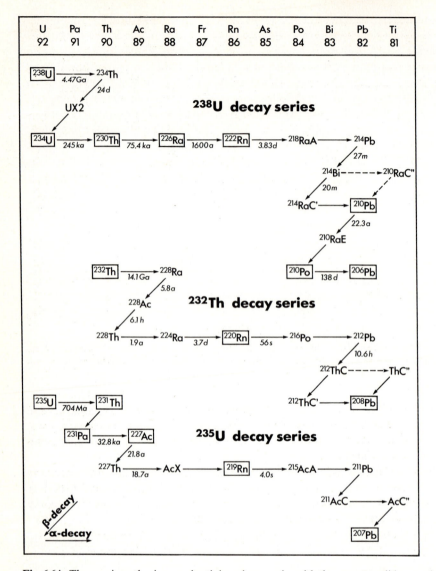

**Fig. 6.64.** The uranium, thorium, and actinium decay series with the parent nuclides uranium-238, thorium-232, and uranium-235, respectively: The geochronologically most important members of the three decay series are enclosed in *boxes*; the others have half-lives that are too short for geological dating

decrease in their activity during the dating range of the radioactive disequilibrium methods is usually neglected. In systems that are in radioactive equilibrium, the number of atoms of any specific daughter nuclide within the series is proportional to its half-life.

Geochemical processes (e.g., weathering, precipitation of lacustrine carbonates, and formation of speleothem), geophysical processes (e.g., deep-sea sedimentation

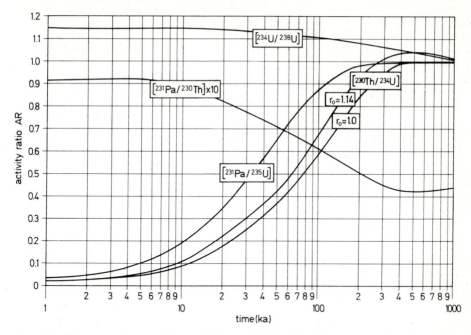

**Fig. 6.65.** Changes in the geochronologically most important activity ratios of members of the uranium decay series in radioactive disequilibrium as a function of time; $r_0$ is the initial activity ratio

and growth of crystals), and biological processes (e.g, growth of foraminifera and molluscs) lead to enrichment or depletion of parent or daughter members as a result of isotopic fractionation, upsetting the radioactive equilibrium and producing radioactive disequilibrium.

If the new system is a closed one, i.e., neither diagenesis nor migration disturbs the isotopic composition, radioactive equilibrium is gradually reestablished. The state of disequilibrium between the activities of the daughter and parent isotopes is then a measure of the age of the system (Fig. 6.65), i.e., the time since the system was disturbed. The change in the activity of the daughter isotope $[A_d]$ with respect to the parent isotope $[A_p]$ as a function of time $t$ is given by the following equation:

$$[A_d] = \frac{\lambda_d}{\lambda_d - \lambda_p}\,[A_p]_0(e^{-\lambda_p t} - e^{-\lambda_d t}) + [A_d]_0 e^{-\lambda_d t} \tag{6.73}$$

where $[A_p]_0$ and $[A_d]_0$ are the initial activities of the parent and daughter nuclides, respectively.

Secular equilibrium is established if (as in the normal case) the half-lives of the daughter products $\tau_d$ are smaller than those of the parent isotopes $\tau_p$. For $\tau_d \ll \tau_p$ and $\lambda_p t \sim 0$, Eq. (6.73) is simplified, as is also the case for systems that at time zero contained only the parent isotope, since the second term is then zero.

Processes that lead to a disturbance of the radioactive equilibrium usually involve water. During the leaching or weathering of rocks, in which there is generally radioactive equilibrium, uranium is enriched in the water relative to thorium owing to the difference in their solubilities. Uranium-234 is preferentially in the $+6$ oxidation state and forms water-soluble complexes, for example, with carbonate, sulfate, and chloride. This is in contrast to uranium-238, which is preferentially in the $+4$ state and forms mainly insoluble compounds (Petit et al. 1985). Therefore, $^{234}$U is more easily leached than $^{238}$U (Cherdyntsev 1955) (Sect. 6.3.4). This behavior explains why the $^{234}$U/$^{238}$U activity ratio of groundwater and the water in rivers and the oceans is usually larger than 1 (far from radioactive equilibrium). This is also the case for travertine and speleothem because they derive their uranium content from the surrounding waters.

The mean residence time of uranium in sea water (mostly present as uranyl carbonate complexes) is about 250 ka; radium, thorium, and protactinium are taken up by plankton or adsorbed on sediment particles within a few decades. This process is very efficient: More than 99% of the $^{230}$Th and $^{231}$Pa of the marine environment is in the top layer of sediment. The concentrations of uranium, radium, and thorium in sea water differ correspondingly (100:12:2). Biological processes enrich uranium in foraminiferal tests, mollusc shells, and corals by a factor of 1000 or more relative to the environment. For this reason, such samples generally have high uranium and low thorium concentrations and are, therefore, particularly suitable for U/Th dating. High uranium enrichments have also been found in strongly reducing environments, e.g., in peat deposits, occluded in iron hydroxide precipitates, and fossil bones.

*Prerequisites.* Dating on the basis of radioactive disequilibrium requires the following conditions:

– The entry of the parent or daughter isotope into the system must be rapid relative to the half-life of the decay product.
– At time zero, the activity of the daughter isotope used for the age determination must be zero, negligibly small, or known.
– The system has been a closed one since its formation, i.e., no parent or daughter radionuclides have been gained or lost, for example, during fossilization or diagenesis. Only when this is the case can radioactive equilibrium be reestablished without disturbance. The dating range is approximately four times the half-life of the daughter isotope used for dating.

Numerous possibilities for dating, several of which have been verified, are provided by the long-lived daughter isotopes, e.g., $^{234}$U, $^{230}$Th, and $^{231}$Pa, and the short-lived radionuclides $^{210}$Pb, $^{228}$Th, and $^{226}$Ra. The dating ranges of these methods (foldout table) depend on the half-lives and abundances of these radionuclides in the system to be dated. Excellent descriptions of these methods have been published by Ku (1976) and Ivanovich and Harmon (1982).

*U/Th Methods.* The methods can be divided into two groups:

– *Daughter isotope deficit methods:* In systems for which these methods can be used, the daughter radionuclide is not present initially and is produced gradually by radioactive decay of the parent isotope. The $^{230}$Th/$^{234}$U (Sect. 6.3.1) and $^{231}$Pa/$^{235}$U (Sect. 6.3.2) methods belong to this group.

– *Daughter isotope excess methods:* In systems for which these methods can be used, the activity of the daughter radionuclide is initially greater than possible for a system in radioactive equilibrium. This excess activity decreases during the period of aging due to radioactive decay. The method described in Sections 6.3.5 to 6.3.10 belong to this group.

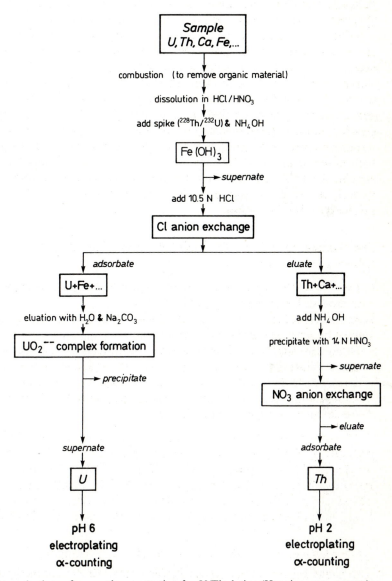

**Fig. 6.66.** General scheme for sample preparation for U/Th dating (Henning, pers. comm.)

## Sample Treatment and Measurement Techniques

The specific activities of suitable, mostly $\alpha$-emitting parent or daughter isotopes or their activity ratios (which are written in square brackets) are measured for the age determination [Eq. (6.73)]. Although there are many dating methods based on disequilibrium of the radioactive disintegration series, the procedures for preparing and measuring the samples are similar for the most commonly used $\alpha$-spectrometric analyses (Fig. 6.66).

**Chemical Treatment.** The first step is acid digestion with HCl or $HNO_3$. A $^{232}U/^{228}Th$ or $^{236}U/^{229}Th$ spike and $FeCl_3$ are then added for the isotope dilution analysis (Sect. 5.2.4.1) so that even at very low isotopic abundances in the ppm to ppb range it is not necessary to use quantitative methods. Uranium, thorium, and protactinium are precipitated together with $Fe(OH)_2$. The precipitate is removed by centrifugation, dissolved in HCl and the U, Th, and Pa isotopes separated by ion exchange. The uranium and thorium are then deposited in very thin layers on stainless steel or platinum discs electrolytically or by evaporation.

If conglomerates containing both autochthonous and allochthonous components are being dated, additional steps are often made to separate the autochthonous component, usually without great success. Pelagic sediment that contains terrestrial detritus appears to be an exception; autochthonous thorium has been successfully extracted with Na-EDTA (Mangini and Dominik 1978). Separate analysis of several grain-size fractions and applying specific evaluation procedures (isochron method; Schwarcz 1980; Ku and Liang 1984) appears to be generally more successful than trying to separate the autochthonous component.

**Measurement Techniques.** *$\alpha$-spectrometric techniques:* The activities of $^{238}U$, $^{234}U$, $^{232}Th$, $^{230}Th$, $^{231}Pa$, $^{228}Ra$, and $^{226}Ra$ (about $10^{-5}$ g U) are measured with an $\alpha$-spectrometer, whose most important components are a silicon-surface-barrier detector (Sect. 5.2.2.3) with high energy resolution, a low-noise amplifier, and a multi-channel analyzer. However, scintillation counters (Sect. 5.2.2.2) or grid chambers are also adequate. Moreover, particle-track analysis (Sect. 6.4.7) is another accurate and inexpensive method (Crawford et al. 1985). The measurement takes 1 to 7 days, depending on the desired precision. The results for unconsolidated sediments are given in dpm/g, for hard rocks they are given as isotope ratios. In addition, the $^{234}U/^{238}U$ atom ratio can also be determined mass spectrometrically (Sect. 5.2.3.1).

Several international projects have been conducted to compare uranium series dates in order to identify and eliminate systematic errors. These projects have demonstrated a reproducibility of at least $\pm 2\%$ for the activity ratios (Ivanovich et al. 1984). This does not include, however, the uncertainty resulting from the lack of knowledge of the exact values of the half-lives.

Several methods have been devised to avoid the time-consuming and elaborate radiochemical separation procedures for the traditional $\alpha$-spectrometric analysis.

*$\gamma$-spectrometric techniques:* An elegant way is to measure the $\gamma$-activities of suitable short-lived members of the decay series. This is done with a silicon surface barrier

detector or a well-type intrinsic germanium detector, for which no sample preparation is necessary (Reyss et al. 1978). Since the probability for $\gamma$-decay and the sensitivity of the instrument are generally less than those for $\alpha$-decay, larger samples (15–1000 g) are needed than for the measurement of $\alpha$-activity. Instead of measuring the uranium and thorium directly, the activities of $\gamma$-emitter descendants are measured, such as $^{214}$Bi (coincident 0.609 and 1.120 MeV) or $^{214}$Po for $^{230}$Th and $^{208}$Tl (coincident 0.583 and 2.614 MeV) or $^{212}$Pb for $^{232}$Th. This substitution assumes that radioactive equilibrium existed between the parent and descendant isotopes. But this is not always the case, for example, between mobile radium-226 and bismuth-214, and therefore, the specific activity of $^{226}$Ra must also be measured. When a very high-sensitivity well-type intrinsic germanium detector is used, the low-energy $\gamma$-ray peaks of the isotopes of interest can be measured directly (Kim and Burnett 1983). This is possible even for $^{210}$Pb (Joshi 1987).

To avoid the complicated isotope dilution method using a $^{233}$Pa spike in the $^{231}$Pa dating method (Sect. 6.3.2), the activity of $^{227}$Th, a $\beta$-emitting descendant ($\tau = 18.6$ d) of $^{231}$Pa, can be measured (Mangini and Sonntag 1977), whereby an additional extraction step is also avoided.

*Neutron activation:* A still more sensitive, although more complicated method is neutron activation (Sect. 5.2.4.2). Irradiation of $^{231}$Pa with neutrons converts it to $^{232}$Pa, which decays in a few days to $^{232}$U, which can be easily measured. The simultaneous production of undesired radionuclides impedes the wide use of this method.

The uranium concentration is also often determined by neutron activation, measuring the $\gamma$-activity of $^{239}$Np ($\tau = 2.35$ d, $E_\gamma = 228.278$ keV) that is formed from $^{238}$U. The concentration of $^{235}$U is obtained from the constant ratio $^{238}$U/$^{235}$U $= 137.88$. If high precision is not required, fluorometric analysis (Sect. 5.2.4.3) is suitable.

*Mass spectrometric analysis:* The development of mass spectrometric techniques for the analysis of $^{234}$U, $^{232}$Th, $^{238}$U, and $^{230}$Th has considerably improved $^{234}$U/$^{230}$Th dating. Sample size has been reduced to gram and milligram amounts (about $10^{-8}$ g U), depending on the age and type of material of the sample. This technique has improved dating precision, for example, to $\pm 1\%$ for 100-ka samples and has extended the dating range to about 500 ka. Dating precision for the Holocene is comparable to that of the radiocarbon method (Sect. 6.2.1; Edwards et al. 1986/87). This improvement may result in a revolution in U/Th dating similar to that resulting from the development of AMS (Sect. 5.2.3.2) for dating with cosmogenic isotopes (Sect. 6.2).

*Emanation measurement:* Heye (1975) dated manganese nodules using a proportional counter (Sect. 5.2.2.1) to measure the $\alpha$-emission of $^{222}$Rn and $^{220}$Rn, daughter products of $^{226}$Ra and $^{232}$Th, respectively. Before the measurement, the samples were kept in a hermetically sealed chamber until radioactive equilibrium was attained. At equilibrium, saturation activity is proportional to the activities of the parent members.

### Scope and Potential, Limitations, Representative Examples

The development of the dating methods based on radioactive disequilibrium of the uranium, thorium, and actinium series began with the work of Joly (1908), who assumed that sediments immediately after deposition contain only uranium and that the $^{226}$Ra produced from the $^{238}$U is immobile and thus suitable for age determinations. In the following decades, however, it was found that neither condition is fulfilled and that 25% of the radium in sea water, for example, is derived from redissolution from marine sediments (Kröll 1953).

***Scope and Potential.*** With the advent of sensitive α-spectrometric techniques, and more recently the impact of mass spectrometric analysis, widely applicable and reliable dating methods have been developed for the uranium and thorium decay series. These are today among the most important instruments for chronological studies in Quaternary geology. The best summary is by Ivanovich and Harmon (1982). The application of U/Th analysis to volcanic rocks not only provides information about their geochronology but also about mobilization of uranium and weathering. The economic geologist is finding increasing use for U/Th dates for uranium ore deposits, marine and insular phosphorite, hydrothermal sulfides and oxides, and sulfate precipitates. Detailed climate records for the late Pleistocene have been obtained from U/Th dating of speleothem, spring deposits, pedogenic carbonate, and coral, as well as peat, lignite, and similar organic materials. Studies of late Pleistocene sea-level history have profited from such records, too. U/Th datings of travertine, mollusc shells, bones, and teeth have made significant contributions to archeological and anthropogenetic studies. The marine environment is another field in which U/Th analysis finds considerable application, e.g., growth rates and formation periods of manganese nodules and pelagic sediments, as well as migration of radium in sediments, scavenging of thorium from ocean water, sediment redeposition and accumulation, oceanic currents and upwelling. Similar studies are made on shallow-marine and river sediments. Contributions to hydrogeology are limited to the chemistry of uranium. The radiogenic gases radon and helium also provide only limited geochronological information. Citations are given in the respective chapters.

*Terrestrial Pleistocene chronology:* The first rather reliable chronology of late Quaternary travertine in Central Europe is based on U/Th dates (Brunnacker et al. 1983). This chronology indicates shorter formation periods for travertines than for speleothem (Gascoyne al. 1983). However, both fit well with the globally most representative time record of marine sediments (Sect. 7.2; Fig. 6.67). An important result is the correlation of the terrestrial Eem interglacial with stage 5e of the marine $\delta^{18}$O record (Mangerud et al. 1979). For research on Middle Pleistocene hominid records, the U/Th methods are indispensible (Cook et al. 1982). The Late Pleistocene eustatic sea-level fluctuations in the Bermudas over the past 250,000 years have been reconstructed on the basis of U/Th ages of speleothem and coral, as well as amino-acid racemization analysis (Sect. 8.1) (Harmon et al. 1983).

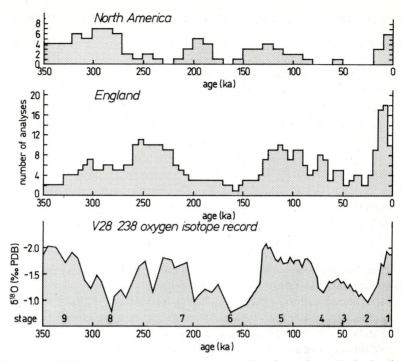

**Fig. 6.67.** Histograms of U/Th dates from uncontaminated speleothem in northwest England and North America (after Ivanovich and Harmon 1982) and correlation with the $\delta^{18}O$ record of planktonic foraminifera from core *V 28-238* (after Shackleton and Opdyke 1973). The histograms reflect the fact that speleothem are deposited mainly during interglacial periods (stages 1,5,7, and 9)

*Marine sediment and oceanographic studies:* Although pelagic sediments are being dated increasingly by the $\delta^{18}O$ method (Sect. 7.2), U/Th dates are still needed for calibration; they are also of interest for studies of erosion and redeposition, upwelling, oxygen consumption, and circulation patterns.

***Limitations.*** *Complex or open systems:* The requirements of the models for the decay-series dating methods are often insufficiently fulfilled or not at all. In complex marine systems the source of $^{230}Th$ (for dating purposes, one of the most important members of the $^{238}U$ series) is not always well known. It enters pelagic sediments, for example, not only from the sea water overlying them, but also via terrestrial, detrital particles, whose $^{230}Th$ activity is unknown. This results in unknown, non-zero initial specific activities of the $^{230}Th$ in samples taken from different cores. It is not always possible to separate the terrestrial component from the marine component (Mangini and Dominik 1978). In such cases, it is necessary to apply other models (e.g., Sect. 6.3.13) which assume, for example, a constant input of radioisotopes in a constant ratio of radioisotope to carrier phase. If the input of radioisotopes or the sedimentation rate changed, the cores must be treated in sections.

Open systems present another complication. Thorium-230 and protactinium-231 may migrate by diffusion, for example, in very slow-growing manganese nodules. This can lead to U/Th ages that are too low by a factor of up to 3 (Ku et al. 1979a). In sediments in a reducing environment, uranium may diffuse too (Yamada and Tsunogai 1984). Mobilization of uranium in rocks and conglomerates has also been detected (e.g., Thiel et al. 1983). When an open system was present, other dating methods should be applied concurrently. In addition, special models must be employed for the evaluation of the data (Rosholt 1967; Szabo and Rosholt 1969; Hille 1979; Szabo 1979). The combined application of ESR analysis (Sect. 6.4.3) with U/Th dating has potential for evaluating the history of the uranium uptake. A fitting procedure should be possible for this correction (Grün et al. 1988).

*Detrital correction:* Detrital inclusions in speleothem and other calcareous formations, such as caliche, usually cannot be removed completely, hindering geochronological interpretation of the results. However, suitable correction procedures based on special models have been developed (Kaufman and Broecker 1965; Schwarcz 1980; Ku and Liang 1984). The most promising of these assume a two-component mixture, of which a number of suitable extracts with differing detrital content must be analyzed (improperly called "isochron method"). These methods have been successfully applied to caliche formations, impure carbonates, case-hardened limestone, and conglomerates that consist of components of differing age (Ku et al. 1979b; Kulp et al. 1979; Ku and Joshi 1981; Ivanovich and Ireland 1984; Schlesinger 1985). However, as none of these methods is entirely satisfactory, samples should be selected that will yield reliable ages with a high probability.

### 6.3.1   $^{230}$Th/$^{234}$U Method***

(Barnes, Lang and Potratz 1956)

#### Dating Range, Precision, Materials, Sample Size

This standard method yields reliable results in the range from several thousand years to $350 \pm 50$ ka. With mass spectrometric techniques, ages of even 500 ka have been measured (Edwards et al. 1986/87). Dating precision decreases with increasing age and decreasing uranium content of the sample. Up to about 250 ka, the usual standard deviation increases from $\pm 1$ to $\pm 10$ ka; for samples above this age it rapidly increases to $\pm 50$ ka (Gascoyne et al. 1983; Cook et al. 1982; Harmon et al. 1983; Ivanovich and Harmon 1982; Schwarcz and Gascoyne 1984). But the precision of the mass spectrometric analysis is better by a factor of 4 to 8 and the sample size decreases by two to three orders of magnitude.

The most reliable data is obtained from coral (several grams) (Osmond et al. 1965; Thurber et al. 1965; Ikeya and Ohmura 1983; Kaufman 1986), manganese nodules (Ku and Broecker 1967a; Heye 1975; Ku et al. 1979a), and foraminiferal tests, ooides, and oolites (Broecker and Thurber 1965) extracted from pelagic sediments (Ku 1965). Mollusc shells yield reliable dates in only about 50% of the cases (Kaufman et al. 1971; Szabo et al. 1981; Ivanovich et al. 1983). The larger and denser the specimen, the more valid the resulting date. Concordant ages have been obtained for marine phosphorite with the $^{231}$Pa/$^{235}$U method (Sect. 6.3.2) (Veeh 1982). Phosphate deposits on islands may be open systems (Roe and Burnett 1985; Burnett and Kim 1986). Accumulation rates on the order of mm/ka to cm/ka can be determined from such data. In addition, marine hydrothermal sulfides and oxides are datable.

The following terrestrial samples are often dated (in order of decreasing suitability): speleothem, travertine (50–100 g), calcareous tuff, inorganic marl, and lacustrine sediments with uranium concentrations above 50 ppb. Pebbles of pedogenic carbonate, which frequently occur in arid and semiarid regions, have been dated successfully despite their complicated genesis (Ku et al. 1979b; Schlesinger 1985). Volcanic rocks and minerals (several grams) (e.g., phyllite, glass, and hornblende) taken from hand specimens up to 10 kg have been dated successfully (Kigoshi 1967; Taddeucci et al. 1967; Allègre and Condomines 1976; Condomines et al. 1982; Schwarcz et al. 1982). The duration of uranium mobilization in the porous zone of hard rocks is determined in the range of 2–800 a using the uranium-trend procedure; the development of soils and alluvial deposits is also studied in this way (Rosholt 1978, 1980). Several 100 g are needed for uranium concentrations below 1 ppm.

Only a few grams are necessary for biogenic, marine samples with a high uranium concentration. Only the compact outer parts of bones (ca. 0.1–5 g) are usable, if at all (Bischoff and Rosenbauer 1981; Rae and Ivanovich 1986). Enamel is not necessarily the best part of a tooth for dating (Bischoff et al. 1988). However, a $^{231}$Pa/$^{235}$U age determination should be carried out for confirmation (Korkisch et al. 1982). Low-bog peat and lignite also appear to be suitable for dating (Titayeva 1966; Vogel and Konfeld 1980; van der Wijk et al. 1986) since uranium can be bound rather irreversibly in them (Kochenov et al. 1965).

Last, but not least, Antarctic ice containing appreciable amounts of dust has been successfully dated, too (Fireman 1986).

The criteria that a sample must fulfill to yield a reliable age are given in Sect. 6.3 (Thurber et al. 1965).

### Basic Concept

It is assumed that the sample contained uranium and no thorium-230 at the time of its formation. When this is the case, the age $t$ of the sample is determined mainly from the specific activity of thorium-230 ($\tau = 75.2$ ka) produced from uranium-234 ($\tau = 248$ ka) using the following equation:

$$\frac{[^{230}\text{Th}]}{[^{234}\text{U}]} = \frac{\lambda_{230}}{\lambda_{230} - \lambda_{234}} \left( 1 - \frac{[^{238}\text{U}]}{[^{234}\text{U}]} \right) (1 - e^{-(\lambda_{234} - \lambda_{230})t})$$

$$+ \frac{[^{238}\text{U}]}{[^{234}\text{U}]} (1 - e^{-\lambda_{230}t}) \tag{6.74}$$

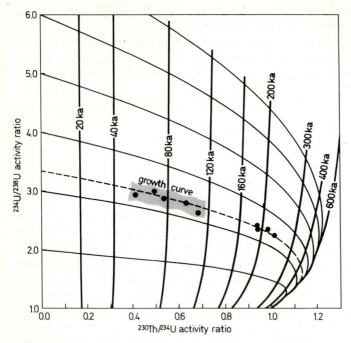

**Fig. 6.68.** Evaluation diagram for U/Th ages: Changes in the activity ratios of $^{234}U/^{238}U$ and $^{230}Th/^{234}U$ as a function of age of a stalagmite (a closed system) containing no initial $^{230}Th$: The values obtained for a stalagmite from a cave in the USA are shown (after Ivanovich and Harmon 1982). The initial $^{234}U/^{238}U$ activity ratio of 3.35 is constant for the entire period of growth, reflecting two hiatuses in speleothem formation (*gray areas*) during glacial periods

Curves derived from this equation for various parameters are shown in Fig. 6.68. The $^{234}U/^{238}U$ activity ratio is plotted versus the $^{230}Th/^{234}U$ activity ratio. The age is read from the intersections of the calculated isochrons. Exact solutions can be obtained by iteration (Bateman equation) or other methods (Catchen 1984). In cold ice, the descendant product $^{226}Ra$ is measured instead of the $^{234}U$ (Fireman 1986).

The graph in Fig. 6.68 is used not only for age determinations, but also for identifying genetically related samples that had the same initial $^{234}U/^{238}U$ activity ratio (Sect. 6.3.4). This is especially important for samples older than 30 ka. Marine carbonates, such as coral, foraminiferal tests, mollusc shells, ooids, and oolites, should have a ratio of $1.14 \pm 0.02$.

Terrestrial carbonates deposited from groundwater show $^{234}U/^{238}U$ activity ratios of up to 10, since $^{234}U$ is preferentially leached out during weathering, but the usual values range from 0.5–3.0 (Kaufmann and Broecker 1965).

When igneous rocks are dated, the model assumes that the initial $^{230}Th/^{232}Th$ activity ratios of all the associated mineral phases were the same and that they are proportional to the $^{232}Th$ concentration measured today (Kigoshi 1967; Schwarcz et al. 1982). But inhomogeneities in the thorium concentration in the melts that formed the minerals place constraints on the reliability of U/Th dating of young volcanic rocks (Capaldi and Pece 1981; Capaldi et al. 1983; see also Hemond and Condomines 1985 and Capaldi et al. 1985).

The model for dating low-bog peat assumes uranium was irreversibly absorbed from the groundwater during growth and that after being covered with new layers, a closed system existed (Kochenov et al. 1965; Titayeva 1966; Vogel and Kronfeld 1980; Vogel 1982).

### Sample Treatment and Measurement Techniques

See Sect. 6.3.

### Scope and Potential, Limitations, Representative Examples

*Scope and Potential.* The $^{230}$Th/$^{234}$U method is widely and routinely used in many laboratories in which analyses for Quaternary geology and anthropology are made (Ivanovich and Harmon 1982; see also Sect. 6.3). The generally high reliability of the ages obtained with this method is demonstrated by the similarity of the chronologies of terrestrial and marine carbonates (Fig. 6.67), which mirror the climatic changes of the last several hundred thousand years (Shackleton and Opdyke 1973). According to the results of an international comparison test, reproducibility is about $\pm 2\%$ (Ivanovich et al. 1984). In spite of this obvious success, there is a great need to internationally standardize the procedure for conventional $^{230}$Th/$^{234}$U ages (Kaufman 1986).

In routine dating such precision can be expected only when appropriate samples are available (Thurber et al. 1965) and when they have been collected properly. A histogram of speleothem data with an apparently greater range of uncertainty (Henning et al. 1983), for example, can be explained by the fact that the samples were incorrectly taken radially instead of along the axis of the stalagmites (Franke 1966).

*Limitations. Open systems:* Even when all of the rules are observed, incorrect data are sometimes obtained. A frequent reason is the presence of an open system, which is often the case with bones, teeth, marine phosphorites (Burnett and Kim 1986), and marine mollusc shells (Kaufman et al. 1971; Ivanovich et al. 1983), all of which acquire uranium in complex, episodic processes. In addition, Rae and Hedges (1989) have demonstrated that under certain circumstances not only uranium but also thorium may become mobile. Cross sampling often yields significantly lower ages than the burial age. For uranium-rich samples, this possibility can be tested by a $^{231}$Pa/$^{235}$U analysis (Sect. 6.3.2) (Hille 1979; Bischoff and Rosenbauer 1981; Ivanovich and Harmon 1982; Korkisch et al. 1982). Even partial mobilization of uranium, e.g., by leaching, accumulation, diagenetic recrystallization, or precipitation of secondary carbonate, may be detected with appropriate procedures for evaluation of the data (Thiel et al. 1983) or should be recognizable in thin section under the microscope.

Leaching of uranium in porous materials is often indicated by anomalously high $^{230}$Th/$^{234}$U activity ratios, anomalously low $^{234}$U/$^{238}$U activity ratios and low uranium concentrations. Under such circumstances, these activity ratios for different samples from the same layer frequently have a negative correlation and the latter ratio has a positive correlation with the uranium concentration.

There have been a number of attempts to develop a model for open systems (e.g., uranium trend dating). The first one to be proposed assumed that all $^{230}$Th and $^{231}$Pa is radiogenic and permanently retained in the sample, whereas the uranium content may change due to leaching by groundwater (Rosholt 1967; Szabo and Rosholt 1969; Szabo 1979; Korkisch et al. 1982, Muhs et al. 1989). Hille (1979) assumed additionally that the radiogenic $^{234}$U is also retained by the sample. Despite several cases in which such models have been successful, it remains questionable whether the models really describe the actual conditions.

*Detrital correction:* Detrital contamination containing allochthonous thorium-230 can usually be recognized by the presence of clay. Neither physical nor chemical separation is found to be successful. It appears to be better to analyze various chemically or physically separated fractions (the latter for the separation of minerals) and then apply the appropriate corrections. This is not necessary, however, if the $^{230}$Th/$^{232}$Th activity ratio is greater than 20 or if no $^{232}$Th is present at all.

Most correction procedures for conglomerates assume a two-component mixture in which (i) the allochthonous component was introduced only at the (short) time of formation and (ii) whose k-value (i.e., the initial $^{230}$Th/$^{232}$Th activity ratio, also called the $^{232}$Th index) is assumed to be the same for all fractions despite the fractions having different allochthonous contents. Common k-values range from 0.5 to 3 for most materials (Kaufman and Broecker 1965). If no detrital uranium is involved and if the mixing occurred before time $t$, the unsupported $^{230}$Th* activity is obtained using

$$[^{230}Th^*] = [^{230}Th_{tot}] - k[^{232}Th]e^{-\lambda_{230}t} \qquad (6.75)$$

In spite of this correction for detritus, U/Th ages obtained for Holocene stalagmites are often too large by several thousand years with no suggestion of any detrital component indicated by the presence of $^{232}$Th (Geyh and Hennig 1986).

*Isochron dating:* More complicated interpretation models employ "isochrons" (mixing lines) to estimate more accurately the detrital contamination with $^{230}$Th (Kigoshi 1967; Bernat and Allègre 1974; Ku et al. 1979b; Schwarcz 1980; Ku and Joshi 1981; Ku and Liang 1984). The applicability of these models has been tested in many comparison trials. Since such corrections can shift the age by as much as 50%, these procedures should be viewed only as a last possibility to obtain an age via multiple analysis of different fractions of one sample layer. These isochron models can be divided into two groups:

– partial dissolution techniques and
– total dissolution techniques.

The partial dissolution techniques are based on a model proposed by Turekian and Nelson (1976). Detritus with a uniform $^{230}$Th/$^{232}$Th activity ratio is admixed

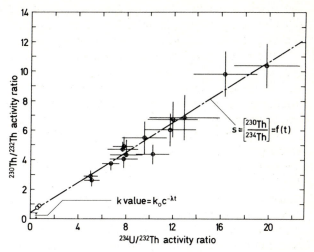

**Fig. 6.69.** Mixing line or "isochron plot" obtained by plotting $[^{230}\text{Th}/^{232}\text{Th}]$ vs. $[^{234}\text{U}/^{232}\text{Th}]$ of coeval samples with an age $t$ (after Ivanovich and Harmon 1982). The slope yields the age (in this case $75 \pm 5$ ka). Samples with a high detritus content yield points to the left and others plot to the right. The y-intercept is a good estimate of the present k-value, which corresponds to the $[^{230}\text{Th}/^{232}\text{Th}]$ of the detritus in the sample. Because the numerator is the same for both axes, the correlation coefficient is always exaggerated

at one time, usually at the time of formation of the sample, in different proportions in the different mineralogical or chemical fractions. Leaching of the uranium can have occurred if this involved no fractionation of the $^{234}$U and $^{238}$U, which of course is questionable (Cherdyntsev 1955).

Combining Eqs. (6.74) and (6.75) and modifying slightly, a graphic solution is obtained which yields an isochron that is actually a simple mixing line (Fig. 6.69). The age is given by the gradient in a plot of $^{230}$Th/$^{232}$Th versus $^{234}$U/$^{232}$Th. The y-intercept corresponds to the $^{232}$Th index at the present time, which can be extrapolated back to time zero using the radioactive decay law (Sect. 5.1). A slope of 0 corresponds to an age of zero and approaches one for an age of infinity.

A plot of $^{234}$U/$^{232}$Th versus $^{238}$U/$^{232}$Th yields the initial $^{234}$U/$^{238}$U activity ratio, which should lie within certain ranges for certain types of carbonates. It is recommended to measure the $^{226}$Ra/$^{232}$Th activity ratio too, as it reflects whether a closed system was present.

The total dissolution techniques involve two assumptions (Ku and Liang 1984). Ku et al. (1979b) assume radioactive equilibrium of $^{238}$U and descendant products in the detrital component and that no fractionation of the thorium isotopes occurred during diagenesis or during the chemical preparation of the sample. Using three equations, the age can then be calculated from the measured activity ratios of the solution and the residue that is insoluble in 2N $\text{HNO}_3$.

Szabo (1979) and Szabo and Butzer (1979) introduced a model in which radioactive equilibrium in the detritus is not assumed. This model, however, does require that the $^{234}$U/$^{230}$Th activity ratio of the solution and that of the acid-insoluble residue have not been changed during chemical handling of the sample. The age is obtained using an isochron plot.

U/Th analysis of $^{14}$C-dated samples can be helpful if the isochron method cannot be applied but the $^{234}$U/$^{238}$U activity ratio must be known for dating (Hillaire-Marcel et al. 1986).

### Non-Chronological Applications

*Igneous rocks:* The $^{230}$Th/$^{234}$U activity ratios of lava samples are used to obtain information about the source rock: whether it originated from the mantle, from subducted oceanic crust, or from terrestrial sediment. The U/Th activity ratios differ greatly among these possibilities (Newman et al. 1983, 1984a, b). Condomines et al. (1982) studied the magmatic evolution of Mt. Etna on the basis of differences in the $^{230}$Th/$^{232}$Th activity ratio of samples representing different stages of development of the caldera during the late Pleistocene (Fig. 6.70). A summary has been published by Condomines et al. (1988).

Another application deals with the mobilization of chemical species in igneous rocks, an important question for the siting of permanent repository sites for radioactive wastes (Schwarcz et al. 1982).

*Marine studies:* Studies of $^{230}$Th in cores of pelagic sediments from the equatorial North Pacific suggest that changes in the bottom currents modified the accumulation pattern of the sediments and that erosion and redeposition both occur within small areas of the ocean floor. Bottom-water flow was obviously faster during glacial and transition periods than during interglacials (Mangini et al. 1982).

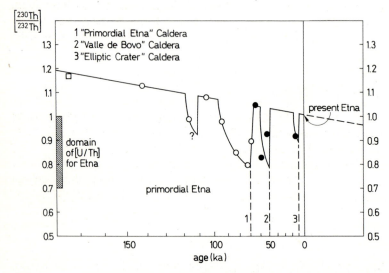

**Fig. 6.70.** Changes in the initial $^{230}$Th/$^{232}$Th activity ratios of lava samples from Mt. Etna as a function of age during the late Pleistocene, reflecting the source of the magma. Two explanations are the mixing of two magmas or the mixing of lava from an alkali magma reservoir with the country rocks (after Condomines et al. 1982)

## 6.3.2   $^{231}$Pa/$^{235}$U Method*

(Sackett 1960)

### Dating Range, Precision, Materials, Sample Size

This method is used less than the $^{230}$Th/$^{234}$U method (Sect. 6.3.1) because $^{235}$U and, therefore, its descendant radionuclide $^{231}$Pa are much rarer by two orders of magnitude. The most successful applications have been the dating of marine carbonate (several grams), such as coral, ooids, and oolites, as well as manganese nodules and marine phosphorites (Veeh 1982), all of which have a relatively high uranium content (Ku and Broecker 1967a). Speleothem and travertine are seldom suitable as the uranium concentration should exceed several ppm. The dense cortex of bones (ca. 20 g) has also been dated successfully (Bischoff and Rosenbauer 1981; Korkisch et al. 1982). Minium ages of volcanic rocks have been determined (Cherdyntsev et al. 1968). The dating range extends from several thousand years to 150 ka, with a maximum of 200 ka. The standard deviations of these ages range from $\pm 10$ to 20% (Sakanoue et al. 1967; Mangini and Sonntag 1977).

### Basic Concept

Analogously to the $^{230}$Th/$^{234}$U method, it is assumed that the system to be dated contained uranium and no protactinium at time zero. Protactinium-231 ($\tau = 34\,300$ years) is formed from uranium-235 ($^{238}$U/$^{235}$U $= 137.5$) after one, very short-lived intervening decay product. The age ($t$) is obtained from the activity of radiogenic $^{231}$Pa using the following equation: $[^{231}\text{Pa}]/[^{235}\text{U}] = 1 - e^{-\lambda_{231}t}$        (6.76)

Protactinium behaves geochemically similarly to thorium and is, for example, almost insoluble in water. For this reason, coral, foraminiferal tests, mollusc shells, ooides, and oolites, as well as secondary terrestrial calcareous rocks (e.g., speleothem) and travertine, initially contain no protactinium and any $^{231}$Pa present really is produced from the initial uranium (Sect. 6.3).

### Sample Treatment and Measurement Techniques

The uranium-235 activity is calculated from the uranium-238 activity, which can be measured in any of various ways (Sect. 6.3). This can be done because the $^{238}$U/$^{235}$U activity ratio in nature is generally a constant 21.7.

Protactinium-231 can be determined with an $\alpha$-spectrometer after addition of a $^{233}$Pa spike for isotope dilution analysis (Sect. 5.2.4.1). Other, more simple procedures are described in Sect. 6.3 and by Ivanovich and Harmon (1982).

### Scope and Potential, Limitations

Scope and Potential. Owing to the generally low activity of uranium-235, $^{231}$Pa/$^{235}$U data are generally less precise than the corresponding $^{230}$Th/$^{234}$U data. For this reason only uranium-rich samples are usually dated, hence, speleothem are not promising samples.

*Limitations.* Bone and mollusc shells, which take up uranium from the groundwater in the course of their fossilization (Sect. 8.5), are suitable only if the process was completed within a few thousand years (Fanale and Schaeffer 1965). According to Szabo et al. (1981), minium ages are obtained for those cases in which diagenetic alteration, e.g., recrystallization, occurs because $^{231}$Pa, in contrast to $^{230}$Th, may be lost. The combined application of both this method and the $^{230}$Th/$^{234}$U method is used to identify the presence of such post-depositional changes on the basis of a difference in the resulting dates (Hille 1979; Bischoff and Rosenbauer 1981; Ivanovich and Harmon 1982; Korkisch et al. 1982).

### 6.3.3  $^{231}$Pa/$^{230}$Th Method**

(Rosholt 1957)

#### Dating Range, Precision, Materials, Sample Size

This method, like the $^{231}$Pa/$^{235}$U method (Sect. 6.3.2), is seldom used (Ku 1976), because only uranium-rich marine carbonate (several grams of coral or mollusc shells) gives counting rates that are high enough to yield a reasonable dating precision. Ages range from several thousand years to 200 ka (Rosholt et al. 1961) with standard deviations of $\pm$ 10–20%.

#### Basic Concept

The $^{231}$Pa/$^{230}$Th method combines the $^{230}$Th/$^{234}$U (Sect. 6.3.1) and the $^{231}$Pa/$^{235}$U (Sect. 6.3.2) methods. It assumes the sample initially contained uranium but no $^{231}$Pa or $^{230}$Th. The age (*t*) is obtained by dividing Eq. (6.76) by Eq. (6.74):

$$\frac{[^{231}\text{Pa}]}{[^{230}\text{Th}]} = \frac{[^{235}\text{U}/^{238}\text{U}](1 - e^{-\lambda_{231}t})}{(1 - e^{-\lambda_{230}t}) + \dfrac{\lambda_{230}}{\lambda_{230} - \lambda_{234}}([^{234}\text{U}/^{238}\text{U}] - 1)(1 - e^{-(\lambda_{230} - \lambda_{234})t})},$$

(6.77)

where $[^{235}\text{U}/^{238}\text{U}] = 21.7$ for natural uranium.

The $^{231}$Pa/$^{230}$Th activity ratio as a function of age is shown in Fig. 6.65. An initial $^{234}$U/$^{238}$U activity ratio of 1.14 is used for marine carbonate.

#### Sample Treatment and Measurement Techniques

See Sects. 6.3 and 6.3.2.

#### Scope and Potential, Limitations

*Scope and Potential.* This method is used as a control to determine whether a system was closed or open. Diagenetic alterations resulting in a gain or loss of uranium cause little change in the $^{231}$Pa/$^{230}$Th activity ratio since geochemically these two elements behave similarly. Only the $^{235}$U/$^{238}$U activity ratio is assumed to be constant, which is quite likely. Since $^{231}$Pa can be lost during recrystallization (Szabo et al. 1981) while $^{230}$Th is retained, the combined application of several U/Th methods provides information on the reliability of the dates, i.e., whether an open or a closed system was present (Hille 1979; Bischoff and Rosenbauer 1981; Ivanovich and Harmon 1982).

**Limitations.** Owing to the low uranium-235 abundance, $^{231}$Pa/$^{235}$U data are generally less precise than corresponding $^{230}$Th/$^{234}$U data. In marine sediments, the $^{231}$Pa content is lower by a factor of about 30 than that of $^{230}$Th. Bone and mollusc shells, which take up uranium after deposition, are suitable only if they formed a closed system during fossilization.

## 6.3.4   $^{234}$U/$^{238}$U   Method**

(Cherdyntsev 1955, Thurber 1962)

### Dating Range, Precision, Materials, Sample Size

This method is sometimes used to date unrecrystallized marine mollusc shells and coral (several grams) (Thurber 1963; Osmond et al. 1965; Veeh 1966), lacustrine and pelagic sediments (50–100 g) (Ku 1965), as well as speleothem (Thompson et al. 1975), with a uranium concentration of more than 10 ppb in the age range of 50 ka to 1.5 Ma. The dating precision is $\pm$ 5–20%. The reliability of the dates is checked using the K/Ar method (Sect. 6.1.1).

Attempts to determine groundwater ages (15–60 L containing at least 0.01 ppb U) from the $^{234}$U/$^{238}$U activity ratio have not yet been successful. But such studies have yielded valuable information on the hydrogeological history of the aquifer, including hydrodynamic mixing (Osmond et al. 1974; Osmond and Cowart 1976; Osmond et al. 1983). New models show promise (e.g., Fröhlich et al. 1984; Fröhlich and Gellermann 1986).

### Basic Concept

Uranium-234 is produced in three steps from uranium-238 by emission of one $\alpha$-particle and two $\beta^-$-particles. In solids, the lattice is damaged at the places where uranium-238 atoms decayed. The $^{234}$U atom is bound less tightly than the $^{238}$U and is in the $U^{6+}$ state more easily dissolved (Petit et al. 1985). This is the main reason why it is enriched in fresh and sea water relative to $^{238}$U, producing radioactive disequilibrium in the oceans, which contain about 3.3 ppb uranium with a worldwide, uniform value of $1.14 \pm 0.02$ for the $^{234}$U/$^{238}$U activity ratio. In groundwater this ratio has a wider range (0.5 to 40) at uranium concentrations of 0.1 to 25 ppb; the most common values are between 0.5 and 6. Often, the

$^{234}$U/$^{238}$U activity ratio and the uranium concentration in groundwater are inversely proportional.

The decrease in excess $^{234}$U activity is given by the following equation:

$$\frac{[^{234}U]}{[^{238}U]} - 1 = \left( \frac{[^{234}U]}{[^{238}U]_0} - 1 \right) e^{-\lambda_{234}t}, \tag{6.78}$$

where zero indicates initial conditions.

### Sample Treatment and Measurement Techniques

Uranium is extracted from groundwater either by precipitation with $Fe(OH)_3$ or aluminum phosphate or by adsorption on a strongly basic anion exchange resin (Gascoyne 1981). The activities of the uranium isotopes are determined alpha-spectrometrically. Details are given in Sect. 6.3.

### Scope and Potential, Limitations, Representative Examples

**Scope and Potential.** This method covers the widest dating range of the radioactive decay series, which is why it is of great interest to anthropologists and Quaternary geologists although relatively inaccurate dates must be accepted.

**Limitations.** The main problem in applying this method to the dating of terrestrial samples is the lack of exact knowledge of the initial $^{234}$U/$^{238}$U activity ratio, which is known only for marine samples. The initial $^{234}$U-excess in lacustrine limestone is locally quite variable and can have changed during the time of formation of the limestone. This can also have occurred during the growth of a stalagmite, although Thompson et al. (1975) did not find any such effect for cases in which growth was continuous. The initial $^{234}$U/$^{238}$U activity ratio of young limestone can be determined by U/Th analysis of $^{14}$C-dated samples (Hillaire-Marcel et al. 1986). Uranium mobilization can result in post-depositional fractionation of $^{234}$U and $^{238}$U (Ku 1985), which causes incorrect dates. Only aragonitic coral that has not been recrystallized seems to be unaffected.

**Representative Examples.** *Limnological applications:* Chalov et al. (1964) estimated the mean residence time of the water in a closed lake and a drainage basin on the basis of the differences in the uranium isotope activity ratios between the lake water and the river water entering the lake. The results were in good agreement with those of conventional methods (Ku 1976).

*Hydrogeological applications:* The dating of groundwater has not been successful, mainly due to complicated and not yet fully understood chemical reactions of uranium with the rocks of the aquifer. Disequilibrium models have been used to explain these processes (Osmond and Cowart 1976; Osmond et al. 1983; Chalov 1983). Obviously, the geochemistry of uranium is determined more by variations in the flow path caused by solid-liquid interface phenomena than by age (Andrews

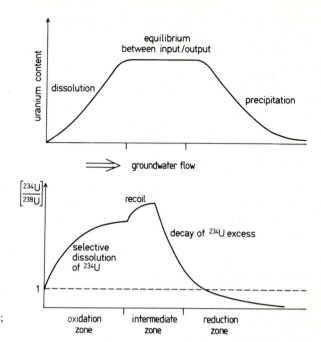

**Fig. 6.71.** Illustration of the changes in the uranium concentration and $^{234}$U/$^{238}$U activity ratio in the oxidizing, transition, and reducing zones of an aquifer (after Osmond et al. 1983; Pearson et al. 1983; Fröhlich et al. 1984)

et al. 1982). Only brines from a salt dome in New Mexico have yielded reasonable $^{234}$U/$^{238}$U ages (Barr et al. 1979).

Three zones of differing uranium chemistry can be distinguished in geologically uniform aquifers if the groundwater movement can be described by the piston-flow model. These zones are determined by the redox potential (Eh) of the environment and the $CO_2$ concentration in the groundwater.

In the oxidizing zone near the catchment area, the uranium is present in the chemically stable, highly soluble + 6 oxidation state and acts as a conservative tracer. Leaching of uranium from the rocks of the aquifer results in a nearly linear increase in uranium concentration and a rather constant $^{234}$U/$^{238}$U activity ratio near unity (Pearson et al. 1983).

In the transition zone (downstream from the oxidizing zone), both this ratio and the uranium concentration pass through a maximum (Fig. 6.71).

Further downstream, in the reducing zone, the $^{234}$U/$^{238}$U activity ratio decreases slowly and linearly as a function of residence time, while the uranium concentration remains relatively constant at a very low level (Pearson et al. 1983). This is due either to α-recoil ejection of $^{234}$Th (from the disintegration of $^{238}$U) at the rock/water interface, which then decays within a short time ($\tau = 24.1$ d) to $^{234}$U (Fleischer 1982) or, more likely, it is due to preferential leaching of $^{234}$U relative to $^{238}$U due to lattice radiation damage (Hussain and Krishnaswami 1980; Andrews et al. 1982). This preferential leaching is explained by recent work as being mainly due to different solubilities of the preferred oxidation states of the two isotopes (Petit et al. 1985). The most important argument for leaching is that the short residence time of the intermediate product $^{234}$Th in the water is not long enough for it to be responsible.

Much more must be known about the details of the uranium adsorption behavior and the $^{234}$U transfer mechanism in an aquifer (e.g., retardation of the uranium, its precipitation, dissolution, and matrix diffusion) before more than the upper limits for the actual age of the water can be estimated (Dickson and Davidson 1985). However, other authors are more optimistic (Pearson et al. 1983; Fröhlich et al. 1984; Fröhlich and Gellermann 1986).

### Non-Chronological Applications

*Hydrologic applications:* The uranium isotope ratios in water samples taken from joints in the rock are used to study mixing processes and flow regimes (Osmond and Cowart 1976; Guttman and Kronfeld 1982). Monitoring of uranium isotope ratios in groundwater for predicting earthquakes has had questionable success (Finkel 1981).

*Lithological applications:* In the case of debris, $^{234}$U/$^{238}$U activity ratios of less than one indicate that its source is weathered rock (Koide and Goldberg 1983). Genetic differences between uranium deposits have also been determined by measuring uranium activity ratios (Osmond et al. 1983).

*Anthropogenic applications:* Anomalous uranium activity ratios (i.e., > 1) were used to identify debris from the aborted Soviet Cosmos 954 satellite in ice. An application that is gaining in importance is concerned with the establishment of permanent repositories for radioactive wastes. Especially uranium adsorption and transfer mechanisms can be studied with this method (Dickson and Davidson 1985).

### 6.3.5  $^{230}$Th-excess Method**

(Pettersson 1937; Piggot and Urry 1942)

### Dating Range, Precision, Substances, Sample Size

This method is used for approximate determination of deep-sea sedimentation rates (several grams of sample) (Ku and Broecker 1967b; Bernat and Allègre 1974; Müller and Mangini 1980; Mangini and Key 1983) and growth rates of manganese nodules (30–300 mg) (Ku and Broecker 1967a; Ku et al. 1979a; Krishnaswami et al. 1982) up to ages of about 300 ka. The precision is better than $\pm 20\%$.

### Basic Concept

Thorium precipitates much more rapidly than uranium from the water column above the sea floor. Thus, thorium-230 ($\tau = 75,200$ a), a descendant radionuclide of

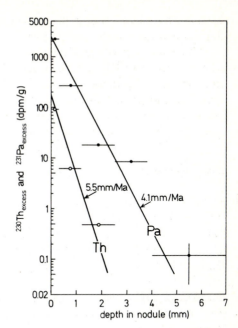

**Fig. 6.72.** Semilog plot of the depth of the specific activities of $^{230}$Th$_{excess}$ and $^{231}$Pa$_{excess}$ in manganese nodule Mn-139 for estimating growth rate; diffusion is neglected. Assuming an effective diffusion coefficient of $^{230}$Th and $^{231}$Pa of $0.95 \times 10^{-8}$ cm$^2$/a, a growth rate of 1.9 mm/Ma is obtained (after Ku et al. 1979a)

uranium-238, is incorporated into pelagic sediments with an excess specific activity (unsupported activity). This $^{230}$Th$_{excess}$ activity decreases faster than the $^{230}$Th is formed from the $^{234}$U (supported activity) in the sediment.

If the sedimentation rate $r$ (cm/a) and $A_0$ (specific activity of $^{230}$Th$_{excess}$) are constant during deposition, Eq. (6.79) describes the specific excess activity ($A^*$) of the unsupported $^{230}$Th [Eq. (6.80)] at a depth $d$ (cm) from the top of the sediment:

$$\ln A^* = \ln A_0 - (\lambda_{230}/r)d \tag{6.79}$$

The age, $t$, of a specific layer is given by $t = d/r$. A semi-log plot of specific activity versus depth yields a straight line whose y-intercept yields $A_0$ and whose slope yields the value of $r$ (Fig. 6.72).

The specific activity of unsupported $^{230}$Th$_{excess}$ is the difference between the total specific thorium-230 activity and the specific activity of the $^{234}$U [Eq. (6.80)], assuming radioactive equilibrium between $^{234}$U, $^{238}$U and $^{230}$Th.

$$[^{230}\text{Th}_{excess}] = [^{230}\text{Th}_{tot}] - [^{234}\text{U}] \tag{6.80}$$

The age is determined independently of the bulk sedimentation rate (Sect. 6.3.6) if the input of thorium-230 can be assumed to be constant. This often seems to be the case for pelagic sediments (Müller and Mangini 1980) and manganese nodules (Ku et al. 1979a).

### Sample Preparation and Measurements

See Sect. 6.3.

*Scope and Potential, Limitations*

*Scope and Potential.* The application of this method to pelagic sediments has been successful in only a few cases because $A_0$ apparently often changes with the rate of sedimentation (Ku 1976).

*Limitations.* The sedimentation process is, in general, more complicated than assumed in this simple model. Post-deposition sediment transport by erosion, turbidity currents, etc. must be taken into account. The $^{230}Th_{excess}/^{232}Th$ method (Sect. 6.3.7) and the $^{231}Pa_{excess}/^{230}Th_{excess}$ method (Sect. 6.3.8) do this, at least partly, for which the preparation and measurement are not significantly more complicated.

In manganese nodules, in addition to changes in $A_0$, $^{230}Th$ can migrate by diffusion (Ku et al. 1979a; Mangini et al. 1986) causing apparent ages that may be too small by up to a factor of 3. Attempts to use diffusion-decay models to correct for this have been made. Moreover, the spatial variation of $^{230}Th_{excess}$ can be large in manganese nodules (Sharma et al. 1984).

### Non-Chronological Applications

*Oceanographic studies:* Records of $^{230}Th_{excess}$ and $^{231}Pa_{excess}$ fluxes into pelagic sediments can be used for studying paleo-upwelling. Glacial periods correlate with maximum fluxes of these nuclides (Mangini and Diester-Haass 1983). Differences in the scavenging activity in the various oceans are reflected in the $^{230}Th_{excess}$ flux into the sediments (Mangini and Key 1983).

Despite large variations in sedimentation rates caused by abyssal currents, the varying content of the terrigenous component in glacial and interglacial sediments has been determined and correlated with the contents of various transition metals (Bacon and Rosholt 1982).

The transport of $^{230}Th$ from the ocean margins to the open sea differs from that of $^{231}Pa$. This is explained by differences in their mixing patterns (Anderson et al. 1983).

*Lithological studies:* The uranium and its descendant isotopes in basalt were found to be not in equilibrium. This disequilibrium has been used to distinguish genetically different magmas (Somayajulu et al. 1966).

### 6.3.6  $^{231}$Pa-excess Method*

(Sackett 1960)

### Dating Range, Precision, Materials, Sample Size

Pelagic sediments (Müller and Mangini 1980), coral (Ku 1968), and manganese nodules (Ku and Broecker 1967a; Ku et al. 1979a) can be roughly dated up to about 150 ka. The precision is generally less than that of $^{230}Th_{excess}$ dates (Sect. 6.3.5).

## Basic Concept

Analogous to the $^{230}$Th$_{excess}$ method, the $^{231}$Pa$_{excess}$ method is based on the difference in precipitation rates of protactinium and uranium from the water column above the seafloor. Protactinium-231 ($\tau = 34{,}300$ a), a descendant of uranium-235, is rapidly incorporated into the pelagic sediments. The specific activity of its unsupported excess decreases according to the radioactive decay law because it is replaced by decay of $^{235}$U at a lower rate. Calculation of $^{231}$Pa$_{excess}$ analogous to Eq. (6.80) is necessary, radioactive equilibrium between $^{235}$U and $^{231}$Pa being here assumed. The enrichment of $^{231}$Pa in manganese nodules is less than that of $^{230}$Th.

The age determination is done analogously to that of the $^{230}$Th-excess method according to Eq. (6.79) (Fig. 6.72). A procedure independent of sedimentation rate which assumes a constant input of $^{231}$Pa$_{excess}$ yields the age from the total unsupported $^{231}$Pa$_{excess}$ activity at the sediment surface.

## Sample Treatment and Measurement Techniques

See Sect. 6.3.

## Scope and Potential, Limitations

The reliability of $^{231}$Pa$_{excess}$ ages is generally relatively low because the input of $^{231}$Pa$_{excess}$ often varies. Moreover, protactinium migrates by diffusion (see also Sect. 6.3.3). More reliable results may be expected from the $^{230}$Th$_{excess}$/$^{232}$Th (Sect. 6.3.7) and $^{231}$Pa$_{excess}$/$^{230}$Th$_{excess}$ methods (Sect. 6.3.8), for which the preparation and measurement are not significantly more complicated.

## Non-Chronological Applications

Sec Sect. 6.3.5.

## 6.3.7   $^{230}$Th-excess/$^{232}$Th or $^{230}$Th/$^{238}$U Method

(Picciotto and Wilgain 1954)

## Dating Range, Precision, Materials, Sample Size

This method is used routinely for determining the age of marine carbonate and manganese nodules (Krishnaswami et al. 1982), glass shards in layers of volcanic ash (Dymond 1969), fish teeth and fish bones, and lacustrine sediments that contain clay minerals (Bernat and Allègre 1974) (several g, respectively) with ages between several thousand years and 300 ka.

Igneous rocks can be dated via the thorium isotope ratios of several cogenetic accessory minerals (e.g., zircon, ilmenite, magnetite, augite, feldspar, and apatite) (Allègre 1968; Capaldi et al. 1983) with ages up to 1 Ma, as well as phosphate deposits (Roe et al. 1983).

### Basic Concept

For sedimentary rocks this method overcomes the difficulty of the $^{230}\text{Th}_{\text{excess}}$ method (Sect. 6.3.5) that arises if the initial $^{230}\text{Th}_{\text{excess}}$ ($\tau = 75,400\,\text{a}$) varied during sedimentation. In this method, the input of $^{230}\text{Th}$ is assumed to correlate with the input of detrital $^{232}\text{Th}$, which, of course, is not always the case (Somayajulu et al. 1983/1984). If this assumption is fulfilled, variations in the sedimentation rate do not influence dating. The two Th isotopes are precipitated in the same proportions. A rather constant U concentration in sea water is another prerequisite of the $^{230}\text{Th}_{\text{excess}}$ method that need not be fulfilled for the $^{230}\text{Th}_{\text{excess}}/^{232}\text{Th}$ method.

Dating of volcanic rocks is possible due to isotopic fractionation of parent and daughter isotopes in a magma (Taddeuci et al. 1967). If the mineral phases with different Th/U ratios had the same initial $^{230}\text{Th}/^{232}\text{Th}$ and $^{234}\text{U}/^{238}\text{U}$ activity ratios at the moment of eruption, the age $t$ can be calculated for a closed system containing two cogenetic minerals $i$ and $j$ (Ku 1976) as follows:

$$t = \frac{1}{\lambda_{230}} \ln \left\{ 1 - \frac{[^{230}\text{Th}_{\text{excess}}/^{232}\text{Th}]_i - [^{230}\text{Th}_{\text{excess}}/^{232}\text{Th}]_j}{[^{238}\text{U}/^{232}\text{Th}]_i - [^{238}\text{U}/^{232}\text{Th}]_j} \right\}^{-1} \tag{6.81}$$

This equation is usually solved graphically (an "isochron" plot, Sect. 6.3.1). The $^{230}\text{Th}_{\text{excess}}/^{232}\text{Th}$ activity ratio is plotted on the $y$-axis and the $^{238}\text{U}/^{232}\text{Th}$ activity ratio along the $x$-axis. The age is given by the slope, which is equal to $(1 - e^{-\lambda_{230}t})$ (Bernat and Allègre 1974). Note that in constrast to the isochrons used for Rb/Sr (Sect. 6.1.3), the isochron rotates around an equipoint. The slope of this line never exceeds one. If it equals one the mineral or rock system is in secular equilibrium ($[^{230}\text{Th}/^{238}\text{U}]$). The range of the slope limits the use of this method to ages between thousand and one million years.

### Preparation of Samples and Measurements. See Sect. 6.3.

### Scope and Potential, Limitations, Representative Examples

The main applications of this method are in oceanography and in the study of recent volcanic rocks. The fact that different sediment fractions often have different origins is, of course, contrary to the prerequisite that the ratio of $^{230}\text{Th}_{\text{excess}}$ and $^{232}\text{Th}$ input has been constant (Bernat and Allègre 1974). In addition to the unsupported $^{230}\text{Th}_{\text{excess}}$, which precipitates rapidly from the sea water (residence time of 5–40 years), there is input of terrestrial detritus, which constains both $^{230}\text{Th}$ and $^{232}\text{Th}$ (Ku 1976). Thus, in each case the extent to which the sediment is genetically uniform must be checked along the entire length of the core; this is the most important prerequisite for a reliable age determination. The presence of detrital components may be reflected by a $^{230}\text{Th}_{\text{excess}}/^{231}\text{Pa}_{\text{excess}}$ ratio exceeding the theoretical value of 10.8 for the oceans (Ku and Broecker 1976b). A case study of volcanic rocks has been published by Roe et al. (1983).

## Non-Chronological Applications

*Lithological studies:* Capaldi et al. (1983) studied coeval mineral phases of lavas from the Lipari Islands (Isole Eolie) in the Tyrrhenian Sea and found that the "felsic Eolian rocks evolved chemically in different magma chambers" and that different mineral phases in these rocks appear to be in radioactive equilibrium with each other. The time of fractional crystallization was calculated on the basis of these findings.

*Manganese and phosphorite nodules:* The dicordant growth rates of manganese nodules derived from $^{230}$Th$_{excess}$ and $^{10}$Be (Sect. 6.2.3) concentration profiles in the nodules document temporal variations in nodule growth rate rather than oceanographic mixing effects (Krishnaswami et al. 1982). A formation rate of millimeters per thousand years has been measured for phosphorite nodules, which is two to four orders of magnitude faster than the accumulation of the underlying sediment (Burnett et al. 1982).

### 6.3.8  $^{231}$Pa-excess/$^{230}$Th-excess Method**

(Sackett 1960; Rosholt et al. 1961)

### Dating Range, Precision, Materials, Sample Size

This method is used for dating pelagic sediments (several grams) up to an age of about 150 ka (Ku and Broecker 1967b). In addition, the alteration between accumulation and erosion resulting from changes in the circulation of deep sea water are reconstructed on the basis of variations in the specific activities of the unsupported excess of $^{230}$Th and $^{231}$Pa along sediment profiles (Mangini et al. 1982; see also Sect. 6.3.7).

### Basic Concept

Because $^{230}$Th and $^{231}$Pa precipitate from sea water after short residence times, chemical and isotopic fractionation can be excluded. Assuming constant concentrations of $^{238}$U, $^{234}$U, and $^{235}$U in the oceans, the ratio of $^{231}$Pa production to $^{230}$Th production is 0.091. The age of a sediment sample is then given by

$$\frac{[^{231}Pa_{excess}]_t}{[^{230}Th_{excess}]_t} = \frac{[^{231}Pa_{excess}]_0}{[^{230}Th_{excess}]_0} e^{-(\lambda_{231} - \lambda_{230}) \cdot t} \tag{6.82}$$

This activity ratio has an effective half-life of 58 ka. The excess activities of the unsupported thorium-230 and protactinium-231 are determined separately using equations analogous to Eq. (6.80). The $^{231}$Pa activity is lower than that of $^{230}$Th by a factor of about 30.

### Sample Treatment and Measurement Techniques. See Sect. 6.3.

## Scope and Potential, Limitations

*Scope and Potential.* The fundamental assumptions of the dating model are not affected by changes in the circulation of the ocean water, by fluctuations in the bulk sedimentation rate, or by the uranium concentration. Only constant isotopic abundances of uranium in the sea water are assumed. As this seems to be the case, reliable ages are to be expected from this method.

*Limitations.* Isotopic fractionation of $^{231}$Pa and $^{230}$Th during precipitation can be a problem, leading to variations in the apparent production rates of the two isotopes. The ratio of these production rates is generally taken to be a constant 0.091 (Broecker and Ku 1969; Yang et al. 1986). In order to check for this, the isotopic activity ratios are plotted analogously to the manner described in Sect. 6.3.7. The average sedimentation rate is given by the slope of the line through the data (Bacon and Rosholt 1982), from which the age is obtained with a relatively small standard deviation.

## Non-Chronological Applications

*Marine sediments:* Areas and periods of increased upwelling can be identified by comparison of the production and deposition fluxes of $^{230}$Th$_{excess}$ and $^{231}$Pa$_{excess}$, which are up to a factor of 4 greater in pelagic sediments than could have been produced in the water column above them (Mangini and Diester-Haass 1983). Maximum fluxes occurred during glacial periods. Moreover, anomalous [$^{231}$Pa-excess/$^{230}$Th-excess] values have been determined for parts of the oceans (Yang et al. 1986).

## 6.3.9  $^{234}$Th-excess Method

(Hussain and Krishnaswami 1980)

### Dating Range, Precision, Materials, Sample Size

This method is an aid in the study of short-term (seasonal) reworking and diagenesis of shallow-water near-shore sediments (1–5 g) in the range of up to 100 days (Coale and Bruland 1985). It is also used to estimate particle mixing rates (Martin and Sayles 1987).

### Basic Concept

Thorium-234 ($\tau = 24.1$ d) is the daughter isotope of uranium-238 and like all other thorium isotopes, is rapidly removed from the water column via scavenging by particulate matter. For this reason, shallow-water sediments show unsupported excess thorium-234 activities, which can be used for age determinations in analogy to the $^{230}$Th-excess method (Sect. 6.3.5).

### Sample Treatment and Measurement Techniques

Details are given in Sect. 6.3 and by Ivanovich and Harmon (1982).

### Scope and Potential, Limitations, Representative Example

*Scope and Potential.* This method has been seldom used by geoscientists owing to the limited interest in very short-term sedimentation processes. However, sedimentation and pollution scavenging studies in estuaries and coastal zones will profit from the use of this method (Coale and Bruland 1985).

*Limitations.* Since the requirement of uniform sedimentation rates is not fulfilled everywhere, the dates can often be used only for orientation.

*Representative Example.* A two-dimensional model of advection and sediment transport has been tested using $^{234}$Th measurements on samples from the continental shelf off the mouth of the Yangtze River. Most of the thorium isotopes are transported in particulate matter. Residence times of 0.3 to 11 days were calculated (McKee et al. 1984). The seasonal variability of both mixing and sediment irrigation rates was estimated for a site in Buzzards Bay, Mass., and biologically driven transport mechanisms were studied (Martin and Sayles 1987).

### Non-Chronological Applications

Short-term particle dynamics near the deep-sea floor has been studied by measuring the specific $^{234}$Th activity of suspended particles. The results show that this radionuclide is a sensitive indicator of particle input and dynamics at the deep-sea floor. The calculated $^{234}$Th mixing rates agree with those obtained with the $^{210}$Pb method (Sect. 6.3.13). Complications are caused by diffusion of dissolved $^{234}$Th (Aller and DeMaster 1984).

## 6.3.10  $^{228}$Th-excess/$^{232}$Th Method

(Goldberg and Koide 1958)

### Dating Range, Precision, Materials, Sample Size

This method is used for determining the age of sediments with high deposition rates in lakes, deltas and estuaries, and along the coast in the range of up to 10 years. Moreover, it is used to determine the time between magma formation and eruption (Cortini 1985).

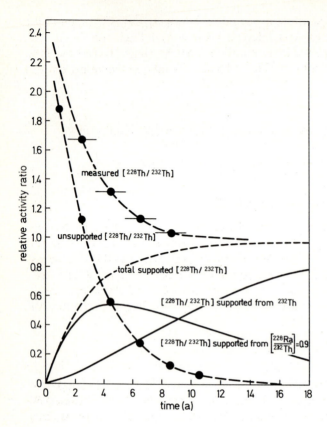

**Fig. 6.73.** Relative contributions to the $^{228}$Th/$^{232}$Th activity ratio from the water (*unsupported*), the $^{228}$Ra (*supported*), and $^{232}$Th (*supported*). The initial $^{226}$Th*/$^{232}$Th at the time of formation of the sediments off Santa Barbara, Calif., is assumed to be 0.9 (after Koide et al. 1973)

## Basic Concept

Thorium-228 ($\tau = 1.913$ a) is the fourth member of the thorium series:

$$^{232}\text{Th} \xrightarrow{1.405 \times 10^{10}\,\text{a}} {}^{228}\text{Ra} \xrightarrow{5.75\,\text{a}} {}^{228}\text{Ac} \xrightarrow{6.13\,\text{h}} {}^{228}\text{Th} \xrightarrow{1.913\,\text{a}} {}^{224}\text{Ra} \xrightarrow{3.66\,\text{d}} {}^{220}\text{Rn}$$

The $^{228}$Th/$^{232}$Th activity ratio in ocean water varies over a wide range (5–30). Thus, the source of the $^{228}$Th cannot be only the $^{232}$Th in the water. The main additional source is $^{228}$Ra, which under reducing conditions migrates from the sediments into the sea water (Koide et al. 1973). Assuming rapid precipitation, the ratio of production from these two sources is about 0.9 (Fig. 6.73). The calculation of the age is done analogously to that described in Sect. 6.3.5.

## Sample Treatment and Measurement Techniques

Details are given in Sect. 6.3 and by Koide et al. (1973).

## Scope and Potential, Limitations, Representative Example

This method has been seldom used; for this reason there is insufficient experience to assess its potential and limitations. A study of $^{228}$Th activity in airborne dust in

Japan showed a pronounced seasonal variation, indicating that the origin of the particulate matter is eastern Asian deserts (Hirose and Sugimura 1984). The formation of the magma involved in the eruption of Mt. St. Helens was estimated to be about 150 a on the basis of the $^{228}$Th/$^{232}$Th activity ratio in the freshly erupted rocks (Cortini 1985).

## 6.3.11 Dating Methods Based on Supported $^{226}$Ra and Unsupported $^{226}$Ra

(Joly 1908; Piggot and Urry 1942; Holmes and Martin 1978)

### Dating Range, Precision, Materials, Sample Size

The age of marine sediments (Krishnaswami and Moore 1973), sedimentation rates in the range of millimeter to centimeter per century (Holmes and Martin 1978), and phosphate nodules (Kim and Burnett 1985) (about 100 g) can be roughly determined up to about 20 ka with this method. Antarctic ice containing dust has been successfully dated using the $^{226}$Ra/$^{230}$Th activity ratio (Sect. 6.3.1) (Fireman 1986). This ratio has also been used to calculate the time between magma formation and eruption (Cortini 1985).

### Basic Concept

Radium-226 ($\tau = 1600$ a) (Fig. 6.64) is a member of the $^{238}$U series. Supported radium-226 would provide a basis for dating analogous to the $^{230}$Th/$^{234}$U method (Sect. 6.3.1) if it were not so mobile in sediments (Kröll 1953; Koide et al. 1973). Only in cold ice is it completely immobilized (Fireman 1986).

Migration of unsupported $^{226}$Ra from the sediment into the sea water results in an excess relative to equilibrium conditions with respect to its ancestor $^{238}$U. This excess radium is enriched further by marine plankton. Sedimentation rates and ages can be estimated by plotting radium-226 activity versus depth (Holmes and Martin 1978).

### Sample Treatment and Measurement Techniques

Details have been given by Ivanovich and Harmon (1982). Direct measurement is possible by monitoring low-energy gamma peaks of $^{238}$U and four peaks of $^{226}$Ra with a well-type germanium detector (Sect. 5.2.2.3) (Kim and Burnett 1983) or by determining the $^{210}$Po activity (Holmes and Martin 1978).

### Scope and Potential, Limitations

Owing to the geochemical mobility of radium and the influence of biological processes on its concentration in sea water, this method is only of historical significance with respect to its application for oceanographic studies. However, together with $^{230}$Th measurements, the time between magma formation and the eruption of Mount St. Helens in 1980 was estimated to be 150 a (Cortini 1985). Moreover, sedimentation rates above the range of the $^{210}$Pb method (Sect. 6.3.13) can be determined (Holmes and Martin 1978).

### Non-Chronological Applications

Radium-226 analyses have been used together with radon measurements to study low-temperature interactions of crystals with water, diffuse hydrothermal flow in aquifers, and sea water mixing on the Galapagos Rift (Dymond et al. 1983). In addition, the $^{234}$U/$^{226}$Ra specific activity ratio provides information for $^{230}$Th/$^{234}$U age determinations (Sect. 6.3.1) about whether the system in question was closed.

## 6.3.12    $^{224}$Ra and $^{228}$Ra Methods

(Moore 1969)

### Dating Range, Precision, Materials, Sample Size

The radium-228 methods is used to date suspensions and bottom sediments from the mixing zone of estuaries, as well as ferromanganese nodules in lakes (Krishnaswami and Moore 1973) and coral (Moore and Krishnaswami 1972) (several grams) for periods of up to 100 a. Marine processes occurring within the past 1 to 30 a are also studied. Lateral mixing of ocean water for periods of up to 10 days can be studied (ca. 700 L) using the descendant radionuclide $^{224}$Ra (Elsinger and Moore 1983; Levy and Moore 1985).

### Basic Concept

Young sediments have as much as a tenfold specific activity for unsupported excess of radium-228 ($\tau = 5.75$ a) (Moore 1969) because radium diffuses from $^{232}$Th-bearing marine sediments into the sea water (Koide et al. 1973). Thus, $^{228}$Ra cannot be considered to be a conservative tracer with only one source, the $^{232}$Th in the sea water. The age determination is carried out as described in Sect. 6.3.5 only with difficulty.

It is possible to study mixing processes in the time range of up to 10 days using radium-224 ($\tau = 3.66$ d), a descendant product of $^{228}$Ra (Elsinger and Moore 1963; Levy and Moore 1985).

### Sample Treatment and Measurement Techniques

Details are given in Sect. 6.3. The measurement is usually done with a semi-conductor gamma-spectrometer (Sect. 5.2.2.3).

## Scope and Potential, Limitations

This method has seldom been used; for this reason insufficient experience is available to assess its potential and limitations. However, in the large drainage system of the Amazon River, $^{228}$Ra was useful for distinguishing input of river water from shield areas from that from other sources (Moore and Edmond 1984).

### 6.3.13 $^{210}$Pb Method***

(Goldberg 1963)

### Dating Range, Precision, Materials, Sample Size

This method is used routinely to determine sedimentation rates of lacustrine (Krishnaswami et al. 1971), fluvial, and coastal marine sediments (Koide et al. 1972; Dominik et al. 1978) in the range of millimeter to centimeter per year and growth rates of coral (samples of several grams) and peat (El-Daoushy et al. 1982) up to ages of 150 years. Ice (ca. 500 kg) has also been dated (Crozaz and Langway 1966). For authentification purposes, the ages of paintings have been determined via white-lead pigments (Keisch et al. 1967), as well as the ages of bronze and tin objects (several mg in each case).

### Basic Concept

Radon-222 ($\tau = 3.825$ d), descendant radionuclide of $^{238}$U (Fig. 6.64), decays via several short-lived daughter products to $^{210}$Pb ($\tau = 22.3$ a). Radium-226, the parent radionuclide of $^{222}$Rn, is present almost universally in crustal material. Radon diffuses continually from the soil into the atmosphere, from which $^{210}$Pb settles out in less than 30 days at an average rate of 10–20 atoms/cm$^2$/min with a distinct continental effect. This lead-210 is incorporated into sedimentary material, resulting in unsupported excess $^{210}$Pb. Hence, a study of sediment or ice profiles can provide information on local differences in the input with time. Lead-210 is bound to organic matter in the soil; in the ocean and lakes, it precipitates together with iron and manganese oxides. But vegetation also acts to remove lead from the soil. The residence time in ocean water is 0.5–50 a.

*Interpretation models:* There are two main models used for age determinations using the $^{210}$Pb dating method. The two models yield the same results if the accumulation rates are constant and not too large (e.g., Oldfield et al. 1978; Chanton et al. 1983). The preferred model (called the CIC model) assumes the initial activity of unsupported excess lead-210 (per unit dry weight) is the same at all depths in the core, independent of the sedimentation rate. The age is calculated with a slight modification of Eq. (6.79). The specific activity of unsupported $^{210}$Pb is derived taking the $^{226}$Ra activity into account analogously to Eq. (6.80).

The $C_{rs}$ model assumes a constant input rate of unsupported $^{210}$Pb from the water to the sediment and that the radionuclides involved in the model do not migrate into the sediment post-depositionally. This model is advantageous for high sedimentation rates. The sedimentation rate $r$ (g/cm$^2$/a) is then obtained from the excess specific activity of lead-210 and the mass $m$ of the core within a depth range $d$ (g/cm$^2$) using Eq. (6.79).

Another possibility, analogous to the $^{231}$Pa-excess method (Sect. 6.3.6), is to estimate the age from the total activity of a core with $n$ subdivisions (dpm/cm$^2$/m) and the flux $\phi$ (dpm/cm$^2$/a) according to the following equation:

$$\sum_{i=1}^{n} [^{210}Pb_{excess}]_i = \phi \int_0^{\infty} e^{-\lambda_{210}t} dt \tag{6.83}$$

### Sample Treatment and Measurement Techniques

The lead is extracted from the sample with hot 6N HCl and then precipitated (Ivanovich and Harmon 1982). The easiest but not the most sensitive way to measure the lead is with a very sensitive $\gamma$-spectrometer (Sect. 5.2.2.3) using the 46.52-KeV line. In this case, no chemical preparation of the sample is necessary (Kim and Burnett 1983).

Another possibility is to wait until radioactive equilibrium between $^{210}$Pb and $^{210}$Po ($\tau = 138$ d) has been established and then measure the $\gamma$-radiation of the polonium-210. The specific activities are generally given in dpm/g dry weight.

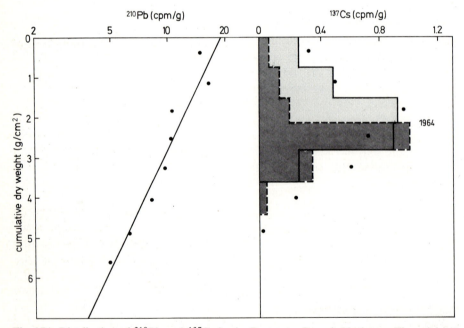

**Fig. 6.74.** Distribution of $^{210}$Pb and $^{137}$Cs in the Pentwater Branch No 1 core: The *solid line* represents the $^{137}$Cs distribution if the inputs are externally integrated (after Kadlec and Robbins 1984)

## Scope and Potential, Limitations, Representative Examples

***Scope and Potential.*** The lead-210 method is used routinely for dating young sediments and to determine sedimentation rates (Krishnaswami et al. 1971; Robbins 1984). Prerequisite is that the sediment has not been disturbed, for example, by bioturbation or physical mixing and that the complete sediment profile is sampled without loss of core. This is checked with a $^{137}$Cs analysis (Sect. 7.5 and Fig. 6.74) (Robbins and Edgington 1975).

***Limitations.*** Incorrect ages are obtained if lead-210 has been mobilized subsequent to sedimentation, e.g., in a reducing environment containing organic matter (Koide et al. 1973), if sorting of the material occurred, or if accumulation did not occur uniformly. In oxidizing environments, which predominate, migration of lead-210 is not a problem (Robbins and Edgington 1975). In systems with varying sedimentation rates, alternative geochronological methods should always be applied to check the results for validity (Fig. 6.75).

***Representative Example.*** White lead-pigments are suitable for the authentification of paintings. However, the samples must be taken very carefully to avoid contamination, especially with gypsum. Such dating is possible because the ancestor radionuclide $^{226}$Ra is separated from the lead during the production of the white lead paint (Keisch et al. 1967).

### Non-Chronological Applications

Non-chronological applications of this method deal with sediment transport and redeposition processes (e.g., Chanton et al. 1983).

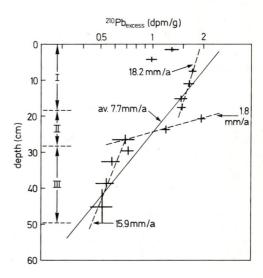

**Fig. 6.75.** Semilog plot of $^{210}$Pb$_{excess}$ activity vs. the depth in the core and an estimate of the sedimentation rates in the different sections of the core (after Dominik et al. 1978)

## 6.3.14  Uranium/Helium (U/He) Method**

(Rutherford 1906; Strutt 1908; Fanale and Schaeffer 1965)

### Dating Range, Precision, Materials, Sample Size

The U/He method gives gas retention ages, so-called He indices, that are usually only minimum values. Non-recrystallized aragonitic Quaternary carbonate (about 100 mg), such as marine fossils (Schaeffer 1976) and coral (Bender 1973; Bender et al. 1979), have been dated. Under favorable circumstances, samples have been dated up to an age of 30 Ma with an uncertainty of $\pm 10$–30%. Bone and other apatite deposits are not suitable (Turekian et al. 1970). Natural hydrocarbons may possibly be dated via the He/Ar ratio (Gerling et al. 1967; Torgersen 1980).

A few successful age determinations have been made on minerals over 50 ka with an undisturbed crystal lattice, particularly ore minerals (0.1–1 g; e.g., magnetite, pyrite, hematite, galena, apatite, and zircon) (Fanale and Kulp 1962; Lippolt et al. 1982; Boschmann 1986), and fine-grained basalt (Ferreira et al. 1975) which have a uranium concentration of 0.1–10 ppm.

The method is also used for studying groundwater with ages from several thousand years up to 400 Ma (Davis and De Wiest 1966; Seifert 1978; Bath et al. 1979; Torgersen 1980; Andrews et al. 1982). An improved model is based on the $^3$He/$^4$He ratio (Andrews 1985).

Another important field of application is the study of extraterrestrial material (Sect. 6.5.5). Olivine and pyroxene extracted from meteorites yielded helium/uranium ages similar to the K/Ar dates (Sect. 6.1.1).

### Basic Concept

The parent nuclides of the natural $^{238}$U, $^{235}$U, and $^{232}$Th decay series and many of the subsequent nuclides (Fig. 6.64) are $\alpha$-emitters, i.e., producers of helium-4. For example, during the decay of $^{238}$U to $^{206}$Pb eight helium atoms are produced; seven are produced by the decay of $^{235}$U to $^{207}$Pb, and six by the decay of $^{232}$Th to $^{208}$Pb.

Production of helium-4 is given by the following expression:

$$^4He = 8 \cdot (^{238}U)(e^{\lambda_{238} \cdot t} - 1)$$
$$+ 7 \cdot (^{235}U)(e^{\lambda_{235} \cdot t} - 1)$$
$$+ 6 \cdot (^{232}Th)(e^{\lambda_{232} \cdot t} - 1) \tag{6.84}$$

where $^{238}$U, $^{235}$U, and $^{232}$Th are the number of parent nuclides per gram rock.

If the units for helium are given in $cm^3$(STP)/g rock and those for the parent nuclides in $\mu g$/g rock, then

$$^4He = 2.2415 \times 10^4 \cdot (0.0336(^{238}U)(e^{\lambda_{238} \cdot t} - 1)$$
$$+ 0.0298(^{235}U)(e^{\lambda_{235} \cdot t} - 1)$$
$$+ 0.0258(^{232}Th)(e^{\lambda_{232} \cdot t} - 1)) \tag{6.85}$$

This equation cannot be solved analytically for time $t$, which must be obtained by iteration. The age calculated in this way is called the "helium index".

When radioactive equilibrium in a rock has been established, $1.19 \times 10^{-13}$ cm$^3$(STP) He/$\mu$g U and $2.88 \times 10^{-14}$ cm$^3$(STP) He/$\mu$g Th are produced annually. Thus, the age of the rock can be calculated as follows:

$$\text{He} = \frac{\rho}{p} \cdot t[1.19 \times 10^{-13}(\text{U}) + 2.88 \times 10^{-14}(\text{Th})] \tag{6.86}$$

where the He concentration is in cm$^3$(STP)/cm$^3$ H$_2$O,
  (U) and (Th) are the uranium and thorium concentrations (in $\mu$g/g rock),
  $\rho$ is the bulk density of the rock (in g/cm$^3$), and
  $p$ is the fractional porosity of the rock (expressed as a fraction of the whole) (Andrews and Lee 1979).

Determination of the uranium and thorium concentrations, as well as the porosity, may be made difficult by, for instance, differences in the diffusivities of the rocks. Hence, a way to avoid having to estimate the unknown parameters in the above equation is to determine the $^{222}$Rn content. It is assumed that $^{222}$Rn is in radioactive equilibrium with uranium. Errors may arise due to decay loss during diffusion within the grains or reequilibrium with a non-uniform uranium distribution (Torgersen 1980; Andrews et al. 1985). Very instructive are $^3$He/$^4$He ratios, which depend mainly on the abundance of lithium in the host formation and the age (Andrews 1985) and, therefore, yield information on crustal diffusion of He.

The helium content increases in a closed system (e.g., in perfect crystals or in a confined aquifer) as a function of uranium concentration and age (Fig. 6.76). If the sample is younger than 1 Ma, a correction that takes the initial radioactive disequilibrium into consideration is necessary (Fanale and Schaeffer 1965).

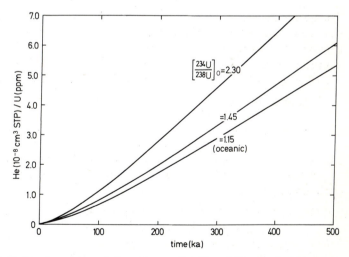

**Fig. 6.76.** Increase in the helium/uranium activity ratio as a function of time in closed systems with different initial $^{234}$U/$^{238}$U ratios. The lines become increasingly parallel as secular equilibrium is approached (after Fanale and Schaeffer 1965)

### Sample Treatment and Measurement Techniques

Solid samples are coarsely ground and pure minerals (i.e., perfect crystals) are handpicked under a microscope. The surface of the grains is removed with acid to eliminate adsorbed helium. Because helium is more easily lost than argon, the samples may not be heated before the measurement, as is usual for the K/Ar method to avoid contamination (Sect. 6.1.1). Because the concentration of helium in the atmosphere is very low, contamination from this source is not a problem.

The helium is driven out of the samples by melting in an ultrahigh vacuum. It is then purified by passing it over hot titanium and adding a $^3$He spike. The very small amount of helium is measured with a very sensitive mass spectrometer (Sect. 5.2.3.1) with a high degree of precision. The limit of detection is $10^{-13}$ mol concentrations are given in units of $10^{-8}$ cm$^3$STP/g sample.

For groundwater dating, the uranium and thorium concentrations are roughly estimated via their α-activities because error due to loss of helium predominates. Determination by isotopic dilution (Sect. 5.2.4.1) would be much more exact. This kind of analysis is used for mineral samples, for example.

The helium content of groundwater is determined using 100-l samples. After degassing and separation of the noble gases, helium is isolated by fractional distillation (Sect. 6.2.8).

### Scope and Potential, Limitations, Representative Examples

**Scope and Potential.** Even the use of very sensitive mass spectrometers cannot make up for the fact that in most cases unreliable ages are obtained due to gas losses and, equally important, the mobility of uranium. This is especially so for terrestrial mineral and rock samples, for which the method has mainly historical significance. This method was used at the beginning of this century for geological age determinations when Rutherford (1906) recognized the possibility for this application.

**Limitations.** The major cause of unreliable U/He dates is the low helium retention, which for most minerals is considerably lower than that of argon. But there are exceptions (nepheline, hornblende, augite, langbeinite; Lippolt and Weigel 1988). In other cases, gas losses of up to 70% are common, even in perfect crystals, in glassy metamict crystals even 99% has been observed. A comparison test of K/Ar ages (Sect. 6.1.1) with U/He ages of whole-rock basalts showed that all of the samples had lost helium. But this makes the analysis valuable for checking whether a closed system was present, which is important for U/He dating (Ferreira et al. 1975). Another cause of false U/He ages is the gain or loss of uranium, e.g., in shells during fossilization; this process, however, usually appears to be completed after a few thousand years (Fanale and Schaeffer 1965). Continual or episodic absorption is thought to be the exception. Open systems (Bender et al. 1979) in which uranium or thorium has been gained or lost yield ages that are too low or too large. Such samples can be identified via the deviation of the aragonite/calcite ratio or $^{230}$Th/$^{226}$Ra from unity. Recrystallization is another common cause of discordant

U/He, $^{234}$U/$^{238}$U (Sect. 6.3.4), and $^{230}$Th/$^{234}$U (Sect. 6.3.1) dates. Loss of radon in such cases exceeds 5%.

Minerals such as beryl, cordierite, and turmaline are not suitable for dating by this method because they accumulate $^4$He already during crystal growth. Even small amounts of initial helium can make an age determination impossible if the uranium concentration is only a few ppm.

*Representative examples. Dating of ores:* A promising possibility for the use of this method is the dating of ore deposits via the minerals hematite, pyrite, magnetite, pyrrhotite, and apatite, whose helium retention seems to be good enough (Fanale and Kulp 1962; Lippolt et al. 1982; Boschmann 1986). This must be accompanied by microscopic examination to determine the presence of mineralogical changes in the rock that could be responsible for helium diffusion and/or uranium migration or are indicative of them.

*Dating of corals:* The results of U/He dating of aragonite fossils, especially of corals, have been mostly satisfactory (Bender 1973) if a correction is made for the helium deficit that occurs when alpha particles leave the domain of intact aragonite. Calcite does not retain helium.

*Groundwater studies:* Radiogenic helium dissolved in groundwater is obtained by subtracting the concentration of atmospheric helium, which is estimated from the argon concentration. An increase in helium content from one point to another in an aquifer may be used to estimate the flow rate. The loss of helium by diffusion is sometimes corrected for by measurement of the $^{20}$Ne or $^{40}$Ar isotopes (Gerling et al. 1967; Pavlov 1970; Tetzlaff et al. 1973; Seifert 1978). The concentrations of uranium and thorium in a rock or groundwater are used to estimate the production rate of helium (Pavlov 1970; Tetzlaff et al. 1973). The $^3$He/$^4$He ratio yields information on the diffusion of helium in a crust, as well as on groundwater mixing (Andrews 1985). Nevertheless, $^4$He concentrations increase due to in-situ production up to 50 ka water age in the aquifer of the Great Artesian Basin. For ages above 100 ka, an additional helium flux equivalent to the entire crustal production must be taken into account (Torgersen and Clarke 1985; Torgersen and Ivey 1985). Therefore, U/He ages calculated using Eq. (6.86) are often to high (Torgersen 1980). The cause of this may be input of He from joints or salt structures (Gerling et al. 1967; Banwell and Parizek 1988), which can be recognized on the basis of the $^3$He/$^4$He ratio. Another source that can explain U/He ages that are too high is the whole crust flux (Heaton 1984; Torgersen and Clarke 1985a, b; Sano 1986; Torgersen and Clarke 1987). Two other sources are pore water that enters the aquifer through the aquitard (Geyh et al. 1984) and diffusion from the underlying rock (Andrews et al. 1984; Heaton 1984). Hence, $^4$He data mirror flow directions rather than quantitative water ages.

### Non-Chronological Applications

Helium isotope geochemistry plays a major role in research on the degassing history of the Earth (e.g., Tolstikhin 1975; Lupton and Craig 1975; Kurz et al. 1985) and on magmatic and geothermal systems (e.g., Torgersen and Jenkins 1982; Torgersen et al. 1982; Craig et al. 1978; Zindler and Hart 1986).

Measurement of $^3$He/$^4$He ratios in marine basalts has established that primordial gases are still escaping from the Earth's mantle. Whereas primordial $^3$He/$^4$He ratios range from $1.4 \times 10^{-4}$ to $3.0 \times 10^{-4}$, the $^3$He/$^4$He ratios for marine basalts range from about $1.4 \times 10^{-6}$ to $4.4 \times 10^{-6}$ (Zindler and Hart 1986), whereas that for crustal helium is $6.6 \times 10^{-8}$. The relatively low ratios in the Earth's mantle are due to radiogenic accumulation of helium-4 from the decay of uranium and thorium over the course of the Earth's history. Recent subcrustal volcanic activity has been detected by helium analysis (Torgersen et al. 1987).

The helium content of groundwater has been found to be helpful in prospecting for uranium and thorium (Clarke and Kugler 1973; Torgersen 1980; Zaikowski and Roberts 1981; Top and Clarke 1981; see also Sect. 6.2.2.2). Combined helium and radon analysis (Sect. 6.3.15) may also be useful as a logging tool in geothermal exploration (Torgersen 1980). Moreover, the helium distribution pattern around oil fields may be applied as an exploration tool (van den Boom 1987).

Analysis of $^3$He/$^4$He ratios in deep-sea sediment cores demonstrated that a significant amount of extraterrestrial helium is present in marine sediments. This may make it possible to estimate sedimentation rates (Takayanaki and Ozima 1987).

## 6.3.15 Radium/Radon Method

(Cherdyntsev 1969)

### Dating Range, Precision, Materials, Sample Size

The radium/radon method has been proposed for determining the mean residence time of groundwater (a few liters) in the range of 30 to 100 days (Gellermann and Gast 1983). Sediment irrigation and solute transport rates can be estimated (10–30 g of wet sediment, Martin and Sayles 1987).

### Basic Concept

This method assumes that groundwater enters an aquifer at a constant velocity, that in part of the aquifer the groundwater and rock have differing radium concentrations, and that exchange of radium and radon occurs between the dissolved and the solid phases. The specific activities of both radionuclides then increase until saturation is reached. The mean residence time (MRT) is given by the following equation (Gellermann and Gast 1983):

$$\text{MRT} = \frac{1}{\lambda_{\text{Ra}}} \ln\left(1 - f \frac{A_{\text{Ra}}}{A_{\text{Rn}} - A_{\text{Ra}}}\right) \tag{6.87}$$

The constant $f$ depends on the production rate of radon from radium. Because it cannot be determined, it is generally set to unity but it may be larger.

The estimation of sediment irrigation and solute transport rates is based on $^{226}$Ra/$^{222}$Rn disequilibrium in sediment pore water (Martin and Sayles 1987).

*Sample Treatment and Measurement Techniques*

See Sect. 6.3 and Ivanovich and Harmon (1982).

*Scope, Potential, Limitations, Representative Examples*

*Scope and Potential.* Mean residence times were determined on groundwater samples taken from springs in the area around Bad Brambach, Germany (Gellermann and Gast 1983). In agreement with the hydrogeological model, the mean residence times (30 d to 3 a) obtained by the radium/radon method were smaller than those derived from the tritium activities (Sect. 7.5). The latter ranged from 9–17 a.

*Limitations.* Very strong Ra adsorption in some aquifers decreases its mobility and may hinder direct groundwater dating but may yield a time-scale of adsorption (Krishnaswami et al. 1981).

*Non-Chronological Applications*

*Hydrothermal systems:* The $^{226}Ra/^{222}Rn$ activity ratio has been shown to be a sensitive indicator of low-temperature interactions of hydrothermal solutions with crustal rocks in a small area along the Galapagos Rift. Information was also obtained about the mixing and flow history of these hydrothermal solutions (Dymond et al. 1983). Variations in the $^{222}Rn$ in well and spring water are monitored as a potential forecaster of earthquakes.

*Hydrological studies:* The interpretation of the $^{222}Rn$ content in stream water is very complicated because this radionuclide is produced by the decay of uranium dissolved in the water, present in the underlying rock but heavily modified by diffusion and transfer processes (Heath 1981). However, radon measurements in river water may be used to estimate the groundwater influx (Ellins 1985).

*Lithological studies:* As the $^{222}Rn$ content in groundwater reaches an equilibrium value after a residence time of about 25 days, this value can be used to estimate the uranium content in the aquifer, as well as the area of rock/water interface (Andrews et al. 1985). Both parameters are important for estimating subsurface production, for example, of $^{36}Cl$ (Sect. 6.2.7) or $^{39}Ar$ (Sect. 6.2.8). Solute transport rates in sediments have been estimated from the $^{222}Rn$ deficit using a molecular diffusion, nonlocal exchange model (Martin and Sayles 1987).

## 6.4 Age Determination Using Radiation Damage

The interaction of non-conducting solids with ionizing alpha, beta, gamma, and cosmic radiation changes their physical and chemical properties (e.g., density, optical parameters, chemical stability and reactivity, defects in the crystal lattice). These changes are called "radiation damage". In minerals, after extreme exposure,

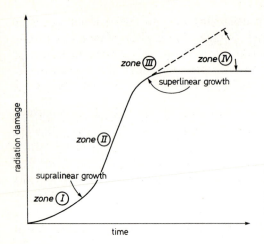

**Fig. 6.77.** Increase in radiation damage (after Singhvi and Wagner 1986); the curve has four distinct regions: *I* supralinear growth; *II* linear growth; *III* superlinear growth; and *IV* saturation or defect creation with change in sensitivity

this process is called metamictization. Most of this damage "heals" immediately after it occurs. A small proportion of it, however, is stable over geological periods of time, increasing quasi linearly with age and continual irradiation within the range of zone II in Fig. 6.77 (Aitken 1978, 1985). The rate at which the density of the radiation damage increases is dependent on the material involved and on the internal and external radiation dose rate.

The age determination methods are based on one of two types of radiation damage:

1) *Electron shell phenomena:* Electrons are detached from the atoms of the material by α, β, γ, and cosmic radiation, some of which are trapped for a quasi infinite time

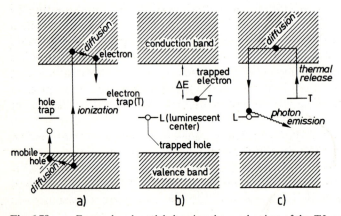

**Fig. 6.78 a–c.** Energy band model showing the mechanism of the *TL* process (after Aitken 1978): **a** Ionization by irradiation promotes an electron from the valence band into the conduction band (about 10 eV energy required); mobile holes move towards the luminescence traps (*L*) and electrons move towards electron traps (*T*). **b** Trapping of an electron at a forbidden energy level (*T*) and a mobile hole forming a luminescence center. If the depths of the electron trap $\Delta E$ exceeds 1.5 eV, the lifetime of the electron in the trap is up to a million years. **c** Recombination of electron and mobile hole after release of the trapped electron by the absorption of thermal energy ($\Delta E$) to the conduction band with the emission of phonons as it falls to the luminescence center (*L*)

(compared to the range of dating) in defects in the crystal lattice. Input of external energy (e.g., heat or electromagnetic radiation) causes some of these electrons to recombine with positive "holes" accompanied by the emission of light (e.g., thermoluminescence; Fig. 6.78). This phenomenon is utilized by the thermoluminescence (Sect. 6.4.1), optical dating (Sect. 6.4.2), and ESR (Sect. 6.4.3) methods.

2) *Lattice phenomena:* Defects in the crystal lattice of nonconducting solids are created by the high-energy particles resulting, for example, from the spontaneous fission of uranium. The traces (several $\mu$m long) of the paths of the fission fragments are made visible under the microscope by etching of the crystal and counted (Fig. 6.92). The fission-track method (Sect. 6.4.7), for example, is based on this phenomenon.

The use of radiation damage for age determinations is based essentially on two assumptions:

1) The radiation damage acquired during the exposure period is stable for the lifetime of the sample; its density is proportional to the absorbed radiation dose ($d_{AD}$) and, thus, to the time the matter has been irradiated.
2) Within the age range being dated, the radiation dose rate ($r$) was constant, e.g., there was no gain or loss of uranium, thorium, or their daughter elements (Sect. 6.3), i.e., the sample was a closed system. This condition may not be met in some cases, e.g., owing to variation in water content or cosmic ray intensity.

If these conditions are fulfilled, two parameters must be determined for the age determination:

1) the *equivalent dose* (also called archeological dose, $d_{AD}$, or accumulated paleodose): The equivalent dose, $d_{ED}$ is the laboratory dose needed to obtain a luminescence signal of the same size as the natural signal. In most cases $d_{AD} = d_{ED}$ if the supralinear correction is zero (see Fig. 6.82). This parameter is a function of the density of the natural radiation damage and a constant that depends on the material involved (radioactive susceptibility). The dose is given in gray (1 Gy = 100 rad). One gray corresponds to one joule of absorbed energy per kg of matter.
2) the *radiation dose rate* ($r$) (in Gy/a): This parameter contains an internal and an external component. The first is a function of the activity of the uranium, thorium, and their descendant elements, as well as potassium in the sample. This is the source of about 95% of the ionizing radiation in rocks. The second component results from these elements in the environment, plus cosmic radiation.

The age is calculated as follows:

$$t = d_{AD}/r \tag{6.88}$$

The application of these dating methods is limited chiefly by four properties of radiation damage:

1) *Thermal fading:* Radiation damage can anneal at low energy input levels (e.g., heat). Hence, only an anomalously small apparent age of the sample is often obtained.

2) *Anomalous fading:* In some substances, the radiation damage has a short lifetime even at ambient temperatures. No dating is possible in this case or special techniques must be used. Anamalous fading is thought to be a quantum mechanical tunneling phenomenon. It could also result from localized transitions between trapping sites (Templer 1986a). A test for the presence of anomalous fading can be made by setting aside irradiated samples for several weeks; anomalous fading is present if the density of the radiation damage decreases by several percent or more and dating becomes difficult. However, if it is assumed that the TL signal from a mineral phase exhibiting anomalous fading consists of the superposition of a stable and an instable component, pretreatment or post-measurement may be developed to overcome this problem (Clark and Templer 1988; Stoneham and Winter 1988). Mejdahl (1989) has shown that TL-age dependent corrections are necessary due to long-term fading estimated by TL-measurements of infinite old samples.

3) *Saturation:* The density of the radiation damage increases quasi-linearly within a certain range (zone II in Fig. 6.77) with age only up to a certain level (saturation level, zone III in Fig. 6.77), which is dependent on the material involved. Under ideal conditions, the density of the radiation damage then remains at a constant level due to steady-state recombination. In some cases it decreases under the influence of further radiation.

4) *Defect creation:* If the material contains $\alpha$-emitting impurities, new traps may be expected to be continually produced (Aitken 1984; Nambi 1984), which increase the possibilities for the recombination of free electrons, lowering the rate at which the amount of radiation damage increases (zone IV in Fig. 6.77). This phenomenon would limit the dating range and is still under discussion.

Various techniques for determining the density of radiation damage are used for age determinations. These techniques give the dating methods their names.

The main applications are related to the age of volcanic eruptions and of important stratigraphic layers, crystallization, the dating of prehistoric obsidian artifacts, as well as determination of tectonic uplift rates and the time since the last exposure to sunlight.

## 6.4.1 Thermoluminescence (TL) Method***

(Daniels, Boyd and Saunders 1953)

### Dating Range, Precision, Materials, Sample Size

The TL method (Aitken 1974; Cairns 1976; Aitken and Mejdahl 1979, 1982; Mejdahl et al. 1983; Wagner 1983; McKeever 1984; Singhvi et al. 1985; Wagner et al. 1985; Aitken 1985) is used routinely for determining the age of archeological objects (e.g., time of firing of ceramics) up to 50 ka. The analyses are done on samples of dielectric solids, often quartz or feldspar fractions ($> 50$ a and $> 500$ a, respectively), also zircon crystals (Templer 1986b; Templer and Smith 1986), extracted from potsherds, from bricks (at least 10 cm from the surface of the brick; dating precision $\pm 2\%$ Kaipa et al. 1988), from terracotta (Fagg and Fleming 1970), and from burned soils or sediments baked by lava, quartz sands, and rocks, silica-rich, porous, crystalline slag from fireplaces and kilns (Elitzsch et al. 1983) as well as mineral pigments (Schroerer et al. 1988). Flint tools that have been heated to above 400°C (Göksu et al. 1974; Göksu-Ögelman 1986), shells (Ninegawa et al. 1988), bones, and dentine (formation age) can be dated with poor precision in the age range of 5 to at least 200 ka. A precision of $\pm 5 - 10\%$ can be attained for the most suitable samples (Aitken 1976). Samples of about 10 g are needed to obtain mineral extracts of about 250 mg from sherds that are, if possible, thicker than 5 mm; 30 mg is sufficient for authenticity testing, since the requirements for accuracy are lower than for precision dating (Fleming et al. 1970; Fleming 1972; Becker and Goedicke 1978).

Geological samples (several tens of grams) are datable with less precision ($\pm 25\%$). The types of samples that are suitable are those whose TL "clock" has been reset by (a) exposure to sunlight (e.g., losse), (b) heating resulting from volcanic or tectonic events, or (c) whose TL "clock" was set at the time of formation. The most suitable mineral extracts are quartz (with ages up to 100 ka), feldspar (5–500 ka), and polymineral fine-grained fine-grained samples ($\geqslant 300$ ka). Special procedures have been developed for quartz. Smoky quartz, igneous and metamorphic rocks, tectites, and impact glass should be datable up to ages of 100 ka. Volcanic glass from pelagic sediments (Huntley and Johnson 1976; Berger et al. 1984) and from tephra (Berger 1985a) are also suitable. The age of lava can be determined between 3 and 300 ka using plagioclase (Hwang 1970; Göksu 1978; Guérin and Valladas 1980).

The dating range of secondary carbonates (travertine and speleothem) with a low clay content (Bangert and Hennig 1979) and gypsum content (Nambi 1982) stretches (with reservations) from 10 ka to possibly 1 Ma (Wintle 1978; Debenham 1983). The dating of eolian sediment (about 1 kg, e.g., loess) is possible up to ages of at least 200 ka, possible 300 ka, with a precision of about $\pm 10\%$ (Wintle 1982; Singhvi and Mejdahl 1985; Zöller and Wagner 1989), some water-laid silts, quartz grains extracted from beach or dune deposits, and silt-sized feldspar from peat (Berger 1986). Prerequisite for success is that the samples have not been diagenetically altered or disturbed by bioturbation (Wintle and Huntley 1982; Wintle et al. 1984). Glacier ice has been dated via the occluded dust (Bhandari et al. 1983). Glaciolacustrine silt can also be dated provided it is in clay-rich annual layers deposited at low rates (Berger 1985b; Berger et al. 1987). Such dating is not without problems (Wintle and Catt 1985). The dating of fossil calcite shells with ages up to 500 ka has been attempted (Ninagawa et al. 1988).

It is a matter of controversy whether terrestrial ages of meteorites (Sect. 6.5.7) can be determined with the TL method (Melcher 1981, 1982; McKeever 1982). Cosmic-ray-exposure ages (Sect. 6.5.6) are also determined (Houtermans and Liener 1966). Reviews have been published by Sears and Hasan (1986) and Sutton and Walker (1986).

Thermoluminescence ages are usually published with two standard deviations. The smaller one (internal variance) corresponds to the actual range of several dates for the same find. The larger one (by a factor of 2 to 3) includes random and systematic errors, for example, calibration errors for the radiation source. The larger standard deviation is to be used when TL dates are compared with ages determined with other methods (Wagner 1983).

Reliable TL dates are guaranteed only when a TL specialist is involved. Sometimes TL dating may not be possible, but thermoluminescence analyses almost always yield valuable stratigraphic information.

### Basic Concept

Crystals contain defects in their structure (e.g., gaps or impurities) which form traps of negative or positive charge deficits relative to the common grid. Defects in crystals create "forbidden" energy levels (traps with activation energies of 0.5–3 eV), which lie in an energy gap between the so-called valence and conduction bands. Electrons can be trapped with high lifetimes in "deep" traps, but are retained long at elevated levels (Fig. 6.78). Electrons and electron deficits are generated when ionizing radiation penetrates the crystal. Most of them recombine immediately, but a few diffuse to the traps, where they become quasi stable. With a constant radiation dose rate, it is assumed that the number of occupied traps increases linearly with age of the sample (Aitken 1967) as a result of internal and external radiation from uranium, thorium and their long-lived daughter elements, as well as potassium, which are

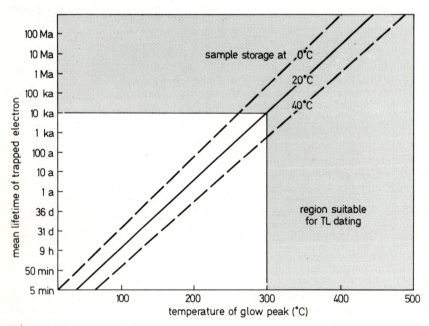

**Fig. 6.79.** Mean lifetime of trapped electrons as a function of the glow peak temperature (after Aitken 1967)

present in all materials, at least in traces. However, the only radiation damage that is suitable for an age determination is that with a lifetime of hundreds of thousands to millions of years at ambient temperatures. The longer the half-life of the trap, the higher the temperature necessary to release the electron from the trap (Fig. 6.79). The lifetimes should be at least ten times longer than the age to be determined (Nambi 1984; Mejdahl 1988).

Thermoluminescence dating assumes that initially (the time of formation or the last heating to above 500°C) none of the sufficiently "deep" traps in the sample were occupied by electrons. This condition is fulfilled by potsherds, natural glasses, magmatic and contact metamorphic rocks (cooling age) because any previous radiation damage is annealed by the high firing temperatures or the high temperatures of the magma. Immediately after the formation of speleothem and siliceous ocean sediments, some of their "deep" traps may be occupied (Huntley and Johnson 1976). For these types of samples, the techniques developed for sediments must be used.

Eolian sediments (light sensitive TL) (especially losse usually loses a part of its primary TL energy by exposure to sunlight (Wintle 1982; Wintle and Huntley 1982). The rate at which it is lost is a function of the duration of the exposure to light, its wavelength, and the type of mineral.

If a crystal with occupied, "deep" traps is heated, the electrons in the traps are elevated to the conduction band from which they fall to luminescence centers with the emission of energy in the visible portion of the spectrum (Sunta et al. 1983). The amount of light emitted above the incandescent glow is due to thermoluminescence and increases with the accumulated paleo-dose and, thus, with age.

Two values must be determined in order to calculate an age using Eq. (6.88)

– the *total accumulated paleo-dose* ($d_{ED}$) during the period of exposure and
– the *annual dose rate* ($r$): the external component from the environment (for potsherds, for example, about 40% of $r$) plus the internal component. For fired clay, $r$ is $< 10 \, mGy/a$; for stalagmites it is only ca. $1 \, mGy/a$.

The techniques for translating TL signals into $d_{ED}$ are described in the following section.

### Sample Treatment and Measurement Techniques

Before any samples are collected, the TL laboratory should be consulted whether a TL expert should visit the site and whether field measurements of the radiation dose rate should be made, which will usually increase the reliability of the dating (Wagner 1983).

*Sampling:* At least six samples (e.g., shards) should be taken from each sampling site. They may not be exposed to sunlight, washed, dried, or treated in any way. They must be taken at depths of at least 20 cm (30 cm would be better). To determine the external radiation dose rate, about 1 kg of fresh, naturally moist material has to be collected about 20 cm from the sampling site; this material should contain pebbles, as far as they are present. It should be kept in mind, however, that inhomogeneous beds usually do not allow very precise TL dating. The samples must be protected from drying out and from heat and sunlight during collection and transport. Special methods of protection must be used if young samples with a low $d_{ED}$ are to be transported by air (Goedicke et al. 1983).

Eolian sediments must be taken at least several centimeters below the surface and may not be exposed to direct sunlight because even a short period of irradiation lowers the TL. It is the best to have a TL expert participate in sample collection. The samples are placed in plastic bags, which are then put in strong paper bags. In the laboratory, the carbonates are dissolved and the 4–11 $\mu$m fraction, consisting mainly of quartz grains, is suspended in acetone and sedimented onto aluminium disks.

*Sample treatment:* In the laboratory, about 2 mm of the surface of the sample is removed to eliminate bleaching by sunlight and the effects of $\beta$-radiation from the soil. The separation of pure mineral fractions (e.g., quartz and feldspar) with grain sizes between 0.1 and 1.0 mm is the next step: After carbonate has been removed by dissolving it in acid, several grams are ground and sieved into several fractions; quartz, plagioclase, and feldspar are separated using heavy-liquids. This is necessary because although quartz has a significantly lower TL sensitivity than feldspar, it shows no anomalous fading, which is often the case for feldspar.

*Glow curve determination:* The TL analysis is usually done in a dark room with a red light. An aliquot, 1–2 mg of the extract, is heated on an aluminium or stainless steel disk under nitrogen or argon to 500°C at a constant rate of 5–10°C/s ($< \pm 0.1\%$). The light emission is measured with a photomultiplier with a quartz window (Fig. 6.80) and recorded together with the temperature. The use of special optical filters to eliminate the red glow and to select for the maximum emission of the mineral phase (e.g., a UV filter for feldspar, a blue filter for quartz) improves the

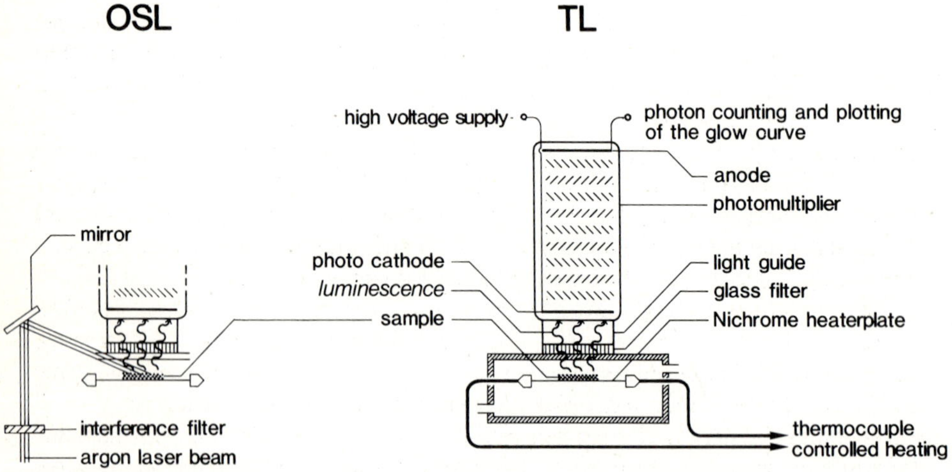

Fig. 6.80. Schematic diagram of a *TL* and *OSL* apparatus (after Aitken 1978)

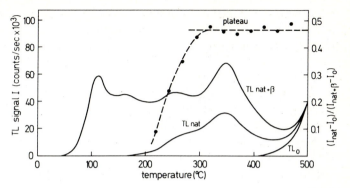

**Fig. 6.81.** "Natural" glow curve ($TL_{nat}$) for an unirradiated aliquot and a "natural and artificial" glow curve for an irradiated (with a dose $\beta$) aliquot of a sample ($TL_{nat+\beta}$), as well as the incadescence curve ($TL_0$): the *dashed line* is the plateau of the dating range derived from the ratio of increments of these curves. A natural dose of 88% of the artificial dose can be calculated from the plateau level (after Aitken 1985)

results, e.g., by suppressing any disturbing thermal radiation above 400°C. Further improvement is possible by pulse height discrimination of the electrical pulses. Computer-controlled TL systems are commercially available.

The term "glow curve" is given to a plot of emitted light versus temperature (Fig. 6.81). The peak areas between 300–450°C are assumed to be proportional to the accumulated paleo-dose $d_{ED}$ in Eq. (6.88). In the low temperature range, the TL is usually instable or distorted, for example, by chemically induced TL.

*Glow growth curve ( Additive radiation technique ):* To determine the total paleo-dose $d_{AD}$, aliquots of several mg are irradiated in increasing steps (e.g., of 5 Gy each) with a carefully calibrated $^{90}Sr$ (a $\beta^-$-emitter of 40–150 mCi) or $^{241}Am$ (an $\alpha$ emitter) source. Each aliquot is exposed to a dose greater than the previous one (additive dose technique, e.g., Aitken 1985), artificially producing a "future" state. The conditions produced by this irradiation are allowed to stabilize at elevated temperature for 1–30 days before the artificial glow curves are recorded. The intensity of the TL signal should not drop more than 15%, otherwise anomalous fading is present and dating is not possible. This annealing can sometimes be produced by very careful heating at low temperature (thermal washing). The thermal stability of the natural TL emission is checked by plotting the ratios of consecutive incremental areas under the glow curves for the artificially irradiated and the unirradiated samples versus temperature. The TL is stable if a plateau is formed (e.g., beginning at ~ 270°C for feldspar and ~ 340°C for quartz (Fig. 6.81).

The areas under the plateaus within certain temperature intervals on the glow curves of the natural and the artificially irradiated aliquots are plotted versus the radiation dose. The intercept of this straight line with the x-axis is equal to the equivalent dose $d_{ED}$ (Fig. 6.82); *first glow growth curve*: Bluszcz 1988).

Less reliable results are obtained with the "*second glow growth curve*" (Fig. 6.82), for which an aliquot of the sample (only one is needed, as opposed to the first glow curve) is annealed at a temperature of more than 500°C in order to remove the

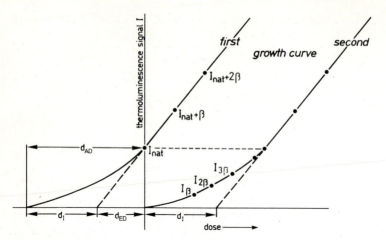

**Fig. 6.82.** Additive radiation technique for estimating the accumulated paleodose $d_{AD}$, the equivalent dose $d_{ED}$, and the supralinearity correction $d_{I}$. Of three aliquots, the first is used to measure the natural $I_{nat}$, the second and third are irradiated with doses of $\beta$ and $2\beta$, respectively, and used to measure $I_{nat+\beta}$ and $I_{nat+2\beta}$. A plot of these three values yields the first glow curve. The second glow curve is obtained by heating and additive irradiation, with measurement of $I_{\beta}$, $I_{2\beta}$, and $I_{3\beta}$

natural TL before the additive irradiation experiment is started. When this is done, however, the TL sensitivity of the sample may be altered by the heat treatment. If this is not the case, the so-called regeneration technique can be used to determine $d_{ED}$ (Wintle and Huntley 1980); it is particularly suitable for very old samples. This technique is independent of the extent to which the TL "clock" was reset to zero by exposure to sunlight; this means that $d_{ED}$ of partially bleached sediments can be obtained directly.

The second glow growth curve is constructed if a supralinearity correction ($d_{I}$) is necessary, e.g., for a young sample (with an age of < 10 ka) with a small $d_{ED}$ value. In the range of low radiation doses (up to 1 Gy), the number of metastable states responsible for the thermoluminescence does not necessarily increase linearly (zone I in Fig. 6.77). The accumulated natural paleo-dose ($d_{AD}$) is then the sum of the equivalent dose ($d_{ED}$) and the supralinearity correction ($d_{I}$) (Fig. 6.82).

*Annual dose rate determination:* The annual dose rate (several mGy/a), including its spatial distribution, is the second parameter that must be determined for an age determination using Eq. (6.88). This parameter consists of terms for $\alpha$, $\beta$, $\gamma$, and cosmic radiation ($r_{\alpha}$, $r_{\beta}$, $r_{\gamma}$, and $r_{cosmic}$, respectively):

$$r_{tot} = k_{\alpha} \cdot r_{\alpha} + r_{\beta} + r_{\gamma} + r_{cosmic}, \tag{6.89}$$

where $k_{\alpha}$ is the $\alpha$-attenuation factor, which ranges from 0.1–0.5 and, theoretically, must be determined for each sample. Alpha-radiation (only U and Th) has a range of only 20 $\mu$m and produces a diminished TL due to its high ionization density, which leads to rapid recombination. Preheated samples show anomalously high $k_{\alpha}$ values. The $\beta$ dose rate (U, Th, K, and Rb) is dependent on grain size because the $\beta$ dose

**Table 6.7.** Specific dose rates of samples in radioactive equilibrium (Bell 1979): The values in parentheses are for 100% loss of radon; see also Nambi and Aitken (1986)

| Dose | $\mu Gy/a$ ppmTh | $\mu Gy/a$ ppmU | $\mu Gy/\%$ $K_2O$ |
|---|---|---|---|
| $\alpha (k_\alpha = 1)$ | 738.0(309) | 2780(1250) | 0 |
| $\beta$ | 28.6 (10) | 146 (61) | 689 |
| $\gamma$ | 51.4 (21) | 115 (6) | 207 |

(range of 2 mm) increases through a 100-$\mu$m grain due to the production of secondary electrons (Mejdahl 1979; Bøtter-Jensen and Mejdahl 1988). With appropriate corrections, it can be determined to $\pm 4\%$ (Haskell 1983; Wagner 1983).

Estimates of the external and internal dose rates are often calculated from the concentrations of uranium, thorium, and potassium using Eq. (6.90) and the factors given in Table 6.7. Uranium and thorium concentrations (ppm) are determined chemically or measured $\alpha$- and $\gamma$-spectrometrically, potassium is measured with a flame photometer or by AA spectrometry (Sect. 5.2.4.3). Sometimes $\alpha$ counting alone will give a good estimate.

Homogeneous distribution of the radionuclides is a prerequisite if their spatial distribution is not taken into account ("isochron" dating). Fission-track analyses (Sect. 6.4.7) can provide information about whether this is the case.

For geologically young samples ($< 300$ ka), it must be taken into consideration that radioactive equilibrium has not yet been attained (Sect. 6.3). The total paleo-dose is then calculated from the $^{234}U/^{238}U$ and $^{230}Th/^{238}U$ activity ratios as follows (Wintle 1978):

$$d_{AD} = \lambda_{238} C \left\{ d_1 t + \frac{d_2}{\lambda_{234}} \left( \left[ \frac{^{234}U}{^{238}U} \right] - 1 \right) (e^{\lambda_{234} \cdot t} - 1) - \frac{d_3}{\lambda_{230}} \left[ \frac{^{230}Th}{^{238}U} \right] \right\}, \qquad (6.90)$$

where $t$ is the age, $c$ is the uranium concentration in uranium atoms/g sample, and $d_1$, $d_2$, and $d_3$ are the specific effective doses per gram sediment and ppm uranium for the $^{238}U$ decay series as far as $^{206}Pb$. For instance, for the realistic value of 0.5 for the $\alpha$-sensitivity factor $(k_\alpha)$, $d_1$, $d_2$, and $d_3$ are ca. 3.4, 3.0, and 2.7 Gy/ppm $U$, respectively.

A more reliable and more exact measurement of the external dose rate is made in the field with a TL dosimeter [$CaSO_4(Tm)$, calcium sulfate dysprosium phosphor $= CaSO_4:Dy$]. Calcium fluoride (natural) dosimeters are probably even better. The dosimeter is buried at least 30 cm deep at the sampling site for several months to a year. Mesurement with a calibrated NaI (T1) scintillation counter (Sect. 5.2.2.2) in a borehole at least 50 cm deep is less accurate but quicker (Mejdahl and Wintle 1984).

For radiochemically homogeneous systems, highly sensitive measurement of the $\gamma$-activity and the $\alpha$-efficiency factor simplifies TL dating (Guibert et al. 1985).

Procedures that do not require determination of the annual dose have been developed (Aitken 1978) and are being used. In the case of microscopic zircon crystals, with their high uranium and thorium levels, the internal alpha-particle dose is greater than the external dose (5–10%). Autoregeneration of the TL signal during storage of at least 6 months makes it possible to carry out TL dating with a precision of 3–10% without the difficult determination of the dose rate. This technique also overcomes the difficulty of zoned TL emission (Templer 1986b; Templer and Smith 1986; Smith 1988).

*Techniques for determining the radiation-dependent paleo-dose:* Because each mineral and grain-size fraction receives different annual dose rates (due to the inhomogeneous distribution of the radioactive isotopes and the differing penetration depths of the different types of radiation), various microdosimetric techniques have been developed to accurately determine the natural paleo-dose $d_{ED}$ for the different types of radiation ($\alpha$, $\beta$, $\gamma$, and cosmic rays) (Cairns 1976). They are based mainly on mineral separations. In practice, several different techniques are usually used for the same sample. If the resulting dates are in agreement, the age is viewed as reliable. Techniques in addition to those described below are in development (Aitken 1978).

Due to the different penetration depths of alpha, beta, and gamma radiation and the inhomogeneity of their spatial distribution, the microdosimetric measurements are carried out on different mineral fractions or aliquots:

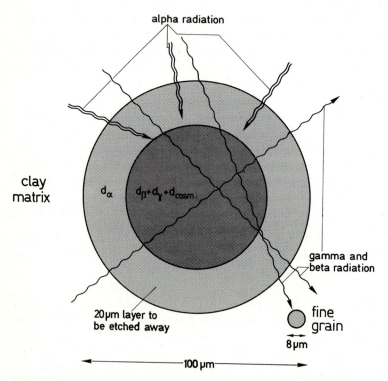

**Fig. 6.83.** Schematic representation of the quartz inclusion technique (after Fleming 1970): The outer 20 $\mu$m of grain surface damaged by $\alpha$-radiation is removed by etching; this restricts the *TL* age determination to the $\beta$, $\gamma$, and cosmic ray doses

*Fine-grain technique* (Zimmerman 1967): This is the most often applied technique. Potsherds about 1 cm thick weighing 5–10 g are used. Since $\alpha$-particles penetrate solid matter only up to 25 $\mu$m, the damage produced by them is limited to a range of several $\mu$m (Fig. 6.83). Thus, the dose-rate calculations are simplified when only fine-grained mineral fractions are used. Undesirable TL is avoided by the use of spectral windows or selective etching. Non-linearity may arise at higher doses (Aitken 1984).

The annual dose rate $r_{tot}$ of the 2–11 $\mu$m feldspar fraction, separated in an acetone suspension, is usually calculated using Eq. (6.89). Considering that $r_{tot}$, needed for Eq. (6.88), can be determined with a precision of no better than $\pm 4\%$, the total error for the age determination is $\pm 5$–10%.

*Quartz inclusion method* (Fleming 1970): 50–100 mg of quartz crystals with grain sizes between 100 and 300 $\mu$m are extracted and about 20 $\mu$m of the surface is removed with hydrofluoric acid. The annual dose rate (Eq. (6.89)] of this fraction does not include the zoned $\alpha$ component (inhomogeneously distributed $\alpha$-emitters) (Fig. 6.83) and therefore the difficult determination of the $\alpha$-attenuation factor is not necessary. A grain-size correction (Mejdahl 1979) for the $\beta$ dose rate is necessary.

The advantages of this technique are that radon losses do not greatly affect the reliability of the results and that quartz shows no anomalous fading. On the other hand, several hundred grains are measured together and the scatter due to a few "bright" grains results in less precise determinations of $d_{ED}$ than obtained with the fine-grain technique. The best precision of TL dating by this techique is $\pm 7\%$. This technique is used, for example, for authentification testing of ancient ceramics (Becker and Goedicke 1978). Because TL saturation occurs earlier in quartz than in feldspar, the upper limit of dating is about 100 ka.

*Subtraction technique* (Fleming and Stoneham 1973): Only the alpha paleo-dose is used for this technique. The $\alpha$ paleo-dose is the difference between the paleo-doses determined using the fine-grain and quartz inclusion techniques. Due to the low alpha decay rate, a precision of better than $\pm 15\%$ is never attained. The technique is limited to dense samples because even very small radon losses lead to very large errors in the age. However, other environmental influences, e.g., secular variations in moisture content, have little effect compared to that on the other techniques.

An improved variant is the *differential dating technique* (Goedicke 1985), which utilizes the dependence of $\alpha$ and $\beta$ absorption on grain size; several quartz fractions between 100 and 1000 $\mu$m are used.

*Zircon inclusion technique* (Zimmerman 1971): This technique is also based only on the alpha paleo-dose. Therefore, it is quasi independent of the environment of the sample because zircon and apatite crystals (40 $\mu$m) have a very high uranium concentration (100–3000 ppm) and thus all other dose components can be neglected. Several crystals of zircon and apatite ($> 100$ $\mu$m across) are handpicked and the TL emission measured. The uranium concentration is derived from the number of tracks resulting from the spallation induced by thermal neutrons (e.g., a dose of 10 Gy) (Sect. 6.4.7).

A handicap of the last two techniques is that the $\alpha$-attenuation factor ($k_\alpha$) changes with increasing dose. But it should be possible to overcome this problem at least partially. High-resolution optical techniques, for which the dating precision is $\pm 10\%$, is one solution, especially for zoned inclusions. An improved version of the zircon technique by Templer (1986b; Smith 1988) removes the need to measure the annual dose and $k_\alpha$. After the natural TL has been measured, the zircon grains are stored for 6 months; the reaccumulated TL is then measured again with a sensitive instrument. Errors can be as low as $4\%$.

*Feldspar inclusion technique* (Mejdahl and Winther-Nielsen 1983): This is a suitable dating method for ceramic objects, heated stones and granite. After grinding of the sample, alkali feldspar and plagioclase ($0.1-0.3\ \mu m$) are separated by suspension in heavy liquids. The $\beta$ dose is obtained via the potassium content, which is especially high in these extracts. Since the TL sensitivity of these minerals is greater than that of quartz, it should be possible to determine ages of up to 500 ka. Tests for anomalous fading are always necessary, however.

*Pre-dose technique* (Fleming 1973): This technique uses quartz extracts whose TL sensitivity at $100°C$ is increased when the aliquots are irradiated with doses of $100-700\ Gy$ before being heated to about $500°C$ for the recording of the glow curves. The procedure is suitable for potsherds that are younger than 1500 years. The dating precision may approach $\pm 7\%$.

*Dating of flint:* The determination of $d_{ED}$ values for heated flint (Göksu et al. 1974; Göksu-Ögelman 1986) and bones was done initially on fragments $300\ \mu m$ thick, because the method is disturbed by triboluminescence and chemoluminescence. In the meantime, it has been learned that $100$-$\mu m$ grains etched with dilute acid yield reliable results. The samples must be stored in the dark and the outer 2 mm removed, since bleaching often occurs.

*Dating of lava and slag:* Plagioclase with a grain size of about 0.1 mm is extracted from the sample and the TL is measured at a high temperature of about $670°C$ (Guérin and Valladas 1980). Crystalline and porous fractions of slag from smelting sites are more suitable for TL dating than the glassy and metallic fractions (Wagner 1983).

*Dating of sediments:* Eolian sediments (e.g., loess, glacier dust, dune and beach sands, glacial and glaciofluvial deposits, as well as water-laid silts) are suitable for dating if the low residual TL energy remaining after bleaching during transport can be determined (Hurford et al. 1986). Deep-sea sediments can also be dated (Wintle and Huntley 1979). Measurements are made along the entire length of a sediment profile. A review of the technique has been published by Wintle (1987).

There are several procedures available (Huntley 1985; Singhvi and Mejdahl 1985; Singhvi and Wagner 1986): The *partial bleach method*, also called the R-$\Gamma$ (*residual-gamma*) *method*, is the most often used. The first glow growth curve is

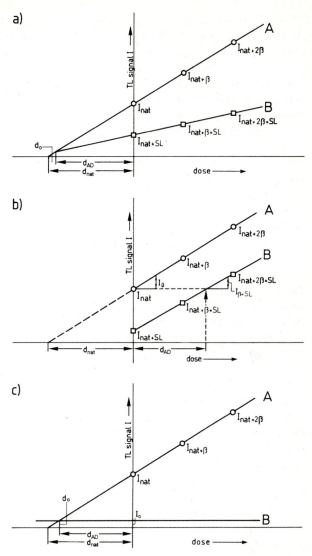

**Fig. 6.84 a–c.** There are three commonly used procedures for dating sediments (after Singhvi and Mejdahl 1985). In the first step of each of these procedures, the natural paleodose $d_{nat}$ of a sediment is obtained by constructing a so-called first growth curve (A). Because a sediment may not have completely lost its initial $TL$ by exposure to sunlight previous to deposition (optical bleaching), the small residual dose $d_0$ must be determined so that the paleo-dose $d_{AD}$ accumulated after deposition can be determined. **a** The most frequent procedure for determining $d_{AD}$ is the R-$\varGamma$ method, in which a second first growth curve (B) is determined after a short period of exposure to the sun or a sunlamp prior to the $TL$ recording. The intercept of the two curves yields $d_0$ and $d_{AD}$. **b** The regeneration method is used if the first growth curve is non-linear. In this case several aliquots are totally bleached, then irradiated, and $I$ is measured. The dose that produces the natural $TL$ signal corresponds to $d_{AD}$, assuming $I_\beta$ is equal to $I_{\beta+SL}$. **c** For the total bleach method, only one sample is exposed to UV light until the $TL$ is no longer lowered ($I_0$). This value is used as $d_0$ for calculating $d_{AD}$. However, "overbleaching" may be a problem

recorded with UV filters as usual and then the second glow growth curve is measured after the aliquots have been exposed to sunlight or a UV lamp for various lengths of time. The intersection of the two straight lines gives a rather imprecise estimate of the equivalent dose (Fig. 6.84) (Wintle 1982). Preheating of the feldspar fraction (which is preferred to the quartz fraction) to 230°C for 1 min (partial thermal washing) increases the precision of the age determination (Wintle et al. 1984). The dating range may be increased to about 300 ka by preannealing at 260°C for 1 min (strong thermal washing) and using a blue filter (Zöller and Wagner 1989). The partial bleach method is also recommended or the feldspar fraction of water-laid deposits using wavelengths longer than 550 nm (Berger 1986).

The *regeneration technique* (Wintle and Huntley 1980) is an alternative method that is particularly suitable for very old sediments. Both methods can be used for sediments, such as loess, that had long exposure to sunlight during deposition (sedimentation rates below 5 mm/a) and where the growth curve is nonlinear. Prerequisite is that the TL sensitivity does not change after bleaching (see also Rendell et al. 1988).

The *total bleach method* (Singhvi et al. 1982) was first applied to sand dunes (Pye 1982). A secondary TL growth curve is regenerated from natural samples that have been bleached artificially for up to 24 h (Fig. 6.84). "Overbleaching" may result in overestimated equivalent doses and, therefore, ages (Berger 1986). The method is preferentially used for young sediments with ages up to at least 30 ka.

Two other procedures have been developed by Mejdahl (1985) for partially bleached sediments. The quartz feldspar procedure for samples of up to 25 ka utilizes differences in optical bleaching characteristics and the dose rates in feldspar and quartz whose dose rate ratio is assumed to be a constant 1.56. Beta/gamma growth curves are constructed for these two fractions. Aliquots of the two minerals are then irradiated with a sunlamp for different periods of time to find the irradiation time that yields the same $d_{ED}$ value for both fractions.

The second procedure uses the feldspar fraction and is suitable for older sediments of ages up to 500–1000 ka because quartz has reached saturation after no more than 70 ka. Several feldspar fractions are irradiated with sunlight or a sun lamp for different periods of time. Only one of them yields a $d_{ED}$ value that is constant over the entire temperature range of the glow curve.

For all procedures, the water content must be known, as it diminishes the dose rate and corrections of up to 50% may be necessary. Moreover, time-dependent changes in the natural dose rate may be caused by post-deposition diagenetic processes like compaction, leaching, or collapse. In the case of volcanic effusives, $^{226}Ra_{excess}$ must be taken into account (Berger 1985a).

## Scope and Potential, Limitations, Representative Examples

*Scope and Potential.* The main application of the TL Method is in the field of archeology (Aitken 1978; Wagner 1983). It permits direct dating of potsherds, which are the basis of the classification of most cultures, owing to the multitude of types. In contrast, radiocarbon dating (Sect. 6.2.1) is often possible only on accessory finds, such as charcoal and food residues, which are not necessarily the same age as the pottery. Because it does not require time calibration, the TL method usually has an advantage over the radiocarbon method, especially for objects from the first millennium BC. Moreover, TL is the only method besides the obsidian hydration method (Sect. 8.6) that can be used for dating heated flint and obsidian, which are usually found without any accessory objects that can be dated by other methods (Göksu-Ögelman 1986). These materials were commonly used during the neolithic period of the Stone Age; they were often heated so that they could be more easily split.

Increasing application of this method is being found for authenticity testing of ceramic objects. For this purpose the precision requirements are relatively low because only a distinction between the period of the style and the last 100 years is necessary, since the latter is the period from which a falsification may be expected.

A new, rapidly expanding utilization in the field of Quaternary geology is the dating of sediments, especially loess (Singhvi and Mejdahl 1985; Berger 1986). It may become more important than archeological applications.

*Limitations.* The main error in TL dating occurs in the determination of the external and internal annual dose rates. Exact and reliable results can be guaranteed only by very time-consuming and personnel-intensive analyses on site and in the laboratory. For example, dating errors of up to 15% have been demonstrated when uranium or thorium has been gained or lost by the sample while it was in the soil (Hille 1979; Ikeya 1982a). So that such errors can be recognized and possibly corrected, $^{210}$Po activity should be measured $\alpha$-spectrometrically (Pernicka and Wagner 1982). This would permit the recognition of any disturbances in the radioactive equilibrium (Wintle 1978; Sect. 6.3). Knowledge of the alpha efficiency is another problem.

The often discussed problem of the influence of secular fluctuations in the moisture content of the samples (ca $\pm$ 20%) and of loss of radon on the dating results appears to be less serious for humid areas than for arid ones. Elevated moisture content decreases the dose rate by increasing absorption by about the same amount as the decreasing loss of radon increases the annual dose. In arid areas this compensation does not always take place because the samples are often completely dried out. Independently of these problems, it will always remain an open question whether the dose rate remained constant the entire time of aging.

Spurious chemoluminescence or triboluminescence, which produce ages which are too large, are caused by chemical reactions or mechanical stress (e.g., grinding, polishing, or pressure).

Part of the radiation damage in some minerals, especially that in feldspar or plagioclase from igneous rocks, is instable (long-term and short-term anomalous fading) and cannot be used for age determinations. The other part may have finite lifetimes, which makes age corrections necessary. For example, Mejdahl (1988, 1989) found lifetimes of 800 to 5400 ka for TL signals of feldspar, which result in ages that are too small by 14 and 6 ka, respectively, in the case of Eemian samples. Suitable pretreatment or UV filters are required to remove the anomalous component. For other kinds of samples, the upper limit of the dating range is reached when all of the electron traps are occupied, i.e., saturation is reached. After this, the TL age does not increase with further radiaton. For example, the saturation dose for calcite is about 1 kGy, for feldspar it is 3–5 kGy, and for quartz only 0.3–0.35 kGy. For many minerals, the upper limit of the dating range corresponds to an $\alpha$-dose of about 1.5 kGy (Nambi 1984).

Apart from eolian sediments, geological samples are seldom suitable for dating if they have been diagenetically altered or have a complex genetic history, e.g., calcrete (Netterberg 1978). Samples that have been heated may have significantly altered TL sensitivities, which results in incorrect ages.

Calcites of the same age can yield results that deviate from each other by up to 200%. This has been explained by inhomogeneous uranium distribution (zoning), which can affect the TL age even in the 20 $\mu$m range. Improvements have been obtained through the use of blue filters (Debenham and Aitken 1984). Because the uranium content of calcite is generally low, the external $\gamma$ dose must be determined especially carefully. As with the analysis of flint, which is disturbed by bleaching, the outer layer must be removed because it may have received an elevated dose of radiation from radon and its daughter elements (Hennig and Grün 1983).

A major problem for TL dating of peat via feldspar grains (100–300 $\mu$m fraction) is the estimation of the dose rate. Compaction may have considerably changed the water content and the uranium content at the boundary layers may have also changed. These problems are discussed by Lamothe and Huntley (1988).

The interpretation of the TL signals for meteorites in terms of radiation ages or terrestrial ages has not yet been solved (Sears and Hasan 1986).


### Non-Chronological Applications

*Stratigraphy:* TL analysis is being increasingly used by sedimentologists for stratigraphic correlations (e.g., Wintle 1987). However, there are many other applications, of which only three examples will be given here:

*Uranium exploration:* The properties of minerals that cause them to act as a TL dosimeter and to change TL capability during metamictization have been utilized for uranium exploration in Australia. Cumulative deposits were traced in this way over distances of several kilometers from the present ore position (Hochman and Ypma 1984). Another study in France used TL to locate uranium occurrences, migration pathways of randon, and uraniferous mineralizing solutions (Charlet et al. 1986).

*Diagenesis:* TL has been successfully applied in a study of crystal defect structures related to diagenesis resulting fron Na, Ca, and Li exchange in clay (Lemons and McAtee 1983). Göksu-Ögelman and Kapur (1982) found that weathering stages in basaltic rocks can be determined quickly and cheaply by TL analysis combined with analysis for kaolinite and smectite.

*Meterorites:* TL analysis of meteorites, especially those found in Antarctica, permits the identification of potentially paired groups, the identification of specimens with unusual pre-fall histories, and a rough determination of terrestrial ages (Sutton and Walker 1986).

## 6.4.2   Optical Dating (OSL) Method*

(Huntley, Godfrey-Smith, Thewalt 1985)

### *Dating Range, Precision, Materials, Sample Size*

The optical dating method, which is in development, is supposed to be suitable for determining the time since sediment grains (e.g., quartz, zircon, and apatite) (several milligrams) were exposed to light for even a short time. This permits the study of sediments deposited at rates greater than 1 cm/a. Dating range: 1–700 ka for quartz and 1–100 ka for zircon with a precision of a few percent (Huntley et al. 1985; Smith 1988).

### *Basic Concept*

The principle of this method is the same as that of the TL method (Sect. 6.4.1), but only the light-sensitive traps are involved, which is why it is called the OSL method (for optically stimulated luminescence). The annual dose rate from environmental radiation and the total radiation dose accumulated since the last short-term exposure to sunlight are determined. The zeroing mechanism is triggered by sunlight and is more effective than that of TL which is why it is especially suitable for dating sediments. An unbleachable residual level does not remain, even after an exposure of only a few seconds. A low-power laser, instead of heat as in the TL method, is used to free trapped, light-sensitive electrons. The induced luminescence is directly proportional to the radiation dose, and thus the age.

### *Sample Treatment and Measurement Techniques*

The beam of a 2-W argon-ion laser (514.5 nm) is passed through a long-pass filter to absorb unwanted light with wavelengths shorter than 500 nm. An energy flux of 0.5–5 mW/cm$^2$ is directed with a mirror onto the sample. The scattered green light of the laser is removed with a filter, resulting in relative attenuation of the optically stimulated luminescence at 400 nm. The luminescence passed by the filter is measured with a photomultiplier. The counts per second per gram are determined. OSL signals show a rapid, non-exponential decay, which is why exposure of the sample to sunlight must be avoided. It is under discussion whether thermal removal of instable components of the OSL signals should be carried out (Smith et al. 1986). Methodical aspects affecting the determination of the equivalent dose are discussed by Rhodes (1988).

Hütt et al. (1988) and Hütt and Jaek (1989a) showed that the use of infra-red laser is physically more appropriate than the use of an argon-ion laser. In addition, the dating equipment should be more simple and the dating results more reliable.

The paleo-dose is determined via an accumulative radiation experiment with a $^{60}$Co gamma source (Fig. 6.82). Rapid, automatic measurements seem to be possible. Another possibility is the TL autoregenerative dating technique for zircon extracted from sediments (Smith 1988).

### Scope and Potential, Limitations

**Scope and Potential.** This method has been applied to grains of quartz and zircon from bleached dune sands and an $A_h$ soil horizon (Huntley et al. 1985; Smith et al. 1986). This application illustrates the great potential of this method, especially as it can also be used for rapidly deposited sediments that cannot be dated with the TL method (Sect. 6.4.1). The agreement with $^{14}$C and TL results was satisfactory. There seems to be no interfering incandescence and the sample properties are not altered during the dating procedure. The high sensitivity and ease of use are the most notable features of this method.

**Limitations.** The main problem is the change in the OSL signal during exposure to light. This makes stringent precautions against bleaching necessary. Instable components may be removed by thermal preheating. However, the minimum age obtainable may be limited by recuperation of the OSL signal following exposure to sunlight or laser beam (Smith et al. 1986). Samples with ages of up to 10 ka may be affected (Aitken and Smith 1988).

### 6.4.3 Electron Spin Resonance (ESR and EPR) Method**

(Zeller, Levy and Mattern 1967)

### Dating Range, Precision, Materials, Sample Size

The ESR method is a young and not yet fully developed dating method with a wide potential for its application (Hennig and Grün 1983; Ikeya and Miki 1985a; Grün 1988, 1989; Ikeya 1988). It is particularly recommended for determining the age of Quaternary aragonitic coral (Ikeya and Ohmura 1983), thick-walled molluscs (e.g., *glycimeris* (Ikeya and Ohmura 1984; Radtke et al. 1985), and planktonic foraminiferal tests (ca. 150 mg) from pelagic sediments (Sato 1981, 1982; Wintle and Huntley 1983; Siegele and Mangini 1985), lunar samples (Ikeya and Miki 1985a), as well as tooth enamel (Grün and Invernati 1985; Schwarcz 1985; Grün et al. 1988). Dating of bones and whole teeth (Ikeya and Miki 1980a) has been less successful. In addition, speleothem and travertine, as well as caliche (Özer et al. 1989) have been studied (Ikeya 1983a), but not always with reliable results (Grün and Hentzsch 1987). The dating of ceramics has been attempted (Maurer et al. 1981). Cooling ages of quartz, feldspar, silicates, glass, apatite, and other minerals and the crystallization age of gypsum have been determined. The carbonate of sediments from Searles Lake in California has been analyzed for the applicability of the method (Ikeya et al. 1987). Petrified wood with high uranium contents (Ikeya 1982b) may also be suitable.

The dating range for these materials stretches from several centuries up to many hundred thousand years, possibly more than 1 Ma or even 100 million years (Odom and Rink 1988). Only 20–250 mg is required for an ESR analysis.

It may be possible to reconstruct earthquake history on the basis of ESR analysis of pressure-stressed minerals (e.g., quartz) in the fracture zone of faults (Miki and Ikeya 1982; Ikeya et al. 1983; Fukuchi et al. 1986) because the ESR signal becomes weaker when the material has been under high pressure.

It may also be possible to determine the production date of foods and the age of other organic matter with the chemical ESR technique (Ikeya and Miki 1980b, 1985a; Ikeya 1988) (Sect. 8.4).

According to the literature, a precision of about $\pm 5\%$ should be obtainable for the total paleo-dose ($d_{ED}$) and about $\pm 10\%$ for the annual dose rate (Hennig and Grün 1983). However, an international comparison test on calcite without any sample treatment yielded $\pm 30$–$150\%$ for the $d_{ED}$ values alone (Hennig et al. 1985). A comparison of U/Th and ESR dates yielded a correlation coefficient of only 0.2 (Fig. 6.86). More realistic values for the precision of $d_{ED}$ determinations have been published by Smith et al. (1985); $\pm 30\%$ for values $\leqslant 10$ kGy and $\pm 10\%$ for values above that.

### Basic Concept

Natural crystals contain $10^{15}$ to $10^{17}$ crystal defects per $cm^3$ (vacant sites in the crystal lattice: sites with missing negative ions are electron traps and sites with missing positive ions are hole traps). These defects are produced during growth or by diagenetic displacement of atoms. Spurious cations in the mineral, especially Mn, Fe, Ni, and Co, also represent electron traps with high lifetimes.

Ionizing radioactive radiation lifts electrons from the valence band to the conduction band (Fig. 6.78). Most of them recombine immediately. A very small number fall into quasi-stable traps in "forbidden" energy levels, some of which may have lifetimes of up to several million years (Fig. 6.79). Prerequisite for the dating method is that the number of traps occupied by electrons is proportional to the total accumulated dose and thus the age.

Traps occupied by a single electron act as paramagnetic centers, whose density can be measured by ESR. In a magnetic field, unpaired electrons have one of two orientations, defining two energy states: the lower corresponding to alignment with the field, the upper one against the field. The difference between these two energy states $\Delta E$ is given by

$$\Delta E = g\beta H, \tag{6.91}$$

where $g$ = the spectroscopic splitting factor, which is characteristic for the energy levels of the electron traps,

$\beta$ = the Bohr magneton ($9.273 \times 10^{-24}$ $Am^2$), and

$H$ = the magnetic field strength (in A/m).

If an electromagnetic field (microwave frequency) is superposed on a static magnetic field, the electron spin vector oscillates between orientation with and against the field. The induced absorption and emission energies theoretically cancel one another. The resonance state is defined as follows:

$$hv = g\beta H \tag{6.92}$$

where $v$ = the frequency of the electromagnetic field (in $\sec^{-1}$), and
$\quad h$ = Planck's constant ($6.625 \times 10^{-34}$ joule·sec).

In practice, some of the energy is lost to the crystal lattice, which is registered as an absorption signal that is almost always the superposition of many individual signals; this is the ESR signal. The amplitude is a function of dose and increases with the number of radiation-induced paramagnetic electrons captured in the traps whose energy levels are defined by the resonance state given by Eq. (6.92). Nearly all substances show several resonance frequencies because the magnetic field vector in the crystal lattice is split into several sublevels due to interaction of the magnetic moments of neighboring atoms. The parameters $h$ and $\beta$ are natural constants, $v$ and $H$ are physical variables, and $g$ is a material-dependent parameter representing the resonance state.

The age determination is based on Eq. (6.88) and includes the determination of the accumulated paleo-dose $d_{ED}$ and the dose rate $r$ (see Also Sect. 6.4.1).

### Sample Treatment and Measurement Techniques

The determination of $d_{ED}$ by the ESR method is similar to that by the TL method (Sect. 6.4.1).

The samples are finely ground and sieved to obtain the 100–400 $\mu$m fraction. This range of grain sizes is chosen to guarantee isotropic distribution of the crystal axes in the ESR spectrometer. To avoid the influence of grinding on the ESR signal, the samples are etched with dilute acid. Several aliquots of one fraction are irradiated in increasing steps (of 1 kGy each), each aliquot with a dose greater than the previous one. Saturation may occur after several steps of 1 kGy. Carefully calibrated $^{60}$Co or $^{90}$Sr sources are necessary. The type of radiation seems to have no influence. After waiting for up to a week for annealing, the instable state has decayed.

An ESR spectrometer consists of a microwave generator, an electromagnet with a highly stable field (in which the sample cuvette is positioned), and an electronic unit for recording the ESR signals. ESR instruments are very expensive, but are often available for age determinations in chemical departments and medical schools of universities.

*ESR measurement:* The untreated and 4–5 irradiated aliquots (several mg) are each placed in a Teflon cuvette (Teflon is not paramagnetic). A tiny ruby crystal built into the wall of this cuvette provides a signal that is used for calibration and stabilization of the microwave generator output of the ESR spectrometer. Another possibility for checking the stability of the system, especially for repeat analyses, is to use on of the

$Mn^{2+}$ lines as internal standard. In samples with a high manganese content these lines are not affected by the radiation. Normally, measurements are made at room temperature; the sensitivity, however, can be increased significantly by cooling the sample with liquid air.

Instead of directly recording the absorption signals, which is assumed to be proportional to the total paleo-dose, the first derivative is plotted by modulation of

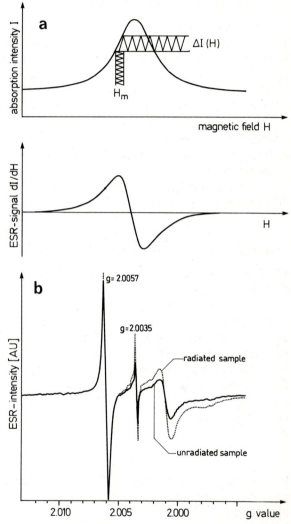

**Fig. 6.85. a** General representation of the procedure for producing an ESR curve: The ESR absorption signal ($I$) is scanned by modulating the magnetic field $H$ with a high-frequency magnetic field $H_m$. This yields the first derivative of the absorption signal $dI/dH$ (after Hennig and Grün 1983). **b** ESR spectra of an unirradiated aliquot and an irradiated aliquot of a aragonitic Pleistocene coral from Antalya, Turkey: Only the peaks at $g = 2.0035$ and $g = 2.100$ increase with radiation dose and thus can be used for dating; the other peak shows no dose effect and hence is not suitable (after Radtke and Grün 1988). *AU* arbitrary units

the magnetic field (Fig. 6.85). The modulation signal (about 100 kHz) should not have an amplitude greater than 0.5 gauss$_{pp}$ (*pp* means "peak to peak") so that sufficiently fine resolution of the ESR signals is attained and interfering signals can be recognized. Such interfering signals may belong to traces of Mn, Fe, REE, and organic radicals. If these signals are too strong, the ESR spectrum may not be usable for age determination. The microwave power may not exceed 2 mW, an order of magnitude less for quartz. However, it is recommended to determine the optimum microwave power (Lyons et al. 1988).

*Natural ESR signals:* ESR signals are described using the following parameters:

1) the *g value* [Eq. (6.92) to four significant figures after the decimal];
2) *line width* (in gauss) or curve form (which also reflects the fine and hyperfine structures);
3) the *thermal stability* of the paramagnetic centers expressed as the mean trap lifetime. This must be at least one order of magnitude greater than the age to be determined if reliable $d_{ED}$ values are to be obtained. The mean trap lifetime may be estimated for each *g* value of interest by heating several aliquots to different temperatures, measuring the equivalent doses, and constructing an Arrhenius plot (Fig. 6.79; see also Sect. 8.3) (Aitken 1978). However, very inaccurate values may be obtained. Another way to estimate the mean trap lifetime is to measure the artificial ESR growth curve of a sample that has already steady-state equilibrium (Barabas et al. 1988a). Incorrect values are obtained if splitting of the signals occurs, for example as a result of the preheating (Grün and Invernati 1985).
4) The *modulation* and *exciter frequencies*, the *microwave energy*, and the *field strength of the magnet* should also be given.

An ESR signal at a particular *g* value is suitable for age determination only if (i) its amplitude is a function of the dose rate, (ii) it is missing in samples with an age of zero, (iii) it is thermally stable, (iv) there is no interference from neighboring signals, and (v) there is no anomalous fading (Templer 1986a).

**Table 6.8.** Common ESR signals in carbonate spectra (after Hennig et al. 1985)

| Label | g value ± 0.0002 | g width × 10⁴ | Half life [a] | Properties |
|-------|------------------|----------------|---------------|------------|
| A h₁ | 2.0057 | 3 | $10^6$ | Induced by organic radicals |
|  | 2.0043 | 4–6 |  | Induced by organic radicals |
| B h₂ | 2.0035 | 3 | $2\text{–}400 \times 10^8$ | Thermally stable, light sensitive, sometimes suitable for dating |
| h₃ | 2.0021 | 3 | 2–3 | Thermally instable |
| C h₄ | 2.0007 | 2 | $0.7\text{–}7 \times 10^6$ | Thermally stable, the most recommended for dating |
|  | 2.0001 | 2 |  | Induced by grinding |
| D h₅ | 1.9996 | 1–2 |  | Interference by $Mn^{2+}$ signal |

The most important properties of the ESR signals characteristic of carbonates (Grün and de Cannière 1984) are compiled in Table 6.8. Not all of these signals are always present. In some publications, various other labels are used instead of $g$ values (Ikeya and Ohmura 1983; Ikeya 1984, 1988).

To determine $d_{ED}$, a suitable signal is selected. For instance, $g = 2.007$ is recommended for forminifers and some travertine, but other signals interfere in the case of coral and molluscs. The ESR peak heights of the unirradiated and irradiated aliquots are plotted versus the radiation dose to obtain a growth curve analogous to Fig. 6.83. The $d_{ED}$ value is given by the x-intercept. For samples like molluscs, non-linear growth curves are obtained (Barabas et al. 1988a); these are extrapolated empirically or an exponential function is used to find the x-intercept (see Singhvi and Wagner 1986). This can be problematic (Katzenberger and Willems 1987; Wintle 1987; Fig. 6.86).

*"Isochron" dating:* It may be possible to date stalagmites more precisely between 0.3 and probably 1 Ma with the "isochron" method (Karakostanoglou and Schwarcz 1983), which can also be called a "mixing line" method. Samples from thin layers of

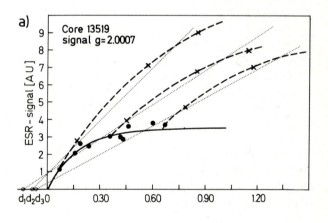

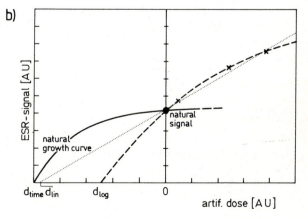

**Fig. 6.86. a** Non-linear natural growth of ESR signals (*dots*) for calcite foraminifers of marine sediments deviating artificial growth of ESR signals (*crosses*). **b** The statement of Barabas et al. (1988b) that a linear fit, instead of a logarithmic fit, yields an ED dose closer to the true one is questionable when the linear extrapolation ($d_1, d_2,$ and $d_3$) in **a** are taken into account; *AU* arbitrary units

nearly the same age but very different uranium contents are analyzed. The $d_{ED}$ values are plotted versus uranium content; the y-intercept of the resulting straight line is equal to the $d_{ED}$ value from which the age is calculated using Eq. (6.88). Unfortunately, the applicability of the isochron method to speleothem is limited because their uranium content is usually very low and varies little.

The "isochron" method should improve dating of tooth enamel (5–10 mg), which seems to be datable up to ages of 5 Ma (Grün and Invernati 1985; Schwarcz 1985). The large variation in the uranium content of the adjoining dentin or cementum limits the dating precision to $\pm 25\%$. Owing to the accumulation of uranium from groundwater during aging, only the inner, protected parts can be viewed as a closed system. The peak suitable for dating purposes is at $g = 2.0018 \pm 0.0004$. Grinding does not affect the results.

*Annual dose rate:* To calculate the age using Eq. (6.88), the internal and external annual doses rates must first be determined (Sect. 6.4.1). This value is the main source of error in ESR age dating. For carbonate samples from caves and deep-sea sediments, $r_{tot}$ is 0.5–2 mGy/a, for corals it is about 0.7 mGy/a. The contribution from cosmic radiation is about 0.15 mGy/a at the Earth's surface. The alpha effectiveness is a major problem (Lyons 1988).

Sometimes, the annual dose is calculated from the uranium and thorium concentrations or from in-situ measurements (Sect. 6.4.1). For young samples, it must be taken into consideration that the uranium and thorium decay series have not yet reached equilibrium [Wintle 1978, Eq. (6.90)]. Large dating errors may result if the uranium distribution is anisotropic.

*Dating independent of the dose rate:* Owing to the difficulties in determining the external and internal dose rate, other possibilities are being searched for. For very young samples, Ikeya and Miki (1980b) have suggested monitoring the in-situ growth of the dose rate with digital ESR spectrometers for several years at the sampling site. A $Mn^{2+}$ line could serve as a reference for the stability of the equipment. Debuyst et al. (1984) proposed that ESR analyses could be made using ESR signals for paramagnetic traps with different, but very exactly known lifetimes. This would provide a direct possibility for dating without the necessity of estimating the annual dose. However, already the first attempt to do this confirmed the great uncertainties resulting from imprecise determination of the lifetimes of the traps (Grün 1984).

### Scope and Potential, Limitations, Representative Examples

**Scope and Potential.** The TL (Sect. 6.4.1) and ESR dating methods are analogous techniques. The latter is better suited for older samples and non transparent ones. The dating range is supposed to cover at least the entire Quaternary. ESR analyses can be repeated many times with the same sample and require little time. The amount of sample needed is very small. Four these reasons the potential of the method seems to be very great. But more time is needed for development of a reliable standard procedure.

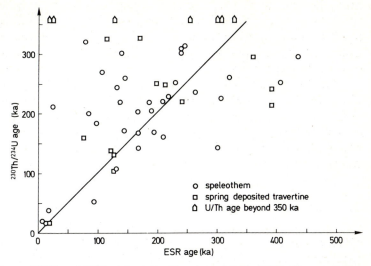

**Fig. 6.87.** Comparison of ESR and U/Th dates (after Hennig and Grün 1983). The correlation coefficient is 0.2. If the ESR and U/Th dates agreed, they would lie along the line

*Limitations.* At the present time, the reliability of ESR dates is often over estimated (e.g., Geyh and Hennig 1986). This is shown for instance, by the poor reproducibility of the $d_{ED}$ values determined in an international comparison test (Hennig et al. 1985). Difficulties are also indicated by a comparison of ESR and U/Th dates taken from the literature (Hennig and Grün 1983; Fig. 6.87). However, processes influencing ESR dating are being increasingly understood (e.g., Barabas et al. 1988a, b; Katzenberger and Willems 1987) and materials formerly believed to be suitable have been found to be unsuitable for dating, e.g., as molluscs (Katzenberger 1989).

*Sample history:* Chemical reactions, heat, pressure, shock, and light attenuate some ESR signals and thus are sources of error. Therefore, the suitability of a sample can be limited by its previous history and hence must be known. For instance, preheated materials yield $d_{ED}$ values, and thus ESR ages, that are too high, recrystallized substances yield ones that are too low. Direct irradiation of the samples with sunlight or a UV lamp, as well as annealing, can cause the transfer of electrons from one trap level to another, resulting in changes in the ESR signals. Moreover, it has been found that preheating results in irreversible annealing of the ESR traps for the ESR signal at $g = 2.0007$ (Barabas et al. 1988a). Anamalous fading has also been observed in carbonates (Templer 1986a). In some rocks the electrons are released from the traps at temperatures below 50 °C. This explains differences in the thermal stability of the ESR signals. Travertine, molluscs, and foraminifers, for example, have very low thermal stabilities (Wintle and Huntley 1983). The growth curve in these cases is nonlinear and nonexponential. making extrapolation to the $d_{ED}$ value very problematic (Wintle 1987; Barabas et al. 1988a, b). But the shape of these saturation growth curves seems to be rather reproducible, at least for the same material (Katzenberger and Willems 1987). Another problem may be that apparently linear growth curves are obtained from traps whose lifetimes are long enough to show no anomalous fading but are much shorter than the dating range of the sample. This can be the case when the signal is the superposition of two or more ESR signals with different lifetimes and saturation behavior. At present it is unknown whether this saturation behavior changes during fossilization (Katzenberger, pers. comm.). Ages corrected for signal lifetime may be as much as twice as large as the signal lifetime, whereas without such a correction only ages up to 0.2 times the signals lifetime are possible. The corresponding correction factors for $d_{ED}$ increase exponentially with increasing $d_{ED}$ and signal half-life (Barabas et al. 1988a; Molodkov 1988; Hütt and Jaek 1989).

Even a small proportion of allochthonous material of very high age in the sample will cause the ESR age to be in error by a large amount, owing to the very wide dating range of the method. The Mg content seems to strongly influence the signals at $g = 2.0007$, which is why the signal for coral is larger than that for molluscs. This is because a weighted mean of the age is determined based on the two components. In contrast, the effect of such very old components is not as large for the radiocarbon method because the age range is narrower.

Anomalous fading of radiation damage at normal temperatures and saturation are additional disturbing effects. Other effects that can also limit the dating range may not even be known yet, as has been shown for the TL method (Nambi 1984).

*Open systems:* As for the TL method, dating errors occur if uranium or thorium is lost or gained during aging. Uranium enrichment leads to ages that are two low and leaching produces ages that are too high. Models for correcting for this are not yet satisfactory (Ikeya 1982a; Ikeya 1985; Grün and Invernati 1985), mainly due to the difficulty of identifying and modeling the presence of an open system. Promising attempts have been made (Grün et al. 1988). However, the errors could be smaller (e.g., for marine sediments) than for uranium-series dating (Sect. 6.3) because the number of traps responsible for the ESR signal increases mainly due to external radiation. The history of the U uptake may be evaluated by combined ESR and U/Th analysis (Sect. 6.3.1) using a fitting procedure for the age estimation (Grün et al. 1988).

*Annual dose rate:* The water content of carbonate samples strongly influences the dose rate and therefore the apparent ESR age (Fig. 6.88). This has serious consequences for an age determination because it has fluctuated by unknown amounts in the past, especially in present-day arid areas. Erroneous estimates of the annual dose rate is another problem of ESR dating, for instance speleothem. In poorly ventilated caves, for example, during glacial and possibly interstadial periods, when a thick cover of snow and ice was present, the radon concentration could have been high enough for radioactive daughter product(s) to be deposited on the surface of the speleothem in sufficient amounts to cause considerable error in an age determination due to underestimation of the annual dose rate (Hennig and Grün 1983). Ages that are too large by a factor of 2 would be obtained for the outer millimeter; in extreme cases the inner zones would also be affected.

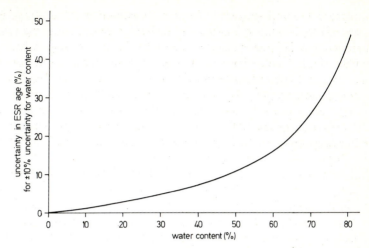

**Fig. 6.88.** Relative uncertainty in the ESR age for a $\pm 10\%$ uncertainty in the water content as a function of the water content of a sample (after Hennig and Grün 1983). The higher the water content, the greater its influence on the ESR age determination

Wintle and Huntley (1983) discuss the complications inherent in determining the annual dose for marine sediments, which show stable signals ($g = 2.0034$) for samples with ages up to 1 Ma (Sato 1982).

*Sample treatment and measurement:* One cause of problems in ESR dating may be the sample preparation, which is sometimes suggested to be unnecessary (Hennig and Grün 1983). Detrimental effects due to grinding, uranium zoning, anomalous fading, and pre-annealing have been demonstrated (e.g., Regulla et al. 1985; Templer 1986a). Analogous problems exist, for example, for the TL method and the fission-track method (Sect. 6.4.7) and are overcome by proper choice of sample and mineral extraction techniques. This is not yet routine for ESR dating. Obviously, measurment parameters also play an important role, for example, the power of microwaves on the accumulated dose (Lyons et al. 1988).

***Representative Examples.*** The ESR method is widely applied in Quaternary research and the spectrum of publications is now so large that it cannot be covered within the scope of this book. A survey is given in a review by Hennig and Grün (1983) and in the Proceedings of TL and ESR specialist seminars (e.g., Ikeya and Miki 1985b; Wagner et al. 1985). As many papers also describe improvements in the method and analytical procedures or compare the results of several dating methods (e.g., Barabas 1988a, b), most of the pioneering applications are already mentioned in this chapter. The citations listed by Ikeya and co-workers (who have been the main promoters of this method) cover more or less the whole spectrum of applications.

Many studies deal with the dating of speleothem at major paleolithic sites (Ikeya 1983a), travertines from river terraces (Grün and Hentzsch 1987), and lacustrine sediments (Ikeya et al. 1987). Direct dating of paleontologic and anthropaleontologic artifacts seems to be possible by ESR analysis of bones (e.g., of mammoths) or tooth enamel (Grün and Invernati 1985; Schwarcz 1985). In connection with studies on global sea-level changes, ESR analyses have been made on fossil shells and coral from coastal terraces in all parts of the world (e.g. Ikeya and Ohmura 1983). The dating of foraminifers from pelagic sediments seems to be promising (e.g., Sato 1981, 1982; Siegele and Mangini 1985). ESR age determinations have been made on volcanic rocks in connection with studies on tectonic movement, earthquake research, and geothermics, mainly in Japan and North America (Ikeya et al. 1983; Fukuchi et al. 1986). A recent study has also used two ESR peaks for quartz grains with different thermal stabilities to obtain the formation age and metamorphic age of volcanic rocks (Shimokawa and Imai 1987). The results are in good agreement with the TL and fission-track ages. Lunar samples have also been analyzed (Ikeya and Miki 1985a).

### Non-Chronological Applications

*Archeological and anthropological applications:* ESR analyses can provide information on whether an archeological object has been subjected to heat treatment (Robbins et al. 1978). This must be known, for example, for the TL dating of flint (Sect. 6.4.1). The temperature to which bones have been heated is supposed to be recognizable from the shift in the $g$ value (Sales et al. 1985). Atomic bomb dosimetry has also been done (Ikeya and Miki 1985a).

*Geological applications:* Geothermal studies have been carried out by Ikeya (1983b). Movement along a fault has been dated by ESR analysis of quartz grains, ages of $0.3 - 2$ Ma were determined (e.g., Fukuchi et al. 1986). The detection of traces of methane trapped in primary quartz is another application. This very sensitive and non-destructive analysis is based on a quartet ESR signal which appears after gamma irradiation of a quartz sample containing traces of methane (Ikeya et al. 1986).

### 6.4.4  Exo-Electron Method (TSEE Method)

(Lewis 1966)

### Dating Range, Precision, Materials, Sample Size

Dating has been done on bones and dentin (ca. 100 mg) with ages of up to 100 ka with $\pm 10\%$ precision (McDougal 1968; Dalton 1972). The sensitivity of the method was reported to be greater than that of the TL method (Sect. 6.4.1). At present, no laboratory is working with the method.

### Basic Concept

Radiation damage produced by radioactive trace elements in bones, as well as environmental radiation, can be detected by the emission of exo-electrons during heating (thermally stimulated exo-electrons, TSEE). The number of exo-electrons is proportional to the absorbed radiation dose (total dose) and, therefore, to the age (Sect. 6.4).

### Sample Treatment and Measurement Techniques

The samples are carefully ground and divided into aliquots. The aliquots are irradiated one after the other increasing the dose in steps of 1000 Gy. The exo-electrons are counted during heating with semiconductor counters (Sect. 5.2.2.3). The total dose is obtained from the growth curve (Fig. 6.82).

The internal and external dose rates are estimated from the uranium and thorium content [Eq. (6.90)], the age is calculated using Eq. (6.88).

Effects comparable to triboluminescence, chemoluminescence, and bioluminescence, which disturb age determinations with the TL method, have not yet been sufficiently investigated.

### Scope and Potential, Limitations

This dating method has not yet been applied in practice, so it is difficult to make an assessment of its potential. One reason for this may be that bones are often open systems with respect to uranium and thorium. Therefore, a constant annual dose rate during aging cannot be assumed (Sect. 6.4.1). In addition, the exo-electron emission is dependent on the content of water of crystallization, which is usually not constant.

### Non-Chronological Applications

The TSEE method is used to determine personal doses in health physics.

## 6.4.5   Thermally Stimulated Current (TSC) Method

(Hwang and Fremlin 1970)

### Dating Range, Precision, Materials

This method has been used only once for dating basalt with ages of 1–2 Ma with ± 25% precision.

### Basic Concept

The number of electrons in thermally stable traps in non-conducting substances as a result of irradiation is proportional to the total absorbed radiation dose ($d_{ED}$) (Sect. 6.4). The electric current that is produced if the sample is heated in an electric field is a linear function of $d_{ED}$. The samples do not need to be transparent or ground up. The principle of the method is analogous to that of the TL method (Sect. 6.4.1).

### Sample Treatment and Measurement Techniques

A thin section, approximately $2\,cm^2$ and 1 mm thick, is prepared; conducting layers are plated to opposite sides. A potential of several hundred volts is placed across these two layers and the changes in the current are recorded while the sample is heated from 90 to 350°C over a period of about 2 min. To determine the total dose ($d_{ED}$), the thin section is irradiated in steps of 10 kGy from 10 to 100 kGy, each time recording the current change (Sect. 6.4.1). Problems with saturation have not been encountered. The $d_{ED}$ value is obtained from the growth curve (Fig. 6.82).

The annual dose is calculated from the uranium and thorium activities [Eq. (6.90)], the age is obtained using Eq. (6.88).

### Scope and Potential, Limitations

The scope and potential of the method cannot yet be assessed because it has been used only once for an age determination (Hwang and Fremlin 1970). Sources of error analogous to black-body radiation or chemoluminescence in TL dating are not expected.

## 6.4.6   Differential Thermoanalysis (DTA)

(Kulp, Volchok and Holland 1952)

### Dating Range, Precision, Materials, Sample Size

This method has been used experimentally for dating a few milligrams of metamict minerals (e.g., zircon, samarskite, microlite, fergusonite, pyrochlore, thorite, ellsworthite and gadolinite) and microlites with ages over 200 Ma. It seems to be possible to date oil paints (ca. 0.2–1 mg) up to 100 years old for authentification tests of suitable paintings (Preuszer 1978, 1979). At present, no laboratory is using the method.

### Basic Concept

The energy stored in a crystal lattice resulting from radiation damage by α-particles can be measured by differential thermoanalysis (DTA). This energy is released during heat-induced recrystallization and measured. It is proportional to the total radiation damage to the crystal lattice. Therefore, with an estimate of the annual dose rate from the uranium and thorium concentrations [Eq. (6.90)], the age is calculated analogously to the TL method using Eq. (6.88) (Sect. 6.4.1).

The cause of the changes with time in the behavior of oil paint with respect to DT analysis is not yet known. However, the principle of this technique is the same as for mineral samples.

### Sample Treatment and Measurement Techniques

The sample and a thermally neutral reference substance (e.g., a metal block) are heated (or cooled) at a constant rate of 5°C/min in two separate, but identical mounts. The temperature difference between the sample and the reference is

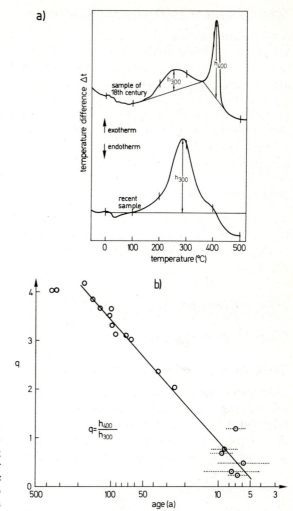

**Fig. 6.89. a** DTA curves for recent and 100-year-old oil paint (after Preuszer 1978. **b** A logarithmic plot shows the age of an oil paint to be a function of peak height ratios

recorded. The measurement is carried out first in a nitrogen atmosphere and then in an oxygen atmosphere so that the influence of oxidation can be observed. Zircon, for example, yields peaks between 700 and 900°C that are characteristic for this mineral and their amplitude is a function of the sample age.

For dating paints, 6–12 samples from the top layer of the painting are analyzed. Dark pigments are more suited for the analysis than light-colored ones. In an oxygen atmosphere, oil paints show two peaks representing highly exothermic reactions between 300 and 400°C. The logarithm of the ratio of the heights of the two peaks at 300 and 400°C is supposed to give the age of the sample (Fig. 6.89).

### Scope and Potential, Limitations

The scope and potential of the method cannot be assessed because it has not yet been applied routinely.

### 6.4.7 Fission Track Method (FT Method)***

(Price and Walker 1962, 1963)

*Dating Range, Precision, Materials, Sample Size*

The fission-track method is a relatively inexpensive standard method for determining the age of minerals (Fleischer et al. 1975; McDougall 1976; Naeser 1979; Wagner 1979, 1981; Storzer and Wagner 1982; Naeser and Naeser 1984), e.g., apatite, zircon, sphene, allanite, epidote, garnet, muscovite, biotite (Price and Walker 1963); obsidian (Yegingil and Göksu 1982; Miller and Wagner 1981); and natural and man-made glass, e.g., tektite, impact glass, volcanic glass (Vincent et al. 1984a), and petrified wood (Srivastava et al. 1986). Even young sediments can be dated via mineral extracts (Hurford et al. 1986). Hence, any etchable, transparent solid whose natural lattice defects produce no disturbing etching grooves and which contain 0.1–1000 ppm U is suitable. Normally, only zircon and glass are used for dating Quaternary rocks. In the case of glass, only minimum ages can be expected. Samples of milligram size are sufficient.

The precision of the method is better than ± 10% (Storzer and Wagner 1982; Green 1985b). With either a "zeta" calibration or calibration on the basis of a best-fit agreement with, for instance, K-Ar ages (Sect. 6.1.1) on well-dated rocks, a standard

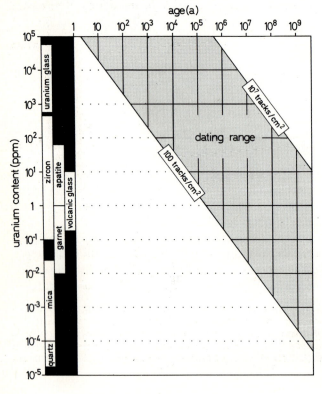

**Fig. 6.90.** Dating range of the fission track method as a function of uranium concentration (*ppm*) in the sample (assuming optical detection limits of *100* and *$10^7$ tracks/cm²*) (after Kaufhold and Herr 1967)

deviation of $\pm 5\%$ is easily obtained. Standard deviations of $\pm 2$–$5\%$, often found in the literature, are based solely on the error in counting the tracks and neglect calibration errors (Poupeau 1981). They correspond approximately to the low standard deviations of TL dates (Sect. 6.4.1). However, the quality of the data is very dependent on the skillfulness and experience of the laboratory staff.

The dating range is highly dependent on the concentration of uranium in the sample, as shown in Fig. 6.90. In a sample with a uranium concentration of 10 ppm, for example, about one track/cm²/ka is formed. Geological samples have been dated in the range of a few ka to 2.7 Ga (Naeser 1979). The age of uranium-rich man-made glass ($> 1\% \, U$) can be determined for ages of more than 20 years (Kaufhold and Herr 1967).

The formation interval (Sect. 6.5.3), the solidification ages, and the track retention ages of meteorites are determined via the minerals they contain (e.g., apatite, whitlockite, pyroxene, and feldspar). The samples must be taken at least 10 cm below the surface of the meteorite. Depending on the technique, several mg of sample or individual grains are sufficient.

## Basic Concept

The fission-track method is based on non-conductors acting as "solid-state track recorders" (SSTR) (Fleischer et al. 1975). Charged particles leave evidence of their passage through such materials in the form of tracks of lattice damage. There are many track types but only fission tracks play a role in the dating of terrestrial

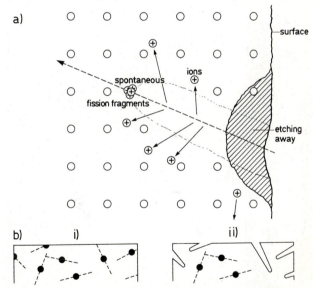

**Fig. 6.91. a** Illustration of the formation of fission tracks in minerals: Along the path of a spontaneous fission fragment, positive ions are produced, which repel each other electrostatically, thus expanding the width of the track. Etching widens the tracks at the surface of the sample to a size visible under a microscope. **b** Cross-sectional view of a sample showing latent fission tracks before etching (*i*) and visible ones after etching (*ii*). Note that the full length of most tracks is not visible

samples. Of the parent isotopes of the natural decay series ($^{238}$U, $^{235}$U, and $^{232}$Th), only $^{238}$U undergoes spontaneous fission with a half-life ($\tau_{sf} = \sim 0.86 \times 10^{16}$ a $\pm 20\%$) short enough to produce a sufficiently large number of spontaneous fission tracks for dating purposes.

The spontaneous fission of a $^{238}$U nuclide produces two fission fragments of approximately the same mass, with multiple positive charges, and a kinetic energy of about 200 MeV. These two fragments recoil in opposite directions (Fig. 6.91), leaving a trail of cations, which repel each other creating tracks (or zones of damage) about 10–20 $\mu$m long and several 0.01 $\mu$m wide in the crystal lattice (Ion Explosion Spike Mechanism; Fleischer et al. 1975). After etching of the sample, these tracks are visible under a normal optical microscope (Fig. 6.92).

At ambient temperatures (up to 80°C), these tracks usually have a lifetime of many million years. Fission tracks are stable only below a certain temperature range (called the effective closing temperature or retention temperature) (Fleischer et al. 1965; Storzer and Wagner 1969) that is specific for each mineral and which depends on the cooling rate. If the effective closing temperature is exceeded, the tracks disappear; this diffusion process is called annealing. The effective closing temperature of apatite, for example, is 105 to 150°C for slowly cooling rocks. The effective closing temperatures for biotite, chlorite, and vermiculite are somewhat higher; those for phlogopite, hornblende, muscovite, and zircon are around 200 $\pm$ 30°C. Allanite, epidote, and sphene have values around 250°C. Besides mineral composition and temperature, the duration of the heating is of great importance (Naeser 1981).

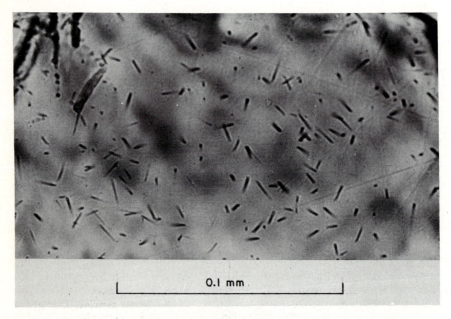

0.1 mm

**Fig. 6.92.** Fission tracks in an apatite from New Mexico (Naeser and Naeser 1984)

The spontaneous-fission-track density ($\rho_{sf}$), obtained by counting the tracks, is a function of the uranium concentration and the age of the sample. The uranium concentration is best determined from the density of artificially induced fission tracks ($\rho_i$) produced by neutron-induced fission of $^{235}U$. The fission-track age, $t$, can then be obtained from the following equation (Price and Walker 1963):

$$t = \frac{1}{\lambda_{238tot}} \ln\left[1 + \frac{\rho_{sf}}{\rho_i} \frac{\lambda_{238} \phi \sigma_{235}}{\lambda_{sf238}}\left(\frac{^{235}U}{^{238}U}\right)\right],$$ (6.93)

where $\lambda_{238tot}$ is the total decay constant for $^{238}U = 1.55125 \times 10^{-10}\,a^{-1}$,

$\lambda_{sf238}$ is the decay constant for the spontaneous fission of $^{238}U$,

$\sigma_{235}$ is the cross-section ($= 580.2 \times 10^{-24}\,cm^2$) for thermal-neutron-induced $^{235}U$ fission tracks,

$\phi$ is the thermal neutron fluence (neutron/$cm^2$) during irradiation, and

$U^{235}/U^{238} = 1/137.88$

Equation (6.93) assumes the same efficiency for the production of both natural and induced tracks, which is only a rough approximation of the actual situation (Gleadow 1981; Storzer and Wagner1982). One of the main problems in solving Equation (6.93) is the choice of $\lambda_{sf238}$. A consensus on which of the values given in the literature should be used has not yet been reached (Storzer and Wagner 1982). Commonly recommended values are

$6.85 \times 10^{-17}$ $a^{-1}$ (Fleischer and Price 1964),
$7.03 \times 10^{-17}$ $a^{-1}$ (Roberts et al. 1968),
$8.46 \times 10^{-17}$ $a^{-1}$ (Galliker et al. 1970), and
$8.57 \times 10^{-17}$ $a^{-1}$ (Thiel and Herr 1976)

*Zeta calibration:* The difficulties in determining the irradiation unknowns and the decay constant $\lambda_{sf}$ in Eq. (6.93) (Hurford and Green 1982; 1985a, b) are avoided if an already dated object or a standard glass (i.e., a monitor with an age $t_m$) (Hurford and Green 1983) is irradiated together with the sample (Fleischer 1975). Geometric corrections become unnecessary. The age is obtained simply from the natural ($\rho_{sf}$) and induced ($\rho_i$) track densities of the sample and the corresponding values ($\rho_{msf}$ and $\rho_{mi}$) of the monitor (the standard):

$$t = t_m \frac{\rho_{sf}}{\rho_i} \frac{\rho_{mi}}{\rho_{msf}} = \xi \frac{\rho_{sf}}{\rho_i} \rho_{mi}$$ (6.94)

The factor $t_m/\rho_{msf} = \xi$, which is why this technique is often called the zeta-correction (Popeau 1981).

*Isochron dating:* The dating precision for minerals with a very heterogeneous uranium distribution (e.g., zoned zircon crystals) is low. In this case, the FT isochron method can be used (Burchart and Kral 1982). Fractions of the same sample with different uranium concentrations are analyzed. The $\rho_{sf}$ values are plotted versus $\rho_n$, yielding a straight line (isochron) with an intercept at $(0,0)$ and a slope proportional to the age of the samples.

### Sample Treatment and Measurement Techniques

In contrast to most of the other hard-rock dating methods (Sect. 6.1), a high-purity mineral concentrate is not necessary. Depending on the techniques used, a mineral fraction or handpicked mineral grains are usually sufficient. Standard mineral separation techniques (e.g., heavy liquids and electromagnets are suitable (Sect. 5.2.1.1). The laboratory procedures are described by Naeser (1976).

*Mineral treatment:* The grains are polished with a diamond paste (finishing with a diamond grain size of $< 1 \mu m$) to remove at least the top $25 \mu m$ of the surface. This polished surface is then etched with a suitable acid or base (hydrofluoric acid for glass and mica, nitric acid for apatite, hydrochloric acid, con. sodium hydroxide or potassium hydroxide for sphene and zircon). This enlarges the tracks so that they can be seen under the microscope. The optimum concentration, as well as period of time and temperature, are determined empirically. A list of the various etching agents and procedures for the FT method has been published by Fleischer et al. (1975).

*Track counting:* The widened tracks (ca. $1 \mu m$ wide after etching) per unit area are then counted under a microscope ($\times 200$ to $\times 2500$). The etched fission tracks differ from other traces by their length ($\leqslant 20 \mu m$), straightness, and their isotropic distribution. Independent counts by two or more persons would reduce subjective error.

There must, of course, be a sufficient number of fission tracks ($> 10/cm^2$) so that the statistical uncertainty is kept small. On the other hand, their density may not be too great ($< 2 \times 10^7/cm^2$), otherwise the individual tracks cannot be distinguished. These limits explain why the dating range is highly dependent on the uranium concentration (Fig. 6.90).

*Efficiency of fission-track formation:* To determine how many fission fragments are represented by each fission track, the sample is irradiated with thermal neutrons in a nuclear reactor to induce fission of $^{235}U$. A standard glass with a known and uniformly distributed amount of uranium is irradiated together with the sample.

The all-important ratio of spontaneous and induced track densities needed for the age determination [Eqs. (6.93) and (6.94)] can be determined in different ways. Gleadow (1981) describes five alternative procedures. The following two are probably the ones most often used (Naeser and Naeser 1984):

- The *population technique* (PM) uses two aliquots of the sample. The aliquots must have a uniform uranium distribution and similar uranium concentrations. One aliquot is irradiated and the natural and induced tracks are counted. The other one is not irradiated. The track density of the irradiated aliquot is subtracted from that of the unirradiated aliquot to obtain the induced track density. It is suitable for apatite and glass but not for sphene and zircon (Fig. 6.93).
- The *external detector technique* (EDM) uses an external detector (also called monitor) during irradiation. This monitor is held against the polished surface of the investigated mineral on which the natural tracks were counted. The

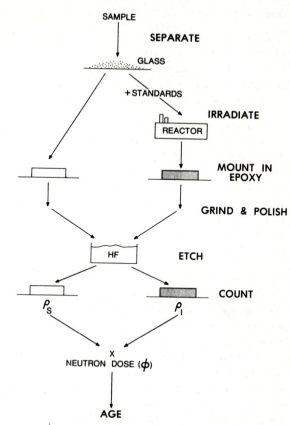

**Fig. 6.93.** Steps involved in obtaining a fission track age using the population method (Naeser and Naeser 1984)

monitor can be a plastic detector or muscovite with a low uranium content. The spontaneous and induced tracks are counted in exactly matching areas of the mineral and the monitor that were next to each other during irradiation (Fig. 6.94). Hence, uranium inhomogeneities, which are usual in zircon, are of no consequence (Green 1985b). Usually, 6–25 crystals are counted.

*Track correction techniques:* Partial annealing of fission tracks, leading to narrower and shorter tracks, or even total track loss, is a gradual process that is often found. This leads to apparent ages that are too low (Wagner 1981; Storzer and Wagner 1982). Partial annealing is caused mainly by heating.

Since the annealing process is both temperature and time dependent, high temperatures that occurred for a short period of time may have the same effect as low temperatures over a long period of time (Green 1988). The result of this annealing is a decrease in track density together with track length. Corrections are made using two independent procedures:

The *plateau-annealing-correction technique* (Storzer and Poupeau 1973; Bertel and Märk 1983) is based on the simple counting of tracks and measurement of their length in aliquots that have been artificially irradiated and then each heated to a different temperature (1 h, 150–250°C; Fig. 6.95a). Up to a certain temperature,

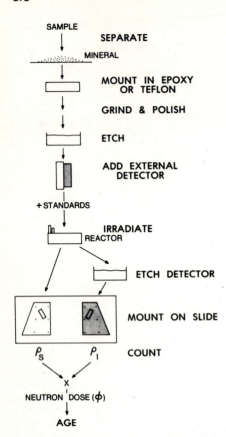

SAMPLE

SEPARATE

MINERAL

MOUNT IN EPOXY
OR TEFLON

GRIND & POLISH

ETCH

ADD EXTERNAL
DETECTOR

+ STANDARDS

IRRADIATE
REACTOR

ETCH DETECTOR

MOUNT ON SLIDE

$\rho_s$     $\rho_i$     COUNT

X

NEUTRON DOSE $(\phi)$

AGE

**Fig. 6.94.** Steps involved in obtaining a fission track age using the external detector method (Naeser and Naeser 1984)

only the number and length of the induced tracks decreases. Above that temperature, the natural tracks also anneal at the same rate but the fission-track ages calculated using Eq. (6.94) remain constant, forming a plateau. This technique has proved useful for glasses, less so for minerals. The dating precision is about $\pm 5\%$.

The *track-size-correction technique* (Storzer and Wagner 1969) is based on temperature-dependent changes in the track lengths and widths and the related decrease in track density (Fig. 6.95b). The dating precision ranges from $\pm 5$ to $\pm 40\%$. This technique is much more time-consuming than the plateau-annealing-correction technique.

The effectiveness of these correction procedures is shown in Fig. 6.96. These techniques are difficult to apply to Quaternary glasses owing to the small number of tracks.

*Scope and Potential, Limitations, Representative Examples*

*Scope and Potential.* Although the FT method is relatively young, it is already one of the few straightforward, standard dating methods used routinely in geology (especially tephrochronology), archeology (reviewed by Wagner 1978), and cos-

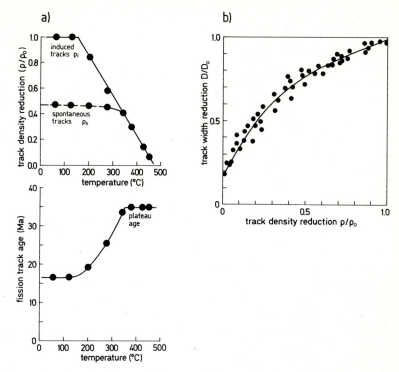

**Fig. 6.95 a, b.** Correction procedures for partially annealed natural tracks: **a** Plateau-annealing-correction technique: Increase in the apparent fission track age with increasing annealing temperature up to the plateau age. The rate of fading of the normalized natural and neutron-induced tracks $\rho_s$ and $\rho_i$ decrease at the annealing temperature for 1 h. Ages affected by the heat of annealing are obtained below 180°C; above 350°C, rather constant fission track plateau ages are obtained (after Storzer and Wagner 1982). **b** Track-size-correction technique: Normalized mean track width $D/D_0$ versus the normalized track density $\rho/\rho_0$ of partially annealed tracks in australite, where $D_0$ and $\rho_0$ are mean values for thermally annealed fission tracks (after Storzer and Wagner 1969)

mology. The method has found application for dating volcanic and metamorphic events, tectonic uplift and mineral deposits, for stratigraphic correlation, and for studies of landform development (see review by Naeser and Naeser 1984). Since each individual fission event is recorded, the method can also be used for geologically young samples. Such samples must contain minerals with a very high uranium concentration (e.g., zircon) and very old (Precambrian) rocks must contain minerals with a very low uranium concentration (e.g., muscovite). As little as a single grain of a mineral can be sufficient for dating. The advantages are a comparatively simple method of measurement with inexpensive equipment (except for the irradiation step) and the fact that the samples do not have to be highly purified. Its potential is considerably increased if it is used in connection with other dating methods (e.g., those of Sect. 6.1).

*Limitations.* The main disadvantage of the FT method is that the results are often difficult to interpret in terms of actual ages. Another problem is the large

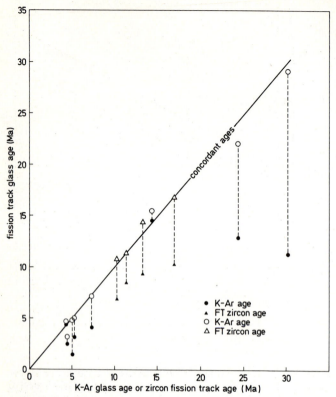

**Fig. 6.96.** Apparent and plateau-annealing fission-track ages of glasses plotted versus the corresponding K-Ar ages (● vs natural state fission-track glass age; ○ vs plateau-annealing fission-track glass age) and zircon fission-track ages of glass (after Naeser et al. 1980)

degree of uncertainty ($\pm 20\%$) in the half-life for spontaneous fission of $^{238}$U; this is overcome to a large degree by the zeta calibration. Moreover, track fading (annealing) is sometimes difficult to correct for. Large errors in dating also occur if naturally induced fission tracks are partially annealed. The problem of anisotropy can be solved, however, by applying the isochron method. Fission-track ages of uranium-rich minerals or extraterrestrial samples are incorrect if the tracks formed by proton- or neutron-induced fission or if the spontaneous fission of uranium or extinct radionuclides (Sect. 6.5.4) are too close together. And in uranium-rich samples, spurious tracks are produced by both proton- and neutron-induced fission.

*Representative Examples. Formation ages:* The fission-track ages obtained for volcanic rock and glass samples usually corresponds to the age of formation if the rock or glass has not been reheated. Apatites and glasses are sometimes affected by fossil track fading (making annealing tests necessary). The formation age of an igneous rock, therefore, can generally be determined from the cogenetic minerals and glasses in it.

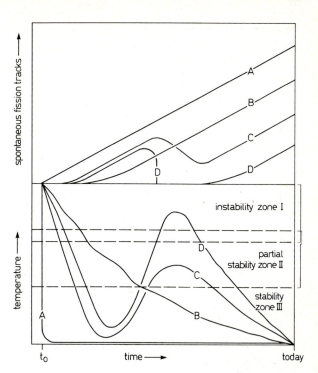

**Fig. 6.97.** Four cooling models for fission tracks (modified after Wagner 1981). *A* Fast-cooling model; *B* slow-cooling model; *C* complex-cooling model with partial annealing of fission tracks; *D* complex-cooling model with total annealing of tracks during burial metamorphism. Only in *zone III* is a complete track record obtained. Partial and total track fading (annealing) occurs in *zones II* and *I*, respectively. The range of temperature for the *zones I–III* depends upon the mineral

*Cooling ages and tectonics:* Since fission tracks in minerals and glasses are stable only below their effective closing temperatures (e.g., Chopin and Maluski 1982), a reconstruction of the temperature history of a rock can be made by analysis of the fission tracks in the various mineral components of the rock (geothermometer, Fig. 6.97) (Naeser 1979; Zeitler et al. 1982; Hammerschmidt et al. 1984). Interpretation of the ages can be very difficult for rocks with a complex thermal history. Four cooling models (selected from the many that have been used) are shown schematically in Fig. 6.97. They illustrate different ways in which fission tracks can develop and be influenced. Thus, it is absolutely necessary to reconstruct the thermal history of the sample to determine its "cooling type". A survey of possibilities for geological interpretation of fission-track ages has been published by Wagner (1981).

For a study of the tectonic history of an area, apatite from plutonic or metamorphic rocks is especially suitable. The fission-track age determined for such a rock indicates when the rock cooled to below about 100°C. Bar et al. (1974) attempted to date tectonic events by fission-track analysis of epidote from fissures.

Calk and Naeser (1973) investigated the effect of a basalt intrusion on the fission-track dates obtained from sphene and apatite in granite from Cathedral Peak in California (Fig. 6.98). The heat developed in the granite by the intrusion of the basalt was sufficient to completely anneal the fission tracks in both the apatite and sphene several meters from the contact, i.e., the fission-track clock was reset to zero at the time of the intrusion. The sphene fission-track age corresponds to the age of the primary granite intrusion at a distance of about 10 m from the basalt-granite contact, the apatite fission-track ages show this correlation for more than a

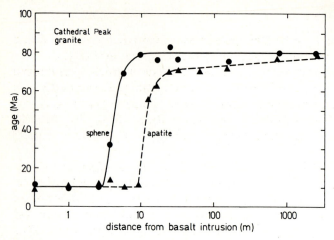

**Fig. 6.98.** Change in fission-track ages as a function of the distance from the intrusion of basalt into granite at Cathedral Park in California (after Calk and Naeser 1973)

kilometer from the contact. This difference is due to the different annealing behavior of sphene and apatite.

The thermal history of sedimentary basins has been studied by FT analysis of apatite and zircon annealing temperatures in many areas. An instructive case study is the publication by Naeser (1986) on data obtained from a well in the Wagon Wheel No. 1 area of Wyoming. It was found that the last cooling 2–4 million years ago occurred with a rapid drop in the temperature (20°C or more).

*Sediment ages:* Sediments in a stratigraphic sequence have been dated by FT analysis of the zircon and apatite in tuffs and bentonites (Storzer and Wagner 1982) and microtektites in deep-sea sediments (Gentner et al. 1970). Fission-track ages for zircon from bauxite and terra rossa provide an indication of the source areas (Comer et al. 1980).

*Archeological application:* The use of the fission-track method in archeology is limited to a very few materials such as obsidian tools (Miller and Wagner 1981) and uranium-rich man-made glass (Kaufhold and Herr 1967). Hearths on rock containing minerals such as apatite and zircon, as well as ceramics containing impurities of uranium-rich minerals, can also be dated.

*Meteorites:* The solidification or track retention age of extraterrestrial samples can be determined not only on the basis of fission tracks from $^{238}$U, but also from the fission tracks of the extinct radionuclides $^{244}$Pu and $^{248}$Cm. Prerequisite for this is a knowledge of the initial ratio $^{244}$Pu/$^{238}$U (Sect. 6.5.4). To avoid errors due to tracks produced by cosmic particle radiation, the samples are taken at least 10 cm beneath the surface of the meteorite or the lunar sample. But an annealing technique has been developed to correct the fission-track ages of samples with a high cosmic-radiation background (Carver and Anders 1976).

*Non-Chronological Applications*

The geochronological and geothermal applications of the FT method are always very close and thus a non-chronological division may appear arbitrary.

Fission tracks open up far-reaching possibilities for reconstructing the thermal and thus the tectonic history of sedimentary basins and mineralizations (Garwin 1984). By analyzing several different minerals in a rock sample (e.g., apatite, zircon, and sphene provide geothermometers for the temperatures 105 to 150, 200 $\pm$ 30, and ca. 250°C, respectively), statements can be made on rates of cooling or uplift (Wagner et al. 1977).

The effects of annealing has been studied primarily via short-term experiments in the laboratory. Measurement of fission-track ages of samples from deep boreholes yields in-situ data on how long-term heat flow has influenced the behavior of the fission-track system. A number of drilling cores have been analyzed (Naeser and Forbes 1976; Zaun and Wagner 1985). An instructive example of the use of such data for the reconstruction of the thermal history of a rock is provided by a study of the geothermal anomaly at Urach, FRG. According to the apparent ages, the rock has cooled at a rate of 0.8°C/Ma; its apatite closure temperature is 106°C (Fig. 6.99) (Hammerschmidt et al. 1984).

To reconstruct the cooling and tectonic history of the Alps, Wagner et al. (1977) used K-Ar (Sect. 6.1.1) and Rb-Sr (Sect. 6.1.3) ages from mica and fission-track ages from apatite. The apatite fission-track ages increase systematically with elevation in the different parts of the central Alps, yielding the rate of uplift. Similar studies have been carried out by Zeitler et al. (1982) on rocks from the Himalayas. Uplift and erosion rates of the Andes in Bolivia have been estimated using FT ages from apatite (Crough 1983).

Cooling rates of meteorites have also been determined (Crozaz 1981).

The method is also very useful for geothermal exploration (Gutierrez-Negrin et al. 1984).

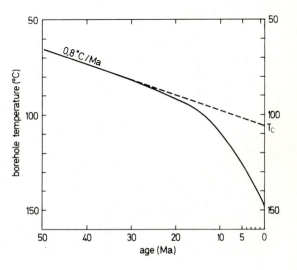

**Fig. 6.99.** Temperature in the Urach III borehole plotted vs. the apatite fission-track ages: The closure temperature $T_c$(106°C) is extrapolated from the linear part of the curve and the slope of this part of the curve yields the cooling rate (0.8°C/Ma) (after Hammerschmidt et al. 1984)

## 6.4.8   Alpha-Recoil Track Method

(Huang and Walker 1967)

### Dating Range, Precision, Materials, Sample Size

This method, which has not been used routinely, has been recommended for the determination of crystallization and rock formation ages of common minerals (several mg), especially mica, as well as the ages of bones and dentin older than 100 ka. In contrast to the fission-track method (Sect. 6.4.7), samples with a low uranium concentration should be datable. The precision of the method has been given as $\pm 5\%$ (Huang et al. 1967).

### Basic Concept

Recoil of the descendent nuclides from $\alpha$-decay of uranium, thorium, or their daughter products in non-conductors produces lattice defects after more than four events. Etching reveals tracks that can be counted under a microscope. Such $\alpha$-recoil tracks are considerably shorter ($2\,\mu m$) and wider than fission tracks. Although $\alpha$-decay of $^{238}U$ occurs $10^6$ times more often than spontaneous fission, the sensitivity of the $\alpha$-recoil track method is expected to be only three orders of magnitude greater than that of the fission-track method because many tracks are not identifiable after etching.

The procedure for this method is analogous to that of the fission-track method.

### Sample Treatment and Measurement Techniques

The rock, mineral, or bone is cut or split, polished and etched for several hours with a suitable acid. The tracks are then counted under a phase-contrast microscope. The track formation rate by uranium and thorium is obtained from the number of fission tracks induced by irradiation with thermal neutrons (Sect. 6.4.7). The age is calculated using Eq. (6.93) or the slightly modified Eq. (6.94).

### Scope and Potential, Limitations

**Scope and Potential.** Although this method is supposed to offer large advantages over the fission-track method (e.g., $\alpha$-recoil tracks are more stable and more frequent than fission tracks), it has not yet been applied. Because a minimum of four $\alpha$-decay events per lattice position are necessary to yield on track, the samples must be at least 100,000 years old. In the $^{238}U$ decay series, for example, not until the seventh daughter product is the fourth $\alpha$-particle ejected (Fig. 6.64).

**Limitations.** The main handicap is the difficult identification of the $\alpha$-recoil tracks. As with the fission-track method, loss or gain of uranium, thorium, or their daughter products leads to incorrect ages.

## 6.4.9  Age Determination Using Pleochroic Haloes

(Mügge 1907; Joly 1907)

### Dating Range, Precision, Materials, Sample Size

Very approximate ages of 100 to 2000 Ma have been estimated with this oldest of the radiometric dating methods (Deutsch et al. 1956; Ramdohr 1960). Only a few milligrams of mineral crystals (e.g., zircon, mica, hornblende, fluorite, monazite, cordierite, apatite, xenolite, coal, phosphates, and carbonates) are needed.

### Basic Concept

Thin sections of many old minerals show microscopic, concentric, colored rings 20–40 $\mu$m in diameter. These "haloes" are radiation damage caused by $\alpha$-particles from uranium, thorium, and their descendant products. The diameters of these haloes is a function of the specific energy of the $\alpha$-radiation, but the intensity of the color is a function of the number of $\alpha$-particles and, thus, the age.

### Sample Treatment and Measurement Techniques

The sample is lightly polished and the intensity of the color of the individual rings is measured under a polarization microscope using a secondary electron multiplier. The uranium and thorium concentrations are determined for each halo by autoradiography. A photographic film is laid on the sample for several days and the intensity of the blackening of the film, which is proportional to the U and Th contents, is measured. For calibration, the sample is then irradiated with a known $\alpha$ dose and the intensity of the color of the induced halo is measured. The age is calculated using Eq. (6.88).

### Scope and Potential, Limitations, Representative Examples

This oldest of the radiometric dating methods is hardly used any more because of the thermal instability of the haloes and the fact that very old samples often reach a saturation point (Fig. 6.77) and thus yield apparent ages that are too small. The intensity of the color may even decrease when saturation is exceeded.

The intensity of the color of the rings depends on the number of absorbed $\alpha$-particles as well as the type and density of the mineral. A lack of knowledge of the exact influence of these parameters hampers the transformation of the measured data into correct ages. In addition, the intensity differs from the inner to the outer zones of the crystal and is also dependent on the size of the crystal. Moreover, anisotropic distribution of the uranium in the micrometer-range is quite common, leading to incorrect dates. Thus, many geologically unacceptable results have been obtained with this method.

*Non-Chronological Applications*

This method is occasionally used to identify metamorphosed samples that must be excluded from K/Ar dating (Sect. 6.1.1).

## 6.5 Dating Meteorites and Lunar Rocks

The dating of meteorites and lunar rocks exhibits special aspects not associated with terrestrial rock samples. These special characteristics arise from the origin and history of the extraterrestrial samples. Although the methods used for dating these samples are much the same as those applied to terrestrial ones, the interpretation of the results is different, yielding other kinds of ages (Fig. 6.100) (Kirsten 1978; Wasson 1985). For this reason, this chapter is organized in terms of interpretation rather than method.

### 6.5.1 Introduction

Radiometric dating of meteorites and lunar rocks has contributed considerably to our understanding of the origin and evolution of the solar system (e.g., Dermott 1978) and the galaxy for almost 40 years. For example, isotopic analysis of meteorites (Geiss et al. 1962; Reedy et al. 1983) has helped to determine past fluctuations in cosmic-ray flux and its solar modulation. These fluctuations must be known for dating with cosmogenic isotopes (Sect. 6.2). Terrestrial rocks are generally not

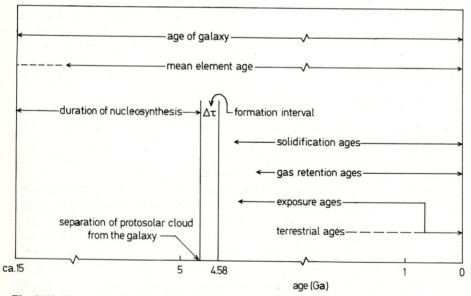

**Fig. 6.100.** Types of ages of the chemical elements and meteorites (after Kirsten 1978)

suitable for this purpose because most of them, for example, have been diagenically altered. Thus, great efforts are justified to locate the impact sites of meteorites already during their flight through the atmosphere or to find old meteorites preserved, for example, in Antarctic ice.

About 95% of all meteorites are stony meteorites, which unfortunately are weathered within a few millenia. Most of them are chondrites, which contain rounded silicate inclusions called chondrules. Stony meteorites that contain no chondrules are called achondrites.The rare, carbonaceous chondrites—the best known is the Allende meteorite—are especially informative about the origin of the solar system.

Less than 2% of all finds are stony-iron meteorites. The remaining 6% are iron meteorites, which consist mainly of iron and nickel and are quite resistant to weathering.

With a few exceptions, all radiometric methods using long-lived radionuclides (Sect. 6.1) to date terrestrial rocks and minerals are also suitable for determining the various types of ages of meteorites and lunar rocks.

The objective of geochronology in the study of meteorites is to decipher the history of the universe by determining the various kinds of ages of stony and iron meteorites and lunar rocks (Fig. 6.100), which range from several years to several billion years (Milliman 1969; Hohenberg 1969; Kirsten 1978; see also the "Proceedings of the Lunar Science Conferences," which were initially published in *Geochim. Cosmo. Acta, Suppl.*, later in the *Journal of Geophysical Research*. Beginning in 1987, monographs have been published).

The history of interstellar matter consists of several stages (Fig. 6.100). Space, time, and matter is theorized to have begun with a gigantic fireball, called the "Big Bang", 15–20 billion years ago, leaving behind radiation and matter in a ratio of about $10^{-9}$ (e.g., Hainebach et al. 1978). At this point in time (called "Hubble time") all matter existed within a few centimeters of space at quasi-infinite density and temperature. In the young, expanding universe, nucleosynthesis was the dominant process (van den Bergh 1981). The light elementary particles were formed during this time. The model developed to estimate the length of this period of nucleosynthesis will not be discussed here because the results are still too uncertain. Evidence for an expanding universe after a "big bang" is the shift into the red wavelengths of the light from the other galaxies and the 2.7-K microwave background radiation interpreted as residual radiation. There are also new hypotheses on a periodic contraction and expansion of the universe.

Our solar system was formed at about $4.57 \pm 0.03$ Ga by the collapse of an interstellar cloud, which was neither chemically nor isotopically uniform. A supernova near the protosolar gas is hypothesized as the initiating factor for this event. The immense amount of energy released by the supernova would have created shock waves that caused local compaction of the interstellar gaseous matter leading to the formation of the heavy elements and later to condensation, first to dust particles within a few million years, which then consolidiated to suns, planets, moons, and comets. Some cosmic dust remained (Barsukov 1981). Short-lived radionuclides, e.g., aluminum-26 (Kuznetsova and Lavrukhina 1981), which decays to magnesium-26 with a half-life of 716 ka (Sect. 6.2.5), provided the heat for melting the interiors of the newly formed bodies (Levin 1982), concentration of the heavy elements within the iron-nickel core and the light elements in the mantle and crust.

Evidence for the initiating cause of a supernova is seen in the existence of heavy elements in our solar system since such elements are formed only at very high neutron fluxes that are never otherwise reached in interstellar clouds. Finally, the presence of pure $^{22}$Ne in meteorites is explained by the formation of $^{22}$Na in a supernova, which later condensed with the dust.

It was only a few million years between the end of nucleosynthesis and the onset of the formation of the first small planets (about 100,000 years). This consolidation of the major planets close to the Sun took about $10^6$ years and those further away about ten times as long. This relatively short period is called the formation interval ($\Delta t$). It is mainly estimated using the daughter products of extinct radionuclides like $^{244}$Pu and $^{129}$I or the fission tracks left by their spontaneous fission.

Comparatively large meteorite parent bodies cooled relatively slowly at a rate of 1–100 K/Ma; this has been derived from the crystalline structure of meteorites. During this time, the silicate phase differentiated from the metal (iron–nickel) phase. The length of this time has been determined especially using the $^{39}$Ar/$^{40}$Ar (Sect. 6.1.1.2) method, but also using the conventional K/Ar (Sect. 6.1.1.1), Rb/Sr (Sect. 6.1.3), U/Th/Pb (Sect. 6.1.9), and Re/Os (Sect. 6.1.8) methods. This period is called the rock age or solidification age of the meteorite. The mean value of $4.57 \pm 0.03$ Ga corresponds to the age of the Earth (e.g., Tilton 1973; Stacey and Kramers 1975).

Not until the planetary bodies of our solar system had cooled down to their blocking temperatures were the volatile substances, especially the primordial noble gases, retained in the solidifying rocks. When the blocking temperatures of helium, argon, krypton, and xenon were reached, their concentrations began to increase with time as a result of radioactive decay, fission, and spallation (Bauer 1948). This is the basis of the determination of gas-retention ages, also called cooling ages (Begemann et al. 1957).

The meteoroid fragments that were formed by the breakup of the meteorite parent bodies are so small that they have been constantly penetrated by cosmic radiation, especially energy-rich protons and alpha-particles. This radiation caused spallation and cascading, forming a great variety of new nuclides (e.g., Fig. 6.101; Wasson 1963), on the order of $10^{-10}$ g per gram of meteorite. Assuming a constant cosmic ray flux, the activities of the radionuclides produces in this way increased until constant saturation activities [Eq. (3.8)] were reached. The stable nuclides were produced at a constant rate. Therefore, the ratio of a stable product and a radioactive cosmogenic product can be used to derive the cosmic ray exposure age, also called radiation age (i.e., the time since the meteoroid was fragmented from the meteorite parent body), povided the ratio of their production rates is known.

The terrestrial age of a meteorite is the time since its impact, i.e., the termination of nuclide production by cosmic rays. It is determined using cosmogenic radio-nuclides (Sect. 6.2), e.g., radiocarbon (Sect. 6.2.1), tritium (Sect. 6.2.2), chlorine-36 (Sect. 6.2.7), argon-39 (Sect. 6.2.8), and manganese-53 (Sect. 6.2.10).

The main problem when isotope abundances in meteorites, particularly those of the noble gases (e.g., Alaerts et al. 1979), are to be converted into ages is the complex

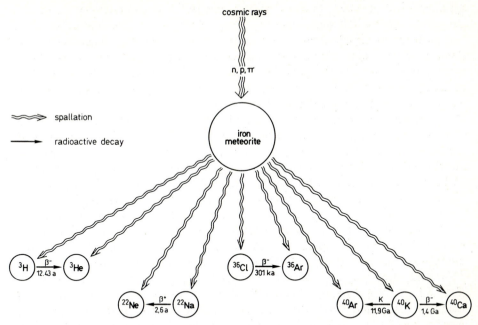

**Fig. 6.101.** An example of production of isotopes via both spallation and radioactive decay: decay products of iron-56 and their daughter products in iron meteorites

mixture of components of differing origin (Shukolyukov 1983). The isotopic spectra of the various components, however, differ significantly (Marti 1967). The most important ones are the

– primordial component (trapped gases from the pre-solar cloud),
– cosmogenic component (produced by solar flares and cosmic radiation; these nuclides can have also been produced directly by spallation or via radioactive progenitors),
– solar component (captured from the solar wind by the parent body regolith),
– atmospheric component (contamination from the air during and after entry),
– radiogenic component (argon and xenon isotopes, for example).

   The abundances of the components of interest (Table 6.9) are estimated (although not very exactly) by subtraction of the other specific isotope concentrations.

## 6.5.2   Sample Preparation and Measurement

The most often applied methods for determining meteorite ages are based on the absolute abundances of noble gas isotopes (Schultz and Kruse 1981). Since the contents of these gases are very low, the slightest contamination distorts the results. Therefore, great care must be taken to avoid contact of the samples with atmospheric gases. Only a few laboratories are skilled enough for this work.

**Table 6.9.** Example of the isotopic composition of
neon in samples from different sources

| Origin | $^{20}$Ne | $^{21}$Ne | $^{22}$Ne |
|---|---|---|---|
| Primordial | 8.2 | 0.025 | 1 |
| Cosmogenic | 0.8 | 0.9 | 1 |
| Solar wind | 12.5 | 0.03 | 1 |
| Atmospheric | 9.8 | 0.03 | 1 |

After removal of the surface in an ultrahigh vacuum, the samples (50 mg to several grams) are heated in a crucible in steps from 1800 to 2000°C. The gases are purified with activated carbon, zircon/titanium getters, and repeated fractional distillation. The isotopic abundances of the gas obtained at each temperature are measured mass spectrometrically (Sect. 5.2.3.1) using isotope dilution analysis (Sect. 5.2.4.1). Non-gaseous, long-lived radionuclides and their daughter isotopes, e.g., Rb/Sr (Sect. 6.1.3), Sm/Nd (Sect. 6.1.6), Re/Os (Sect. 6.1.8), and U/Th/Pb (Sect. 6.1.9), are also measured mass spectrometrically.

"Plateau" ages are determined, for example, from $^{39}$Ar/$^{40}$Ar and Xe isotope ratios by degassing of the samples in ultrahigh vacuum while increasing the temperature stepwise. In this way, $^{40}$Ar is released from lattice sites of differing activation energies. If only partial gas losses occurred, the $^{39}$Ar/$^{40}$Ar ages (Sect. 6.1.1.2) obtained at each step remain constant from a certain temperature onwards, remaining on a plateau. In cases where a catastrophic event occurred during the history of the meteorite, e.g., collision with another meteorite, multi-method studies have often revealed information about the event. Gas retention ages are susceptible to the slightest heating of the meteorite, whose effect is different for each gas. In the extreme case of melting and homogenization of the strontium isotopes, even the Rb/Sr age is affected.

In the conventional methods for measuring the activity of radionuclides, samples of several hundred grams are ground and individual dual mineral fractions separated. These are digested in acid in an ultrasonic bath. The elements of interest are isolated using ion exchangers. Chlorine-36 is precipitated as silver chloride. Aluminum-26 and berylium-10 are purified with acetylacetone and precipitated as the oxide. $\beta^-$-Emitters are measured using low-level proportional counters (Sect. 5.2.2.1) and anticoincidence circuits (Sect. 5.2.2). $\beta^+$-Emitters, e.g., $^{22}$Na and $^{26}$Al, are detected with $\gamma$-$\gamma$ coincidence spectrometry (Sect. 5.2.2.3). Manganese-53 (Sect. 6.2.10) is determined by neutron activation analysis (Sect. 5.2.4.2).

The ultra-sensitive AMS method (Sect. 5.2.3.2) is increasingly replacing the traditional radiometric method because a significantly smaller sample suffices, e.g., 25 mg for $^{10}$Be, 10 g for $^{129}$I, 5 mg for $^{14}$C (e.g., Brown et al. 1984), and 25 mg for $^{26}$Al, and because the method is significantly more sensitive and much faster. Resonance ionization by laser (Sect. 5.2.3.3) to selectively ionize the sample in the mass spectrometer makes it possible to avoid isobaric interference and is applicable, for example, to lutetium, thorium, bismuth, uranium, and plutonium with extremely high sensitivity.

## 6.5.3 Formation Interval

The formation interval ($\Delta t$) is the time between the separation of the protosolar system from the sources of nucleosynthesis and the consolidation of first planetary bodies from the solar nebula (Kirsten 1978). Because the term "age" was already used for the periods of the Earth's history, the term formation interval was introduced.

As the precision of rock dating is about $\pm 50$ Ma for ages around 4.5 Ga, special methods had to be developed to determine formation intervals up to 100 Ma with a precision of $\pm 1$ Ma. Methods based on the daughter isotopes of extinct radio-nuclides with relatively short half-lives or the density of the fission tracks they produced have proven to be useful for this purpose. The isotopes most used for determining $\Delta t$ are as follows:

| Mode of decay | Half-life | Reference |
|---|---|---|
| $^{244}Pu \xrightarrow[\text{(tracks)}]{\alpha,\text{sp,sf}} {}^{131-136}Xe$ | 82.6 Ma $\lambda_{sf}/\lambda_\alpha = 1.25 \times 10^{-3}$ | Kuroda (1967) Hudson et al. (1982) |
| $^{129}I \xrightarrow{\beta^-} {}^{129}Xe$ | 15.7 Ma | Sect. 6.2.12; Kuroda (1967) |
| $^{107}Pd \xrightarrow{\beta^-} {}^{107}Ag$ | 6.5 Ma | Kelly and Wasserburg (1978) |
| $^{26}Al \xrightarrow{\beta^-} {}^{26}Mg$ | 716 ka | Sect. 6.2.5; Lee et al. (1978) |

Other extinct radionuclides are used to explain the discrepancy that still exists between values for the formation interval:

| | | |
|---|---|---|
| $^{92}Nb \xrightarrow{EC} {}^{92}Zr$ | $36 \pm 1.5$ Ma $130 \pm 1.5$ Ma | Makino and Honda (1977) Minster and Allègre (1982) |
| $^{146}Sm \xrightarrow{\alpha} {}^{142}Nd$ | 103 Ma | Notsu et al. (1973) |
| $^{147}Sm \xrightarrow{\alpha} {}^{143}Nd$ | 106 Ga | Sect. 6.1.6 |
| $^{205}Pb \xrightarrow{EC} {}^{205}Tl$ | 15 Ma | Kirsten (1978) |
| $^{247}Cm \xrightarrow{3\alpha} {}^{235}U$ | 15.6 Ma | Chen and Wasserburg (1979) |
| $^{248}Cm \xrightarrow{\alpha,\text{sf}} {}^{131,136}Xe$ | 340 ka | Blake and Schramm (1973) |
| $^{250}Cm \xrightarrow{\text{sf}} {}^{131,136}Xe$ | 11.3 ka | |

The simplest model (the Yoni model) for the formation interval assumes continual, uniform nucleosynthesis in the galaxy (Wasserburg et al. 1969) in which stable ($s$) and radioactive ($r$) nuclides are formed. At the time of separation of the solar nebula from the rest of the galaxy, these nuclides had concentrations of $r_0$ and $s_0$. While cooling during the formation interval, high-temperature condensates formed first, accompanied by chemical fractionation. During this cooling period, stable isotopes, $s_A$, were formed by radioactive decay. A closed system has existed since the end of this separation phase.

According to this model, the noble gases in the meteorites consist mainly of (1) a primordial component $s_0$ and $s_A$ (trapped gases) and (2) a radiogenic component $s*$ formed in the meteorite (Alaerts et al. 1979).

A more complicated model, the so-called multi-spike model (Trivedi 1977), assumes multiple, discrete production of "spikes" of equal intensity and a quasi-steady-state equilibrium between formation and decay of radionuclides.

Because they yield different values for the formation interval, only one or neither of these models corresponds to the truth (Kirsten 1978; Lee et al. 1978).

One problem related to this is the estimation of saturation activities of radioactive isotopes and initial abundances of stable isotopes during the formation interval. It is not known whether the isotopes fractionated during separation from the solar nebula or from the galaxy. A definite answer has not yet been obtained as gases extracted from meteorites are almost always contaminated by terrestrial gases.

Some of these problems are avoided if the relative instead of the absolute formation interval is determined. This can be done with greater precision, but at least two cogenetic meteorites must be available that had the same initial abundance ($s_0$ and $s_A$) and saturation activity ($r_0$) of the isotopes to be measured. The trapped and radiogenic gases can be distinguished by degassing the sample by stepwise heating in an ultrahigh vacuum and measuring the isotope abundances mass spectrometrically (Sect. 5.2.3.1). The results are interpreted analogously to the $^{39}Ar/^{40}Ar$ method (Sect. 6.1.1.2).

The $^{129}I/^{129}Xe$ method (Brown 1947) assumes that one part of the initial $^{129}Xe$, which is uniformly distributed in the meteorite, resulted from the decay of $^{129}I$ ($\beta = 15.7$ Ma) formed during galactic nucleosynthesis. The other part resulted from the decay of primordial $^{129}I$ that was concentrated in iodine condensation centers in the meteorite after its temperature dropped below the xenon retention temperature. This radiogenic $^{129}Xe*$ correlates with the present $^{127}I$ content. The ratios of isotopic abundances measured for two meteorites $i$ and $j$ yield formation interval $\Delta t$ with a precision of several Ma:

$$\Delta t = \frac{1}{\lambda_{129}}\left[\ln\left(\frac{^{129}Xe*}{^{127}I}\right)_i - \ln\left(\frac{^{129}Xe*}{^{127}I}\right)_j\right] \tag{6.95}$$

The abundance of the radiogenic component $^{129}Xe*$ is obtained from the xenon isotope spectrum after subtracting the following components: the trapped gas component $^{129}Xe_0$; the xenon isotopes $^{129}Xe$, $^{131}Xe$, $^{132}Xe$, $^{134}Xe$, and $^{136}Xe$ (formed by spontanous fission of $^{238}U$ and $^{244}Pu$); $^{129}Xe$ (formed by spallation); and finally the solar wind component.

The most exact measurements of the various xenon isotopes are obtained using neutron activation analysis (Sect. 5.2.4.2) and stepwise degassing (see Sect. 6.1.1.2). This procedure converts iodine-127 to xenon-128 by the following reaction:

$$^{127}I(n,\gamma)^{128}I \xrightarrow{\beta^-} {}^{128}Xe$$

Because the trapped primordial gases are uniformly distributed in the meteorite whereas the radiogenic $^{129}Xe^*$ occurs only near iodine condensation centers, stepwise degassing of the sample makes it possible to correlate $^{129}Xe^*$ with $^{128}Xe$, whose abundance relative to $^{132}Xe$ is a measure of the spallation component. The trapped gas component is obtained from the linear relationship between $^{129}Xe$ and $^{130}Xe$ or between $^{129}Xe$ and $^{132}Xe$ (e.g., Drozd and Podosek 1976).

The relative formation interval determined using the $^{129}I/^{129}Xe$ method is always less than 20 Ma (Kirsten 1978).

The $^{244}Pu/^{136}Xe$ *method* (Kuroda 1960; Alaerts et al. 1979; Jost et al. 1981) is based on the fact that $^{136}Xe$ is formed only by spontaneous fission of $^{244}Pu$ ($\tau = 82.6$ Ma). The relative formation interval is calculated analogously to Eq. (6.95) or in combination with the $^{129}I/^{129}Xe^*$ method:

$$\left(\frac{^{136}Xe_{sp}}{^{136}Xe^*}\right) = \left(\frac{^{244}Pu_0}{^{129}I_0}\right)\frac{\lambda_{244}}{\lambda_{129}}\xi_{244}\cdot\xi_{136}e^{\Delta t\cdot(\lambda_{129}-\lambda_{244})}, \tag{6.96}$$

where $\xi_{244} = 0.006$, the probability of spontaneous fission of $^{244}Pu$ and
$\xi_{136} = 0.00012$, the production ratio of $^{136}Xe^*$ from $^{244}Pu$ and $^{129}Xe^*$ from $^{129}I$.

Because the initial isotopic abundances of $^{244}Pu_0$ and $^{129}I_0$ are not known, a relative formation interval must be calculated using the isotopic abundances from two meteorites $i$ and $j$ with the same initial $^{244}Pu$ and $^{129}I$ concentrations:

$$\Delta t = \frac{\ln(^{136}Xe^*/^{129}Xe^*)_i - \ln(^{136}Xe^*/^{129}Xe^*)_j}{\lambda_{129} - \lambda_{244}} \tag{6.97}$$

Instead of the $^{136}Xe^*$ abundance, sometimes the $^{238}U$ activity is used, assuming a constant $^{244}Pu/^{238}U$ ratio in the protosolar cloud of 0.015 or even only 0.005, which follows from the constant ratio of $^{244}Pu/Nd = 1.5 \times 10^{-4}$ in several meteorites (e.g., Drozd et al. 1977; Jones 1982).

## 6.5.4 Solidification Ages

The age of the meteorite (solidification age) is defined as the time since the stony and metal phases differentiated or the meteoroid parent body consolidated. If this process occurred stepwise, the age refers to the last phase. Nearly all meteorites have been found to have a solidification age within the narrow limits of $4.57 \pm 0.03$ Ga, which is considered to be the time of the formation of our solar system. This observation is an important piece of evidence for the thesis that all meteorites are fragments of only a few meteorite parent bodies that cooled rather rapidly.

The most reliable dates are obtained with the *Rb/Sr method* (Sect. 6.1.3), even though the primordial $(^{87}Sr/^{86}Sr)_0$ ratio of meteorites is not known. This difficulty is overcome by analyzing two or more genetically linked meteorites that had the same initial isotopic composition. The "internal" isochrons from different mineral phases of a meteorite (mainly olivine, pyroxene, and plagioclase) tend to be less reliable than whole-rock isochrons obtained from several meteorites. The Rb/Sr values do not form isochrons if the meteorite has been metamorphosed sufficiently for the strontium isotopes to have been even partially homogenized, i.e., reequilibrated.

The range of application of the *Re/Os method* (Sect. 6.1.8) makes it especially suitable for the dating of iron meteorites and lunar rocks now that the half-life of rhenium has been determined exactly by AMS. A value of 0.83 is used for the primordial isotopic ratio $(^{187}Os/^{186}Os)_0$.

The $^{207}Pb/^{206}Pb$ *method* (Sect. 6.1.11) and the *U/Th/Pb methods* (Sect. 6.1.9) (Tera and Wasserburg 1974) have the advantage that the isotopic abundances of only one element need be measured. However, the low uranium and thorium concentrations of meteorites increase the probability of terrestrial contamination, caused by weathering or isotopic exchange, for example.

The concentration of primary lead and its isotopic composition in a meteorite need not be known if two meteorites (*i*) and (*j*) of the same age (*t*) are analyzed. A graphic solution of the following equation is common:

$$\frac{(e^{\lambda_{235}t} - 1)}{137.88(e^{\lambda_{238}t} - 1)} = \frac{(^{207}Pb/^{204}Pb)_i - (^{207}Pb/^{204}Pb)_j}{(^{206}Pb/^{204}Pb)_i - (^{206}Pb/^{204}Pb)_j} \tag{6.98}$$

The greater the difference between the $\mu$ values (Sect. 6.1.10) of the meteorites being analyzed, the lower the standard deviation of the determined age.

Meteorites are also dated using the $^{147}Sm/^{143}Nd$ *method* (Sect. 6.1.6) (e.g., Notsu et al. 1973; Lugmair et al. 1975).

Still another method used for obtaining solidification ages of meteorites is the counting of *fission tracks* (Sect. 6.4.7) resulting from the spontaneous fission of uranium-238 and the extinct radionuclides $^{244}Pu$ and $^{248}Cm$. Cooling ages can be determined for samples taken at least 10 cm below the surface of a meteorite or a lunar rock. Samples taken from less than this depth contain spurious tracks made by cosmic particle radiation. The initial $^{244}Pu$ and $^{238}U$ concentrations, which must be known for the age determination, are estimated by iteration.

Olivine containing little or no uranium is occasionally found in melted stony meteorites. Fission tracks in this olivine result only from the matrix material. In such rare cases, the "contact age" at which the mineral was formed in the meteorite can be determined.

### 6.5.5  Gas Retention Ages

Until the meteorite had cooled below the blocking temperature, which is specific for each gas (e.g., ca. 570 K for xenon and about 470 K for argon), the gas was all lost by diffusion through the crystal lattice. The gas retention age (also called the cooling age) represents the time since the gas began to be retained in the crystal matrix. This assumes later reheating did not occur, e.g., by collision with another meteoroid or owing to a path near the sun.

The multitude of possible events that can influence the gas retention age explains the wide range of values that have been obtained (0.3–4.5 Ga). Attempts to explain gas retention ages that were apparently too small by continual loss of gas by diffusion have not been successful because differences between the U/He (Sect. 6.3.14) and K/Ar (Sect. 6.1.1) ages that would be expected in this case were not found. The assumption is now made that gas retention ages date catastrophic events that raised the temperature of the meteorite above the blocking temperature.

Although the potassium concentration of stony meteorites is not very high (about 1‰) and the autochthonous argon is often contaminated with atmospheric argon, K/Ar ages have generally proved to be reliable (Schaeffer and Schaeffer 1977). This has especially been the case when the $^{39}Ar/^{40}Ar$ method (Sect. 6.1.1.2) was used. This method makes the difficult potassium determination in the microgram range unnecessary and primordial, cosmogenic, and radiogenic argon can be distinguished. But $^{39}Ar$ recoil losses during the neutron activation (Sect. 5.2.4.2) may cause anomalous presolar ages (e.g., Villa et al. 1983).

The K/Ar method cannot be used for iron meteorites because they contain too little potassium.

The use of the U/He method for dating meteorites led to the discovery of noble gases produced by spallation. The precision of U/He ages is generally low because the content of radiogenic He is often less than 65%, especially in iron meteorites, due to their low uranium concentrations. The rest of the helium is cosmogenic or primordial. Somewhat more favorable conditions are found in stony meteorites, in which the radiogenic component is sometimes as high as 90%.

In some cases, gas retention ages obtained using the $^{39}Ar/^{40}Ar$, K/Ar, and U/He methods are so high that they must be interpreted as the formation age of the parent body.

## 6.5.6   Cosmic Ray Exposure Ages

The relative and absolute cosmic ray exposure ages (radiation ages; Bauer 1947, 1948) of meteorites range from 200 ka to 2.4 Ga and can be determined with a precision of $\pm 10$–30% (Cressy and Bogard 1976).

On their paths through interplanetary space, meteoroids are irradiated with high-energy solar and galactic cosmic particles, mainly protons and alpha-particles. These penetrate up to about 500 g/cm$^2$, e.g., up to 0.7 m into iron meteorites, causing spallation, which produces both stable and radioactive nuclides (e.g., Fig. 6.101). the latter with half-lives of up to $10^9$ a. Many of these spallation products are gases (Gentner and Zähringer 1955). Their concentrations are on the order of $10^{-10}$ g/g meteorite. The cosmic ray exposure ages determined from these isotope abundances thus correspond to the time since the meteorite became smaller than twice the penetration depth of cosmic protons. Relatively large meteorite parent bodies presumably still exist in the asteroid belt. The cosmic ray exposure ages of iron meteorites (4 Ma–2.3 Ga with a mean of about 500 Ma) are greater than those of stony meteorites (20 ka to 80 Ma with a mean of about 10 Ma), presumably due to their greater chemical and mechanical stability. But all exposure ages are small compared to the rock ages of meteorites and thus date collisions resulting in the breaking up of the parent body.

The cosmic ray exposure age of lunar rocks gives the time the rock has lain on the moon's surface. Events that brought the rocks to the surface, e.g., crater formation by meteorite impact, can thus be dated.

Cosmic ray exposure ages are usually calculated using a one-step irradiation model according to which the spallation products are produced at a continuous rate. This requires that the self-shielding (i.e., the size of the meteorite) and the intensity of the cosmic radiation remain the same during the entire exposure period. This model also assumes that "space erosion" can be neglected. Sometimes, two-step irradiation models explain the isotopic composition more effectively (e.g., Heusser et al. 1985).

In the one-step irradiation model, the cosmic ray exposure age ($t$) is derived from the abundance ($c_s$) of a cosmogenic, stable nuclide ($s$) and the saturation abundances of the radionuclide ($c_r$) of a cosmogenic, radioactive nuclide ($r$) with a decay constant $\lambda$:

$$a = \frac{1}{\lambda} \frac{c_s}{c_r} \frac{P_r}{P_s} (1 - e^{-\lambda t}) \simeq \frac{1}{\lambda} \frac{c_s}{c_r} \frac{P_r}{P_s}, \tag{6.99}$$

where $P_s$ and $P_r$ are the production rates of the stable and radioactive isotopes, respectively, at constant irradiation intensity.

$P_s$ and $P_r$ are rather equal if the nuclides $s$ and $r$ are isobars and are produced from one and the same target isotope. For isobar nuclide pairs (e.g., $^3$H and $^3$He or $^{36}$Cl and $^{36}$Ar) for which the stable nuclide is also produced by radioactive decay, Eq. (6.99) modifies to

$$t = \frac{1}{\lambda} \frac{\sigma_r^*}{\sigma_r^* + \sigma_s^*} \frac{c_s}{c_r} \tag{6.100}$$

Exposure ages are derived from isotope pairs such as $^{21}$Ne/$^{26}$Al, $^{36}$Ar/$^{36}$Cl, $^3$He/$^3$H, $^{38}$Ar/$^{39}$Ar, $^{22}$Ne/$^{22}$Na, $^{41}$K/$^{40}$K (Voshage 1984), $^{53}$Mn/$^{26}$Al (up to 10 Ma, Herpers and Englert 1983), and $^{81}$Kr/$^{83}$Kr (Table 6.10).

**Table 6.10.** Examples of ratios of the production rates of often measured cosmogenic nuclides in meteorites and lunar rocks

| Stony meteorites | | | | | | |
|---|---|---|---|---|---|---|
| $^3$He/$^{36}$Cl | $^{26}$Al/$^{21}$Ne | $^3$He/$^{21}$Ne | $^{45}$Sc/$^{21}$Ne | $^{36}$Cl/$^{36}$Ar | $^{81}$Kr/$^{83}$Kr | $^{53}$Mn/$^{26}$Al |
| 34 | 0.35 | 85.2 | 4.2 | 0.83 | 0.63 | 1.48 |

| Iron meteorites | |
|---|---|
| $^3$He/$^4$He | $^{39}$Ar/$^{38}$Ar |
| 0.257 | 0.52 |

| Lunar rocks | | |
|---|---|---|
| $^{26}$Al/$^{21}$Ne | $^{36}$Ar/$^{38}$Ar | $^{38}$Ar/($^{39}$Ar + $^{37}$Ar) |
| 1.0 | 0.66 | 1.0 |

**Table 6.11.** Half-lives and examples of specific saturation activities of radionuclides used for determining cosmic ray exposure ages of meteorites .

| Nuclide | Half-life | $A_r$ (dpm/kg Fe) iron meteorite | $A_r$ (dpm/kg Si) stony meteorite | Ref. |
|---|---|---|---|---|
| $^3$H | 12.43 a | | | (a) |
| $^{10}$Be | 1.6 Ma | | 16– 26 | (b) |
| $^{22}$Na | 2.6 a | | 70–120 | |
| $^{26}$Al | 716 000 a | | $300 \pm 55$ | (b) |
| $^{36}$Cl | 301 ka | 8–23 | | (c, d) |
| $^{41}$Ca | 103 ka | 6.9 | | |
| $^{53}$Mn | 3.7 Ma | $478 \pm 80$ | | (e) |
| $^{60}$Co | 5.3 a | | 60–230 | (b) |
| $^{129}$I | 15.7 Ma | | 4– 25 | (e) |

(a) Begemann et al. (1957); (b) Ehmann and Kohman (1958); Herpers and Englert (1983); Sarafin et al. (1984); (c) Sammet and Herr (1963); (d) Vilcseck and Wänke (1963); Goel and Kohman (1963); (e) Nishiizumi et al. (1983a, b).

In this case, the effective cross-sections $\sigma_s^*$ and $\sigma_r^*$ are used instead of the absolute production rates. However, $\sigma^*$ is a function of the unknown mean flux, the energy spectrum of the cosmic radiation impinging on the meteoroid, and the self-shielding. Moreover, the energy-dependent production cross-sections of all the spallation processes must also be known, which is generally not the case.

Therefore, empirically determined values of $\sigma^*$ are often used. These values are obtained by comparing cosmic ray exposure ages determined for the same sample using different isotopes; or they can be obtained from neutron and proton simulation experiments (e.g., Englert et al. 1984). But ratios of the production rates of stable noble gases normalized to the chemical composition of the sample are also used, although the precision is often as low as $\pm 30\%$ (Marti 1967; Eugster et al. 1967a). Recently, production rates have been calculated assuming oxygen, magnesium, silicon, aluminium, and calcium as the target nuclides for the production of $^3$He, $^{21}$Ne, and $^{38}$Ar in stony meteorites; iron and nickel are assumed to be the target nuclides in iron meteorites. The cross-sections for the production of $^{26}$Al, $^{22}$Na, and $^{39}$Ar from argon and $^9$Be and $^{10}$Be from nitrogen have been estimated (e.g., Reyss et al. 1981; Michel et al. 1982).

Cosmic ray exposure ages are calculated using Eq. (6.99) and (6.100) using the radionuclides listed in Table 6.11. They are sometimes roughly estimated from the absolute production rates of stable isotopes (Table 6.12). The actual rates, like the relative production rates, are dependent on the type and composition of the meteorite, the self-shielding and the path in space (the latter two determine the intensity of the interstellar particulate radiation).

Errors due to diffusion losses are avoided by measuring the isotopic ratio of a single element, e.g., $^{81}$Kr/$^{86}$Kr or $^{78}$Kr/$^{86}$Kr (Marti 1967; Eugster et al. 1967a; Schultz 1986; Freundel et al. 1986; Eugster 1988). Such errors can occur when absolute isotope abundances are used to determine cosmic ray exposure ages.

**Table 6.12.** Values for absolute production rates of gaseous nuclides in meteorites found in the literature

| Nuclide | $P$ ($10^{-8}$ cm$^3$ STP/g/Ma) |
|---------|---------------------------------|
| $^3$H | 0.5 |
| $^3$He | 1.2–2.00 |
| $^{21}$Ne | 0.12–0.30 ($\pm$ 7%) |
| $^{38}$Ar | 0.10–0.14 |
| $^{126}$Xe | 0.13 |
| $^{130}$Xe | 0.00025 |

Very reliable results have been obtained using the $^{41}$K/$^{40}$K/$^{39}$K method (Voshage and Hintenberger 1963; Voshage 1984).

"Space erosion" amounting to 0.02–0.05 cm/Ma may have occurred as a result of the constant bombardment of the meteorite with cosmic particles. Reliable evidence for this has not yet been obtained. An incorrect estimate of the space erosion would lead to incorrect cosmic ray exposure ages, which are calculated on the assumption of a constant production rate of the spallation products. To test whether the self-shielding changed during the flight of the meteoroid through space, the absolute concentrations of $^3$He and $^4$He or the ratios $(^{22}$Ne/$^{21}$Ne$)_{\text{spall}}$, $(^3$He/$^{21}$Ne$)_{\text{spall}}$, and $^3$He/$^{38}$Ar are plotted versus depth in the meteorite (Eberhardt et al. 1966). Isotopic depth profiles of $^{26}$Al and $^{10}$Be also yield information on the former size and shape of the meteorite because the production of the various isotopes results in different depth profiles (e.g., Schultz and Signer 1976; Wright et al. 1973; Reedy et al. 1983). Last, but not least, nuclear track records may also be helpful, especially for estimating the depth of ablation (Bhandari et al. 1980). Moreover, loss of gas by diffusion can be estimated (Eberhardt et al. 1966).

The "trapped gases" component, which must be eliminated to determine the cosmic ray exposure age, is estimated from the changes in the isotope ratios that are measured with a mass spectrometer (Sect. 5.2.3.1) while the sample is being degassed in ultrahigh vacuum accompanied by a stepwise increase in temperature. A plot of $^i$Xe/$^{136}$Xe versus $^j$Xe/$^{136}$Xe yields a straight line whose slope is a function of the proportion of the spallation component and whose y-intercept corresponds to the proportion of the trapped gases. Analogous corrections are possible for $^i$Kr/$^{86}$Kr versus $^j$Kr/$^{86}$Kr, $^{41}$K/$^{40}$K versus $^{39}$K/$^{40}$K, and $^{20}$Ne/$^{22}$Ne versus $^{21}$Ne/$^{22}$Ne.

Thermoluminescence analysis (Sect 6.4.1) may also yield cosmic ray exposure ages (Houtermans and Liener 1966).

Cosmic ray exposure ages derived from tritium concentrations (Sect. 6.2.2) are often unreliable owing to $^3$He losses in space.

### 6.5.7 Terrestrial Ages of Meteorites

The terrestrial age of meteorites is defined as the length of time it has spent on Earth. It depends partly on the terrestrial weathering "half-life", which can range from a few

**Table 6.13.** Cosmogenic radionuclides used for determining terrestrial ages of meteorites

| Nuclide | Half-life | Dating range | Section | Ref. |
|---|---|---|---|---|
| $^3$H | 12.43 a | 100 a | 6.2.2.1 | |
| | | | 6.2.2.2 | |
| $^{10}$Be | 1.6 Ma | 12 Ma | 6.2.3 | (a) |
| $^{14}$C | 5730 a | ~40 ka | 6.2.1 | (b) |
| $^{22}$Na | 2.6 a | 20 a | 6.2.4 | |
| $^{26}$Al | 716 ka | 4 Ma | 6.2.5 | (c) |
| $^{36}$Cl | 301 ka | 2 Ma | 6.2.7 | (d) |
| $^{39}$Ar | 269 a | ~1500 a | 6.2.8 | |
| $^{40}$K | 1469 Ma | 4.5 Ga | 6.1.1 | |
| $^{41}$Ca | 103 ka | 500 ka | 6.2.9 | (e) |
| $^{53}$Mn | 3.7 Ma | 15 Ma | 6.2.10 | (c) |
| $^{81}$Kr | 210 ka | 1 Ma | 6.2.11 | (f) |
| $^{129}$I | 15.7 Ma | ~100 Ma | 6.2.12 | (g) |

(a) Tuniz et al. (1984); Kohmann and Goel (1963); Sarafin et al. (1984); (b) Brown et al. (1984); (c) Herr et al. (1981); (d) Nishiizumi et al. (1983a); (e) Kubik et al. (1986); (f) Schultz (1986); (g) Nishiizumi et al. (1983b).

**Table 6.14.** Examples of ratios of the saturation activities ($A_r$) of several cosmogenic radionuclides found in meteorites

| $[^{39}\mathrm{Ar}/^{36}\mathrm{Cl}]$ | $[^{10}\mathrm{Be}/^{36}\mathrm{Cl}]$ | $[^{26}\mathrm{Al}/^{36}\mathrm{Cl}]$ | $[^{10}\mathrm{Be}/^{26}\mathrm{Al}]$ |
|---|---|---|---|
| 0.89 | $0.63 \pm 0.03$ | $0.32 \pm 0.02$ | $1.95 \pm 0.08$ |

thousand years for stony meteorites in a humid climate to as much as 100,000 years in Antarctica. The terrestrial age can be determined with nearly all of the known long-lived cosmogenic radionuclides (Sect. 6.2; Table 6.13).

Terrestrial ages ($t$) of meteorites are calculated using Eq. (6.58) with the specific saturation activity $A_r$ instead of the initial activity $A_0$. $A_r$ is either estimated theoretically or derived from the corresponding specific isotope activity of a meteorite with a known terrestrial age.

Another possibility for determining the terrestrial age ($t$) of a meteorite is the measurement of the activities of two isotopes ($i$) and ($j$) with similar atomic masses but different half-lives (Table 6.14). In this case,

$$t = \frac{1}{\lambda_j - \lambda_i} \ln\left(\frac{A_r^j A^i}{A_r^i A^j}\right) \tag{6.101}$$

Presupposition for this is that both radionuclides were formed by similar mechanisms. This means that the depth profiles are similar and that correction for a difference in unnecessary.

The saturation activity according to Eq. (5.8) is equal to the ratio of the production rate to the decay constant. The production rate is proportional to the cosmic radiation flux, the effective cross-section of the isotope in question, and the number of target atoms per gram of meteorite. The production rate is also affected by the self-shielding, which is estimated from the depth distribution of the noble gas isotopes in the meteorite (Sect. 6.5.6).

Ambiguous results may be obtained if multi-step irradiation occurred, as is often the case with lunar rocks. When this happens the one-step irradiation model cannot be applied (Tuniz et al. 1983).

A sample size of 10–40 g is needed for the conventional analysis of iron meteorites and 100–400 g for stony meteorites (Kohman and Goel 1963; Sammet and Herr 1963). The quantities needed for the modern AMS technique (Sect. 5.2.3.2) are three orders of magnitude less than these amounts. The isotopes should be measured in the metal phase of the meteorite because only in these parts of the meteorite are the production rates constant. The thermoluminescence method (Sect. 6.4.1) has been used with controversial success for Antarctic meteorites with ages of up to $10^6$ years (Melcher 1981, 1982; McKeever 1982; McKeever and Townsend 1982; Sears and Hasan 1986; Sutton and Walker 1986). Anomalous fading at a constant rate is assumed.

# 7  Chronostratigraphic Methods Using Global Time Markers

Several widely applied chronostratigraphic dating methods are based on globally distributed strata that contain physical or chemical anomalies, climatic rhythms or seasonal cycles resulting from global events (Walliser 1983/1984). To yield time marks, these globally present anomalies or stratigraphic records must be very accurately dated by suitable methods. Paleomagnetic remanence in magmatic rocks and sediments (Sect. 7.1) recording changes in the Earth's magnetic field is an example of time markers. Variations in the stable oxygen isotope ratio in planktonic foraminiferal tests from pelagic sediments caused by variations in climate (Sect. 7.2) are another example of a natural stratigraphic record.

Anthropogenic time markers have appeared worldwide during the last several decades due to the introduction of new industrial chemicals (Sect. 7.6), fallout from nuclear weapons, and emissions from nuclear power plants into the environment (Sect. 7.4), for which the input dates are often rather well known.

These methods are mostly applied for cases in which no other method yields a usable stratigraphic differentiation and when a worldwide correlation with other stratigraphic records is needed.

The dating methods of palynology, dendrochronology (a review of this relatively young field has been published by Eckstein et al. 1984), lichenometry, tephrochronology, sedimentology, and other similar fields will not be discussed here since either they are not based primarily on physical and chemical effects or their application is not valid worldwide.

## 7.1  Paleomagnetic Dating Methods***

(Folgheraiter 1899; Thellier and Thellier 1959)

*Dating Range, Precision, Substances, Sample Size*

*Archeomagnetic studies:* This method is routinely used for dating ceramics, bricks, burned flint, burned soil under former firepits, slag from smelters, and oven linings up to ages of about 15 ka. The standard deviations range from $\pm 40$ to 200 a, depending on how good the calibration of the reference time scales are (e.g., Tarling 1975; Fleming 1976; Aitken 1978). The three-dimensional, in-situ orientation of the sample (at least several cm$^3$) must be recorded when it is collected. This is not necessary if the position of the sample during firing can be reconstructed from the shape, e.g., of ceramics.

*Magnetochronostratigraphic studies* are made on fine-grained sediment cores. Careful attention must be given to the nature of the magnetized material and to the magnetization process ("second-generation studies") (Creer and Tucholka 1982a, b, 1983). Master curves valid for areas from a hundred to several thousand km² are necessary. They are available for parts of North America (Creer and Tucholka 1982a), Europe (Creer and Tucholka 1982b), South America (Creer et al. 1983), and

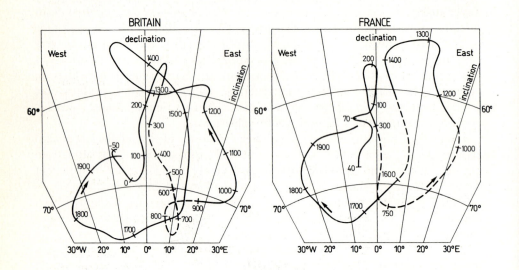

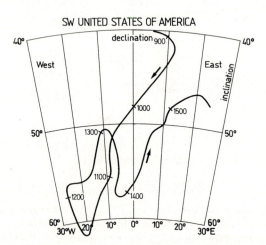

**Fig. 7.1.** Paths of the virtual geomagnetic poles (VGP) represented in declination-inclination diagrams for Britain, France, and the Southwest USA for up to two thousand years (after Kovacheva 1982)

Australia (Constable and McElhinny 1985). Stalagmites have also been dated by this method (Latham et al. 1979, 1982).

*Paleomagnetic dating* is done for rough correlations. It is based on geomagnetic reversals recorded in samples with an age of at least 400 Ma (Harland et al. 1982; Craig et al. 1986; Obradovich et al. 1986). Undisturbed iron-bearing sediments and igneous rocks are often suitable objects for this method. The dating precision is

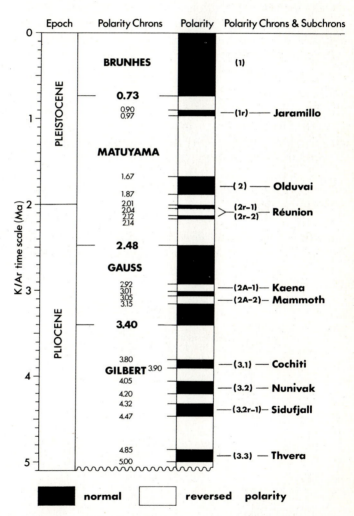

**Fig. 7.2.** Paleomagnetic polarity time scale for the Pleistocene and Pliocene (after Harland et al. 1982), portions of which have been modified (Craig et al. 1986; Obradovich et al. 1986) and will continue to be subject to modification; the numbers refer to the sequence numbers of marine paleomagnetic reversals. The main subdivisions of time based on polarity resersals are now called "polarity chrons", replacing the earlier term polarity epoch; the term subchron is used instead of the term "polarity event" for polarity intervals shorter than 100 ka

dependent on the resolution of the regional (Fig. 7.1) or global (Fig. 7.2) reference time scale (Cox 1969; Kennett 1980; Palmer 1983). The best precision is $\pm 1-2\%$.

### Basic Concept

*Magnetostratigraphy:* The paleomagnetic records used for dating reflect changes in the Earth's magnetic field, which have occured numerous times during the Earth's history for reasons that are not completely understood. Various phenomenological models provide explanations (e.g., Cox 1975; Creer 1983).

The magnetic field of the earth can be viewed as the superposition of a dominant dipole field and a non-dipole field (5–40%), the resultant of which can be described by the following three parameters (Fig. 7.3):

1) *declination* (D): the difference between the geographic and magnetic north poles (in degrees);
2) *inclination* (I), also called dip angle: the difference between the horizontal and the resultant (in degrees);
3) *geomagnetic field intensity* (F): in dating, given as the ratio between former and present intensities $F/F_0$.

Various phenomena are reflected in archeomagnetic and paleomagnetic records:

*Secular variation of the direction of the geomagnetic field:* This is caused by a westward drift of the non-dipole component of the geomagnetic field and by the

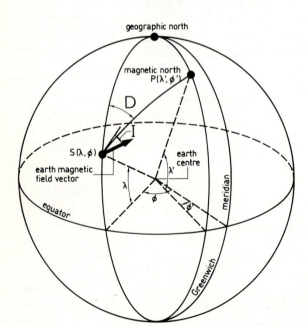

**Fig. 7.3.** Vector diagram of the geomagnetic field: inclination *I* and declination *D* at sampling site *S* with a geographic latitude $\lambda$ and longitude $\phi$

growth and decay of the regional non-dipole anomalies (Creer 1981; Thompson 1983). This drift differs for different regions and is recorded as changes in inclination and, less accurately, as changes in declination in many regions over the last several centuries (Fig. 7.1) (e.g., Kovacheva 1982). The data for one region can, with a certain amount of uncertainty, be extrapolated to other magnetic latitudes. Maximum variation in inclination and declination is about 20 and 40°, respectively.

*Secular variations in the intensity of the geomagnetic field:* The validity of the time records of the intensity of the geomagnetic field determined, for example, by Bucha (1973) (Fig. 6.50) for Central Europe, previously considered to be valid worldwide, has been increasingly questioned (Barton et al. 1979; Beer et al. 1984). Dates based on changes in geomagnetic field intensity are generally less precise than those based on direction measurements, and are, therefore, suitable only for authenticity tests of ceramics and art objects.

*Reversal of polarity:* The polarity of the geomagnetic field has changed at irregular intervals during the history of the earth (Fig. 7.2). By convention, the polarity is called normal when the field is directed towards the north and the inclination is downwards in the northern hemisphere and upwards in the southern hemisphere. The polarity is considered reversed when the direction of the field is the opposite. The transition periods from one stage to another lasted only several hundred years, the periods of stability (Raisbeck et al. 1985; Prevot et al. 1985) lasted from several hundred thousand to several tens of millions of years. The cause(s) of polarity reversals are not yet understood (e.g., Jacobs 1984). They are thought to be due to convective instabilities in the outer fluid core of the Earth.

In addition to pole reversals, changes of 40–90° in the direction (called excursions), as well as the intensity, of the geomagnetic field occurred over periods of 1–100 ka. It has not yet been established whether they are of global extent and, therefore, may be of only regional significance. Their existence is even questioned (Verosub 1982).

*Apparent pole wandering curves:* The apparent wandering of the poles (APW) is a process that occurs over a period of 10–100 Ma (e.g., Evans 1983). It is due to a relative wandering (drift) of the continental plates, which is why it differs on the different continents (e.g., Piper 1982). True polar wander (TPW) is still subject do debate.

*Magnetostratigraphy and magnetochronology:* To convert a paleomagnetic stratigraphic master curve to a paleomagnetic time scale, prominent markers are dated as precisely as possible with the most suitable radiometric method (Chap. 6). Regional paleomagnetic stratigraphic master curves are based on measurements on undisturbed drill cores, sections or profiles, or suitable archeological finds for which the orientation was recorded when the sample was collected or for which the orientation could be reconstructed. These master curves must be determined for each particular geographic region (e.g., Creer and Tucholka 1982a, b). The best example of a well-dated stratigraphic record is the magnetic polarity time scale (MPTS) (Fig. 7.2), which has a precision of $\pm$ 2–5 ka for the last 3.5 Ma (Cox 1969; Harland et al. 1982; Palmer 1983); for the time up to 100 Ma it is also relatively well known (Lowrie and Alvarez 1981; Harland et al. 1982). The polarity reversals, determined on marine sediments, are numbered consecutively up to 33 back to 79 Ma (Watkins 1976; Harland et al. 1982). For Quaternary studies, the most important time marker is the Brunhes-Matuyama reversal, which occurred 730,000 years ago (Mankinen and Dalrymple 1979).

Ages derived from thermoremanent magnetic records refer to the last time the rock cooled below the blocking temperature or new minerals formed during sedimentation, metamorphism, or after dehydration. Ages derived from detrital remanent magnetic records may date either the time of the sedimentation or post-depositional events.

*Thermoremanent magnetic records:* The archeomagnetic and paleomagnetic dating methods make use of the property of ferromagnetic minerals by which they receive a thermoremanent magnetization (TRM) when they cool below a certain temperature (variously called the blocking temperature, the Curie temperature or Curie point). Because the thermoremanent magnetization is limited to magnetic domains in the micrometer range, there is a strong dependence on grain size and the kind of mineral. In materials that have not been heated, the magnetic domains are oriented at random. On heating to temperatures above the blocking temperatures, a small percentage of these domains become oriented with the prevailing magnetic field and remain (blocked) in this state after cooling. (Remanent magnetization may also be acquired during the formation of geologic materials at temperatures below the blocking temperature.) The strength of the thermoremanent magnetization is proportional to the field intensity at the time the temperature dropped below the blocking temperature. Minerals suitable for dating are present in most igneous rocks.

The blocking temperature of magnetite, for example, is 580°C, that of hematite is 670°C. But it is not the Curie temperatures that are decisive, but the lower blocking temperatures, which cover a range of values. The higher these temperatures, the more stable the TRM. The life time of the TRM ranges from several hundred thousand years to several hundred million years, as long as the materials are subjected only to normal ambient temperatures. If thermoremanent magnetization occurs at temperatures below 150°C (called "viscous remanent magnetization" – VRM), the object is usually unsuitable for dating because the lifetime of the VRM is too short.

Thermoremanent magnetization is also acquired by minerals formed in a magnetic field.

*Detrital remanent magnetic records:* In undisturbed, fine-grained ($\sim$ 100 $\mu$m) lacus-trine sediments (e.g., King et al. 1983), a weak remanent magnetization (detrital remanent magnetization = DRM) is often observed. This is caused by the orien-tation of the detrital magnetic grains parallel to the geomagnetic field during deposition. This kind of magnetic record, however, is sometimes difficult to interpret because its direction may deviate from the natural field direction if the magnetized

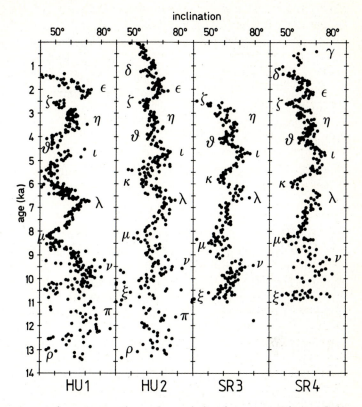

**Fig. 7.4.** Inclination type-curves for geomagnetic secular variation from cores taken at Lake Huron (*HU1* and *HU2*) and Lake Superior (*SR3* and *SR4*) in North America. The principal peaks and lows (*Greek letters*) correlate with each other and with the time scale fo Lake St. Croix and Lake Kylen (after Creer and Tucholka 1982a)

particles rolled owing to an irregular shape when they settled on the bottom of a still body of water or if water currents were present. Therefore, DRM is dependent on particle form and sedimentation rate. In addition the record may be disturbed by diagenetic changes (post-depositional DRM: PDRM), especially when organic matter or bioturbation is present. PDRM begins as soon as the overlying sediment is thick enough to prevent further changes from the surface. Stratigraphic DRM correlations are usually based on curve fitting using as many records as possible from neighboring profiles (Fig. 7.4).

### Sample Treatment and Measurement Techniques

Valuable discussions on selection of sample type, on sample collection, and on measurement techniques are given by Collinson (1983) and Aitken (1974).

*Sample collection:* For the determination of thermoremanent magnetization, the measurement of I, D, and F on a sufficient number of samples (at least 10) from the same site is recommended so that local disturbances can be recognized and a reliably representative value can be obtained. A theodolite (or sun compass), an exact clock, and astronomical tables are used to determine the orientation of the object to the nearest degree. The directions are marked on the sample, for example on plaster seals. Magnetic compass measurements are less suitable because the Earth's magnetic field is often locally disturbed; they are sufficient only if reversals are of interest. No iron tools should be used when collecting samples. The samples are to be protected from strong magnetic fields, elevated temperatures, and solar radiation during transport and storage.

*Pretreatment:* Cubes 0.5–3 cm to a side are cut from the sample in the laboratory. If necessary, the sample is impregnated with water glass (sodium silicate) or reinforced with a plaster cast. The $x$, $y$, and $z$ direction marks must be retained.

*Magnetic cleaning:* Instable viscous remanent magnetization (VRM) acquired during transport and storage or resulting in situ from climatic variations, is removed by "cleaning" in a high-frequency alternating magnetic field of 10–50 mT (after testing several pilot specimens for the proper strength) or by heating to 100–150°C ("thermal cleaning") for at least 15 min, followed by cooling in a zero magnetic room. Generally, VRM hardness increases with the age of the sample.

*Magnetometers:* Astatic, spinner, and squid magnetometers are used for measuring D, I, and F. *Astatic magnetometers,* although still in use, are no longer applied for routine measurements. They have been replaced by the other two types of magnetometers because analyses can be done faster with them and the SQUID type is more sensitive.

A *flux-gate spinner magnetometer* contains one or two specially wound coils (flux gates) for compensating for external magnetic fields. A potential is produced in these coils proportional to the magnetic moment of the sample (ca. 10 cm³) rotated at 300 rotations per minute. This potential is registered at specific positions at each rotation by a computer and converted into a two-dimensional magnetic vector field. Replicate measurements ("stacking") increase the accuracy. The magnetic field in the third dimension is obtained by turning the sample by 90°. The sensitivity of a spinner magnetometer is ten times greater than that of most astatic magnetometers.

A two-axis *SQUID magnetometer* can be up to 100 times more sensitive than a spinner magnetometer. A SQUID (superconducting quantum interference device) is a cryogenic paramagnetic semiconductor device that utilizes the Josephson effect (Walton 1977). At liquid helium temperature, the paramagnetic inductance is a function of the external magnetic field. For archeomagnetic samples, several milligrams may be sufficient. It is also more rapid than the spinner or astatic magnetometers.

*Measurement of D and I:* The values of D and I should be determined to at least $\pm 1°$, if possible to $\pm 0.5°$. The standard deviation in the declination measurement is generally greater than that of the inclination by a factor of 2. For magnetostratigraphic studies, at least three consecutive sequences of sediments (if possible, spanning several chrons) should be taken from the same site. For I/D records in lacustrine sediments, several cores should be taken from the same lake. If the curves are similar, further cores should be taken from two or more other lakes in the region to eliminate artifacts of nature or measurement. The mean of the curves is the master curve, which must then be calibrated by an absolute dating method, e.g., the radiocarbon method (Sect. 6.2.1).

*Measurement of F:*  The value of F is obtained using the double-heating method (Thellier and Thellier 1959; Walton 1977) or similar procedures (Aitken 1978; Shaw 1983). The samples are heated to 600°C in steps of 100°C; the samples are cooled and the magnetization M is determined after each step. This is done twice: the first time they are cooled in a zero magnetic field, the second time in the known magnetic field $F_e$ of the laboratory. In the first case, the value of M decreases with increasing temperature. The second time, the values of $M_e$ increase with increasing temperature. The ratio of the change in M to that of $M_e$ for each 100° temperature interval is proportional to the field F in the past:

$$F = \frac{\Delta M}{\Delta M_e} F_e \tag{7.1}$$

F represents the paleomagnetic field if the same value F is obtained for several neighboring temperature intervals (plateau test). If this is not the case, the thermoremanent magnetism has changed during the aging of the sample. Temperatures above 600°C are to be avoided so that the structure and magnetic properties of the sample will not be altered.

*The age determination:*  Ages derived from D and I records are generally significantly more reliable and precise than those derived from F values (whose confidence interval is seldom better than $\pm 15\%$) because the intensity of the Earth's magnetic field has varied by as much as 50% or more over periods on the order of 50 ka (McElhinny and Senanayake 1982).

The age of archeological samples is taken from the reference curves for the study area (e.g., Fig. 7.2) (Aitken 1974, 1978). The error in the dates may be less than $\pm 20$ years in the periods with the greatest resolution.

The task of paleomagnetic dating is to correlate the magnetostratigraphic polarity record to MPTS. If the top is Holocene, then the first reversal encountered is probably the Brunhes/Matuyama reversal at 730 ka. Otherwise, a constant sedimentation rate is assumed; however, this is seldom true. In special cases, independently determined data from biostratigraphic records, K-Ar dates (Sect. 6.1.1), for example from intercalated lava flows, or fission-track dates (Sect. 6.4.7) to calibrate parts of the time interval of the study.

### Scope and Potential, Limitations

*Scope and Potential.*  The archeomagnetic and paleomagnetic dating methods have a solid place in archeology (Aitken 1978; Tarling 1975), Quarternary research, and geochronology (Opdyke 1972).

Coarse magnetostratigraphic dating based on polarity records can be used to distinguish between rocks formed before and after the last reversal at 730,000 BP. For example, this method was used to revise Canadian glacial stratigraphy (e.g., an area previously assumed to have been glaciated was determined to be permanently ice-free during the Quaternary; Vincent et al. 1984b). Another area was identified as containing no active faulting (important for the siting of nuclear power plants; Davis et al. 1977).

Geoscientists use the method primarily for studying the chronology of tectonic events and for studies of the Earth's magnetic field.

*Limitations.* Objects whose thermoremanent magnetization has changed at any time are unsuitable (Storetvedt 1970; Verosub 1977). Among the causes are chemical processes (giving chemical remanent magnetization – CRM) during diagenesis (e.g., formation of iron oxides), dehydration, weathering, recrystallization (e.g., conversion of magnetite to hematite), redeposition (Dymond 1969), being struck by lightning, or subjection to tectonic processes under high pressure (dynamic magnetization from shearing and shock (SRM)). Changes brought about by heating for the measurement can also make an object unsuitable (Ness et al. 1980).

Thermoremanent magnetization also yields unusable data if the object cooled very slowly over a very long period (Fig. 6.97). This is indicated by the lack of a plateau in the results of the double-heating method. In addition, an object whose orientation was changed after its initial deposition will give incorrect results.

### Non-Chronological Applications

It has also been used to estimate, for example, the degree of water saturation of sediments in the past (Ensley and Verosub 1982). Paleomagnetic studies are made in geophysics mainly to research the causes of the magnetic field of the Earth, changes in the magnetic field, and to find relationships between geomagnetic excursions and paleoclimatological events and orbital parameters (Rampino 1979). The dating of cave paintings is a beautiful, although exceptional example of the possibilities for application of archeomagnetics to prehistoric research (Creer and Kopper 1974).

## 7.2  Chronostratigraphic Time-Scale Using $\delta^{18}O$ Values***

(Urey 1947; Emiliani 1954, 1955)

### Dating Range, Precision, Substances, Sample Size

The stable isotope composition of oxygen (expressed as $\delta^{18}O$) yields very precise ages ($\pm 1500$–5000 years) (Emiliani 1966; Shackleton and Opdyke 1973; Hays et al. 1976; Berger et al. 1984; Imbrie et al. 1984; Martinson et al. 1987) for monospecies planktonic foraminiferal tests (ca. 0.01 g) extracted from undisturbed pelagic sediment cores up to at least 1 million years.

Continental and glacier ice (ca. 5 ml or less) up to 8000 years old has been dated to ± 1‰ (Dansgaard et al. 1982; Dansgaard 1985; Hammer et al. 1986). Continental ice can be dated up to an age of 150 ka with the application of a suitable ice flow model (Dansgaard et al. 1969, 1971; Johnsen et al. 1972; Hammer et al. 1978; Lorius et al. 1979, 1985; Reeh 1989).

Interglacial and interstadial groundwater (Bath et al. 1979) and micropore water from speleothem, preferentially stalagmites (Harmon and Schwarcz 1981; Harmon et al. 1983), can be distinguished on the basis of their stable oxygen isotope and hydrogen isotope records (Rozanski and Dulinski 1987).

Limnic sediments contain stable oxygen isotope records that reflect the transitions from glacial to post-glacial periods (Siegenthaler et al. 1984) and allow stratigraphic correlations.

### Basic Concept

*The basic of oxygen isotope stratigraphy* is the temperature dependence of the ratio of the stable oxygen and hydrogen isotopes in water (Urey 1947; Epstein et al. 1953). Therefore, seasonal temperature changes causes the formation of isotopically distinguishable annual layers, which can function as a time-scale hundreds or even thousands of years long. Long time periods are covered by the oxygen isotope record of the frequent changes between interglacial and glacial periods of the Quaternary (Lorius et al. 1985; Martinson et al. 1987).

The average ratio of $^{16}O$ to $^{18}O$ in nature is about 490 to 1, that of normal hydrogen to deuterium is about 6000 to 1. As a result of the lower vapor pressure of the isotopically heavier water molecules $H_2^{18}O$ and $^1H^2HO$ (ca. 1‰ and 7‰ lower, respectively) relative to the more abundant, lighter $H_2^{16}O$, isotopic fractionation occurs during phase transitions (e.g., evaporation, condensation, and sublimation). The lighter molecules are enriched in the gas phase, the heavier ones in the liquid or solid phase. These isotopic shifts are expressed as $\delta^{18}O$ and $\delta D$ values, analogous to $\delta^{13}C$ values (Eq. 6.59).

*Isotopic variations in precipitation:* The seasonal changes in the $\delta^{18}O$ values for precipitation (Craig 1961a; Dansgaard 1964) are a result of temperature-dependent fractionation (Fig. 7.5). Precipitation in winter is isotopically lighter than in summer. The temperature coefficient of the $\delta^{18}O$ value for continental precipitation is 0.7‰ per °C or less, that of the $\delta^2H$ value is about 5.6‰ per °C. In maritime areas, this gradient may be much smaller, e.g., 0.2‰ per °C for $\delta^{18}O$ (Gat and Gonfiantini 1981). Analogous fractionation occurs between the hydrogen isotopes $^1H$ and $^2H$ and thus the following linear relationship exists between $\delta^2H$ and $\delta^{18}O$ for continental precipitation:

$$\delta^2H = s\,\delta^{18}O + d \tag{7.2}$$

The theory and methods of isotope chronostratigraphy are described by Williams et al. (1988).

A slope of $s = 8$ is characteristic for continental precipitation and defines the global meteoric water line (Craig 1961a; Dansgaard 1964; Gat and Gonfiantini

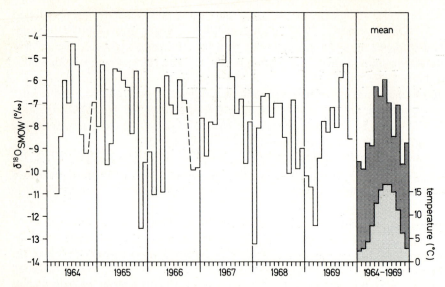

**Fig. 7.5.** Monthly changes in the $\delta^{18}O$ values for precipitation at Groningen and the mean monthly temperatures (after Mook 1970)

1981). The intercept $d$ calculated for $s = 8$ is the deuterium excess, which is about + 10‰ for continental rain, about + 20‰ for Mediterranean rain, about 0‰ for Antarctica. The value of $d$ is chiefly a function of the mean relative humidity of the atmosphere above the oceans (Merlivat and Jouzel 1979). Evaporation from surface water may cause the slope to be as low as 4. The slope $s$ can be as low as 2 for soil water in the unsaturated zone if it evaporates. Thus, groundwater that has been previously subjected to evaporation can be identified on this basis (Allison 1982).

This temperature dependence of $\delta^{18}O$ values for precipitation is also preserved in Pleistocene and Holocene groundwaters, which were recharged under different climatic conditions. For example, in Europe the difference between the $\delta^{18}O$ values for groundwater from these two periods is about 1.2‰ (Bath et al. 1979).

In glacier and continental ice, the seasonal variations in the oxygen isotope ratio in precipitation are "frozen in" as annual layers. These can be counted for at least 1000 years into the past on the basis of the seasonal maxima and minima of the $\delta^{18}O$ values. In older cores, the "isotope varves" may no longer be present due to thinning under the pressure of the overlying ice and to diffusion (e.g., Johnsen 1977), but pronounced isotope time marks reflecting the long-term $\delta^{18}O$ trend at the boundaries of glacial to interglacial periods are still preserved. Flow models for continental ice have been set up to obtain time-scales, which are calibrated by the $\delta^{18}O$ records of suitable sections (Dansgaard et al. 1971; Johnsen et al. 1972; Hammer et al. 1978; Lorius et al. 1979, 1985).

*Oxygen isotopes in carbonates:* At least some solid carbonates formed in water (e.g., foraminiferal tests and mollusc shells) are in isotopic equilibrium with the bicarbonate dissolved in the water. Thus, oxygen isotope analysis ($\delta^{18}O_{PDB}$) of

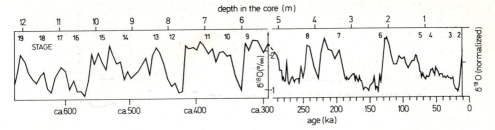

**Fig. 7.6.** Quaternary deep-sea oxygen-isotope record for the Brunhes chron. A final orbitally based chronostratigraphy derived from a stacked oxygen isotope record is available for the last 300 ka (after Martinson et al. 1987). The earlier part of the representation is the $\delta^{18}O$ record of the core V28-238, for which only age estimates are available (after Shackleton and Opdyke 1973)

planktonic foraminifers from pelagic sediment (Fig. 7.6), for example, can be used to reconstruct the $\delta^{18}O$ records ($\delta^{18}O_{SMOW}$) for ocean water during secular climatic fluctuations if the paleotemperature (T) of the deep sea water is known (Epstein et al. 1953):

$$T = 16.0 - 4.2(\delta^{18}O_{PDB} - \delta^{18}O_{SMOW}) + 0.14(\delta^{18}O_{PDB} - \delta^{18}O_{SMOW})^2 \qquad (7.3)$$

The $\delta^{18}O$ values for planktonic foraminifers mainly are governed by the mixing of isotopically heavy ocean water with isotopically light melt water from the polar ice caps and reflect to only a minor extent changes in temperature. At the last glacial maximum, the sea level was 100–150 m lower than at present and the $\delta^{18}O$ value for the ocean water was about 1.5‰ larger, because the isotopically light water was bound in continental ice. The $\delta^{18}O$ record in marine sediments, therefore, is a paleoglacial curve rather than a paleotemperature curve (Shackleton 1967, 1987).

The general trend of the fluctuations in the $\delta^{18}O$ curves for foraminifers (Fig. 7.6) is valid worldwide and reflects global climatic changes, especially between glacial and interglacial periods (Shackleton and Opdyke 1973; Shackleton 1982; Martinson et al. 1987). The peaks of the curve belonging to interglacials have been given odd numbers (Emiliani 1955).

*Marine oxygen-isotope record and time-scale:* The marine oxygen isotope record of equatorical planktonic foraminifers has been dated using the U/Th method (Sect. 6.3) and the paleomagnetic method (Sect. 7.1). This record shows at least 12 glacial periods instead of the four to six of classical Quaternary geology (Würm/Weichsel/Wisconsin – Riss/Saale/Illinonian – Mindel/Elster/Hansan – Günz/ Menapian/Nebraskan – Donau/Eburanonian/? – Biber) within the Brunhes polarity chron, and a possible further 18–20 within the last 2 Ma. Great efforts have been made to correlate the peaks of the Quaternary deep-sea oxygen-isotope record with the biostratigraphically classified interglacials recognized on the continents but only peak 5e could be definitely correlated (with the Eemian, Fig. 7.6; Mangerud et al. 1979).

Since the mid-1970's, the time-scale of the deep-sea oxygen-isotope record for the last 730,000 years (most of the Quaternary) has been "tuned" to within $\pm 1500$–5000 years (Hays et al. 1976, 1984; Berger et al. 1984; Imbrie et al. 1984; Herterich and Sarnthein 1984; Martinson et al. 1987). This precise tuning is based on the idea of Milankovitch (1930) that the climatic change between glacial and interglacial periods is governed by variations in the Earth's orbit with well-known periods which causes changes in the insolation of the Earth: for the last 800,000 years. The parameters that change are obliquity (with a period of 41,000 years) and precession of the equinoxes (with main periods of 19,000 and 23,000 years) and eccentricity of the Earth's orbit (with periods of 413,000, 100,000 and 54,000 years) (Berger 1988). Previously the 41-ka cycle dominated.

### Sample Treatment and Measurement Techniques

*Storage:* Ice should be stored in a deep-freeze; water samples should be kept in tightly closed, completely filled glass bottles. Otherwise, molecular exchange with atmospheric water vapor will lead to contamination of the samples, which easily occurs through the walls of plastic bottles.

*Pretreatment:* Sediment is sieved with distilled water under an inert gas. The foraminiferal tests of a single species (50–100 $\mu$g) are handpicked under the microscope and cleaned in an ultrasonic bath. In the case of benthic foraminifers, at least ten should be analyzed to get an adequate representation of the mixture (Boyle 1984). The organic residue in the foraminifers, which would cause incorrect results, is removed by heating to 400°C in vacuo. Chemical treatment with hydrocarbons may cause incorrect results (Ganssen 1981).

*Measurement:* The $\delta^{18}O$ values are determined by measurements on carbon dioxide with a mass spectrometer (Sect. 5.2.3.1), often automatically for many samples. For this purpose, carbonates are decomposed with 100% water-free phosphoric acid at constant temperature; water samples are equilibrated with $CO_2$ at atmospheric pressure for a few hours so that the $CO_2$ acquires the same oxygen isotope ratio as the water sample. If ice is being dated, at least five oxygen isotope analyses must be made for each annual layer. $\delta^2H$ values are determined mass spectrometrically on hydrogen, obtained by reduction of the water with hot uranium or zinc.

So that the smallest possible differences in the isotope ratios can be measured, relative measurements are made instead of absolute measurements. The errors in the $\delta^{18}O$ and $\delta^2H$ values, which are defined analogously to $\delta^{13}C$ [Eq. (6.59)], are $\leqslant \pm 0.1$‰ and $\pm 1$‰, respectively. The PDB standard (Cretaceous belemnite from the Peedee Formation in South Carolina) was formerly used for carbonates (Craig 1957). Because the PDB standard is no longer available, NBS 19 has been used for the last several decades. The SMOW standard (Standard Mean Ocean Water, Craig 1961b) or derived standards are used for water.

### Scope and Potential, Limitations, Representative Examples

*Scope and Potential.* Oxygen isotope analysis of foraminifers is indispensible for stratigraphic and geochronological studies in paleo-oceanography (e.g., Ruddiman and McIntyre 1981, 1984). For reliable dates, it is important to obtain carbonatic cores of sediments that contain little or no disturbance by turbidity currents, erosion, redeposition, recrystallization and that had high deposition rates so that errors due to bioturbation are kept small (Berger and Johnson 1978).

Oxygen isotope analyses are also important for continental ice studies (e.g., Robin 1983; Hammer et al. 1986). The great analytical effort that is necessary and the exponential thinning of the annual layers of ice with increasing depth (Johnsen 1977) limits the application of this method to ice no older than about 8000 a. For older ice, the accumulation rates are determined only for suitable parts of the cores and these are used to calibrate the time-scales of ice flow models (Dansgaard et al. 1971; Johnsen et al. 1972; Hammer et al. 1978; Lorius et al. 1979, 1985; Reeh 1989) Chemostratigraphic methods may also be employed (Sect. 7.6).

The mean residence time of freshly recharged groundwater can be estimated up to several years on the basis of the damped annual $\delta^{18}O$ variations of precipitation (IAEA 1969–1986; Stichler and Herrmann 1983). In tropical areas, the strong negative correlation between $\delta^{18}O$ and intensity of the rainfall (heavy rainfall results in increasingly negative $\delta^{18}O$ values: amount effect) is another seasonal variation that can be used for studies of residence times of groundwater. In the case of fossil groundwater, Holocene and Pleistocene samples may sometimes be distinguished (Bath et al. 1979). $\delta^{18}O$ values for paleogroundwater appear to correlate with those for bone apatite and phosphate collected in the same area (Longinelli 1984; Luz et al. 1984).

**Limitations.** *Marine carbonates:* The resolution of the deep-sea record is limited due to bioturbation (e.g., Broecker et al. 1984), which reaches depths of up to 8 cm in deep-sea cores (Berger and Johnson 1978), sedimentation rates that are too low, erosion and lateral transport, or other, still unknown effects (Pisias 1983). These efforts have been reviewed by Butzer (1983) and Nicolis and Nicolis (1984). The carbonate of most species of benthic foraminiferal tests is not in isotopic equilibrium with the carbonate dissolved in the sea water (Duplessy et al. 1970; Dunbar and Wefer 1984) and, therefore, cannot yield reliable age records. Detrital marine carbonate is obviously allochthonous material and, therefore, is not suitable.

*Ice:* Wind scouring is a serious problem for the interpretation of stable isotope records of ice cores (Fisher et al. 1983). Diagenetic processes, such as vapor diffusion, may also mask the isotope record. The dating range of temperate ice is often very short, because percolating melt water and molecular diffusion cause a rapid disappearance of $\delta^{18}O$ stratification. In the Alps, complex accumulation processes governed by microclimate make chronological interpretation difficult (Schotterer et al. 1977; Haeberli et al. 1983).

*Groundwater:* Holocene and Pleistocene groundwater may not always be distinguishable on the basis of their $\delta^{18}O$ values because they may be masked by dispersion effects in the aquifer or by changes in meterologic trajectories in the past (continental effect). As the heavier isotopes are always enriched in the condensate, precipitation becomes increasingly isotopically heavier along the path of a storm across a continent (e.g., Sonntag et al. 1980).

## Non-Chronological Applications

The important role of stable oxygen and hydrogen isotope analysis for geothermal (Javoy 1977) and hydrological (Gat and Gonfiantini 1981; IAEA 1983) studies, as well as studies on the genesis of rocks (Fritz and Fontes 1980, 1986) has been the subject of so many textbooks that only a few representative examples should be given here.

*Oceanographic studies:* Sources of error in the $\delta^{18}O$ record are lateral transport of sediment, bioturbation (Broecker et al. 1984), and delays in the mixing of fresh meltwater with the sea water (Berger et al. 1977) and deep oceanic circulation (Barnola et al. 1987). These and other problems for the interpretation of the $\delta^{18}O$ data have been discussed in detail by Shackleton (1982). Specimens that have been altered from aragonite to calcite are also unsuitable objects for analysis (e.g., Stahl and Jordan 1969).

Coastal upwelling (Ganssen and Sarnthein 1983) and the circulation of deep water (Shackleton et al. 1983a; Ruddiman and McIntyre 1984; Barnola et al. 1987), as well as sea-level fluctuations, have been studied using simultaneous $\delta^{18}O$ and $\delta^{13}C$ analyses (see Berger 1981; Shackleton 1987). The results imply that a large part of the observed variations represent global changes in the carbon distribution between the biosphere and the oceans (Labeyrie and Duplessy 1985). There is growing evidence that $CO_2$ variations (Neftel et al. 1982) are also involved in the mechanisms of climatic response to orbital variations (Broecker 1982; Shackleton et al. 1983b; Pisias and Shackleton 1984; Barnola et al. 1987).

*Limnic studies:* The $\delta^{18}O$ record in lacustrine deposits is much more complex than that of marine sediments. Lakes dry out during droughts, sediment is removed during glacials. During interglacials, the $\delta^{18}O$ values of the lake water and carbonate represent a superimposed record from various sources. The isotopic composition of precipitation predominates with its positive temperature gradient. Variation is damped by the $\delta^{18}O$ shift with a negative temperature coefficient between the lake water and carbonate in isotopic equilibrium. The isotopic composition is modified further by evaporation of the lake water (Eicher et al. 1981). Owing to these effects, lacustrine sediments contain only regional isotope markers at the transitions from glacial to interglacial periods (Siegenthaler et al. 1984). $\delta^{18}O$ studies on lacustrine sediments in semi-arid areas should be accompanied by the analysis of calcite, Mg-calcite, aragonite, and dolomite, which reflect salinity trends. Autochthonous and diagenetic minerals must also be separated for the analysis (Talbot and Kelts 1986).

Water balance studies of lakes have successfully applied oxygen isotope analysis (Zuber 1983); the precision is not better than $\pm 20\%$.

*Brackish and fresh water studies:* Paleotemperatures of coastal water have been derived from $\delta^{18}O$ values for the carbonate in sea shells (Mook 1971); $\delta^{18}O$ values for snail shell carbonate have documented sea-level fluctuations (Covich and Stuiver 1974). Paleotemperatures have also been determined using fresh-water carbonates (Stuiver 1970).

*Groundwater studies and brines:* Seasonal, diurnal, or artificial fluctuations in the mixing of groundwater from various sources yield numerous examples of the application of stable oxygen isotopes. When the groundwater in two aquifers is isotopically different, the best method for studying relationships between the aquifers is the concurrent analysis of deuterium and oxygen. The mixing of bank infiltrate with groundwater (Stichler et al. 1986) and the composition of the runoff of glaciated areas (direct run-off, snow meltwater, ice meltwater, long-retained meltwater, and groundwater) (Behrens et al. 1979; Stichler and Herrmann 1983; Obradovic and Sklash 1986) have been differentiated. The content of meteoric water in brines in mines (Geyh 1969; Zuber et al. 1979), the formation of deep brines on the Canadian Shield (Frape and Fritz 1982), and in the deep sea (Schoell and Faber 1978) have been studied by isotopic analysis. The elevation of the recharge areas for springs has been estimated from orographic $\delta^{18}O$ gradients, which are between $-0.15$ to $-0.4\%$ per 100 m (Gat and Gonfiantini 1981). $\delta^{18}O$ and $\delta^2H$ analyses of soil moisture provide a means of estimating evapotranspiration in the range of 4–700 mm/a, e.g., in depressions in the eastern Sahara (Christman and Sonntag 1987). Changes in the $\delta^{18}O$ values of groundwater seem to be associated with earthquakes (O'Neil and King 1981).

*Speleothem studies:* It was expected that absolute paleotempertures could be determined under favorable conditions by isotopic analysis of the stable oxygen and hydrogen of micropore water in stalagmites (Gascoyne et al. 1980; Harmon and Schwarcz 1981; Harmon et al. 1983). However, more recent studies have shown that in-situ isotopic exchange with the host calcite during aging and during extraction for preparation of the sample may mask the paleoclimatic record (Rozanski and Dulinski 1987). This may partly explain the discrepancy between the results of these studies and those based on other isotopic records (Friedman 1983).

The $\delta^{18}O$ values for speleothem do not give a global isotope record because kinetic-controlled isotope fractionation during carbonate precipitation often masks the considerably smaller amount of temperature-dependent equilibrium-controlled fractionation (Fantidis and Ehhalt 1970). An additional complication is that stalagmite growth is not continuous. It is interrupted during glacial periods, and possibly also during interstadials, which is why a long time-scale cannot be expected (Geyh and Hennig 1986).

*Biological and paleoclimatic studies:* Ambiguous results have been obtained from isotopic analysis of carbon-bound, non-exchangeable hydrogen and oxygen in cellulose in tree rings (Tans et al. 1978; Burk and Stuiver 1981; Hughes et al. 1982) because kinetic isotopic fractionation during metabolism masks temperature-dependent equilibrium-controlled fractionation (e.g., Stiller and Nissenbaum 1980). The same is true for other organic materials, such as peat (e.g., Schiegl 1972; Brenninkmeijer et al. 1982; Dupont and Brenninkmeijer 1984). But the results may yield new insight on the metabolic processes of plants (Burk and Stuiver 1981; de Niro and Epstein 1981). They may also assist in the reconstruction of the isotopic content of paleoprecipitation, which is governed by temperature and humidity

(Buchardt and Fritz 1980; Yapp and Epstein 1982a,b; Lawrence and White 1984; Gray and Se Jong Song 1984; Edwards and Fritz 1986).

*Archeological studies* are aided by $\delta^{18}O$ analyses in the reconstruction of prehistoric ecological conditions (Shackleton 1970) and the seasons in which prehistoric settlements were occupied (Shackleton 1973; Bailey et al. 1983).

*Meterological studies:* Stable isotope analysis of precipitation and humidity has provided information on their spatial and temporal distribution, their origin, and the transport of water masses in the troposphere (Hübner et al. 1979; Förstel and Hützen 1983; Schoch-Fischer et al. 1984). Such information has also been obtained for the past (Gat and Carmi 1970; Lorius et al. 1979; Jouzel et al. 1982; Rozanski 1985).

## 7.3 Chronostratigraphic Time-Scale Using $\delta^{34}S$ and $\delta^{13}C$ Values and $^{87}Sr/^{86}Sr$ Ratios**

(Nielsen and Ricke 1964)

### Dating Range, Precision, Substances, Sample Size

The method is useful for dating marine evaporites (several 100 mg) with ages of 0–650 Ma (Nielsen 1979; Claypool et al. 1980), in some cases also for the sulfate in leachate waters (Pearson and Rightmire 1980) and in crude oil (Thode and Rees 1970).

### Basic Concept

Sulfur has four stable isotopes: $^{32}S$ (95.02%), $^{33}S$ (0.75%), $^{34}S$ (4.21%), and $^{36}S$ (0.02%). The ratio of the first two is given as $\delta^{34}S$ values, defined analogously to the $\delta^{13}C$ value [Eq. (6.59)]. Iron sulfide from the troilite phase of the Diablo Canyon iron meteorite (DCT) (with a $^{32}S/^{34}S$ ratio of 22.220) is used by convention as standard.

Fractionation of sulfur isotopes of up to 50‰ occurs mainly by reduction of sulfate to $H_2S$ by anaerobic bacteria and by isotope exchange in which the heavier sulfur isotope is preferentially enriched in compounds with the higher oxidation state. Input of $^{34}S$ into the oceans occurs via weathering and erosion products; removal occurs via precipitation either as sulfide (in muds) or as sulfate (in evaporites).

There are three main reservoirs of sulfur: evaporite sulfate (with $\delta^{34}S$ values of + 10 to + 30‰, mean + 17‰), and dissolved sulfate in the oceans ($\delta^{34}S$ value of + 21‰), balanced by the largest of the three, the sedimentary sulfides (roughly − 12‰).

During the course of the Earth's history, the $\delta^{34}S$ value of the world's oceans, and consequently, of marine evaporite sulfate has varied between + 10 (Permian)

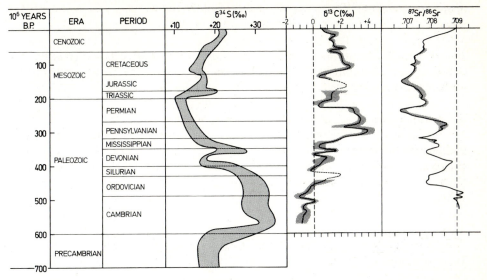

**Fig. 7.7.** Composite $\delta^{34}$S curve for sulfates in marine evaporites (after Claypool et al. 1980) and corresponding $\delta^{13}$C values, as well as $^{87}$Sr/$^{86}$Sr values from carbonates and apatite in marine sediments (after Holser et al. 1986)

and $+35‰$ (Cambrian) (Fig. 7.7), possibly correlated with the sulfate content (François and Gerard 1986). Sulfur and carbon isotope fractionation ($\delta^{34}$S and $\delta^{13}$C values) appear to correlate inversely with one another over the long term (Ohmoto 1972; Veizer et al. 1980). There are several models (e.g., Nielsen 1965; Claypool et al. 1980; Lein 1985) that attempt to explain these changes. The $^{87}$Sr/$^{86}$Sr ratio correlated with the $\delta^{13}$C values (Holser et al. 1986; Figs. 6.19 and 7.7).

In addition to long-term changes (Fig. 7.7), several "catastrophic geochemical events" (Sect. 7.4) associated with sharp rises in $\delta^{34}$S values produced stratigraphic markers. The causes might have been the mixing of brines with ocean water. The most important events occurred in the late Early Triassic (ca. 215 Ma), early Late Devonian (ca. 355 Ma), and late Proterozoic (ca. 635 Ma) (Holser 1977).

### Sample Treatment and Measurement Techniques

For a $\delta^{34}$S analysis, the sulfur must first be converted quantitatively to sulfate. The samples (0.5–1.5 g) are treated successively with $H_2O_2$ and aqua regia, evaporating to dryness after each treatment. The sulfate is reduced to sulfide by heating with iron powder and some zinc powder in a muffle oven under an inert gas and then precipitated as BaS, CdS, AgS, or ZnS. This is oxidized to $SO_2$ by fusing it with $V_2O_5$ or CuO in a sealed, evacuated quartz glass tube or heating with oxygen at 100°C. The isotopic composition of the $SO_2$ is measured mass spectrometrically (Sect. 5.2.3.1) to about $\pm 0.3‰$. To obtain information about the natural variability, several samples are always analyzed from the same strata.

*Scope and Potential, Limitations, Representative Examples*

*Scope and Potential.* Due to the wide distribution of sulfate on the Earth, interest in dating via sulfur isotopes is great (Nielsen 1979), but only marine sulfates and mineral water are suitable (Pearson and Rightmire 1980). Crude oil has been dated via $\delta^{34}S$ values or at least its source rock identified, since the $\delta^{34}S$ value of the sulfur compounds in oil is $+15‰$ greater than that of the cogenetic marine sulfate (Thode and Rees 1970).

*Limitations.* Reasons for the sometimes very large differences in the $\delta^{34}S$ values for different materials have been discussed by Holser and Kaplan (1966). The method cannot be applied for hydrogeological studies if the sulfate in the water includes anthropogenic sulfate (from industry, households, or agriculture, i.e., artificial fertilizers) or sedimentary sulfate. Their presence is certain if $\delta^{34}S$ values greater than $+8‰$ are found. Another interference is bacterial sulfate reduction, which is associated with the presence of $H_2S$. For these reasons, groundwater with a high sulfate concentration is more suitable than that with a low sulfate concentration.

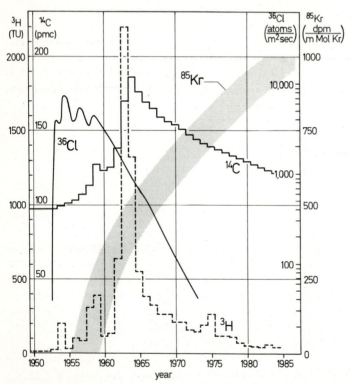

**Fig. 7.8.** Changes in the radiocarbon concentration (*pmc*) in atmospheric $CO_2$ and changes in the tritium concentration (*TU*) in precipitation in Central Europe caused by nuclear weapons tests. Present concentrations are due to emissions from nuclear reactors and nuclear processing plants. The $^{85}Kr$ content of the atmosphere derives mainly from emissions from plants reprocessing spent fuel from nuclear power plants (after Rozanski and Florkowski 1979; Salvamoser 1986). The chlorine-36 fall-out rate is estimated from Dye-3 data from Greenland (after Bentley et al. 1986a)

### Non-Chronological Applications

Sulfur isotope analysis has been a part of numerous geochemical and hydrochemical studies (Fritz and Fontes 1980; IAEA 1987). These have included studies on the sedimentary sulfur cycle, the sulfur compounds in the atmosphere, and those dissolved in groundwater and ocean water. Moreover, sulfur isotope studies have been applied for genetic differentiation of ore deposits (e.g., magmatic, hydrothermal, or biogenic) (Nielsen 1979).

*Rock studies:* Attempts to distinguish volcanic sulfate deposits from sedimentary ones have not been successful because the isotopic sulfur compositions have a too wide range. Genetic information can be obtained, however, in many cases (Faure 1977; Krouse 1980), also for crude oil (Guffney et al. 1980; Venkatesan et al. 1982).

*Cave studies:* Sulfur and oxygen isotope data, together with hydrochemical data, have yielded at least a partial explanation of the formation of two caves in Banff National Park in Alberta, Canada: oxidation of $H_2S$ in spring water to $H_2SO_4$, which leached out the caves (van Everdingen et al. 1985).

*Meteorological studies:* According to meteorological studies on the origin of sulfate in rain carried out in Israel (Wakshal and Nielsen 1982), the air masses of the lower troposphere carry aerosol particles of mainly marine origin with higher $\delta^{34}S$ values and the upper troposphere contains volatile organic compounds containing isotopically lighter sulfur of biogenic origin.

## 7.4   Artificial Radionuclides as Time Markers***

(Crane 1951)

### Dating Range, Precision, Substances, Sample Size

Artificial radionuclides are used for dating lacustine, fluvial, and marine sediments (several grams) (Delaune et al. 1978; Robbins 1984), as well as cold and temperate ice (ca. 250 g) (Wilgain et al. 1965; Ambach and Dansgaard 1970; Schotterer et al. 1977; Elmore et al. 1982). Such deposits contain time markers (mainly $^{137}Cs$) formed by fallout from the nuclear weapons tests after 1955, the most prominent of these appeared immediately after the end of the tests in 1963/64 (Fig. 7.8). A new time marker was formed in Europe by the Chernobyl accident.

Radiocarbon, tritium, $^{85}Kr$, and $^{36}Cl$ have been used to determine mean residence times (MRT) of up to 150 years for the long-term components of spring water (Eriksson 1958; Rozanski and Florkowski 1979; Bentley et al. 1982), for groundwater in unconfined aquifers, for carbon dioxide in the atmosphere and carbon in the biosphere and the oceans (Oeschger et al. 1975; Bentley et al. 1982; Nydal and Lövseth 1983; Nydal et al. 1984). Multi-isotope analysis ($^3H$, $^{14}C$, $^{36}Cl$, and $^{85}Kr$ have been used) yields the most accurate results, or in some cases is even necessary, especially if mixtures of water are being investigated (e.g., Salvamoser 1986; Smethie et al. 1986). Iodine-129 from the Chernobyl reactor accident may become a tracer for global environmental processes (Paul et al. 1987). The production date of foodstuffs can be tested (Walton et al. 1967) and supporting data for ethnological studies obtained (Tamers 1969).

Cesium-137 analyses can be made in the laboratory on sediment samples of several grams or in the field on borehole cores and in the borehole itself.

For $^{14}$C analyses, 5 mg–5 g carbon is necessary; this amount is contained in 0.1–300 L groundwater. For a tritium analysis, 15–2000 mL is sufficient; for a $^{85}$Kr analysis, 200–300 L must be degassed in the field (Rozanski and Florkowski 1979). For $^{36}$Cl using AMS (Sect. 5.2.3.2), 1–2 mg AgCl is analyzed, which can be obtained from a few liters of water or several kilograms of ice, depending on the Cl content.

### Basic Concept

*Fallout in solids:* From the beginning of the 1950's to 1963/64, large amounts of short-lived, radioactive isotopes were produced by nuclear weapons tests, e.g., $^{22}$Na ($\tau = 2.6$ a) (see also Sect. 6.2.4), $^{55}$Fe ($\tau = 2.7$ a), $^{60}$Co ($\tau = 5.27$ a), $^{90}$Sr ($\tau = 28.5$ a), $^{137}$Cs ($\tau = 30.17$ a), as well as $^{238}$Pu, $^{239}$Pu, $^{240}$Pu, $^{242}$Pu, $^{241}$Am, and $^{243}$Am. A considerable portion of these radionuclides was blown into the stratosphere, where they formed solids that were rapidly adsorbed on aerosols. Gradually, with a prominent maximum in early summer, when the tropopause is relatively low, they have returned via precipitation to the Earth's surface, where they have been deposited on ice and sediments. In favorable circumstances, the fallout has formed resistent annual layers (Delaune et al. 1978). Since the limited nuclear weapons test ban treaty of 1963, only a few, difficult-to-identify stratigraphic markers have been formed, which correlate with low-energy atomic bomb tests and sporadic emissions (extreme cases) of the nuclear industry (Hardy et al. 1980). The most extreme case of emissions from a nuclear power plant resulted from an accident at Chernobyl in the Soviet Union in 1986 (Paul et al. 1987). This fallout can be distinguished from others on the basis of its known initial $^{134}$Cs/$^{137}$Cs activity ratio of 0.5. The method, however, is based mainly on the fallout marker of 1963/64. Resolution to within a year is possible only for samples formed within a single year, e.g., foodstuffs, annually varved ice and sediments.

*Fallout in liquids and gasses:* Radiocarbon (in carbon dioxide) and tritium (in water) from the nuclear tests (about 500 kg was injected in the stratosphere) both followed the same path from the stratosphere (mean residence time of tritium is about 2 years) into the hydrosphere and biosphere. From 1950 to 1963/64, the $^{14}$C content of atmospheric $CO_2$ increased by a factor of almost 2 and the tritium content of rain in the northern hemisphere increased by three orders of magnitude, in Canada by four orders of magnitude (Fig. 7.8) (Nydal and Lövseth 1983). In the ocean the increase was smaller (Nydal et al. 1984). The radiocarbon level has continued to decrease since 1963. After 1964 the tritium level decreased exponentially; since 1973 it has remained at a nearly constant level of 30–50 TE in the northern hemisphere, which is why it has become increasingly difficult to use for dating. The loss of tritium due to radioactive decay is compensated for by tritium emissions from nuclear reactors, nuclear fuel reprocessing plants, laboratories and industries producing and using tritiated products. Since the mid-1980s the proportion from these sources has become greater than that from the atomic weapons tests (Eisenbud et al. 1979). The tritium level is subject to pronounced continental and seasonal effects. The tritium level in the southern hemisphere peaked about two orders of magnitude lower than it did earlier in the northern hemisphere. At present, it approaches the probable pre-nuclear levels. The $^{14}C$ variations in the oceans may be reconstructed by analysis of corals (Druffel 1982), as well as narwal tusks (Bada et al. 1987).

Anthropogenic $^{14}C$ and $^3H$ have marked the hydrosphere, making it possible to determine the mean residence time of long-term components of karst spring water and groundwater in unconfined aquifers up to 150 years (e.g., Geyh 1972a, b); hydrodynamic models (Eriksson 1958) are used for this purpose. Cold ice up to 30 years old has been dated to within a year. Complex box models based on long-term isotopic records have been developed to estimate the mean residence time of $CO_2$ in the various geophysical reservoirs (Oeschger et al. 1975; Oeschger and Siegenthaler 1978; Nydal and Lövseth 1983; Nydal et al. 1984).

The concentration of $^{85}Kr$ ($\tau = 10.76$ a, $E_{max} = 687$ keV) in the atmosphere has been increasing worldwide since the beginning of the 1950's (Fig. 7.8). Initially it increased by 40 Bq/L Kr annually, in the last 15 years by only about 15 Bq/L Kr per year. Its concentration in rainwater in 1982 was about 710 Bq/L Kr or about 0.07 Bq/m$^3$ water or 1 Bq/m$^3$ air. It appears that the $^{85}Kr$ level in the atmosphere is approaching steady-state conditions and thus a constant level. Because $^{85}Kr$ enters the troposphere directly, rather than via the stratosphere, seasonal fluctuations and the continental effect are absent (Weiss et al. 1983). The primary source of $^{85}Kr$ is the industry that extracts plutonium from spent fuel elements from nuclear power plants. However, efforts are being made to avoid release of $^{85}Kr$ from the reprocessing plants. The cosmogenic portion can be neglected.

Krypton-85 serves as a time marker in the hydrosphere in the same way as $^{14}C$ and $^3H$. As a chemically inert gas, it has nearly ideal properties for studying hydrodynamic movement and mixing of groundwater and ocean water into which it has diffused (Salvamoser 1986).

Chlorine-36 was formed by the nuclear weapons tests primarily from $^{35}Cl$ in the ocean (Bentley et al. 1986a) (Fig. 7.8). It was distributed irregularly across the surface of the Earth via the stratosphere, where it had a maximum residence time of 3 years. In the meantime, the $^{36}Cl$ level has returned to its very low natural level, in contrast to tritium and radiocarbon (Bentley et al. 1982). Anthropogenic $^{36}Cl$ is the best tracer for soil moisture in semiarid zones worldwide.

Anthropogenic iodine-129 is released by nuclear weapons tests and is present in gaseous emmisions from nuclear reactors and reprocessing plants.

### Sample Treatment and Measurement Techniques

*Radiocarbon:* The carbon dioxide in groundwater is present mainly as bicarbonate. It is extracted for $^{14}C$ analysis from 0.1–300 L in the field. This is usually done by precipitation with barium hydroxide or a $SrCl_2/NaOH$ solution. Ion exchangers have also been successfully used (Fröhlich et al. 1974). In another technique, the groundwater is acidified, a stream of nitrogen is recycled through it, the $CO_2$ is absorbed in an alkaline trap (Linick 1980).

Carbon dioxide is extracted from air by absorption in sodium hydroxide solution in open pans over a period of 4–7 days; molecular sieves are used for sampling air at high altitudes (Nydal and Lövseth 1983).

Wood and other organic matter are burned; the $CO_2$ thus produced is processed as described in Sect. 6.2.1 and measured using proportional counters (Sect. 5.2.2.1) or scintillation counters (Sect. 5.2.2.2). For the alcohol distilled from wine or whisky, only a scintillator need be added and the measurement can be made directly.

*Tritium:* The analysis of tritium (15–2000 mL groundwater samples) is described in Sect. 6.2.2.1, where recommendations for sample collection and storage are also given. In the meantime, tritium can be detected down to levels as low as 0.003 TU (Sect. 6.2.2.2).

*Krypton-85:* For the analysis of $^{85}Kr$, 200–300 L of groundwater are sprayed under pressure at a rate of several $m^3/h$ into a vacuum and the released gases are compressed in steel cylinders. In the laboratory, $CO_2$ and water are removed in dry-ice traps, oxygen and nitrogen are removed with barium or lithium at 400–600°C. Krypton is separated chromatographically from the other noble gases using hydrogen as carrier gas and then adsorbed on activated charcoal. The isotopic spectrum of krypton is measured mass spectrometrically (Sect. 5.2.3.1). The $^{85}Kr$ activity is measured on 10–50 $\mu L_{STP}$ krypton after mixing with argon and methane in a very small proportional counter in anticoincidence for 1–6 days. One of two man-days are required for the processing of one sample.

*Chlorine-36* is determined using the AMS technique (Sect. 5.2.3.2) on milligram samples extracted from a few liters of water (Bentley et al. 1982).

*Solid fallout:* The $\gamma$-emitters, e.g., $^{137}Cs$ and $^{22}Na$, in fallout are measured either in situ with a well-type NaI(T1) scintillation counter (Sect. 5.2.2.2) or in the laboratory with a semiconductor counter (Sect. 5.2.2.3). The best resolution is obtained by measurement of whole cores. Concurrent $^{210}Pb$ measurements (Sect. 6.3.13) are usually made as a check on sediment-rate estimates (Oldfield et al. 1978; Chanton et al. 1983; Robbins 1984). Iodine-129 is determined by AMS analysis (Sect. 6.2.12).

### Scope and Potential, Limitations, Representative Examples

**Scope and Potential.** The most extensive use of dating on the basis of anthropogenic radionuclides is in the fields of hydrology, hydrogeology, and sedimentology. The detection of $^3H$ and $^{85}Kr$ in groundwater is proof that it was recharged after 1950, if only partially. In many cases these isotopes can be used to estimate mean residence times of groundwater. Recently developed techniques apply $^{36}$Cl for dating ice (Elmore et al. 1982) and groundwater (Bentley et al. 1982). Owing to the short length of time in which solid fallout in sediments and ice has been available for dating purposes, the use of fallout radionuclides often can do no more than indicate whether the recovered core satisfactorily includes the top deposits and whether there was bioturbation or other mixing (Robbins and Edgington 1975). Results of studies of lake dynamics have not always been very precise.

**Limitations.** The anthropogenic tritium content of precipitation varies greatly with season and from place to place. Summer rain generally contains more tritium than snow; less tritium is measured on the coast than inland (continental effect). These effects cause the tritium input to vary considerably from place to place. The IAEA recognized early the necessity of determining input curves for different areas and established a global network of about 125 stations to collect precipitation for isotopic analysis. The measured isotopic abundances have been published regularly in the IAEA Technical Report Series since 1969 (IAEA 1969–1986). These tables provide sufficiently reliable input curves for extrapolation to nearly any site on the Earth. The tritium concentrations in the 1950's are estimated from the $^{90}$Sr concentrations measured at many places during that decade.

$^{137}$Cs, $^{55}$Fe, $^{90}$Sr, $^{239}$Pu, $^{240}$Pu, and $^{241}$Am from fallout are not always completely immobile in sediments (e.g., Henzel and Strebel 1967). Cesium is transported by molecular diffusion through pore water, plutonium transport is mainly due to its affinity to sediment particles (Santschi et al. 1983). The Cs/Pu and $^{238}$Pu/$^{240}$Pu ratios are also not always constant. Upwards diffusion, mixing, recycling of the sediment and other effects, such as changes in the sedimentation rate, acidification, organic decomposition, and scarcity of clay minerals for adsorption, can interfere (IAEA 1981b). The trapping of fallout by vegetation and organic components of the soil may result in delayed input of these short-lived radionuclides into groundwater and sediments with lag times of decades, leading to failure of the dating method (Davis et al. 1984). In the case of ice from temperate glaciers, the fallout is transported with the meltwater and enriched in dust layers. Only $^{210}$Pb seems to be suitable for determining sedimentation rates.

**Representative Examples.** *Groundwater and spring water studies:* An exponential model has been developed for spring water (Eriksson 1958). This has also been used for groundwater in unconfined aquifers. This model is also used to estimate mean residence times and the initial $^{14}$C concentration (Sect. 6.2.1) of groundwater using previously determined input curves for radiocarbon and tritium from local precipitation (Geyh 1972a,b). It is assumed that spring water consists of water of different ages whose proportions decrease exponentially with increasing age. By analyzing several isotopes, it can be tested whether the model is valid for describing a specific case (Rozanski and Florkowski 1979; Grabczak et al. 1982). However, dates can be interpreted by suitable models (Eriksson 1958; Nir 1964; Zuber 1986) only if the type of age distribution (e.g., exponential or linear) in the groundwater is known (Egboka 1985).

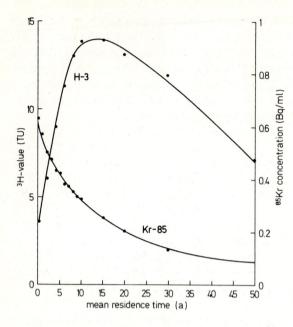

**Fig. 7.9.** Relationship between the mean residence times (MRT) of $^3$H and $^{85}$Kr calculated with an exponential ground-water model for southern Germany (after Salvamoser 1986). Most of the tritium values yield two MRT values and each $^{85}$Kr value corresponds to only one, even more precise MRT

For mean residence times in the range of 10 to 30 years, $^{85}$Kr yields at the present time more precise and less ambiguous values than $^3$H (Salvamoser 1986) (Fig. 7.9). If the $^{85}$Kr saturation level is reached, a dating method analogous to the radiocarbon method becomes possible.

*Unsaturated zone:* Tritium records for soil moisture in the unsaturated zone are used to determine groundwater recharge rates because groundwater moves laterally at depth. The simplest method is based on the water column above the prominent peak of 1963/64 (Zimmermann et al. 1967; Smith et al. 1970; Andres and Egger 1985; Athavale 1984). A more sophisticated, more reliable procedure includes the saturated zone of the upper aquifer (Atakan et al. 1974; Herweijer et al. 1985). The applicability of all these techniques is diminishing rapidly as the 1963/64 peaks are disappearing with the groundwater runoff (Fontes 1980). But in irrigated areas, the differing values of the rain and the irrigation water result in annual layers in the tritium content of the pore water. This allows the water flow rate to be estimated, as well as the percentage of immobile water, the effective dispersion, the dispersivity, and tortuosity (Gvirtzman and Magaritz 1986).

*Global carbon cycle:* Complex models have been used to estimate the mean residence time of carbon in the various geophysical reservoirs, e.g., the atmosphere, the biosphere, and the hydrosphere (Oeschger et al. 1975; Oeschger and Siegenthaler 1978).

*Miscellaneous applications:* The age of wine and whisky has been tested using the $^{14}$C concentration of the alcohol (Walton et al. 1967). Tamers (1969) applied the method for ethnological studies in Brazil. Harkness et al. (1986) determined carbon turnover times of different fractions in soil by monitoring the $^{14}C$ in the top 15 cm of soil over a period of 15 years. The application of artificial radionuclides has contributed to the understanding of the growth of human gallstones (Druffel and Mok 1983) and to the determination of the ages of monkeys by studies of proteins from eye lenses (Bada et al. 1987).

### Non-Chronological Applications

The annual records of radiocarbon, tritium, and krypton-85 in the atmosphere and hydrosphere for the last two decades have extended considerably man's knowledge and understanding of ecological relationships. Especially oceanographic (e.g., Bien et al. 1960; Stuiver and Oestlund 1980; Oestlund and Stuiver 1980; Robinson 1981; Broecker and Peng 1982; Smethie et al. 1986) and meteorological large-scale meridional, interhemispheric, and vertical mixing studies (Weiss and Roether 1980; Weiss et al. 1983) have profited. Fallout analyses have also contributed to studies on the transport of large particles through the oceanic water column (e.g., Livingston and Anderson 1983). The operation time of the Chernobyl reactor was estimated from the $^{129}I/^{131}I$ in the fallout from the reactor accident (Paul et al. 1987).

*Pollution studies:* Radiocarbon analyses have made it possible to identify the source of smoke and other pollution (e.g., Lodge et al. 1960; Rosen and Rubin 1965; Vogel and Uhlitzsch 1975; Baxter and Harkness 1975; Court et al. 1981; Currie et al. 1983) and to study the influence of meteorological conditions on the distribution of radioactive emissions around nuclear power plants (Levin et al. 1980; Otlet et al. 1983b; Segl et al. 1983).

*Erosion and deposition studies:* Studies on mass transport of sediments, in addition to their geochronology, are routine. Erosion and deposition rates are determined either by the decrease or increase in the depth of the 1964 layer or by the mean $^{137}Cs$ activity of the redeposited sediments (e.g., Brown et al. 1981; de Jong et al. 1983; Lance et al. 1986). Wind and water erosion cannot be distinguished. Soil loss from farmland can be quantified (Martz and de Jong 1987).

*Studies of the hydrosphere using tritium:* The wide range of applications for tritium analysis in environmental studies is illustrated by the papers published in the Proceedings of an IAEA symposium (e.g., IAEA 1979). These papers deal with the production and emission of tritium at present and in the future and its distribution in the environment: Tritium in aquatic systems are dealt with, as well as the hazards of tritium for man. Typical applications of tritium deal with lake dynamics, circulation of water in fissured and fractured rocks, and groundwater recharge in arid and semi-arid regions. Tritium has also aided ice studies, e.g., on the formation of ground ice in permafrost areas (Chizhov et al. 1983).

Anthropogenic radionuclides are also useful for checking and regulating the amount and rate of infiltration water during the development of gas and oil field for exploitation (Barnov et al. 1983) or for estimating the proportion of meteoric water in brine from salt mines (Geyh 1969; Zuber et al. 1979).

## 7.5   Geochemical Time Markers*

(Alvarez, Alvarez, Asaro, and Michel 1979)

### Dating Range, Precision, Substances, Sample Size

Geochemical time markers are found in marine sediments (Holser 1977) at the Cretaceous/Tertiary boundary at about 65 Ma (Alvarez et al. 1979, 1980), at the Eocene/Oligocene boundary at 34 Ma (Ganapathy 1982), and at 2.3 Ma (Kyte et al. 1981) on the basis of anomalies of iridium and other siderophilic elements whose ages have been precisely determined by other methods and correlated with well-based geological time-scales.

The seasonal or climatological fluctuations of the input of continental dust, volcanic debris, or salt can be used to date ice from several years to more than 1000 years old (Fig. 7.10) (Dansgaard 1981; Herron 1982; Finkel and Langway 1985; Hammer 1989).

### Basic Concept

Anomalously high concentrations of iridium and other siderophilic elements were first found in an Apennine limestone section from the Cretaceous/Tertiary boundary at about 65 Ma. The usual content is exceeded in these rocks by two orders of magnitude and is best explained as a result of an extraterrestrial event (Luck and Turekian 1983). Since then, other anomalies have been found at the

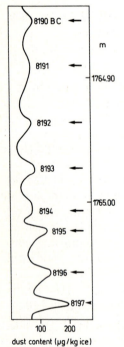

**Fig. 7.10.** Annual variation of the dust content of the deep part of ice core Dye-3 from Greenland. Dating only on the basis of seasonal $\delta^{18}O$ fluctuations becomes increasingly difficult in this part of the core (after Hammer et al. 1986).

Eocene/Oligocene boundary at about 34 Ma (Ganapathy 1982) and also at the Pliocene/Pleistocene boundary at about 2.3 Ma (Kyte et al. 1981).

Anomalies of siderophilic elements could have different causes. The impact of large asteroids, presumably iron meterorites, or the impact of a comet (Alvarez et al. 1979; Hsü 1980; Kyte et al. 1980) are possibilities that have been discussed. It has been pointed out that at least the iridium anomaly at the Cretaceous/Tertiary boundary coincides with a period in which 75% of the animal species living at that time became extinct (Alvarez et al. 1979, 1980). Urey (1973) had previously suggested that a collision of the Earth with a comet could have been the cause of the extinction of the dinosaurs, initiating the Tertiary division of geologic time. The coincidence of the iridium anomaly with the extinction of the dinosaurs, however, could be coincidental (Walliser 1983/1984). The results of Alvarez et al. figure prominently in the papers presented at a conference on this topic (Silver and Schultz 1982).

Seasonally or periodically distributed materials represent another type of time marker. They from strata in ice that can be counted (Herron 1982; Hammer et al. 1986; Hammer 1989) or reflect paleoclimatic trends (Finkel and Langway 1985); examples are airborne continental dust and biological material, volcanic ash, and sea salt (Fig. 7.10).

### Sample Treatment and Measurement Techniques

Because the concentrations of iridium and other siderophilic elements are very low, very sensitive techniques must be used, such as instrumental or radiochemical neutron activation analysis (INAA and RNAA, see Sect. 5.2.4.2) (Barker and Anders 1968; Kyte et al. 1980).

To date ice on the basis of natural contaminants, nearly all of the usual analytical methods used in glaciochemistry (Sect. 5.2.4) can be used.

### Scope and Potential, Limitations, Representative Examples

Anomalies of iridium and other siderophilic elements have been detected in a terrestrial sediment core (from New Mexico) and about 50 pelagic sediment cores (from Europe, the Pacific, and the Antarctic). Extension to other areas is necessary in order to develop a widely applicable dating method. It may be assumed that still other geochemical or isotopic anomalies will be found (e.g., Luck and Turekian 1983) that are suitable for dating.

Within the scope of ecological and paleoclimatological studies, natural contaminant records from the last century are needed. These can be obtained from ice studies (Finkel and Langway 1985). Changes in the contaminant input are obtained in addition to the age data.

## 7.6 Chemical Pollution as Time Markers*

### Dating Range, Precision, Substances, Sample Size

Chemical pollutants are suitable for dating lacustrine and marine sediments, cold ice (Hammer et al. 1978, 1986), and groundwater that have been formed or recharged since the beginning of the industrial age (Erlenkeuser et al. 1974; Schell et al. 1983; Ortlam 1983). The precision of the ages depends on how accurately the time of introduction of a specific pollutant or the production of a new industrial chemical is known.

### Basic Concept

Since the beginning of the industrial period in the middle of the last century, pollution by heavy metals, industrial dust (e.g., soot), etc. has increased in the atmosphere, hydrosphere (groundwater, river and sea water, ice), and lithosphere (sediments). In the present century, the introduction of industrial chemicals that do not easily decompose (e.g., pesticides, preservatives, and organohalogen compounds) has formed time marks that can be traced worldwide. So that sediments, etc. can be dated in the future, information on the beginning and course of production of these substances will have to be compiled. Gaseous pollutants in the atmosphere and hydrosophere provide a means of estimating mean residence times as described in Sect. 7.4. Liquid pollutants can be used as tracers in the hydrosphere. Solid pollutants are deposited on sediments or ice, frequently as annual layers, which can be counted.

### Sample Treatment and Measurement Techniques

Sample preparation and measurement depends on the kind of trace substance and its concentration. Nearly all of the chemical analytical methods (Sect. 5.2.4) are used (e.g., INAA, IDMS, AA, and chromatography).

### Scope and Potential, Representative Examples

Dating via chemicals will increase in importance for studies of groundwater and the atmosphere, especially for the clarification of environmental pollution.

**Representative Examples.** Studies of the method have been conducted primarily on coastal and lacustrine sediments (Erlenkeuser et al. 1974; Goldberg et al. 1977; Schell et al. 1983) that reflect the history of pollution. As a check, concurrent $^{210}$Pb determinations (Sect. 6.3.13) are often made (Dominik et al. 1978).

One application is related to the stepwise increases in the salinity of the Weser River due to discharges by the potash industry of the GDR; its use made it possible to predict when the drinking water resources for the city of Bremen will become too saline to be used (Ortlam 1983).

### Non-Chronological Applications

The pollution records since the beginning of the industrial age preserved in cold ice (e.g., southern Greenland, Mayewski et al. 1986) may aid our understanding of the impact of industrial pollutants on the global ecology.

# 8 Chemical Dating Methods

Chemical age determinations are based primarily on the assumption of reaction rates (diffusion, exchange, oxidation, hydration, etc.) that are at least nearly constant. The age is estimated from the initial and end concentrations of suitable reactants or products.

The main problem for dating is that chemical reaction rates are highly dependent on temperature and often also on other environmental factors, e.g., pH, eH, and moisture. Thus, chemical dating usually yields only relative ages, whose usefulness for stratigraphic classification (Oakley 1980; Miller et al. 1979; Wehmiller and Belknap 1982; Wehmiller 1986), however, is of great value. Absolute ages are obtained only after calibration for *each* sampling site with samples that have been reliably dated with other methods. Calibration also requires an estimate of the effective ambient temperature (Eq. 8.6; Lee 1969; Michels et al. 1983a, b). Mean values for reaction rates from the literature yield only very rough estimates of the absolute ages.

These circumstances explain why most results of chemical dating methods based on the decomposition of organic substances or the sorption of trace elements must be viewed more critically than those obtained using physical, mainly isotopic, dating methods. Their results usually cannot be used without critical consideration of the geochemical and diagenetic processes that may have changed the material to be dated. In spite of this, modern analytical methods have steadily increased the value of chemical dating (Tite 1981) because they can be carried out relatively quickly and inexpensively. Suitable techniques are chromatography, atomic absorption analysis (Sect. 5.2.4.3), amino acid analysis (Sect. 8.1), X-ray fluorescence spectrometry (Sect. 5.2.4.5), and ESR spectrometry (Sect. 6.4.3).

Chemical dating methods are applied especially by anthropologists, paleontologists, and geomorphologists. Another field is the reconstruction of the thermal history of a region, made possible by the temperature dependence of chemical reaction rates. Paleotemperatures are obtained using samples that have been reliably dated by other methods (Bada et al. 1973; Kvenvolden et al. 1981).

## 8.1 Amino-Acid Racemization Method (AAR)***

(Abelson 1954, 1955)


### Dating Range, Precision, Materials, Sample Size

This method has been used routinely for dating fossil matter that contains amino acids (Wehmiller 1984a, 1986). The first datings with the AAR method were done on dense, compact bone fragments and tooth enamel (a few grams) (Bada 1985) in the age range of several thousand to several hundred thousand years. An object that has been treated with chemical preservation agents, and thus is unsuitable for radiocarbon dating (Sect. 6.2.1), can usually be dated. Manganese nodules have been dated using enclosed bone nuclei (Kvenvolden and Blunt 1979).

The major application of the method is the dating of foraminifers extracted from pelagic sediments (Bada and Schroeder 1975) and coprolites (several 100 mg), corals (results are reasonable in less than 50% of the cases) and mollusc shells (5–150 mg) (Szabo et al. 1981; Wehmiller 1982, 1986), as well as land snails (Goodfriend 1987b). Apparently reliable ages of up to 200 ka have been determined (Bada and Schroeder 1975). With improvements in the method, it may be possible to determine ages of up to 20 Ma. Dating precision may be $\pm$ 5000 a up to ages of 120 ka, $\pm$ 60 ka up to 500 ka, and about $\pm$ 500 ka for older samples (Müller 1984).

Marine phosphorites (Cunningham and Burnett 1985), Pleistocene tuff deposits, carbonate muds, and oolites, as well as speleothem have yielded encouraging results. Freshwater sediments (ca. 25 g samples) and molluscs (Scott et al. 1983), as well as terrestrial molluscs from archeological sites (Masters and Bada 1979), have been dated with limited success (Schroeder and Bada 1978). The use of proline and hydroxyproline to date wood with an age of up to 100 ka is under development (Lee et al. 1976; Rutter and Crawford 1984).

The racemization calibration curve for a study area is obtained with independently and reliably dated samples (Bada 1985). Holocene samples can then be dated with a precision of about $\pm$ 200 a, older samples to $\pm 2$–10% (Wehmiller 1984a) when data from different laboratories are taken into consideration. The results from a single laboratory may look more consistant. Multiple analyses from one site increase the precision, which should be within 2 to 10%.

A measure of the age of humans and other mammals is the recemization of aspartic acid in enamel (Helfman and Bada 1975) or better in dentine (Helfman and Bada 1976; Bada et al. 1983) at a rate of about $8 \times 10^{-4}$ $a^{-1}$.


### Basic Concept

About ten amino acids are found in fossil skeletal material, since only 40–70% is decomposed during fossilization. Recent proteinaceous substances contain about 20 relatively simple amino acids. Except for the simplest amino acid, they all have at least two stereoisomeric forms, i.e., two configurations that are mirror images of each other owing to the presence of an asymmetric C-atom. Amino acids that have

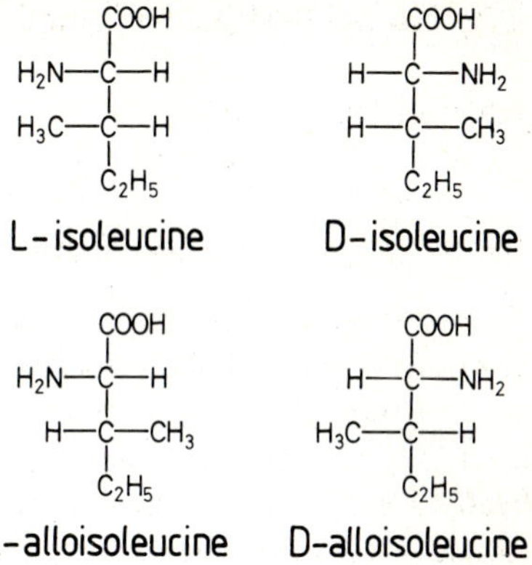

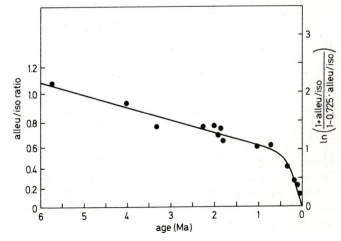

**Fig. 8.1.** Isoleucine and alloisoleucine are examples of stereoisomeric forms of amino acids; L- and D-isoleucine, as well as L- and D-alloisoleucine, are enantiomers; L-isoleucine and D-alloisoleucine, as well as D-isoleucine and L-alloisoleucine, are epimers

two asymmetric carbon atoms have an additional pair of isomers, which are called diastereoisomers of the first pair.

The amino acid method is based on the racemization of amino acids with one asymmetric C-atom (e.g., aspartic acid, alanine, leucine, and proline) or the epimerization of amino acids with two asymmetric C-atoms (e.g., hydroxyproline and isoleucine, Fig. 8.1), as well as of proteins and peptides. When formed in the living body, natural amino acids are L-isomers (L for levo), which are optically active, i.e., they rotate plane polarized light. The mirror image form, the D-isomer (D for dextro), rotates plane polarized light by the same amount but in the opposite direction.

Once the organism has died, the amino acids slowly convert reversibly into D-isomers, ultimately leading to an equilibrium mixture of L- and D-isomers. This process is called racemization and for intact collagen follows first-order kinetics (if the sample formed a closed system, e.g., no leaching occurred), but not for the amino acids in marine sediments (Kvenvolden et al. 1981; Müller 1984; Fig. 8.2). According to Julg et al. (1987), this process occurs spontaneously at temperatures below 40°C. According to Müller (1984), curves for total isoleucine epimerization

**Fig. 8.2.** Epimerization of isoleucine in micropaleontologically dated foraminiferal tests from DSDP cores (after Bada and Schroeder 1975). The curve can be approximated by two rather linear components (after Kvenvolden et al. 1981)

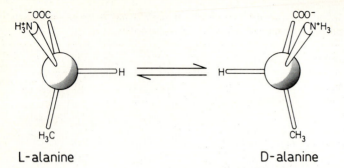

L-alanine                                                      D-alanine

**Fig. 8.3.** Mechanism of amino acid racemization (after Bada and Schroeder 1975)

can be separated in three rather linear segments of decreasing apparent rate with increasing age of the foraminifera.

The ratio of D- and L-isomers (enantiomeric ratio) determines the optical activity of an amino acid. During the course of aging this ratio increases from 0 to 1 and the optical activity decreases to zero. Both are thus a direct measure of the degree of racemization that has taken place and, therefore, for the age.

The racemization rate, analogous to radiactive decay, may be expressed in terms of the half-life ($\tau$). This is the time in which the ratio of enantiomers increases from 0 to 0.33. Since a sample with an age of zero contains no D-isomer and one with an age of infinity contains 50%, an enantiomeric ratio of $25/75 = 0.33$ is obtained when half of the molecules have been racemized.

For the reversible first-order reaction (Fig. 8.3) leading to the racemization of amino acids, e.g., aspartic acid, from intact collagen,

$$L_{ASP} \underset{k_D}{\overset{k_L}{\rightleftharpoons}} D_{ASP},$$

the following differential equation describes the disappearance of the L-form during racemization:

$$-\frac{d(L_{ASP})}{dt} = k_L \cdot L_{ASP} - k_D \cdot D_{ASP} \tag{8.1}$$

where $k_L$ and $k_D$ are the first-order rate constants for the forward and reverse reactions. The equilbrium constant $K_{DL} = k_D/k_L$ or the D/L ratio at equilibrium. Integration of Eq. (8.1) yields

$$\ln \frac{1 + (D)/(L)}{1 - K_{DL} \cdot (D)/(L)} - \text{const} = (1 + K_{DL}) \cdot k_L \cdot t \tag{8.2}$$

The constants $k_L$ and $k_D$ are equal if the activation energies of the forward and reverse reactions are the same, which is the case for many amino acids. The constant of integration is the initial enantiomeric ratio, which is not zero, and for aspartic acid, for example, is about 0.14, since the samples either already contain some of the D-isomer or some is formed by hydrolysis during extraction of the amino acid from the sample.

**Table 8.1.** Half-lives (in ka) for the racemization of several amino acids at pH 7.6 in various materials at various temperatures (Bada and Schroeder 1975; Bada 1984)

| Amino acid | Material | 0 | 10 | 20 | 25 | 37°C |
|---|---|---|---|---|---|---|
| Isoleucine | Marine sediment | 6000 | | | 48 | 6.5 |
| | Bone | | | 100 | | |
| | Wood | | | | 480 | |
| | Shells | | 300 | | ca 100 | |
| Alanine | Bone and teeth | 1400 | | 50 | 12 | 1.5 |
| Aspartic acid | Bone and teeth | 430 | 40 | 15 | 3.5 | 0.46 |
| Proline | Wood | | | | 115 | |

The racemization rate depends on many factors, e.g., the amino acid, the temperature of reaction, (often called the effective diagenetic temperature (EDT) for samples with a complex thermal history), the moisture content, type of chemical bonding (interior position in a protein or free (both slow rates), or terminal (rapid rate) (Müller 1984), the state of preservation, and various environmental factors (e.g., pH) during the time the sample was in the ground. Anatomically different parts of a shell (intrashell variations) yield slightly different D/L ratios (Brigham 1983). The large range of variation in half-life, $\tau$, can be seen in Table 8.1 (half-lives are given instead of racemization rates). For example, the half-life for racemization at ambient temperatures is about 3500 years for peptide-bound aspartic acid and 12,000 years

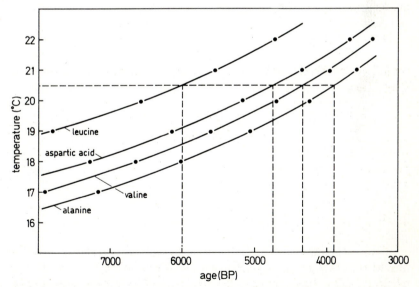

**Fig. 8.4.** Relationship between the apparent age of a tendon from a 6000-year-old Egyptian mummy and the diagnetic temperature for various amino acids (after Dungworth 1976). Using the current mean annual temperature of 20.5°C, apparent ages of 3900, 4300, 4700, and 6000 years are obtained from alanine, valine, aspartic acid, and leucine, respectively

for alanine, at 0°C both are more than 100 times longer. Thus a 1°C variation in the effective diagenetic temperature changes the racemization rate by about 25%, resulting in a corresponding shift in the calculated age (Fig. 8.4). This is a great disadvantage for this method because, especially during the Quaternary, temperatures have varied widely. An exception is the dating of pelagic sediments, which is not seriously affected by the temperature dependence of amino-acid racemization (Dungworth 1976) because the temperature of the deep ocean hardly changed during the ice ages. However, due to the curvature of the epimerization curve, the dating precision decreases rapidly with increasing age (Müller 1984).

The effective racemization temperature (also effective diagenetic temperature) for the Holocene is estimated using the chemical temperature integration equation of Lee (1969). A temperature called the exponential temperature ($T_{exp}$) is obtained from the CMAT (current mean of the mean annual air temperature, $T$) and the maximum difference of the mean monthly values, $\Delta T$:

$$T_{exp} = \frac{1}{0.973}(T + 0.226\,\Delta T - 2.23)$$

for areas with annual snowfall

and

$$T_{exp} = \frac{1}{1.065}(T + 0.161\,\Delta T + 1.23) \tag{8.3}$$

for areas with no annual snowfall

The values for $T_{exp}$ obtained with Eq. (8.3) are in reasonably good agreement with the effective diagenetic racemization temperatures (EDT) or effective Quaternary temperature (EQT) obtained from samples dated with other methods (Bada et al. 1973, 1979; Michels 1986).

Since the rate constants for racemization are not invariable constants like those of radioactive decay, they must be determined anew for each site and material. For this purpose, samples dated by other methods are necessary to obtain the in-situ amino acid racemization rate (Goodfriend 1987b). Ages that are not based on such site-specific calibrations can deviate by several orders of magnitude from the actual ones (Dungworth 1976). For bone samples, an Arrhenius plot [Eq. (8.6), Sect. 8.6] is also used to estimate the racemization rate (Bada 1985).

### Sample Treatment and Measurement Techniques

Summaries of the methods for analyzing amino acids have been published by Hare et al. (1985) and Engel and Hare (1985).

To avoid contamination of the samples, they should not be touched with the fingers when they are being collected. They should be taken from within an outcrop rather than from the surface, where they would have been subjected to environmental effects. They should also be stored frozen. The degree of preservation of the sample is checked by microscopic or X-ray methods. Criteria for the suitability of a sample for dating with this method include luster, porosity, and texture.

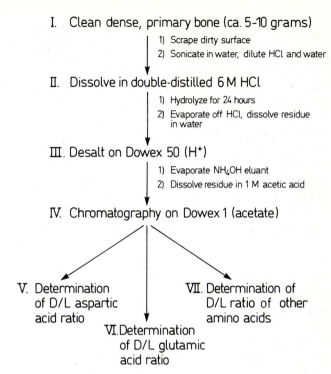

I.  Clean dense, primary bone (ca. 5-10 grams)

    1) Scrape dirty surface
    2) Sonicate in water, dilute HCl and water

II.  Dissolve in double-distilled 6 M HCl

    1) Hydrolyze for 24 hours
    2) Evaporate off HCl, dissolve residue
      in water

III.  Desalt on Dowex 50 (H⁺)

    1) Evaporate NH₄OH eluant
    2) Dissolve residue in 1 M acetic acid

IV.  Chromatography on Dowex 1 (acetate)

V. Determination          VII. Determination of
of D/L aspartic               D/L ratio of other
acid ratio                    amino acids
      VI. Determination
      of D/L glutamic
      acid ratio

**Fig. 8.5.** An outline of a procedure used in amino acid racemization dating of bone (after Bada et al. 1979)

An outline of the procedures used in AAR dating is shown in Fig. 8.5. The first step in the preparation of the samples (Bada et al. 1979) is ultrasonic cleaning in double-distilled water. Julg et al. (1987) suggest that various steps be included to remove inorganic ions, free amino acids, and degraded collagen fragments, e.g., peptides with molecular weights smaller than 3500. The dried samples (0.1–1 g) are then pulverized under sterile conditions and digested in double-distilled, amino-acid-free acid (e.g., 6N HCl) for about 24 h at 100°C to obtain free amino acids. After desalting using a standard cation exchange resin (e.g., Dowex 50), the amino acids are separated in an automatic amino acid analyzer by one of two chromatographic methods: conventional anion-exchange liquid chromatography (rapid, sensitive, economic, but suitable for only one amino acid at a time) or gas chromatography (more time consuming and more expensive, but applicable to many amino acids) (Bada 1985; Engel and Hare 1985). Internationally accepted reference samples of powdered fossil mollusc shells are now available as standards for enantiometric ratios (Wehmiller 1984b).

Proline and hydroxyproline, which have very low rates of racemization, have been extracted from wood for analysis (Rutter and Crawford 1984).

Amino acid contamination of the samples, which can lead to ages that are too low, are recognized by deviation from the following sequence of the D/L ratios of six to eight amino acids. The following sequence is characteristic for pure fossil bones

(Dungworth 1976):

aspartic acid > alanine ≅ glutamic acid > isoleucine > alloisoleucine = leucine

The following information is necessary to assess the reliability of a D/L age (Bada et al. 1979):

- the D/L ratio and the method used to determine it;
- the $K_{DL}$ value and how it was obtained;
- a comparison of the present mean temperature (CMAT) of the sampling site with the one obtained from the $K_{DL}$ value, for which the temperature dependence must be known;
- the racemization rates of the amino acids used for the dating; and
- the range of the results of repeated analyses.

### Scope and Potential, Limitations, Representative Examples

**Scope and Potential.** The AAR method is quick and less expensive than most of the other dating methods. It is used worldwide for local and regional chronostratigraphic studies in glaciology, paleoclimatology, paleoecology, marine stratigraphy, sea-level changes, and tectonics. Wehmiller (1984a, 1986) has published a summary of AAR studies of Quaternary molluscs from coastal sediments in North America and other parts of the world. Miller and Mangerud (1985) have reported on the use of the method for European marine interglacial deposits. In an international comparison test, however, it was shown that the D/L ratios of aspartic acid and alanine are now reproducible to 3–8%, that of leucine to 5–10%, those of isoleucine and proline to 10–18% (Wehmiller 1984b).

The racemization method is also attractive for anthropologists (for example, those studying hominid evolution) owing to the low amount of bone sample required, the large range that can be dated, and the direct applicability to bones. Bada (1985) has given a summary of the applications to fossil bone and tooth samples collected in Africa and North America. However, absolute dating is possible only if properly dated samples are available for calibration (Bada 1985; Ennis et al. 1986).

Future research will be concentrated on increasing knowledge on the geochemistry of racemization in various materials.

**Limitations.** Material to be dated must be examined for sample-specific effects. For example, racemized amino acids can be leached from porous bone during weathering, in which case small samples or parts near the surface may give different ages than large samples or parts towards the middle. Metal ions (e.g., copper and magnesium) seem to accelerate racemization. Water content and temperature are the decisive factors. A further difficulty is that the rate of racemization may change during fossilization (Dungworth 1976). The pH of the material around the bone is not important within the range of 3–8 (Bada and Schroeder 1975) because the minerals in the bone effectively buffer the system. The precision of AAR dating of mollusc shells is affected by (a) the precision of the multiple analyses of shells from the same site, (b) differences in the racemization kinetics of different mollusc genera, and (c) temperature effects (Wehmiller 1982, 1986). Moreover, epimerization studies of fossil molluscs show that the racemization kinetics of isoleucine can be either first or second order, depending on whether the isoleucine is bound in a peptide (interior or COOH-terminal) or is free. Therefore, the extent of racemization of the total isoleucine may not yield very reliable ages (Kriausakul and Mitterer 1983; Müller 1984). Evidence for reversal of aspartic acid racemization in shells has also been found (Kimber et al. 1986; Kimber and Griffin 1987).

These and other AAR sources of error can explain racemization ages that are too large by several thousand to several ten thousand years (Dungworth 1976; Williams and Smith 1977). However, often the problems are apparent and arise from poor dating of samples used for calibration (Bischoff and Rosenbauer 1981; Bada 1985; Ennis et al. 1986). Julg et al. (1987) showed that more reliable results can be obtained if the AAR dating is based on intact collagen and samples that were never exposed to temperatures above 40°C (where "chemical" rather than "spontaneous" racemization occurs). The reliability of AAR is highly dependent on the diagenetic status of the organic components.

*Representative Examples.* The AAR method is widely used to establish a chrono-stratigraphic framework based mainly on analysis of mollusc shells (Miller et al. 1979; Wehmiller 1982; Miller and Mangerud 1985; Wehmiller 1986; Hearty and Aharon 1988). Two examples from the large number of papers: A comparison study of AAR, U/Th dating (Sect. 6.3), biostratigraphic and paleomagnetic data from 22 sites of Quaternary deposits on the North American coastal plain from Florida to Nova Scotia revealed a generally consistent chronostratigraphy. AAR analysis reflects most of the depositional events. Major conflicts between model ages indicate that the basic temperature assumptions for the amino-acid stratigraphy may be incorrect. Various kinetic models for epimerization were considered (Wehmiller and Belknap 1982).

In another study it became apparent that the generally complex temperature history of Europe need not be known exactly for AAR dating. However, the use of the same genera with the same temperature history, e.g., in the case of molluscs, is unavoidable.

Using D/L isoleucine ratios in mollusc shells, Miller and Mangerud (1985) have made the most serious correlation between the marine oxygen isotope record (Sect. 7.2) and the classical land-based sequence of interglacials in Europe. Their results correlate the age of the classical Eemian sites in Europe with that of sub-stage 5e of the marine isotope record, whereas the time of the Holsteinian inter-glacial most likely correlates with substage 7c, possibly with that of stage 9, but certainly later than stage 11 (Fig. 8.6). Amino acid epimerization analysis of alloisoleucine/isoleucine in Holocene land snail shells taken from rodent middens and from fluvial and colluvial deposits in the Negev Desert yielded a very strong correlation between A/I and radiocarbon ages. The combined use of both methods is quite promising (Goodfriend 1987c).

Another application is the determination of the age of humans and other mammals (Helfman and Bada 1975, 1976; Bada et al. 1983).

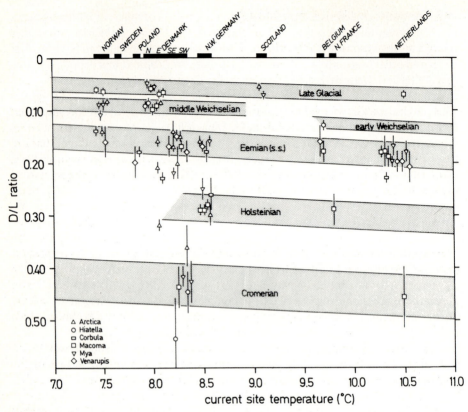

**Fig. 8.6.** Mean D/L ratios and $1\sigma$ intervals for six dominant taxa in the moderate-rate group from four Pleistocene interglacials as a function of the current mean site temperature: More rapid epimerization at the warmer sites produces higher D/L ratios within equivalent time periods than at cooler sites (after Miller and Mangerud 1985)

### Non-Chronological Applications

*Heat flux and paleotemperature:* A very important non-chronological application of the amino acid racemization method is the estimation of average heat flux since the time of sample deposition. This has been done on multi-specie foraminiferal assemblages extracted from selected Deep-Sea Drilling Project (DSDP) cores. The results suggest that for the last 4 million years there has been no significant change in the hydrothermal circulation pattern in the crust (Katz et al. 1983). Another example is the determination of the effective diagenetic racemization temperature (Table 8.1) as representative of late Quaternary thermal history (Bada et al. 1973; Bada and Schroeder 1975; Mitterer 1975; Dungworth 1976; Wehmiller 1982, 1986). In this case, well-dated samples must be available.

*Relative dating:* D/L dating of paleosols is one of the many examples of relative dating in the geosciences (Mahaney et al. 1986).

## 8.2   Amino-Acid Degradation Method

(Hare and Mitterer 1967; Denninger 1971)

### Dating Range, Precision, Materials, Sample Size

This method was developed using mollusc shells of up to Miocene age (Hare and Mitterer 1967). Foraminifers (0.1–0.5 g) extracted from pelagic sediments between 0.1 and 2 Ma have also been dated (Bada et al. 1978). For authenticity tests of paintings, it has been used to determine the age of protein (several milligrams) in paint binders up to 2000 years old. The dating precision seems to be independent of the age of the albumin sample (Table 8.2).

### Basic Concept

The determination of the age of foraminifers is based on the natural degradation (mainly dehydration) of the hydroxy amino acid threonine into $\alpha$-amino-n-butyric acid (ABA), whereby metal ions catalyze the reaction. The age is a linear function of natural logarithm of the ratio of serine to leucine (ser/leu) or threonine to leucine (thr/leu). Leucine is chosen as the reference amino acid because it is the most stable one.

The best correlation with age ($t$) was obtained with the ratio of the ABA and threonine concentrations. This ratio is rather independent of the species of the foraminiferal tests and should not be susceptible to contamination.

$$\frac{C_{ABA}}{C_{threonine}} = 0.019 + 2.3 \times 10^{-7} t \tag{8.4}$$

For dating proteins in paint binders, this method is a modification of the collagen or nitrogen method (Sect. 8.3). Since the rates of degradation of the different

**Table 8.2.** Relationship between the number of remaining amino acids, the age, and the precision of dating paint binders containing protein (Denninger 1971)

| Remaining amino acids | Age (BP) | Standard deviation (a) |
|---|---|---|
| 10 | 5–   10 | ±   5 |
| 8 | 10–   30 | ±   10 |
| 7 | 30–   40 | ±   10 |
| 6 | 40–   80 | ±   20 |
| 5 | 80–  120 | ±   20 |
| 4 | 120–  300 | ±   50 |
| 3 | 300–  600 | ± 100 |
| 2 | 600–1200 | ± 200 |
| 1 | 1200–1800 | ± 300 |
| 0 | >1800 | |

amino acids differ, the number of different amino acids remaining in the sample decreases during aging nearly independently of temperature and other environmental conditions (Table 8.2). The cause of this time-dependent decrease in the number of amino acids is not yet known.

### Sample Treatment and Measurement Techniques

The foraminiferal tests are extracted using a 62-$\mu$m stainless steel sieve and cleaned in an ultrasonic bath using deionized, double-distilled water. Samples of 0.1–0.5 g are hydrolyzed in an excess of double-distilled 6N HCl at 100°C for 6–24 h. The solution is then evaporated to dryness under reduced pressure, redissolved, and the salts removed with a cation exchange resin. The amino acids are analyzed using an automatic amino acid analyzer.

Paint samples are digested in concentrated acid and the amino acids determined by one-dimensional ascending paper or thin-layer chromatography.

### Scope and Potential, Limitations, Representative Examples

Only a few studies have applied this method since it was developed. Thus, no statements can be made about its scope, potential, or limitations. Homogeneous assemblages of species and single species samples extracted from two well-dated deep-sea cores were used to prove irreversible first-order kinetics for the decomposition of serine and threonine. This finding has not yet been used for dating deep-sea sediments. The situation is similar for its application to proteins in tests of the authenticity of paintings.

## 8.3  Dating of Bones Using the Nitrogen or Collagen Content*

(Oakley 1949)

### Dating Range, Precision, Materials, Sample Size

This empirical method is occasionally used for rough, and often only relative dating of untreated Quaternary bones, antlers, and teeth (ca. 100 mg), mostly in connection with the fluorine method (Sect. 8.9), up to an age of 100 ka, perhaps older. The precision of the method is very poor, making only the distinction between old and young samples possible (Oakley 1980; Ortner et al. 1972; Buczko et al. 1978; Hille et al. 1981).

### Basic Concept

Proteins, especially collagen, forms the matrix for the phosphatic mineral matter in bones and antlers. These proteins decompose with the loss of nitrogen (initial concentration $\sim 4.5\%$) at a rate that is highly dependent on the physical, chemical, and bacteriological conditions (Hare and Mitterer 1967), mainly temperature, moisture, and pH (Ortner et al. 1972). The proteins in the bones are better preserved in clay and other environments in which air (i.e., oxygen) is excluded than in porous, acid soils, where bacteria rapidly decompose them. Only during climatically stable periods can a linear relationship be expected between the collagen content and the age of the bone.

The amino acid degradation method (Sect. 8.2) is a sophisticated modification of this method.

### Sample Treatment and Measurement Techniques

Since the nitrogen and collagen contents of bones correlate with each other, only one of them need be measured. The fluorine content of the sample should be determined at the same time (Sect. 8.9) (Hille et al. 1981).

The bone is first cleaned with distilled water in an ultrasonic bath. It is then ground, for example, in a centrifugal ball mill. Instead of the chemically complicated extraction of the collagen, a micro-Kjeldahl nitrogen determination is preferred. For this analysis, at least 10 mg of bone is heated in concentrated sulfuric acid to 70°C; on cooling the solution is neutralized with sodium hydroxide. The ammonium that is produced can be determined gas chromatographically to ± 0.1%. Multiplication of the percentage of nitrogen in bones, antlers and teeth by 2.5 to 3 yields the carbon content, which is useful for estimating the sample size needed for a $^{14}$C dating (Sect. 6.2.1). Another technique is to determine the amino acids in the hydrolysate by chromatographic analysis

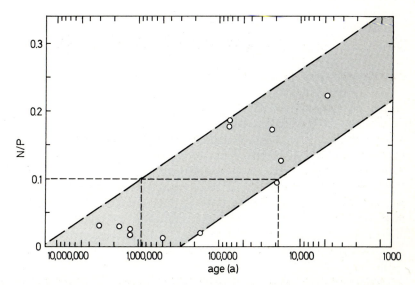

**Fig. 8.7.** Nitrogen/phosphorus wt/wt ratio ($N/P$) in dated bones from Austrian caves as a function of age (after Hille et al. 1981), showing the great uncertainty of about two orders of magnitude in dating such samples. For example, N/P ratio of 0.1 corresponds to an age range of 20,000 to one million years

A technique that does not require destruction of the sample is based on the activation of nitrogen with high-energy neutrons. The radionuclide produced in the ensuing nuclear reaction decays by $\beta^+$ emission producing annihilation radiation, which is measured with a Ge(Li) $\gamma$-spectrometer (Sect. 5.2.2.3), yielding the nitrogen content. This would not be worth the effort if the "fluorine" age (Sect. 8.9) were not obtained at the same time (Hille et al. 1981), which appears to be more reliable. The N/P ratio correlates weakly with age of the sample (Fig. 8.7). A calibration curve between nitrogen content or the N/P ratio and the age of the sample must be determined for each site.

### Scope and Potential, Limitations, Representative Example

***Scope and Potential, Limitations.*** The collagen method is of interest due to the small amount of sample necessary. Naturally, the reliability of the results cannot compete with the results of the absolute dating methods. The uncertainty of $\pm 10$ ka that is sometimes given is hardly realistic since constant rates of decompostion cannot be assumed (Heizer and Cook 1952).

The combined application of the N, U, and F methods (Sect. 8.9) for dating is termed the FUN method. It has been successfully applied in studies conducted in Europe, Asia, Africa, and the USA, especially for identifying fraud in apparent finds of new genera (Oakley 1980).

***Representative Example.*** An example for the restricted potential of this method is provided by the correlation of the N/P ratios with the ages of bone samples from Austrian caves (Hille et al. 1981). The dates have a large scatter for the Late and Middle Pleistocene but are nearly constant for ages greater than 100 ka.

### Non-Chronological Applications

*Studies on paleodiet:* Information on the diet of humans and other animals is provided by the $\delta^{13}C$ values (Sect. 6.2.1) of bone collagen. Collagen from terrestrial animals yields different values from that from marine animals (van der Merve 1982). But there are also differences between terrestrial samples from different climatic regions owing to differences in the metabolism of the plants. Plants that have the C3-cycle predominate in humid areas, resulting in a lower $^{13}C$ abundance in them than in plants arid areas, where the C4-cycle predominates. Tauber (1981) used this method to show that the mesolithic inhabitants of Denmark lived primarily from fish, but that the neolithic population lived from farming. Application of nitrogen and strontium isotopes provides further clues to prehistoric diets (Lewin 1983; Nelson et al. 1986).

The carbon of the hydroxyapatite in very old bones that no longer contain any collagen yields the same information when it is taken into consideration that its $\delta^{13}C$ value is 13‰ greater than that of collagen (Sullivan and Krueger 1981). However, there are also contrary observations (Schoeninger and DeNiro 1982).

## 8.4   Chemical Electron-Spin-Resonance (ESR) Dating

(Ikeya and Miki 1980b)

### Dating Range, Precision, Materials, Sample Size

This empirical method is suggested for determining the age of organic materials such as foodstoffs (e.g., sugar, oils, grains, and potato chips), blood, furs, and wool, silk, and cotton within a range of days to weeks and paper up to 500 a (Ikeya and Miki 1980b,c; Ikeya 1988). ESR signals of a motor oil also reflect the time a motor has been run (Ikeya and Miki 1985b).

### Basic Concept

The aging of foodstuffs is associated with chemical conversions, e.g., oxidation and the accumulation of thermally very stable free radicals with increasing dose, which leads to a constant increase in the ESR signal (Sect. 6.4.3). Acetyl and methyl derivatives in wood behave similarly (Ikeya 1983a).

### Sample Treatment and Measurement Techniques

The amplitude of the ESR signal is read continuously over a period of several days with a digital ESR spectrometer. In a graph of the amplitude of the ESR signal (e.g., $g = 2.0055$) versus time, the intercept with the $x$-axis gives the approximate time of production (Fig. 8.8).

### Scope and Potential, Limitations

As this method is not yet widely applied, its scope and potential cannot yet be assessed nor can its limitations be given. Problems with reproducibility may occur as discussed for the ESR method (Sect. 6.4.3) (Hennig et al. 1985).

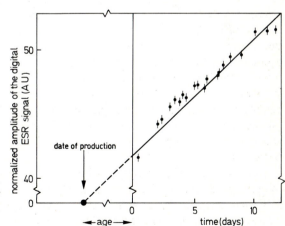

**Fig. 8.8.** Increase in the normalized amplitude of the digital ESR signal of potato chips with time used for the determination of the date of production (after Ikeya and Miki 1980b)

### Non-Chronological Applications

The maximum temperature to which cereal grains at archeological sites had been heated is reflected by the $g$-values of ESR signals. A temperature range of 50°C to about 500°C is covered (Hillman et al. 1983).

Alanine free-radical amino-acid ESR dosimeters are used in accelerator physics routinely for heavy-ion and fast-neutron high-level dosimetry. Exposure of humans to radiation is measured via dosimetry of tooth material.

## 8.5  Molecular (Protein and DNA) Clocks

(Zuckerkandl and Pauling 1962; Sarich and Wilson 1967)

### Dating Range, Precision, Materials, Sample Size

Molecular (protein and DNA) clocks have been proposed to determine the times of branching within the evolution of humans and other mammals during the last 100 Ma. Optimistic estimates of $\pm 10\%$ for 10 Ma and $\pm 20\%$ for 2 Ma have been given for the precision of this method. At present, work is being done to calibrate the molecular clocks.

### Basic Concept

Phylogeny is associated with biomolecular evolution, which is reflected by replacement of amino acids in proteins and by nucleotide substitution in genes. The two basic assumptions are that (a) some significant proportion of the amino acids in the primary structure of a protein is replaced by other amino acids without

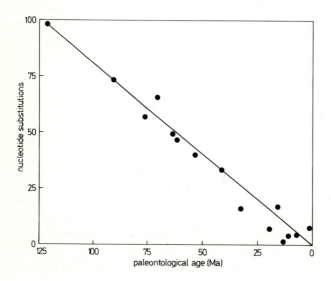

**Fig. 8.9.** Nucleotide substitutions vs. paleontological age (after Fitch 1977)

significant alteration of the biological function of the protein and (b) replacements occur at rates that can be measured in units of geologic time (Lovejoy et al. 1972). Such replacements are quite distinct from mutations within individuals and accumulate at a stochastically uniform rate over geological time (null hypothesis). Various studies have yielded evolutionary rates on the order of $10^{-9}$ substitutions per site per year (Fig. 8.9). The calibration of the molecular clock of mitochondrial DNA, for example, assumes the divergence between primates and ungulates occurred at the Cretaceous/Tertiary boundary (Hasegawa et al. 1985).

### Sample Treatment and Measurement Techniques

Amino acid sequences in orthologous proteins (e.g., globins, cytochromes, fibrinopeptides, albumins) are determined for the molecular clock. Indirect determinations are based on "immunological distances" of homologous serum proteins (Sarich and Wilson 1967). The DNA clock is based on a statistically significant number of nucleotide substitutions in DNA (Hasegawa et al. 1985).

### Scope and Potential, Limitations

**Scope and Potential.** The validity of the basic concepts of the molecular and DNA clocks is still in discussion (e.g., Goodman et al. 1982). Although there is general agreement that the evolution of individual amino acid replacements yields less accurate time information than the nucleotide substitutions in DNA. However, it is already obvious that these studies on molecular clocks have aided the dating of the time of divergence of the hominoid primates from the other mammals, as well as the reconstruction of geneological trees.

**Limitations.** It is not yet clear whether molecular changes within a lineage are cummulative and gradual (at a constant rate) or discontinuous (at varying rates) (Cronin et al. 1981). Moreover, the generation length is a critical parameter for the calculation of the rate of molecular evolution (Lovejoy et al. 1972).

## 8.6   Obsidian Hydration Method***

(Friedman and Smith 1960)

### Dating Range, Precision, Materials, Sample Size

The absolute age of little weathered artifacts made from obsidian, ignimbrite, basaltic glass, fused shale, slag, vitrophyre, and other natural glasses can be determined quickly and relatively inexpensively with this method. This type of object cannot be dated by other methods and is generally of great interest in terms of cultural prehistory. The dating precision, a low $\pm 100$ years, is comparable with that of the radiocarbon method (Sect. 6.2.1) if the calibration was carried out properly, but its dating range is considerably larger: from at least a few hundred years to possibly as much as 1 Ma (Michels 1986; Michels et al. 1983a, b). For other chipped lithic materials, see Sect. 8.9 (Taylor 1975).

The time of a volcanic eruption can be determined from silica-rich lava, as well as basaltic and volcanic glass from pelagic sediments (Friedman and Obradovich 1981) up to about 20 Ma. The precision is comparable to that of the K/Ar method (Sect. 6.1.1). Manganese nodules have also been dated by this method (Landford 1978). The structural water content of ceramics appears to be a function of the age (Zaun 1982).

## Basic Concept

Obsidian and other glasses, including man-made glass, absorb water on the surface, where it becomes chemically bound, forming a hydrated layer with a water content of ca. 3.5%, which is about ten times that of the original silicate. Since the process is diffusion controlled, the hydrated layer grows very slowly: several $\mu$m per century to several $\mu$m per millennium. These layers can be up to 50 $\mu$m thick.

The density and optical properties (e.g., index of refraction) of this hydrated layer differ from the original glass. As the diffusion front of the hydrated layer is a sharp boundary, it can usually be easily recognized in polarized light (Fig. 8.10).

The relationship between the thickness ($d$) of the hydrated layer and age ($t$) is derived from the law of diffusion:

$$d^2 = kt \tag{8.5}$$

where $k$ is the hydration rate constant, usually given in $\mu m^2/ka$. The value of $k$ ranges from 0.4–300 $\mu m^2/ka$, the most frequent values are around 10. It depends mainly on the chemical composition of the obsidian (Friedman and Long 1976) and the temperature. Moisture content and pH of the surrounding environment seem to have no influence. The relationship between $k$ and temperature is given by the Arrhenius equation:

$$k = Ae^{-E/RT} \tag{8.6}$$

where  T = effective hydration temperature in K,
        E = activation energy in J/mol,
        R = universal gas constant (8.317 J/K/mol), and
        A = frequency factor in $sec^{-1}$.

E varies from sample to sample. Combining Eqs. (8.5) and (8.6), the following equation (Michels et al. 1983a, b) is obtained:

$$d^2 = Ate^{-E/RT} \tag{8.7}$$

To solve Eq. (8.7) for the age of the sample, the energy of activation E and the value of A must be determined for each obsidian source (Michels 1986). For this purpose, an induced hydration experiment is carried out in a pressurized reaction vessel on chemically identical obsidian flakes with a fresh surface. These are placed in 500 mL deionized water in 1-L flasks at temperatures of 150–200°C for various lengths of time up to several days. For example, five flakes are hydrated at one temperature of 0.5, 1, 2, 4, and 6 days, respectively; four flakes are hydrated for 4 days at 150, 175, 225, and 250°C, respectively. The thickness of the hydration layer $d$ is

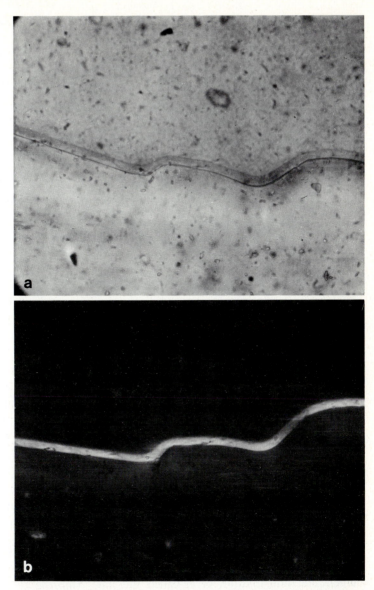

**Fig. 8.10 a, b.** A 4-$\mu$m-thick hydration rim in **a** plain and **b** plane-polarized light (after Trembour and Friedman 1984)

then measured. The rate constant for a single temperature is obtained from a plot of d versus $\sqrt{t}$ [Eq. (8.5); Fig. 8.11]. The slope ($-$ E/R) of the Arrhenius plot of ln $k$ versus 1/T yields the activation energy [Eq. (8.7)].

The hydration rate k* at the temperature T* of the study site is derived from the k value obtained from the induced hydrolysis experiment as follows:

$$k^* = k e^{E/R(1/T - 1/T^*)} \tag{8.8}$$

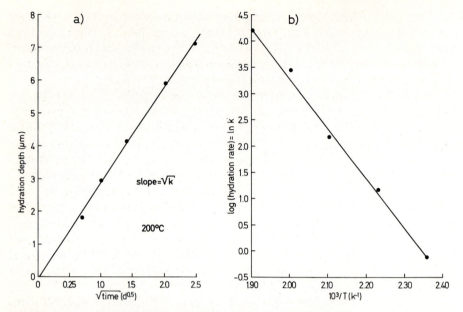

**Fig. 8.11. a** Induced hydration rate at 200°C for obsidian from Valle of Mexico. **b** Arrhenius plot for the same obsidian obtained using induced hydration rates at different temperatures (after Michels et al. 1983b)

The Mohlab obsidian dating laboratory in Pennsylvania publishes the results of induced hydration experiments on obsidian and other volcanic glasses from around the world. By the end of 1986, 71 such reports were available.

The effective hydration or diagenetic temperature (EHT) is determined in different ways. One possibility is the use of Eq. (8.3), proposed by Lee (1969). An error of ± 0.1°C results from an uncertainty of 30% in the value of E. This is negligible for Holocene samples. For Pleistocene samples with a complex temperature history, the geologic period involved should be subdivided into intervals $i$ that can be assumed to have a constant mean temperature. The $k_i$ values are determined for each of these intervals and then averaged.

In order to take into account fluctuation of the temperature around the annual mean and the variance between the soil temperature at various depths and the air temperature, implantation of so-called thermal "Ambrose cells" for 1 year is recommended as a second possibility (Michels 1986).

It is particularly important to know the effective hydration temperature when dating volcanic glass (Friedman and Obradovich 1981).

### Sample Treatment and Measurement Techniques

Information on location and elevation of the site, the local air temperature, and possible obsidian sources must be given to the laboratory making the age determination. There is seldom more than one source for a site.

It must be established for each sample whether it is chemically identical with the one used for induced hydration; if this is the case, the published hydration data can be used. XRF (Sect. 5.2.4.5), AA (Sect. 5.2.4.3), and neutron activation analysis (Sect. 5.2.4.2) are suitable procedures for determining chemical identity. Optical inspection of thin sections instead of instrumental analysis is often sufficient.

For optical techniques to determine the thickness of the perlite rim (i.e., the hydrated layer) (Friedman and Smith 1960; Michels 1986), 2-mm-thick slices are cut with a diamond saw to a depth of 4 mm perpendicular to the surface of the object to be dated or a small wedge is removed from each artifact. These slices are mounted on a microscope slide with Canada balsam, ground to a thickness of 75 $\mu$m, washed, dried, and sealed with a cover glass. A technician can prepare 3–6 mounts per hour.

The thickness of the hydrated layer is determined to $\pm 0.1$ $\mu$m in cross-polarized light using a x45 or x100 oil-immersion lens and an image-splitting eyepiece. It is sometimes necessary to use optical filters to improve the contrast (Michels et al. 1982; Michels 1986). About four measurements at each of two widely separated points must be made. The wavelength of the light limits the resolution of the optical method and thus the dating range is restricted to old samples whose hydrated layer is thicker than the wavelength of the light.

The nuclear resonance technique (Bird et al. 1983), which uses $^{19}$F ions accelerated to 16–22 MeV (Lee et al. 1974), is more sensitive than the optical method, but more complicated. Resolution is about 0.02 $\mu$m, hence this technique can be used for young samples, for example antique glass. The hydration profiles are measured directly since the reaction $^{1}$H $(^{19}$F$,\alpha\gamma)$ $^{16}$O produces $\gamma$-rays, which can be measured with a NaI (T1) scintillation detector (Sect. 5.2.2.2). The thickness of the hydrated layer is derived from a plot of the counts versus energy.

The $^{15}$N nuclear-resonance depth-profiling technique (Lanford 1978) uses accelerated nitrogen ions to react with the hydrogen atoms of the glass to produce resonance radiation: $^{1}$H $(^{15}$N$,\alpha\gamma)^{12}$C.

A modification of this technique uses accelerated argon ions to detect the depletion of potassium resulting from the hydration (Tsong et al. 1978; Michels 1986). Duoplasmatron ion sources with a focusing lens are used to produce ions with acceleration energies of 15–25 keV and a current of about 20 $\mu$A/mm$^2$. This removes the surface at a rate of about 10 $\mu$m/h (sputtering). After several hours at the most, the diffusion front is reached, detected by a sharp drop in the sputter-induced optical emission. The optical emission is detected with a photomultiplier.

For young glass samples, a non-destructive technique by exchange of tritiated water into the water of the hydrated layer and back exchange has been recommened (Lowe et al. 1984) for those cases where the hydrated layer is not adequate for accurate optical measurement.

### Scope and Potential, Limitations, Representative Examples

**Scope and Potential.** Although originally used as a relative dating technique, this method has achieved full operational capability as a chronometric technique in recent years (Michels 1986).

*Archeologic application:* This method is widely used in the USA, in Europe it is still almost unknown. Tens of thousands of artifacts, e.g., projectile points, knives, scrapers, and choppers, found in North, Central, and South America, Easter Island, Japan, the Near East, and Africa have been dated, classified, or identified using it. This has revealed typological transformations and innovations over long periods. In some cases, however, only the relative chronology was determined, based on the clustering of hydration thicknesses (Friedman and Smith 1960; Michels 1986). But the good agreement between hydration and K/Ar ages is encouraging (Friedman and Odradovich 1981). Due to the worldwide distribution of obsidian artifacts, the obsidian hydration method may become one of the important dating methods for archeologists.

**Limitations.** Dating errors occur when the obsidian artifact has been subjected to heat because this changes the rate of hydration. Moreover, erosion of the surface, which is difficult to detect, also leads to incorrect ages.

   Michels (1986) reports that the effective diagenetic hydration temperature determined with Lee's method is in many cases of sufficient precision to yield obsidian ages of high precision. The hydration velocity increases by about 10% per °C.

**Representative Examples.** *Deep-sea sediments:* Basaltic glass from pelagic sediment shows banded hydration layers about 50 $\mu$m thick, which can be easily counted. Because the temperature of the deep sea has varied little, even during the Quaternary, a constant hydration rate can be assumed: 1 $\mu$m/ka at 5°C, 7 $\mu$m/ka at 17°C, and 11 $\mu$m/ka at 24°C.

*Terrestrial sediments:* For terrestrial samples, the maximum age that can be determined using this method is set by the peeling off of the hydrated rim when it is 40–50 $\mu$m thick, less when subjected to heat or mechanical stress. But usually at least a small part of the surface of the sample remains intact, which can then be used for dating.

### Non-Chronological Applications

*Paleothermal studies:* The thermal history of a sampling site can be reconstructed on the basis of the thickness of the hydration layer of samples dated by other methods (Friedman and Obradovich 1981). The results, of course, need not correspond to the mean ambient temperature, since the rate of hydration does not increase linearly with temperature [Eq. (8.6)].

## 8.7 Dating of Man-Made Glass

(Brill and Hood 1961)

### Dating Range, Precision, Materials, Sample Size

This empirical non-destructive method is recommended for dating antique glass with ages of up to several hundred years. The chronometer starts at the time of deposition in the soil rather than the time of production.

### Basic Concept

Man-made glasses are amorphous, isotropic silicates that are subject to weathering when they lie in moist soil (Brewster 1863). Alkali cations are leached out during weathering forming silicon-rich crusts $0.3-20\,\mu m$ thick (Sect. 8.6). They reflect cyclic or periodic changes in its environment that are interpreted for dating as seasonal variations.

### Sample Treatment and Measurement Techniques

Polished sections are made (cut perpendicular to the surface) from several protected and well-preserved parts of the surface of the object to be dated. Ancient glasses show layers that are counted under a microscope with about x350 magnification and interpreted as age.

### Scope and Potential, Limitations

Owing to the large number of glass finds, archeologists are greatly interested in the dating of these objects. Application of the method is limited, however. Apart from the many finds that have no layered weathering crusts (e.g., glass objects from Byzantium or ancient Rome, which are scarcely weathered) and others that are completely weathered, the number of layers in many objects does not correlate with the age of the find (Brill 1969). Laboratory experiments in autoclaves have shown that by just a single heating several layers can form. Since the mechanism of the formation of the weathering layers is not yet understood, doubts have been raised about the suitability of this phenomenon for dating (Newton 1971). However, some objects may be suitable for dating with the obsidian hydration method (Sect. 8.6).

## 8.8 Calcium Diffusion and Cation-Ratio Methods

(Waddell and Fountain 1984; Dorn et al. 1986)

### Dating Range, Precision, Materials, Sample Size

This method has been used once to date riverbed clay, adobe, and fired clay bricks (and other objects containing little or no calcium) embedded in calcium-rich cement, plaster or mortar, as well as rock varnish ubiquitous on stable natural and cultural rock surface (Dorn et al. 1986). Ceramics coated with a lime-based slip or wash after firing are another example of objects that may be datable using this method. The age range is from several decades to about 100,000 years. The precision is expected to be about $\pm 10\%$ (Waddell and Fountain 1984; Dorn et al. 1986).

### Basic Concept

Calcium migrates at a nearly constant rate across clay–cement interfaces. The effective diffusion coefficient is $\sim 10^{-20}$ m$^2$/s, corresponding to an average migration rate of about 9 $\mu$m/century. This rate is, of course, controlled not only by diffusion, but also by ion exchange and adsorption. It is essentially independent of moisture content and temperature. Hence, the age of the sample may be estimated from the distance the calcium has migrated from the cement into the clay (Fig. 8.12).

Another basic concept is based on differences in the rate at which minor chemical elements are leached out of rock varnish ($K^+$, $Ca^{2+}$, $Ti^{4+}$). These ratios plotted versus the logarithm of age yield a straight line for an age range of 100,000 a (Dorn et al. 1986).

### Sample Treatment and Measurement Techniques

An area of undisturbed contact between the clay matrix and the calcium-rich coating is selected under a microscope. A tungsten carbide saw is used without lubricant to cut a $5 \times 5 \times 9$ mm perpendicular section from this part of the sample. This is embedded in epoxy resin, polished by hand with a block coated with 10-$\mu$m diamond particles, and then coated with carbon. The samples are analyzed using an ETEC electron microprobe.

The cation ratios are determined by proton-induced X-ray emission (PIXE). Several traverses are examined for each area of contact. With a calibration curve (Fig. 8.12), the mean depth of calcium migration, which can be determined to $\pm 2\,\mu$m, gives the age by extrapolation.

### Scope and Potential, Limitations

The method has not yet been used enough to be able to make a statement on its reliability. But it does seem to have potential due to its quick, inexpensive, and simple procedures. Especially the work on rock varnish is very impressive (Dorn et al. 1986).

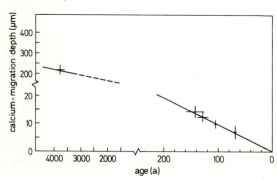

**Fig. 8.12.** Non-linear relationship between age and depth of calcium migration from a carbonate coating into a clayey material (after Waddell and Fountain 1984)

## 8.9   Dating of Bones Using the Fluorine or Uranium Content*

(Middleton 1844; Oakley 1949)

### Dating Range, Precision, Materials, Sample Size

This empirical method is used for determining relative ages of Pliocene and Pleistocene skeletal material, e.g., bones, antlers, and tooth enamel (Oakley 1980), up to an age of several million years (Hille et al. 1981). As the fluorine and uranium concentrations depend on local conditions and their secular variations, the precision of the method is rather poor (greater than several 100,000 years). A number of samples should be analyzed for each site so that the precision of the obtained ages can be estimated. Only 10–200 mg are needed for the analysis. Chipped lithic materials with an age of up to several thousand years may also be datable (Taylor 1975).

### Basic Concept

Bones, antlers, and teeth in groundwater continually take up fluorine (which occurs in trace amounts in percolating groundwater) by irreversible ionic exchange. During the course of fossilization the rate of uptake gradually decreases. The acquired fluorine transforms hydroxyapatite into the more stable, insoluble fluorapatite. Saturation, which is dependent on several parameters, e.g., species, is reached at about 3.8 wt.% in bones.

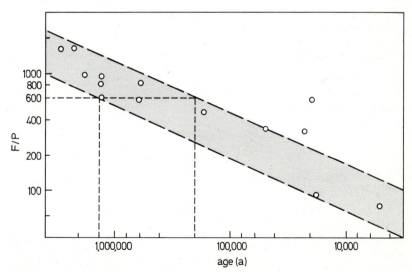

**Fig. 8.13.** Fluorine/phosphorus wt/wt ratio ($F/P$) in dated bones from Austrian caves as a function of age (after Hille et al. 1981), showing the large uncertainty in the F/P dates obtained. For example, an F/P ratio of 600 corresponds to an age range of 200,000–1.35 million years

Analogously, the uranium age determination is based on the adsorption of uranium from the groundwater. Because this adsorption is generally very slow, only Pleistocene samples can be distinguished from Holocene ones.

Due to the pronounced climate fluctuations during the Quaternary, paleohydrogeological conditions changed and thus fluorine and uranium concentrations in the groundwater may be expected to have also varied. Since constant concentrations cannot be assumed, calibration curves (Fig. 8.13) must be made to take variations in time and locality into account. Samples reliably dated by other methods are used for this purpose (Hille et al. 1981). Bone, antlers, dentin, and enamel all yield different calibration curves.

Fluorine in groundwater can diffuse into a lithic matrix analogous to the hydration of obsidian (Sect. 8.6); this phenomenon is also used for dating (Taylor 1975).

### Sample Treatment and Measurement Techniques

After ultrasonic cleaning, 10–200 mg of bone is ground to a fine, homogeneous powder and digested in concentrated sulfuric acid at 80°C for several days. The hydrofluoric acid thus produced is determined quantitatively by gas chromatography (Tite 1981). A modern technique uses a calibrated fluoride-sensitive electrode. Phosphate is determined by normal volumetric analysis.

The fluorine can also be determined simply and quickly by X-ray powder fluorescence diffraction analysis (Sect. 5.2.4.5), particularly if limited amounts of sample are available. The fluorine content is determined by the distance between a suitable pair of peaks in the XRF pattern. If an already dated sample is used, only distance between these two X-ray lines needs to be measured (Niggli and Overweel 1953). However, because the signal consists of three variable components, the fluorine content cannot be reliably determined (McConnell 1962).

A non-destructive technique is the activation of fluorine, nitrogen, calcium, and phosphorus with high-energy (14 MeV) neutrons (Sect. 5.2.4.2). The following reactions yield $\beta^+$-emitting nuclides, whose 511 keV annihilation radiation is detected by Ge(Li) $\gamma$-ray spectrometers (Sect. 5.2.2.3):

$$^{19}\text{F} \quad (n, 2n) \; ^{18}\text{F}$$
$$^{14}\text{N} \quad (n, 2n) \; ^{13}\text{N}$$
$$^{31}\text{P} \quad (n, 2n) \; ^{30}\text{P}$$
$$^{44}\text{Ca} \, (n, p) \quad ^{44}\text{K}$$

The activities of the individual emitters can be determined on the basis of their different half-lives; $^{44}\text{K}$ is measured using its characteristic gamma line (Hille et al. 1981).

The estimate of the age is based on the ratio of fluorine to phosphorus (Fig. 8.13) or nitrogen to phosphorus (Fig. 8.7), taking into account the different densities of the mineral matter of the bones. Bones that have been diagenetically altered or contaminated, and thus are unsuitable for dating, can be identified by deviation of the ratio of Ca/P from 1.67.

Uranium concentrations are determined radiometrically (Sect. 6.3) or by the fission-track method (Sect. 6.4.7). For this non-destructive method, only a few grams of sample are necessary.

The fluorine diffusion profile in chipped lithic material is determined by measuring the resonance energy from the nuclear reaction $^{19}F$ $(p,\alpha\gamma)$ $^{16}O$ with an ion microprobe mass analyzer (Taylor 1975).

### Scope and Potential, Limitations, Representative Example

*Scope and Potential.* The advantages of this empirical method are its simplicity and the small amount of sample necessary to distinguish contemporary from ancient skeletal materials with similar matrixes. Its use, however, is limited because (with a few exceptions) only relative ages with great uncertainty are obtainable, even if the samples were taken from similar geological and paleohydrogeological conditions. In spite of this, paleontologists and paleoanthropologists are still applying this method for relative dating or stratigraphic correlation studies. Improvements in the method are desirable because it is the only method available for the direct dating of bones older than te age range of the U/Th methods (Sect. 6.3). Promising results have been obtained from the inner parts of the hard tissue of enamel and dentin using SIMS (Fischer et al. 1986).

*Limitations.* Differences in the N and F contents are sometimes found even between porous and dense samples and between parts near and far from the surface of a sample. Therefore, when a series of samples is measured, only samples of the same kind should be selected. Samples from sites in which the groundwater has an excessively high fluorine concentration are unsuitable, e.g., in most tropical volcanic areas. Arid regions are also unsuitable.

It is not yet clear whether uranium is absorbed uniformly over a long time or only for a short time after the sample was deposited in the soil (Hennig and Grün 1983). According to the results of U/Th age determinations (Sect. 6.3), both cases appear to be possible. In addition, post-deposition leaching must also be considered.

*Representative Example.* A modern example for the application of this method is a study of bone samples from Austrian caves for which the logarithms of the F/P ratios correlate rather linearly with age (Fig. 8.13) (Hille et al. 1981). The combined application of the F, U, and N (Sect. 8.3) methods (called the FUN method) is recommended, which has been used, for example, to prove that the Piltdown fossils are modern (Oakley 1980).

# 9 Phanerozoic Time-Scale

## 9.1 Objectives and History of Geochronology

The field of geochronology involves more than just age determination. It also includes the chronostratigraphy of the Earth, in a wider sense, the cosmochronology of the universe. On the basis of the development of the biological world found in them, chronostratigraphic sequences reflect the history of the Earth in a qualitative way. Their significance lies in the possibility widely scattered natural records, often no longer in their original formation, and assign them to a previously reconstructed time sequence. In chronostratigraphic studies, attributes that can be assigned to a specific period of time, e.g., a certain isotopic composition (Chap. 6), are required. In biostratigraphic studies, key fossils that permit global correlations are of importance for correlating geological sequences whose biofacies and lithofacies differ completely from place to place.

Chronometry is a subdivision of geochronology that deals with the quantitative determination of the times and time spans of geological events and conditions. Such studies provide the time-scales for chronostratigraphy. This book covers mainly this topic, but also discusses the main chronostratigraphic methods (Chap. 7), which yield the most precise time-scales for the Quaternary (Fig. 7.6).

Not until the development of modern natural science was interest in geology rekindled. For many centuries the Earth was considered to be static without change. The first thoughts about evolution were made in antiquity. The Greek philosopher Xenophanes in the sixth century B.C. recognized the organic origin of fossil mollusc shells. Herodot in the fifth century B.C. saw relics of past times in the sediments of the Nile. The Roman Strabo (63 B.C. to 19 A.D.) described uplift and subsidence of the earth as an explanation for the presence of marine fossils in areas that are high mountains today. Leonardo da Vinci and Giordano Bruno recognized in the fifteenth and sixteenth centuries that the division between land and sea was continually changing. A primary contribution to modern geology was made by the Danish physician Niels Stensen (1638–1687), who in 1669, on the basis of stratified rocks in Tuscany, formulated the stratigraphic principle: The closer a layer of rock is to the surface, the younger it is.

Although this principle, which assumes uniform deposition and unchanging petrography of the rocks, provides a starting point for organizing the history of the Earth, it was the biostratigraphic principle of William Smith (1769–1839) that provided a more certain differentiation of stratigraphic sequences. Further developments led to fine stratigraphic subdivisions based on changes in microfossils with time (e.g., pollen, foraminifers).

With the determination of the first stratigraphic sequences arose the question of absolute ages. Various attempts to answer this question were made. An early estimate placed the age of the oceans around 360 Ma on the basis of the salt load in the rivers and the salt content of the oceans. Darwin in 1859 gave 300 Ma for the time since the end of the Mesozoic. The age (20–40 Ma) calculated by Lord Kelvin on the basis of the rate of cooling of the Earth was much too low. Goodschild (1897) estimated

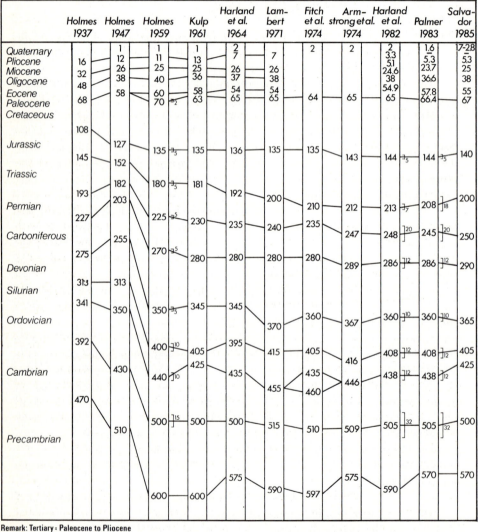

Remark: Tertiary = Paleocene to Pliocene

**Fig. 9.1.** Current adjustment of the Phanerozoic time scale during the last 50 years (see, e.g., Menning 1989). Time marks are given in Ma. The *light bars* indicate assumed confidence intervals

704 Ma for the time since the Cambrium on the basis of the thickness of sediments and the observed sedimentation rates. This is very close to the value accepted today (Fig. 9.1). However, these values could not be trusted because the processes governing the events of the Earth's history are not static but occur in a constant alternation between active phases and pauses.

Thus, that geological time is not in thousands of years, but in millions, or even billions of years was not recognized until the middle of the last century. It is quite interesting that in the mythology of India (the Veda) the age of the Earth is given as 4 billion years, which corresponds quite well with the value accepted today.

After the discovery of radioactivity by Bequerel in 1896, Rutherford formulated the idea of determining the age of minerals by the accumulation of radiogenic products. He calculated the first U/He age (Sect. 6.1.12) in 1905. His data showed that the Earth had to be older than 2 billion years. With the discovery of the exothermic character of radioactive decay, the source of heat was found that caused Lord Kelvin's calculation for the age of the Earth to be too low.

The radiometric ages determined in the first several decades were still rather imprecise and explain the initially only gradual acceptance by geologists. At first it was only a small, but enthusiastic group of geochronologists, with Arthur Holmes as mentor, that devoted themselves to this new field. They explored almost all of the possibilities for using uranium isotopes (Sect. 6.1) for age determination. Especially the U/He and the U/Th/Pb methods should be mentioned here, but also the now obsolete method using pleochroic haloes (Sect. 6.4.9).

The birth of modern geochronology was presaged by the invention of the mass spectrometer by Nier and Mattauch at the end of the 1930s. For the first time isotope abundances could be measured with sufficient accuracy, making it possible to distinguish non-radiogenic from radiogenic components. The potential of this instrument was expanded very rapidly and further developments are still being made today. This has been accompanied by the development of the model concepts for interpretation (e.g., the common lead concept and isochron and concordia diagrams). These developments made possible the "standard" methods of today (Sect. 6.1: U/Pb, Pb/Pb, K/Ar, and Rb/Sr methods).

After World War II, the range of methods available for geochronological studies expanded extremely rapidly, particularly the new radiocarbon (Sect. 6.2.1), U/Th (Sect. 6.3.1), and fission track (Sect. 6.4.7) methods. With the development of very precise instruments and improved techniques, new possibilities were opened up: Re/Os (Sect. 6.1.8), K/Ca (Sect. 6.1.2), and the La methods (Sects. 6.1.4 and 6.1.5). The Sm/Nd method (Sect. 6.1.6), although was introduced only in 1974, is already one of the main methods of geochronology and isotope geochemistry.

The spectrum of geochronological applications has been expanded to new types of samples. For example, the dating of sediments (e.g., Hurford et al. 1986), ore deposits, and tectonic events. An example is the dating of glauconite and clay mineral fractions, which permit the dating of the diagenesis of certain sediments and the determination of chronostratigraphic sequences. The dating of single crystals in the micro range has recently caused excitement. The use of long-lived cosmogenic isotopes as chronometer (Sect. 6.2) has been made possible by laser and ion microprobes (Sect. 5.2.4.4) and the accelerator mass spectrometer (Sect. 5.2.3.2).

**Table 9.1.** Phanerozoic time-scale of the American Geological Society (Palmer 1983). The dates assigned to the boundaries are being continually adjusted according to the most recent data.

| Era | Period | Epoch | Age | Age of boundaries Ma |
|---|---|---|---|---|
| **Cenozoic** | *Quaternary* | Holocene | | |
| | | | | 0.01 |
| | | Pleistocene | Calabrian | |
| | | | | 1.6 |
| | *Neogene* (Tertiary) | Pliocene | Piacenzian | |
| | | | | 3.4 |
| | | | Zanclean | |
| | | | | 5.3 |
| | | Miocene | Messinian | |
| | | | | 6.5 |
| | | | Tortonian | |
| | | | | 11.2 |
| | | | Serravallian | |
| | | | | 15.1 |
| | | | Langhian | |
| | | | | 16.6 |
| | | | Burdigalian | |
| | | | | 21.8 |
| | | | Aquitanian | |
| | | | | 23.7 |
| | *Paleogene* (Tertiary) | Oligocene | Chattian | |
| | | | | 30.0 |
| | | | Rupelian | |
| | | | | 36.6 |
| | | Eocene | Priabonian | |
| | | | | 40.0 |
| | | | Bartonian | |
| | | | | 43.6 |
| | | | Lutetian | |
| | | | | 52.0 |
| | | | Ypresian | |
| | | | | 57.8 |
| | | Paleocene | Thanetian (Selandian) | |
| | | | | 60.6 |
| | | | Unnamed (Selandian) | |
| | | | | 63.6 |
| | | | Danian | |
| | | | | 66.4 |
| **Mesozoic** | *Cretaceous* | Late | Maastrichtian | |
| | | | | 97.5 |
| | | Early | Albian | |
| | | | | 113 |
| | | | Aptian | |
| | | | | 119 |
| | | | Barremian | |
| | | | | 124 |
| | | | Hauterivian | |
| | | | | 131 |
| | | | Valanginian | |
| | | | | 138 |
| | | | Berriasian | |
| | | | | 144 |
| | *Jurassic* | Late | Tithonian | |
| | | | | 152 |
| | | | Kimmeridgian | |
| | | | | 156 |
| | | | Oxfordian | |
| | | | | 163 |
| | | Middle | Callovian | |
| | | | | 169 |
| | | | Bathonian | |
| | | | | 176 |
| | | | Bajocian | |
| | | | | 183 |
| | | | Aalenian | |
| | | | | 187 |
| | | Early | Toarcian | |
| | | | | 193 |
| | | | Pliensbachian | |
| | | | | 198 |
| | | | Sinemurian | |
| | | | | 204 |
| | | | Hettangian | |
| | | | | 208 |
| | *Triassic* | Late | Norian | |
| | | | | 225 |
| | | | Carnian | |
| | | | | 230 |
| | | Middle | Ladinian | |
| | | | | 235 |
| | | | Anisian | |
| | | | | 240 |
| | | Early | Scythian | |
| | | | | 245 |

| Era | Period | Epoch | Age | Age of boundaries Ma |
|-----|--------|-------|-----|----------------------|
| **Paleozoic** | *Permian* | Late | Tatarian | |
| | | | Kazanian | 253 |
| | | | Ufimian | |
| | | Early | Kungurian | 258 |
| | | | Artinskian | 263 |
| | | | Sakmarian | 268 |
| | | | Asselian | |
| | *Pennsylvanian* (Late Carboniferous) | | Gzelian | 286 |
| | | | Kasimovian | |
| | | | Moscovian | 296 |
| | | | Bashkirian | |
| | *Mississippian* (Early Carboniferous) | | Serpukhovian | 320 |
| | | | Visean | 333 |
| | | | Tournaisian | 352 |
| | Devonian | Late | Fammenian | 360 |
| | | | Frasnian | 367 |
| | | Middle | Givetian | 374 |
| | | | Eifelian | 380 |
| | | Early | Emsian | 387 |
| | | | Siegenian | 394 |
| | | | Gedinnian | 401 |
| | *Silurian* | Late | Pridolian | 408 |
| | | | Ludlovian | 414 |
| | | Early | Wenlockian | 421 |
| | | | Llandoverian | 428 |
| | *Ordovician* | Late | Ashgillian | 438 |
| | | | Caradocian | 448 |
| | | Middle | Llandelian | 458 |
| | | | Llanvirnian | 468 |
| | | Early | Arenigian | 478 |
| | | | Tremadocian | 488 |
| | *Cambrian* | Late | Trempealeauan | 505 |
| | | | Franconian | |
| | | | Dresbachan | |
| | | Middle | | 523 |
| | | Early | | 540 |
| | | | | 570 |

## 9.2   Geological Time-Scales

The setting up of geological time-scales for different regions and for the world as a whole requires the use of both chronostratigraphic and chronological data. Only when both are used can geochronological dates be interpreted in the proper geological context. Attempts to assign times to the chronostratigraphic sequences that were being developed began in 1911 with Arthur Holmes. He devoted his entire life to this task, continually correcting the time-scales (Table 9.1). Whereas until the end of the 1930s, these changes were the result of improvements in instrumentation, since then they have been the result of finding better samples for defining geological boundaries. It is still necessary to check and correct the Phanerozoic time-scale, and will continue to be so in the future. A number of international groups are working on this task (e.g., Cohee et al. 1978; Harland et al. 1982; Odin 1982a; Snelling 1985; Menning 1989).

There are a number of problems associated with the assignment of age values to stratigraphic boundaries. The first problem is the exact biostratigraphic definition of the geological time boundaries and their correlation between continents. Where this problem has been solved, rocks associated with these boundaries and which can be reliably dated must be found (e.g., tuff layers or bentonite). This, however, is a matter of happenstance and is just barely fulfilled if large regions are taken into consideration. Extrapolations to locally defined chronostratigraphic boundaries in other regions are often necessary. One difficulty is that dissimilar time-scales often must be correlated, time-scales that are based on dates obtained with different methods or different rocks and minerals (e.g., Fig. 4.3). The precision of the dates plays a role in their comparison. For a long time uncertainties in the half-lives of many of the nuclides relevant to geochronological studies were a handicap. Not until the fixing of conventional values by the Subcommission for Geochronology of the IUGS at the 24th Geology Congress in 1976 was this problem overcome for the standard dating methods (K/Ar, Rb/Sr, and U/Pb). In the meantime, international laboratory comparisons have been more or less successfully conducted for other methods: radiocarbon (International Study Group 1982), U/Th (Ivanovich et al. 1984), and the amino-acid racemization method (Wehmiller 1984b). On the other hand, disappointing results for other methods (Henning et al. 1985) have led to the discontinuation of such necessary efforts.

The dates assigned to the Phanerozoic time-scale by the Geological Society of America (Palmer 1983) are given in Table 9.1.

## 9.3   The Future

There have been great improvements in methods and techniques in recent years that have revolutionized geochronology anew. These have been accompanied by advances in the interpretation of the data. This development is continuing, also for methods that today receive little attention, opening up a great potential and range of applications in all fields of the geosciences, as well as for many ecological problems.

The analyses are being increasingly simplified in terms of the necessary manipulations; the instruments (e.g., mass spectrometers, accelerators, scintillation instruments) are being increasingly on-line computer controlled and automated. This allows increasingly routine application. The methods that still require complicated laboratory manipulations, e.g., the K/Ca, Re/Os, Lu/Hf, and also that La/Ce and La/Ba methods, will likely attain the importance that the Sm/Nd method already has.

This apparent simplication of the application of geochronological methods leads to certain problems, with which Quaternary chronologists are already being confronted. Increasingly, geochronological laboratories are being set up by insufficiently qualified scientists, who produce large amounts of data that do not stand up to careful inspection (Geyh 1988). The production of new chronologies and the accompanying publication of the authors' names appear to have priority to international comparison tests (e.g., Henning et al. 1985) and efforts to learn how to assess the reliability of the dates on the basis of scientifically reproducible data. It is urgently necessary to set standards on an international level, to conduct international laboratory comparisons (e.g., Wehmiller 1984b; Ivanovich et al. 1984; International Study Group 1982), and possibly even quality control tests.

The dating of sediments, ore deposits, and tectonic events present problems at the present time. The gap between 10 Ma and 50 ka is being gradually closed. Important contributions to this are being made by the complicated $^{39}$Ar/$^{40}$Ar method (Sect. 6.1.1.2) and recently also by the conventional K/Ar method (Sect. 6.1.1.1). Important dates are expected from the thermoluminescence method (Sect. 6.4.1) and possibly some day from the ESR method (Sect. 6.4.3). Recent advances in mass spectrometry make it possible to apply the U/Th method (Sect. 6.3.1) to this time period (Edwards et al. 1984).

The greatest changes and innovations in geochronology, however, are to be expected in dating in the micro range: Dating of single crystals or even parts of minute crystals (e.g., zircon and sanidine). Tephrochronology has already profited from $^{39}$Ar/$^{40}$Ar dating of individual grains using a laser microprobe (van den Bogaard et al. 1987).

Quaternary research has profited from the introduction of the AMS technique (Sect. 5.2.3.2): For example, from the dating of very small samples via cosmogenic isotopes that were not previously applicable for geochronological problems. Such studies will aid the work of archeologists and anthropologists, as well as Quaternary geologists, for whom the extraction of very pure components of samples will minimize the problem of contamination in Pleistocene samples that occurs with the radiocarbon method (Sect. 6.2.1). The AMS technique has also led to breakthroughs for methods that utilize very long-lived isotopes (e.g., $^{10}$Be, $^{36}$Cl, and $^{129}$I). Improvements in the methods are subject of current work. Major new applications, however, are being found already, for example, in hydrology. It extends the dating range of groundwater to above 50 ka, as well as to very recently regenerated groundwater. Any problems associated with hydrochemical reactions should be solvable. AMS analyses are providing significant contributions to oceanography and marine sedimentology with respect to paleoclimatology and oceanic circulation, well as variations in the Earth's magnetic field and trace gases. The influence of this technique could extend to microbiology and medicine. Studies of continental ice may expand our understanding of the role of extraterrestrial radiation on climate, particularly when the inputs of various cosmogenic isotopes are compared and the glaciogenic processes are understood. A further major task will be to learn to assess the reliability of the time-scales obtained with the various methods. Especially in the late Pleistocene part of the time-scale of terrestrial samples, large corrections are to be expected.

It is obvious that the work of international stratigraphic correlation programs will not decrease in significance. The Phanerozoic time-scale can still be improved. Many of the time marks used today as boundaries for geological formations are still based on only a small amount of data. It will be necessary to obtain samples for dating from as many parts of the world as possible.

These and many other, as yet unrecognized, questions will occupy geochronological laboratories involved in geoscientific and ecological research. The objective to correlate dissimilar time-scales will be attained bit by bit, deciphering the Earth's history, which is so important for our future.

# 10 Literature

From the immense amount of geochronologically relevant literature, the attempt was made in this book to survey, on the one hand, the primary literature, on the other hand, the most recent publications. Some of the reviewers, quite correctly, criticized that this has resulted in neglecting important publications that reflect the development of the individual methods. We have attempted to avoid this in the following compilation, which also includes secondary literature that seems to us to give a particularly valuable presentation or contain an extensive compilation of references. We hope that we have found an acceptable compromise.

## 10.1 Journals that Frequently Publish Geochronological Papers

Ancient TL, Oxford
Archaeometry, Oxford
Canadian Journal of Earth Sciences, Ottawa
Chemical Geology (Isotope Geoscience Section), Amsterdam
Contributions to Mineralogy and Petrology, Heidelberg
Earth and Planetary Science Letters, Amsterdam
Economic Geology, Lancaster
Geochemistry International, Washington
Geochimica et Cosmochimica Acta, Fairview Park, Elmsford
Geokhimiya, Kiev
Geological Society of American Bulletin, Boulder
Geology, Boulder
Geophysical Research Letters, Washington
IAEA Publications, Vienna
Journal of Geophysical Research, Washington

Journal of Hydrology, Amsterdam
Journal of Petrology, Oxford
Journal of Volcanology and Geothermal Research, Amsterdam
Meteoritics, Tempe, Arizona
Nature, London
Naturwissenschaften, Heidelberg
Nuclear Instruments and Methods in Physical Research, Amsterdam
Nuclear Tracks and Radiation, Oxford
PACT, Brussel
Palaeogeography, Palaeoclimatology, Palaeoecology, Amsterdam
Physical Review, New York
Precambrian Research, Amsterdam
Quaternary Science Reviews, Oxford
Quaternary Research, New York
Radiocarbon, New Haven
Science, Washington
Tectonophysics, Amsterdam
Terra Cognita, Paris
U.S. Geological Survey Bulletin, Washington

## 10.2  Geochronology Textbooks

Aitken MJ (1974) Physics and Archaeology. University Press, Oxford
Bates RL, Jackson JA (eds) (1980) Glossary of Geology: 751 pp, Am Geol Instr; Falls Church, Virginia
Berger A, Imbrie J, Hays J, Kukla G, Saltzman B (1984) Milankovitch and Climate, Vol I and II. 895 pp Reidel Publ Comp, Dordrecht
Bishop WW, Miller JA (eds) (1972) Calibration of Hominoid Evolution. 487 pp; Scottish Academic Press
Craig H, Miller SL, Wasserburg GJ (eds) (1964) Isotopic and Cosmic Chemistry. 553 pp; North-Holland, Amsterdam
Currie LA (ed) (1982) Nuclear and Chemical Dating Techniques. Interpreting the Environmental Record. Am Chem Soc Symp Ser 176. 516 pp, ACS; Washington DC
Dalrymple GB, Lanphere MA (1969) Potassium-Argon Dating. Principles, Techniques and Applications to Geochronology. 258 pp; WH Freeman; San Francisco
Doe Br (1970) Lead Isotopes. 137 pp; Springer, Berlin, Heidelberg, New York
Faul H (1966) Ages of Rocks, Planets and Stars. 109 pp; McGraw-Hill, New York
Faure G (1986) Principles of Isotope Geology (2nd edn). 589 pp; Wiley, New York
Faure G, Powell JL (1972) Strontium Isotope Geology. 188 pp; Springer, Berlin, Heidelberg
Fleischer RL, Price PB, Walker RM (1975) Nuclear Tracks in Solids: 605 pp; University of California, Berkeley, Los Angeles, London.
Fleming SJ (1976) Dating in Archaeology. A Guide to Scientific Techniques. London
Fritz P, Fontes J CH (eds) (1980) Handbook of Environmental Isotope Geochemistry, The Terrestrial Environment A, Vol 1: 546 pp; Elsevier, Amsterdam
Fritz P, Fontes J CH (eds) (1986) Handbook of Environmental Isotope Geochemistry, The Terrestrial Environment B, Vol 2: 557 pp, Elsevier, Amsterdam
Geyh MA (1980b) Einführung in die Methoden der physikalischen und chemischen Altersbestimmung. 276 pp; Wissenschaftliche Buchgesellschaft, Darmstadt
Hamilton EI (1965) Applied Geochronology. 267 pp; Academic Press, London, New York
Hamilton EI, Farquhar (eds) (1968) Radiometric Dating for Geologists. 506 pp; Interscience, London
Handbook of Chemistry and Physics, ca 2500 pp; CRC Press, London (annually new edition)
Harper CT (ed) (1973) Geochronology: Radiometric Dating of Rocks and Minerals. 469 pp; Dowden, Hutchinson and Ross, Stroudsburg, Pa
Hoefs J (1980) Stable Isotope Geochemistry (2nd edn). 208 pp, Springer Berlin, Heidelberg, New York
Hurford AJ, Jäger E, Ten Cate JAM (1986) Dating Young Sediments. COOP Techn Secr: 393 pp; (Proc workshop Beijing, China, Sept. 1985)
Ivanovich M, Harmon RS (eds) (1982) Uranium Series Disequilibrium. Applications to Environmental Problems. 571 pp; Clarendon Press, Oxford
Jäger E, Hunziker JC (1979) Lectures in Isotope Geology. 329 pp; Springer, Berlin, Heidelberg, New York
Mahaney WC (ed) (1984) Quaternary Dating Methods. 431 pp; Elsevier, Oxford
Michels JW (1973) Dating Methods in Archaeology. 240 pp; Academic Press, New York, London
Moser H, Rauert W (1980) Isotopenmethoden in der Hydrologie. 400 pp; Gebrüder Borntraeger, Berlin Stuttgart
Odin GS (ed) (1982) Numerical Dating in Stratigraphy (2 vol). 1040 pp; John Wiley, Chinchester, UK
Rankama K (1963) Progress in Isotope Geology. 705 pp; Interscience, London
Roth E, Poty B (eds) (1985) Methods de datation par les phenomènes nucleaires naturels— Applications. 642 pp; Masson, Paris
Roth E, Poty B (eds) (1989) Nuclear Methods of Dating. Dordrecht (Kluwer)
Russell RD, Farquhar RM (1960) Lead Isotopes in Geology. 243 pp; Interscience, John Wiley, New York
Rutter N, Brigham-Grette, Catto N (eds) (1989) Applied Quaternary Geochronology. Quatern Int 1:1–166
Schaeffer OA, Zähringer J (eds) (1966) Potassium Argon Dating. 234 pp; Springer, Berlin, Heidelberg

Tite MS (1981) Methods of Physical Examinations in Archaeology. 385 pp; Seminar Press, London, New York
York D, Farquar RM (1972) The Earth's Age and Geochronology. 178 pp, Pergamon Press. Oxford, New York, Toronto, Sydney, Braunschweig
Zimmerman MR (ed) (1986) Dating and Age Determination of Biological Materials. 300 pp; Croom Helm, Beckenham, UK

## 10.3  References

Abelson PH (1954) Amino acids in fossils. Science 119:576
Abelson PH (1955) Paleobiochemistry. Carnegie Inst Wash Yearbook 54:107–109
Åberg G, Bollmark B (1985) Retention of U and Pb in zircons from shocked granite in the Siljan impact structure, Sweden. Earth Planet Sci Lett 74:347–349
Ahrens LH (1951) The feasibility of a calicium method for the determination of geological age. Geochim Cosmochim Acta 1:312–316
Aitken MJ (1967) Thermoluminesce. Science Jour 1:32–38
Aitken MJ (1974) Physics and Archaeology. University Press, Oxford
Aitken MJ (1976) Thermoluminescent age evaluation and assessment of error limits: revised system. Archaeometry 18:233–239
Aitken MJ (1978) Archaeological Involvements of Physics. Physics Reports. A Review of Physics Letters 40C:277–351
Aitken MJ (1984) Non-linear growth: allowance for alpha particle contribution. Ancient TL 2:2
Aitken MJ (1985) Thermoluminescence Dating. Academic Press, London, 359 pp
Aitken MJ, Mejdahl V (coeds) (1979) A specialist seminar on thermoluminescence dating— Oxford, Research Laboratory for Archaeology and History of Art, July 1978. PACT 2 & 3, Brussels
Aitken MJ, Mejdahl V (coeds) (1982) Second specialist seminar on thermoluminescence dating, Oxford, 1980. PACT 6:562 p; Brussels
Aitken MJ, Smith BW (1988) Optical dating: recuperation after bleaching. Quatern Res Rev 7:387–393
Alaerts L, Lewis RS, Anders E (1979) Isotopic anomalies of noble gases in meteorites and their origins III. LL-chondrites. Geochim Cosmochim Acta 43:1399–1415
Albarède F (1982) The $^{39}Ar/^{40}Ar$ technique of dating. In: Odin GS (ed.) Numerical dating in stratigraphy. Wiley, Chichester pp 181–197
Alburger DE, Harbottle G, Norton EF (1986) Half-life of $^{32}Si$. Earth Planet Sci Letters 78:168–176
Aldrich LT, Nier AO (1948) Argon-40 in potassium minerals. Phys Rev 74:876–877
Alder B, Oeschger H, Wasson JT (1967) Aluminium-26 in deep-sea sediments. In: Radioactive Dating and Methods of Low-Level Counting: 189–198; IAEA, Vienna
Aleinikoff JN (1983) U-Th-Pb systematics of zircon inclusions in rock-forming minerals: a study of armoring against isotopic loss using the Sherman granite of Colorado-Wyoming, USA. Contrib Mineral Petrol 83:259–269
Alexander EC Jr (1978) Noble gases. In: Wedepohl KH (ed) Handbook of geochemistry, vol II-I. Springer, Berlin, Heidelberg, New York
Alexander EC Jr, Coscio MR Jr, Dragon JC, Pepin RO, Saito K (1977) K/Ar dating of lunar soils III: comparison of $^{39}Ar$-$^{40}Ar$ and conventional techniques; 12032 and the age of Copernicus. Proc Lunar Sci Conf 8th, pp 2725–2740
Allègre CJ (1968) $^{230}Th$ dating of volcanic rocks: a comment. Earth Planet Sci Lett 5:209–210
Allègre CJ (1969) Comportement des systèmes U-Th-Pb dans le manteau superieur et modèle d'evolution de ce dernier au cours des temps géologiques. Earth Planet Sci Lett 5:261–269
Allègre CJ (1987) Isotope geodynamics. Earth Planet Sci Lett 86:175–203
Allègre CJ, Condomines M (1976) Fine chronology of vulcanic processes using $^{234}U$-$^{230}Th$ systematics. Earth Planet Sci Lett 28:395–406
Allègre CJ, Ben Othman D (1980) Nd-Sr isotopic relationship in granitoid rocks and continental crust development: a chemical approach to orogenesis. Nature (Lond) 286:335–342
Allègre CJ, Luck JM (1980) Osmium isotopes as petrogenetic and geological tracers. Earth Planet Sci Lett 48:148–154

Allègre CJ, Rousseau D (1984) The growth of the continent through geological time studied by Nd isotope analysis of shales. Earth Planet Sci Lett 67:19–34

Allègre CJ, Albarède F, Grünenfelder M, Köppel V (1974) $^{238}U/^{206}Pb$-$^{235}U/^{207}Pb$-$^{232}Th/^{208}Pb$ zircon geochronology in alpine and non-alpine environments. Contrib Mineral Petrol 43: 163–194

Allègre CJ, Ben Othman D, Polve M, Richard P (1979) Nd-Sr isotopic correlation in mantle materials and geodynamic consequences. Phys Earth Planet Inter 19:293–306

Aller RC, DeMaster DJ (1984) Estimates of particle flux and reworking at the deep-sea floor using $^{234}Th/^{238}U$ disequilibrium. Earth Planet Sci Lett 67:308–318

Allison GB (1982) The relationship between $^{18}O$ and deuterium in water in sand columns undergoing evaporation. J Hydrol 55:163–169

Allsopp HL (1961) Rb-Sr measurements on total rock and separated mineral fractions from the Old Granite of the Central Transvaal. J Geophys Res 66:1499—1508

Alvarez W, Alvarez LW, Asaro F, Michel HV (1979) Experimental evidence in support of an extra-terrestrial trigger for the Cretaceous-Tertiary extinctions. EOS 60:734

Alvarez LW, Alvarez W, Asaro F, Michel HV (1980) Extraterrestrial cause for the Cretaceous–Tertiary extinction. Science 208:1095–1108

Ambach W, Dansgaard W (1970) Fall-out and climatic studies in firn cores from Carreflow, Greenland. Earth Planet Sci Lett 8:311–316

Amin BS, Kharkar DP, Lal D (1966) Cosmogenic $^{10}Be$ and $^{26}Al$ in marine sediments. Deep-Sea Res 13:805–824

Amin BS, Lal D, Somayajulu BLK (1975) Chronology of marine sediments using the $^{10}Be$ method: Intercomparison with other methods. Geochim Cosmochim Acta 39:1187–1192

Andersen T, Taylor PN (1988) Pb isotope geochemistry of the Fen carbonatite complex, S.E. Norway: age and petrogenetic implications. Geochim Cosmochim Acta 52:209–215

Anderson DL (1982) Isotopic evolution of the mantle: the role of magma mixing. Earth Planet Sci Lett 57:1–12

Anderson EC, Libby WF, Weinhouse S, Reid AF, Kirshenbaum AD, Grosse AV (1947) Natural radiocarbon from cosmic radiation. Phys Rev 72:931–936

Anderson RF, Bacon MP, Brewer PG (1983) Removal of $^{230}Th$ and $^{231}Pa$ at ocean margins. Earth Planet Sci Lett 66:73–90

André L, Deutsch S (1985) Very low-grade metamorphic Sr isotopic resettings of magmatic rocks and minerals: evidence for a late Givetian strike-slip division of the Brabant Massif, Belgium. J Geol Soc Lond 142:911–923

Andree M, Moor E, Beer J, Oeschger H, Stauffer B, Bonani G, Hofmann HJ, Morenzoni E, Nessi M, Suter M, Wölfli W (1984) $^{14}C$ dating of polar ice. Nucl Instrum Methods 233 (B5):385–388

Andree M, Oeschger H, Broecker WS, Beavon N, Mix A, Bonani G, Hofmann HJ, Morenzoni E, Nessi M, Suter M, Wölfli W (1986) AMS radiocarbon dates on foraminifera from deep sea sediments. Radiocarbon 28(2A):424–428

Andres G, Egger R (1985) A new tritium interface method for determining the recharge rate of deep groundwater in the Bavarian Mollasse basin. J Hydrol 82:27–38

Andres G, Geyh MA (1970) Paläohydrogeologische Studien mit Hilfe von $^{14}C$ über den pleistozänen Grundwasserhaushalt in Mitteleuropa (Südliche Frankenalb). Naturwissen-schaften 59:418

Andrew A, Godwin CI, Sinclair AJ (1984) Mixing line isochrons: A new interpretation of galena lead isotope data from southeastern British Columbia. Econ Geol 79:919–932

Andrews JN (1985) The isotopic composition of radiogenic helium and its use to study groundwater movement in confined aquifers. Chem Geol 49:339–351

Andrews JN, Kay RLF (1982) Natural production of tritium in permeable rocks. Nature (Lond) 298:361–363

Andrews JN, Lee DJ (1979) Inert gases in groundwater from the Bunter Sandstone in England as indicators of age and palaeoclimatic trends. J Hydrol 41:233–252

Andrews JN, Giles IS, Kay RLF, Lee DJ, Osmond JK, Cowart JB, Fritz P, Barker JF, Gale J (1982) Radioelements, radiogenic helium and age relationships of groundwaters from the granites at Stripa, Sweden. Geochim Cosmochim Acta 46:1533–1543

Andrews JN, Balderer W, Bath AH, Clausen HB, Evans GV, Florkowski T, Goldbrunner JE, Ivanovich M, Loosli H, Zojer H (1984) Environmental isotope studies in two aquifer systems:

A comparison of groundwater dating methods. In: Isotope Hydrology 1983:535–576. IAEA, Vienna

Andrews JN, Goldbrunner JE, Darling WG, Hooker PJ, Wilson GB, Youngman MJ, Eichinger L, Rauert W, Stichler W (1985) A radiochemical, hydrochemical and dissolved gas study of groundwaters in the Mollasse basin of Upper Austria. Earth Planet Sci Lett 73:317–392

Andrews JN, Fontes JC, Michelot J-L, Elmore D (1986) In-situ neutron flux, $^{36}$Cl production and groundwater evolution in crystalline rocks at Stripa, Sweden. Earth Planet Sci Lett 77:49–58

Armstrong RL, Mac Dowell WG (1974) Proposed refinement of the Phanerozoic time scale. Abstr. Meeting Geochron Cosmochron and Isotope Geology, August 1974, Paris

Arnal GB, Audrieux P (1986) Origins of carbon in potsherds. Radiocarbon 28(2A):711–718

Arnold JR (1956) Beryllium-10 produced by cosmic rays. Science 124:584–585

Atakan Y, Roether W, Münnich KO, Matthess G (1974) The Sandhausen shallow groundwater tritium experiment. In Isotope Techniques in Groundwater Hydrology 1974 (I):21–43. IAEA, Vienna

Athavale RN (1984) Nuclear tracer techniques for measurement of natural recharge in hard rock terrains. In: Int Workshop on Rural Hydrogeology and Hydraulics in Fissured Basement Zone. Univ of Roorkee, pp 71–80

Athavale RN, Chand R, Rangarajan R (1983) Groundwater recharge estimates for two basins in the Deccan Trap basalt formation. Hydrol Sci 328:525–538

Baadsgaard H (1987) Rb-Sr and K-Ca isotope systematics in minerals from potassium horizons in the Prairie Evaporite Formation, Saskatchewan, Canada. Chem Geol Isot Geosci Sect 66:1–15

Baadsgaard H, Lerbekmo JF (1982) The dating of bentonite beds. In: Odin GS (ed) Numerical dating in stratigraphy, part I. Wiley, Chichester, pp 423–440

Bachmann G, Grauert B (1986) Dating by means of $^{87}$Sr/$^{86}$Sr-disequilibrium profiles. Terra Cognita 6:148

Backus MM (1955) Mass spectrometric determination of the relative isotopic abundances of calcium and the determination of geological age. Ph D Thesis, Department of Geology and Geophysics, Massachusetts Institute of Technology, Boston USA

Bacon MP, Rosholt JN (1982) Accumulation rates of Th-230, Pa-231, and some transition metals on the Bermuda Rise. Geochim Cosmochim Acta 46:651–666

Bada JL (1984) In vivo racemization in mammalian proteins. Methods Enzymol 106:98–115

Bada JL (1985) Amino acid racemization dating of fossil bones. Ann Rev Earth Planet Sci 13:214–268

Bada JL, Schroeder RA (1975) Amino acid racemization reactions and their geochemical implications. Naturwissenschaften 62:71–79

Bada JL, Protsch R, Schroeder RA (1973) The racemization reaction of isoleucine used as a palaeotemperature indicator. Nature (Lond) 241:394–395

Bada JL, Ming-Yung Shou, Man EH, Schroeder RA (1978) Decomposition of hydroxy amino acids in foraminiferal tests; kinetics, mechanism and geochronological implications. Earth Planet Sci Lett 41:67–76

Bada JL, Masters PM, Hoopes E, Darling D (1979) The dating of fossil bones using amino acid racemization. In: Berger R, Suess HE (eds) Radiocarbon Dating. Univ California Press, Los Angeles, pp 740–756

Bada JL, Mitchell E, Kemper B (1983) Aspartic acid racemization in narwhal teeth. Nature (Lond) 303:418–420

Bada JL, Vrolijk CD, Browns S, Druffel ERM, Hedges REM (1987) Bomb radiocarbon in metabolically inert tissues from terrestrial and marine mammals. Geophys Res Lett 14:1065–1067

Bagge E, Willkomm H (1966) Geologische Altersbestimmung mit $^{36}$Cl. Atomkernenergie 11:176–184

Bailey GN, Deith MR, Shackleton NJ (1983) Oxygen isotope analysis and seasonality determinations: limits and potential of a new technique. Am Antiq 48:390–398

Baksi AK (1982) $^{40}$Ar/$^{39}$Ar incremental heating studies on a suite of "disturbed" rocks from Cape Breton Island, Canada: detection of a deformational episode. Geol Soc India 23:267–276

Bangert U, Hennig GJ (1979) Effects of sample preparation and the influence of clay impurities on the TL-dating of calcite cave deposits. PACT 3:281–289

Banwell GM, Parizek RR (1988) Helium-4 and radon-222 concentrations in groundwater and soil gas as indicators of zones of fracture concentration in unexposed rock. J Geophys Res 93:355–366

Bar M, Kolodny Y, Bentor YK (1974) Dating faults by fission track dating of epidotes—an attempt. Earth Planet Sci Lett 22:157–162

Barabas M, Bach A, Mangini A (1988a) An analytical model for the growth of ESR signals. Nucl Tracks Radiat Meas 14 (1/2):231–235

Barabas M, Mangini A, Sarnthein M, Stremme H (1988b) The age of the Holstein interglaciation: a reply. Q Res 29:80–84

Barker F, Millard HT Jr, Hedge CT, O'Neil JR (1976) Pikes Peak batholith: Geochemistry of some minor elements and isotopes, and implications for magma genesis. In: Epis RC, Weimer RJ (eds) Studies in Colorado field geology. Colorado School of Mines Prof Contrib 8:44–56

Barker JL Jr, Anders E (1968) Accretion rate of cosmic matter from iridium and osmium contents of deep-sea sediments. Geochim Cosmochim Acta 32:627–645

Barnes IL, Murphy TJ, Gramlich JW, Shields WR (1973) Lead separation by anodic deposition and isotope ratio mass spectrometry of microgram and smaller samples. Anal Chem 45:1881–1884

Barnes JW, Lang EJ, Portratz KA (1956) Ratio of ionium to uranium in coral limestone. Science 124:175–176

Barnola JM, Raynaud D, Korotkevich YS, Lorius C (1987) Vostok ice core provides 160,000-year record of atmospheric $CO_2$. Nature (Lond) 329:408–414

Barnov VA, Kartvelishvili II, Laliev AG, Tsetskhladze TV (1983) Determination of natural tritium in formation waters of the Samgari-Patardzeuli Oil field. Water Res 9:546–549

Barr GE, Lambert SJ, Carter JA (1979) Uranium isotope disequilibrium in ground waters of south eastern New Mexico and implications regarding age-dating of waters. In: Isotope Hydrology 1978 (II): 645–660. IAEA, Vienna

Barsukov VL (1981) Comparative planetology and the earth's early history. Geokhimiya 11:1603–1614

Barton CE, Merrill RT, Barbetti M (1979) Intensity of the earth's magnetic field over the last 10 000 years. Phys Earth Planet Inter 20:96–110

Barton JC, Watson AH, Wright AG (1982) A direct measurement of the distribution in depth of $^{26}Al$ in the Estacado meteorite. Geochim Cosmochim Acta 46:1963–1967

Basu AR, Tatsumoto M (1980) Nd-isotopes in selected mantle-derived rocks and minerals and their implications for mantle evolution. Contrib Mineral Petrol 75:43–54

Bath AH, Edmunds WM, Andrews JN (1979) Palaeoclimatic trends deduced from the hydro-chemistry of a Triassic sandstone aquifer, United Kingdom. In: Isotope Hydrology 1978 (II): 545–566. IAEA, Vienna

Bauer CA (1947) Production of helium in meteorites by cosmic radiation. Phys Rev 72:354–355

Bauer CA (1948) Rate of production of helium in meteorites by cosmic radiation. Phys Rev 74:501–502

Baxter MS, Harkness DD (1975) $^{14}C/^{12}C$ ratios of urban pollution. In: Isotope Ratios as Pollutant Source and Behaviour Indicators:135–140. IAEA, Vienna

Becker K, Goedicke C (1978) A quick method for authentication of ceramic art objects. Nucl Instrum Methods 151:313–316

Beer H, Walter G, Macklin RL. Patchett PJ (1984) Neutron capture cross sections and solar abundances of $^{160,161}Dy$, $^{170,171}Yb$, $^{175,176}Lu$, and $^{176,177}Hf$ for the s-process analysis of the radionuclide $^{176}Lu$. Phys Rev 30:464–478

Beer J, Giertz V, Möll M, Oeschger H, Riessen T, Strahm C (1979) The contribution of the Swiss lake-dwellings to the calibration of radiocarbon dates. In: Berger R, Suess HE (eds) Radiocarbon Dating. Univ California Press, Los Angeles, pp 566–584

Beer J, Andree M, Oeschger H, Siegenthaler U, Bonani G, Hofmann H, Morezoni E, Nessi M, Suter M, Wölfli W, Finkel R, Langway C Jr (1984) The Camp Century $^{10}Be$ record: implications for long-term variations of the geomagnetic dipole moment. Nucl Instrum Methods Phys Res 233 (B5): 380–384

Beer J, Bonani G, Hofmann HJ, Suter M, Synal A, Wölfli W, Oeschger H, Siegenthaler U, Finkel RC (1987) $^{10}Be$ measurements on polar ice: comparison of arctic and antarctic records. Nucl Instrum Methods Phys Res B 29:203–206

Beer J, Siegenthaler U, Blinov A (1988a) Temporal $^{10}Be$ variations in ice: information on solar activity and geomagnetic field intensity. In: Stephenson FR, Wolfendale AW (eds) Secular Solar and Geomagnetic Variations in the Last 10,000 Years. Kluwer Academic Publ. Dordrecht, pp 297–313

Beer J, Siegenthaler U, Bonani G, Finkel RC, Oeschger H, Suter M, Wölfli W (1988b) Information on past solar activity and geomagnetism from $^{10}$Be in the Camp Century ice core. Nature (Lond) 331:675–679

Begemann F, Geiss J, Hess DC (1957) Radiation age of a meteorite from cosmic-ray produced He$^3$ and H$^3$. Phys Rev 107:540–542

Behrens H (1982) New insights into the chemical behavior of radioiodine in aquatic environments. In: Environmental Migration of Long Lived Radionuclides: 27–40. IAEA, Vienna

Behrens H, Moser H, Oertner H, Rauert W, Stichler W, Ambach W, Kirchlechner W (1979) Models for the runoff from a glaciated catchment area using measurements of environmental isotope content. In: Isotope Hydrology 1978: (II) 829–845. IAEA, Vienna

Bell K, Powell JL (1970) Strontium isotopic studies of alkalic rocks: The alkalic complexes of Eastern Uganda. Geol Soc Am Bull 81:3481–3490

Bell K, Blenkinsop J, Cole TJS, Menagh DP (1982) Evidence from Sr isotopes for long-lived heterogeneities in the upper mantle. Nature (Lond) 298:251–253

Bell WT (1979) Thermoluminescence dating: revised dose-rate data. Archaeometry 21:243–245

Bender ML (1973) Helium-uranium dating of corals. Geochim Cosmochim Acta 37:1229–1247

Bender ML, Fairbanks RG, Taylor FW, Matthews RK, Goddard JG, Broecker WS (1979) Uranium-series dating of the Pleistocene reef tracts of Barbados, West Indies. Geol Soc Am Bull Part I 90:577–594

Bennett CL, Beukens RP, Clover MR, Gove HE, Liebert RP, Litherland AE, Purser KH, Sondheim WE (1977) Radiocarbon dating using electrostatic accelerators: negative ions provide the key. Science 198:508–510

Bentley HW, Phillips FM, Davis SN, Gifford S, Elmore D, Tubbs LE, Gove HE (1982) Thermonuclear $^{36}$Cl pulse in natural water. Nature (Lond) 300:737–740

Bentley HW, Phillips FM, Davis SN (1986a) Chlorine-36 in the terrestrial environment. In: Fritz P, Fontes JCh (eds) Handbook of Environmental Isotope Geochemistry, Vol 2. The Terrestrial Environment. Elsevier, Amsterdam, pp 427–480

Bentley HW, Phillips FM, Davis SN, Habermehl MA, Airey PL, Calf GE, Elmore D, Gove HE, Torgersen TL (1986b) Chlorine-36 dating of very old groundwater 1. The Great Artesian Basin, Australia. Water Resour Res 22(13):1991–2001

Berger A (1988) Milankovitch theory and climate. Rev Geophys 26:624–657

Berger A, Imbrie J, Kukla G, Saltzman B (eds) (1984) Milankovitch and Climate, I and II. Reidel, Dordrecht, p 510 and p 385

Berger GW (1985a) Thermoluminscence dating of volcanic ash. J Volcanol Geotherm 25:333–347

Berger GW (1985b) Thermoluminscence dating applied to a thin winter varve of the late glacial South Thompson silt, south-central British Colombia. Can J Earth Sci 22:1736–1739

Berger GW (1986) Dating Quaternary deposits by luminscence – recent advances. Geosci Canada 13:15–21

Berger GW, York D (1981) Geothermometry from $^{40}$Ar/$^{39}$Ar dating experiments. Geochim Cosmochim Acta 45:795–812

Berger GW, Huntley DJ, Stipp JJ (1984) Thermoluminescence studies on a $^{14}$C-dated marine core. Can J Earth Sci 21:1145–1150

Berger GW, Clague JJ, Huntley DJ (1987) Thermoluminescence dating applied to glaciolacustrial sediments from central British Columbia. Can J Earth Sci 24:425–434

Berger R, Suess HE (eds) (1979) Radiocarbon Dating. Univ California Press, Los Angeles, 789 pp

Berger WH (1981) Oxygen and carbon isotopes in foraminifera: an introduction. Palaeogeogr Palaeoclimat Palaeoecol 33:3–7

Berger WH, Johnson RF (1978) On the thickness and the $^{14}$C age of the mixed layer in deep-sea carbonates. Earth Planet Sci Lett 41:223–227

Berger WH, Johnson RF, Killingley JS (1977) "Unmixing" of the deep-sea record and the deglacial meltwater spike. Nature (Lond) 269:661–663

Bernat M, Allègre CJ (1974) Systematics in uranium-thorium dating of sediments. Earth Planet Sci Lett 21:310–314

Bertel E, Märk TD (1983) Fission tracks in minerals: annealing kinetics, track structure and age correction. Phys Chem Mineral 9:197–204

Bhandari N, Lal D, Rajan RS, Arnold JR, Marti K, Moorce CB (1980) Atmospheric ablation in meteorites: a study based on cosmic ray tracks and neon isotopes. Nucl Tracks 4:213–262

Bhandari N, Gupta DS, Singhvi AK, Nijampurkar VN, Vohra CP (1983) Thermoluminescence dating of glaciers. PACT 9:513–521

Bibron R, Chesselet R, Crozaz G, Leger G, Mennessier JP, Picciotto E (1974) Extra-terrestrial $^{53}$Mn in Antarctic ice. Earth Planet Sci Lett 21:109–116

Bickford ME, Harrower KL, Hoppe WJ, Nelson BK, Nusbaum RL, Thomas JJ (1981) Rb-Sr and U-Pb geochronology and distribution of rock types in the Precambrian basement of Missouri and Kansas. Geol Soc Am Bull 92:323–341

Bien GS, Rakestraw NW, Suess HE (1960) Radiocarbon concentration in Pacific ocean water. Tellus 12:436–443

Birck JL, Minister JF, Allègre C (1975) $^{87}$Rb-$^{87}$Sr chronology of achondrites. Meteoritics 10:364–365

Bird JR, Duerden B, Wilson DJ (1983) Ion Beam Techniques in Archaeology and the Arts. Nucl Sci Appl Sec B: 483–513. Harwood Academic, New York

Bischoff JL, Rosenbauer RJ (1981) Uranium series dating of human skeletal remains from the Del Mar and Sunnyvale Sites, California. Science 213:1003–1005

Bischoff JL, Rosenbauer RJ, Tavoso A, de Lumley H (1988) A test of uranium-series dating of fossil tooth enamel: results from Tournal cave, France. Appl Geochem 3:145–151

Black LP, Fitzgerald JD, Harley SL (1984) Pb isotopic composition, colour, and micro-structure of monazites from a polymetamorphic rock in Antarctica. Contrib Mineral Petrol 85:141–148

Black LP, Williams IS, Compston W (1986) Four zircon ages from one rock: the history of a 3930-Ma-old granulite from Mount Sones, Enderby Land, Antarctica. Contrib Mineral Petrol 94:427–437

Blake JB, Schramm DN (1973) $^{247}$Cm as a short-lived r-process chronometer. Nat Phys Sci 243:138–140

Blümel WD (1982) Calcretes in Namibia and SE-Spain relations to substratum, soil formation and geomorphic factors. CATENA Suppl 1:67–82

Bluszcz A (1988) The Monte-Carlo experiment with the least squares methods of line fitting. Nucl Tracks Radiat Meas 14:355–360

Bøtter-Jensen L, Mejdahl V (1988) Assessment of beta dose rate using a GM multicounter system. Nucl Tracks Radiat Meas 14:187–191

Bogard DD, Husain L, Nyquist LE (1979) $^{40}$Ar-$^{39}$Ar age of the Shergotty achondrite and implications for its post-shock thermal history. Geochim Cosmochim Acta 43:1047–1055

Bogard DD, Nyquist LE, Johnson P (1984) Noble gas content of shergottites and implications for the Martian origin of SNC meteorites. Geochim Cosmochim Acta 48:1723–1739

Boltwood BB (1907) On the ultimate disintegration products of the radioactive elements. Part II. The disintegration products of uranium. Am J Sci 23(4):77–78

Bonani G, Balzer R, Hofmann H-J, Morenzoni E, Nessi M, Suter M, Wölfli W (1984) Properties of milligram size samples prepared for AMS $^{14}$C dating at ETH. Nucl Instrum Methods 233(B5):284–288

Bonhomme MG (1982) The use of Rb-Sr and K-Ar dating methods as a stratigraphic tool applied to sedimentary rocks and minerals. Precambrian Res 18:5–25

Bonhomme MG, Lucas J, Millot G (1966) Signification des determinations isotopiques dans la géochronologie des sediments. Actes du 151$^e$ Coll Int CNRS, Nancy 1965:541–565

Bonhomme MG, Bühmann D, Besnus Y (1983) Reliability of K-Ar dating of clays and silicifications associated with vein mineralizations in Western Europe. Geol Rundsch 72:105–117

Bonner FT, Roth E, Schaeffer OA, Thompson SO (1961) Chlorine-36 and deuterium study of Great Basin lake waters. Geochim Cosmochin Acta 25:261–266

Boschmann W (1986) Uran und Helium in Erzmineralien und die Frage ihrer Datierbarkeit. Heidelberger Geowissenschaftl Abh 4:234 pp

Boudin A, Deutsch S (1970) Geochronology: recent development in the lutetium-176/hafnium-176 dating method. Science 168:1219–1220

Bourles D, Raisbeck GM, Yiou F, Loiseaux JM, Lienoin M, Klein J, Middleton R (1984) Investigation of possible association of $^{10}$Be and $^{26}$Al with biogenic matter in the marine environment. Nucl Intrum Methods 233 (B5):365–370

Bourles D, Raisbeck GM, Yiou F (1989) $^{10}$Be and $^{9}$Be in marine sediments and their potential for dating. Geochim Cosmochim Acta 53:443–452

Boyle EA (1984) Sampling statistic limitations on benthic foraminifera chemical and isotopic data. Mar Geol 58:213–224

Brenninkmeijer CAM, van Geel B, Mook WG (1982) Variations in the D/H and $^{18}$O/$^{16}$O ratios in cellulose extracted from a peat bog core. Earth Planet Sci Lett 61:283–290

Brereton NR (1970) Corrections for interfering isotopes in the $^{40}$Ar/$^{39}$Ar dating method. Earth Planet Sci Lett 8:427–433

Brévart O, Dupré B, Allègre CJ (1982) Metallogenesis at spreading centres: lead isotope systematics for sulfides, manganese-rich crusts, basalts, and sediments from the Cyamex and Alvin areas (East Pacific Rise). Econ Geol 77:564–575

Brévart O, Dupré B, Allègre CJ (1986) Lead-lead age of komatiitic lavas and limitations on the structure and evolution of the Precambrian mantle. Earth Planet Sci Lett 77:293–302

Brewer MS, Lippolt HJ (1974) Petrogenesis of basement rocks of the Upper Rhine Region elucidated by rubidium-strontium systematics. Contrib Mineral Petrol 45:123–141

Brewster D (1863) On the structure and optical phenomena of ancient decomposed glass. Trans R Soc Edinbourgh 23:193–204

Brigham JK (1983) Intrashell variations in amino acid concentrations and isoleucine epimerization ratios in fossil 'Hiatella arctica'. Geology 11:509–513

Brill RH (1969) The scientific investigation of ancient glass. 8th Int Congr Glass, pp 47–68

Brill RH, Hood HP (1961) A new method for dating ancient glass. Nature (Lond) 189:12–14

Broecker WS (1981) Glacial to interglacial changes in ocean and atmospheric chemistry. In: Berger A (ed) Climate Variations and Variability: Facts and Theory. Reidel, Holland, pp 111–121

Broecker WS (1982) Ocean chemistry during glacial time. Geochim Cosmochim Acta 46:1689–1705

Broecker WS (1987) Paleo-ocean circulation rates as determined from accelerator radiocarbon-measurements on hand-picked foraminifera. Terra Cognita 7(1):43-44

Broecker WS, Ku TL (1969) Caribbean Cores P6304-8 and P6304-9: New analysis of absolute chronology. Science 166:404–406

Broecker WS, Peng TW (1982) Tracers in the sea. Eldigio, Columbia Univ New York

Broecker WS, Thurber DL (1965) Uranium-series dating of corals and oolites from Bahaman and Florida key limestones. Science 149:58–60

Broecker W, Mix A, Andree M, Oeschger H (1984) Radiocarbon measurements on coexisting benthic and planktonic foraminifera shells: potential for reconstructing ocean ventilation times over the past 20 000 years. Nucl Instrum Methods Phys Res 233 (B5):331–339

Bronk CR, Hedges REM (1987) A gas ion source for radiocarbon dating. Nucl Instrum Methods Phys Res B 29:45–49

Brookins DG, Krueger HW, Bills TM (1985) Rb-Sr and K-Ar analyses of evaporate minerals from southeastern New Mexico. Isochron/West 43:11–12

Brooks C, Wendt I, Harre W (1968) A two-error regression treatment and its application to Rb-Sr and initial Sr$^{87}$/Sr$^{86}$ ratios of younger Variscan granitic rocks from the Schwarzwald Massif, southwest Germany. J Geophys Res 73:6071–6084

Brown H (1947) An experimental method for the estimation of the age of the earth. Phys Rev 72:348

Brown L (1987) $^{10}$Be as a tracer of erosion and sediment transport. Chem Geol Isot Geosci Sect 65:189–196

Brown L, Klein J, Middleton R, Sacks IS, Tera F (1982) $^{10}$Be in island-arc vulcanoes and implications for subduction. Nature (Lond) 299:718–720

Brown RB, Kling GF, Cutshall NH (1981) Agricultural erosion indicated by $^{137}$Cs redistribution: II. Estimates of erosion rates. Soil Sci Soc Am J 45:1191–1197

Brown RM, Andrews HR, Ball GC, Burn N, Imahori Y, Milton JCD, Fireman EL (1984) $^{14}$C content of ten meteorites measured by tandem accelerator mass spectrometry. Earth Plan Sci Lett 67:1–8

Brown TA, Nelson DE, Southon JR, Vogel JS (1985) The extraction of $^{10}$Be from lake sediments leaching versus total dissolution. Chem Geol Isot Geosci Sect 52:375–378

Brunnacker K, Jäger K-D, Hennig GJ, Preuss J, Grün R (1983) Radiometrische Untersuchungen zur Datierung mitteleuropäischer Travertinvorkommen. EAZ Ethnogr-Archäol Z 24:217–266

Bruns M, Münnich KO, Becker B (1980a) Natural radiocarbon variations from AD 200 to 800. Radiocarbon 22 (II):273–277

Bruns M, Levin I, Münnich KO, Hubberten HW, Fillipakis S (1980b) Regional sources of vulcanic carbon dioxide and their influence on $^{14}$C content of present-day plant material. Radiocarbon 22(II):532–536

Bucha V (1973) Archaeomagnetic dating. In: Michael HN, Ralph EK (eds) Dating Techniques for the Archaeologist. MIT, Cambridge, MA, London, pp 57–117

Buchardt B, Fritz B (1980) Environmental isotopes as environmental and climatological indicators. In: Fritz P, Fontes JC (eds) Handbook of Environmental Isotope Geochemistry, vol 1, The Terrestrial Environment A. Elsevier, Amsterdam, p 473–504

Buczko CM, Borbely A, Ilkov NI (1978) Fossil bones and the paleoclimatology. Radiochem Radioanal Lett 36:175–179

Burchart J, Kral J (1982) Application of fission-track isochrons method to accessory minerals of the crystalline rocks of the West Carpathians. Geol Carpathica 33:141–146

Burk RL, Stuiver M (1981) Oxygen isotope ratios in trees reflect mean annual temperature and humidity. Science 211:1417–1419

Burke WH, Denison RE, Hetherington EA, Koepnick RB, Nelson HF, Otto JB (1982) Variation of seawater $^{87}Sr/^{86}Sr$ throughout Phanerozoic time. Geology 10:516–519

Burnett WC, Kim KH (1986) Comparison of radiocarbon and uranium-series dating methods as applied to marine apatite. Q Res 25:369–379

Burnett WC, Beers MJ, Roe KK (1982) Growth rates of phosphate nodules from the continental margin off Peru. Science 215:1616–1617

Burwash RA, Krupicka J, Basu AR, Wagner PA (1985) Resetting of Nd and Sr whole-rock isochrons from polymetamorphic granulites, northeastern Alberta. Canad J Earth Sci 22:992–1000

Butler WA, Jeffery PM, Reynolds JH, Wasserburg GJ (1963) Isotopic variations in terrestrial xenon. J Geophys Res 68:3283–3291

Butzer KW (1983) Global sea level stratigraphy: an appraisal. Q Sci Rev 2:1–15

Cadogan PH (1977) Palaeoatmospheric argon in Rhynie chert. Nature (Lond) 268:38–40

Cairns T (1976) Archaeological dating. Anal Chem 48:266A–280A

Calk LC, Naeser CW (1973) The thermal effect of a basalt intrusion on fission tracks in quartz monzonite. J Geol 81:189–198

Cameron AE, Smith DH, Walker RL (1969) Mass spectrometry of nanogram-size samples of lead. Anal Chem 41:525–526

Campbell DA, Paul EA, Rennie OA, McCallum KJ (1967) Factors affecting the accuracy of the carbon dating method in soil humus studies. Soil Sci 104:81–85

Canalas RA, Alexander EC, Manuel OK (1968) Terrestrial abundance of noble gases. J Geophys Res 73:3331–3334

Capaldi G, Pece R (1981) On the reliability of the $^{230}Th$-$^{238}U$ dating method applied to young volcanic rocks. J Volcan Geotherm Res 11:367–372

Capaldi G, Cortini M, Pece R (1983) U and Th decay-series disequilibria in historical lavas from the Eolian Islands, Thyrrhenian Sea. Isot Geosci 1:39–55

Capaldi G, Cortini M, Pece R (1985) On the reliability of the $^{230}Th$-$^{238}U$ dating method applied to young volcanic rocks–reply. J Volcanol Geotherm Res 26:369–376

Carl C, Dill H (1985) Age of secondary uranium mineralizations in the basement rocks of northeastern Bavaria, F.R.G. Chem Geol Isot Geosci Sect 52:295–316

Carlson RW, Lugmair GW, Macdougall JD (1981) Columbia River volcanism: the question of mantle heterogeneity or crustal contamination. Geochim Cosmochim Acta 45:2483–2499

Carver EA, Anders E (1976) Fission-track ages of four meteorites. Geochim Cosmochim Acta 40:467–477

Cassignol C, Gillot PY (1982) Range and effectiveness of unspiked potassium-argon dating: experimental groundwork and applications. In: Odin GS (ed) Numerical dating in stratigraphy, Part I. Wiley, New York, pp 159–173

Castagnoli G, Lal D (1980) Solar modulation effects in terrestrial production of carbon-14. Radiocarbon 22 (II):133–158

Catanzaro EJ (1967) Triple-filament method for solid-sample lead isotope analysis. J Geophys Res 72:1325–1327

Catanzaro EJ (1968) The interpretation of zircon ages. In: Hamilton EI, Farquhar RM (eds) Radiometric dating for geologists. Wiley, New York, pp 225–258

Catanzaro EJ, Murphy TJ, Garner EL, Shields WR (1969) Absolute isotopic abundance ratio and atomic weight of terrestrial rubidium. J Res US Nat Bur Stand Sect A 73:511–516

Catchen GL (1984) Application of the equations of radioactive growth and decay to geochronological models and explicit solution of the equations by Laplace transformation. Isot Geosci 2:181–195

Cattell A, Krogh TE, Arndt NT (1984) Conflicting Sm-Nd whole-rock and U-Pb zircon ages for Archaean lavas from Newton Township Abitibi Belt, Ontario. Earth Planet Sci Lett 70:280–290

Cavazzini G (1988) Linear correlation between pairs of Rb-Sr isochron ages from coexisting metamorphic micas. Chem Geol Isot Geosci Sect 72:29–36

Cerling TE, Brown FH, Bowman JR (1985) Low-temperature alteration of volcanic glass: hydration, Na, K, $^{18}$O and Ar mobility. Chem Geol Isot Geosci Sect 52:281–293

Cerveny PF, Naeser ND, Zeitler PK, Naeser CW, Johnson NM (1988) History of Uplift and Relief of the Himalaya During the Past 18 Million Years: Evidence from Fission-Track Ages of Detrital Zircons from Sandstones of the Siwalik Group. In: Kleinspehn KL, Paola C. Frontiers in Sedimentary Geology. New Perspectives in Basin Analysis: 43–61; New York, Berlin, Heidelberg (Springer)

Chalov PI (1983) Uranium disequilibrium as an indicator of process in the hydrosphere. Water Res 9:466–479

Chalov PI, Tuzova TV, Musin YA (1964) The $U^{234}/U^{238}$ ratio in natural waters and its use in nuclear geochronology. Geochim Inter 1:402–408

Chanton JP, Martens CS, Kipphut GW, (1983) Lead-210 sediment geochronology in a changing coastal environment. Geochim Cosmochim Acta 47:1791–1804

Chapelle FH, Morris JT, McMahon PB, Zelibor Jr JL (1988) Bacterial metabolism and the $\delta^{13}$C composition of groundwater, Floridan aquifer system, South Carolina. Geology 16:117–121

Chappell J, Polach HA (1972) Some effects of partial recrystallization on $^{14}$C dating Late Pleistocene corals and molluscs. Q Res 2:244–252

Charlet J-M, Quinif Y, Dupuis Ch, Lair P (1986) A case study of thermoluminescence in uranium exploration. Uranium 2:279–285.

Chase CG (1981) Oceanic island Pb: Two-stage histories and mantle evolution. Earth Planet Sci Lett 52:277–284

Chaudhuri S, Clauer N (1986) Fluctuations of isotopic composition of strontium in seawater during the Phanerozoic Eon. Chem Geol Isot Geosci Sect 59:293–303

Chauvel C, Dupré B, Jenner GA (1985) The Sm-Nd age of Kambalda volcanics is 500 Ma too old! Earth Planet Sci Lett 74:315–324

Chen CH, Kramer SD, Allman SL, Hurst GS (1984) Selective counting of krypton atoms using resonance ionization spectroscopy. Appl Phys Lett 44:640–642

Chen JH, Tilton GR (1976) Isotopic lead investigations on the Allende carbonaceous chondrite. Geochim Cosmochim Acta 40:635–643

Chen JH, Wasserburg GJ (1979) Cm/U, Th/U, and $^{235}U/^{238}U$ in meteorites. Meteoritics 16:301

Cherdyntsev VV (1955) 3rd session of the commission of the determination of absolute ages of geologic formation. Izd Acad Nauk SSSR: 175 (in Russian)

Cherdyntsev VV (1969) Uranium. Atomizdat, Moscow, 234 pp

Cherdyntsev VV, Kuptsov VM, Kuz'mina YA, Zverev VL (1968) Radioisotopes and protactinium age of neovolcanic rocks of the Caucasus. Geokhimiya 1:77–85

Chizhov AB, Chizhova NI, Morkovkina IK, Romanov VV (1983) Tritium in permafrost and in ground ice. Proc Int Conf Permafrost 4:147–151

Chopin C, Maluski H (1982) Unconvincing evidence against the blocking temperature concept? Contrib Mineral Petrol 80:391–394

Christmann D, Sonntag C (1987) Groundwater evaporation from east-Saharian depressions by means of deuterium and oxygen-18 in soil moisture. In: Isotope Techniques in Water Resources Development: 189–204. IAEA, Vienna

Church SE, Tatsumoto M (1975) Lead isotope relations in oceanic ridge basalts from the Juan de Fuca-Gorda ridge area, N.E. Pacific Ocean. Contrib Mineral Petrol 53:253–279

Clark PA, Templer RH (1988) Dating thermoluminescence samples which exhibit anomalous fading. Archaeometry 40:19–36 (s.a. Nucl Tracks Radiat Meas 14:139–141)

Clarke WB, Kugler CW (1973) Dissolved helium in groundwater: a possible method for uranium and thorium prospecting. Econ Geol 68:243–251

Clarke WB, Jenkins WJ, Top Z (1976) Determination of tritium by mass spectrometric measurement of $^3$He. Int J Appl Rad Isot 27:515–517

Clauer N (1976) Geochimie Isotopique du strontium des milieux sedimentaire. Application à la géochronologie de la couverture du craton ouest-africain. Mem Sci Geol Strasbourg 45:256

Clauer N (1979) A new approach to Rb-Sr dating of sedimentary rocks. In: Jäger E, Hunziker JC (eds) Lectures in isotope geology. Springer, Berlin Heidelberg New York, pp 30–51

Clauer N (1982) The rubidium-strontium method applied to sediments: certitudes and uncertainties. In: Odin GS (ed) Numerical dating in stratigraphy, part I. Wiley, Chichester, pp 245–276

Clausen HB (1973) Dating of polar ice by $^{32}$Si. J Glaciol 12:411–416

Claypool GE, Holser WT, Kaplan IR, Sakai H, Zak I (1980) The age curves of sulfur and oxygen isotopes in marine sulfate and their mutual interpretation. Chem Geol 28:199–206

Cliff RA, Gray CM, Huhma H (1983) A Sm-Nd isotopic study of the South Harris Igneous Complex, the Outer Hebrides. Contrib Mineral Petrol 82:91–98

Coale KH, Bruland KW (1985) $^{234}$U:$^{238}$U disequilibrium within the California current. Limnol Oceanogr 30:22–23

Cohee GV, Glaessner MF, Hedberg HD (eds) (1978) Contributions to the geologic time scale. Stud Geol Am Assoc Petrol Geol 6:388

Coleman ML (1971) Potassium-calcium dates from pegmatitic micas. Earth Planet Sci Lett 12:399–405

Collins CB, Russel RD, Farquhar RM (1953) The maximum age of the elements and the age of the earth's crust. Can J Phys 31:402

Collinson DW (1983) Methods in Rock Magmatism and Palaeomagnestism. Techniques and Instrumentation. Chapman and Hall, London, 503 pp

Comer JB, Naeser CW, McDowell FW (1980) Fission-track ages of zircon from Jamaican bauxite and terra rossa. Econ Geol 75:117–121

Compston W, Jeffery PM (1959) Anomalous common strontium in granite. Nature (Lond) 184:1792–1793

Compston W, Kröner A (1988) Multiple zircon growth within early Archaean tonalitic gneiss from the Ancient Gneiss Complex, Swaziland. Earth Planet Sci Lett 87:13–28

Compston W, McDougall I, Wyborn D (1982) Possible two-stage $^{87}$Sr evolution in the Stockdale Rhyolite. Earth Planet Sci Lett 61:297–302

Compston W, Williams IS, Black LP (1983) Use of the ion microprobe in geological dating. BMR 82, Yearb Bureau Mineral Resour, Geol Geophys, Canberra

Compston W, Williams IS, Campbell IH, Gresham JJ (1985/86) Zircon xenocrysts from the Kambalda volcanics: age constraints and direct evidence for older continental crust below the Kambalda-Norseman greenstones. Earth Planet Sci Lett 76:299–311

Condomines M, Tanguy JC, Kieffer G, Allègre CJ (1982) Magmatic evolution of a volcano studied by $^{230}$Th-$^{238}$U disequilibrium and trace elements systematics: the Etna case. Geochim Cosmochim Acta 46:1397–1416

Condomines M, Hemond Ch, Allègre CJ (1988) U-Th-Ra radioactive disequilibria and magmatic processes. Earth Planet Sci Lett 90:243–262

Constable CC, McElhinny MW (1985) Holocene geomagnetic secular variation records from north-eastern Australian lake sediments. Geophys J 81:103–120

Cook J, Stringer CB, Currant AP, Schwarcz HP, Wintle AG (1982) A review of the chronology of the European middle Pleistocene hominid record. Yearb Phys Anthropol 25:19–65

Cordani UG, Kawashita K, Filho AT (1978) Applicability of the rubidium-strontium method to shales and related rocks. In: Cohee GV, Glaessner MF, Hedberg HD (eds) Contributions to the Geologic Times Scale. Am Assoc Petrol Geol, Stud Geol 6:93–117

Cortini M (1985) An attempt to model the timing of magma formation by means of radioactive disequilibria. Chem Geol 58:33–43

Court DJ, Goldsack RJ, Ferrari LM, Polach HA (1981) The use of carbon isotopes in identifying urban air particulate sources. Clean Air 1:6–11

Covich A, Stuiver M (1974) Changes in oxygen 18 as a measure of long-term fluctuations in tropical lake levels and molluscan populations. Limnol Oceanogr 19:682–691

Cowan GA, Haxton WC (1982) Solar neutrino production of technetium-97 and technetium-98. Science 216:51–54

Cox A (1969) Geomagnetic reversals. Science 163:237–245

Cox A (1975) The frequency of geomagnetic reversals and the symmetry of the nondipole field. Phil Trans R Soc Lond Ser A 243:67–92

Craig H (1957) Isotopic standards of carbon and oxygen and correction factors for mass-spectrometric analysis of carbon dioxide. Geochim Cosmochim Acta 12:133–149

Craig H (1961a) Isotopic variations in meteoric water. Science 133:1702–1703

Craig H (1961b) Standard of reporting concentrations of deuterium and oxygen-18 in natural waters. Science 133:1833–1834

Craig H, Lupton JE, Horibe Y (1978) A mantle helium component in circum-Pacific volcanic gases: Hakone, the Marianas and Mt. Lassen. Adv Earth Planet Sci 3:3–16

Craig LE, Smith AG, Armstrong RL (1986) A geological time scale. Terra Cognita 6:141

Crane HR (1951) Dating of relics by radiocarbon analysis. Nucleonics 9:16–23

Crawford RW, Trole J, Baxter MS, Thomson J (1985) A comparison of the particle track and alpha-spectrometric techniques in excess thorium-230 dating of eastern Atlantic pelagic sediments. J Environ Radioact 2:135–144

Creer KM (1981) Long-period geomagnetic secular variations since 12,000 yr BP. Nature (Lond) 292:208–212

Creer KM (1983) Computer synthesis of geomagnetic palaeosecular variations. Nature (Lond) 304:695–699

Creer KM, Kopper JS (1974) Paleomagnetic dating of cave paintings in Tito Bustillo cave, Asturias, Spain. Science 186:348–350

Creer KM, Tucholka P (1982a) Construction of type curves of geomagnetic secular variation for dating lake sediments from east central North America. Can J Earth Sci 19:1106–1115

Creer KM, Tucholka P (1982b) Secular variation as recorded in lake sediments: a discussion of North America and European results. Phil Trans R Soc Lond A 306:87–102

Creer KM, Tucholka P (1983) On the current state of lake sediment paleomagnetic research. Geophys J 74:223–238

Creer KM, Valencio DA, Simitro AM, Tucholka P, Vilas JFA (1983) Geomagnetic secular variations 0-14,000 yr BP as recorded by lake sediments in Argentina. Geophys J 74:199–221

Cressy JP Jr, Bogard DD (1976) On the calculation of cosmic-ray exposure ages of stone meteorites. Geochim Cosmochim Acta 40:749–762

Criss RE, Lanphere MA, Taylor HP Jr (1982) Effects of regional uplift, deformation, and meteoric-hydrothermal metamorphism on K-Ar ages of biotites in the southern half of the Idaho Batholith. J Geophys Res 87 (B8):7029–7046

Cronin JE, Boaz NT, Stringer CB, Rak Y (1981) Tempo and mode in hominid evolution. Nature (Lond) 292:113–120

Crough ST (1983) Apatite fission-track dating of erosion in the eastern Andes, Bolivia. Earth Plan Sci Lett 64:396–397

Crozaz G (1981) Fission tracks and cooling rates of meteorites. Proc Earth Planet Sci 90:383–358

Crozaz G, Langway CC (1966) Dating of Greenland firn ice cores with $^{210}$Pb. Earth Planet Sci Lett 1:194–196

Cumming GL, Richards JR (1975) Ore lead isotope ratios in a continuously changing earth. Earth Planet Sci Lett 28:155–171

Cumming GL, Eckstrand OR, Peredery WV (1982) Geochronologic interpretations of Pb isotope ratios in nickel sulfides of the Thompson Belt, Manitoba. Can J Earth Sci 19:2306–2324

Cunningham R, Burnett WC (1985) Amino acid biogeochemistry and dating of offshore Peru/Chile phosphorites. Geochim Cosmchim Acta 49:1413–1419

Currie LA (1968) Limits for qualitative detection and quantitative determination. Anal Chem 40:586–589

Currie LA (1972) The limit of precision in nuclear and analytical chemistry. Nucl Instrum Methods 100:387–395

Currie LA (ed) (1982) Nuclear and Chemical Dating Techniques. Interpreting the Environmental Record. Am Chem Soc Symp Ser 176:516 ACS, Washington DC

Currie LA, Klouda GA, Continetti RE, Kaplan IR, Wong WW, Dzubay TG, Stevens RK (1983) On the origin of carbonaceous particles in American cities: results of radiocarbon "dating" and chemical characterization. Radiocarbon 25 (2):603–614

Currie LA, Klouda GA, Voorhees KJ (1984) Atmospheric carbon: the importance of accelerator mass spectrometry. Nucl Instrum Methods 233 (B5):371–379

Currie LA, Stafford TW, Sheffield AE, Klouda GA, Wise SA, Fletcher RA, Donahue DJ, Jull AJT, Linick TW (1989) Microchemical and molecular dating. Radiocarbon 31(3) (in press)

Czamanske GK, Lanphere MA, Erd RC, Blake MC Jr (1978) Age measurements of potassium-bearing sulfide minerals by the $^{40}$Ar/$^{39}$Ar technique. Earth Planet Sci Lett 40:107–110

Dallmeyer RD (1979) $^{40}$Ar/$^{39}$Ar dating: principles, techniques, and applications in orogenic terranes. In: Jäger E, Hunziker JC (eds) Lectures in isotope geology. Springer, Berlin Heidelberg New York, pp 77–104

Dalrymple GB, Lanphere MA (1969) Potassium-argon dating. Freeman, San Francisco, 258 pp

Dalrymple GB, Lanphere MA (1971) $^{40}$Ar/$^{39}$Ar technique of K-Ar dating: a comparison with the conventional technique. Earth Planet Sci Lett 12:300–308

Dalrymple GB, Lanphere MA (1974) $^{40}$Ar/$^{39}$Ar age spectra of some undisturbed terrestrial samples. Geochim Cosmochim Acta 38:715–738

Dalton P (1972) A new method of dating bone. MASCA Newslett 8:1–2

Damon PE (1970) A theory of "real" K-Ar clocks. Eclogae Geol Helv 63:69–76

Damon PE, Linick TW (1986) Geomagnetic-heliomagnetic modulation of atmospheric radiocarbon production. Radiocarbon 28, 2A:266–278

Daniels F, Boyd CA, Saunders DF (1953) Thermoluminescence as a research tool. Science 117:343–349

Dansgaard W (1964) Stable isotopes in precipitation. Tellus 16:436–468

Dansgaard W (1981) Ice core studies: dating the past to find the future. Nature (Lond) 290:360–361

Dansgaard W (1985) Greenland ice core studies. Palaeogeogr Palaeoclimat Palaeoecol 50:185–187

Dansgaard W, Johnsen SJ, Møller J, Langway CC Jr (1969) One thousand centuries of climatic record from Camp Century of the Greenland ice sheet. Science 166:377–381

Dansgaard W, Johnsen SJ, Clausen HB, Langway CC Jr (1971) 3. Climatic record revealed by the Camp Century ice core. In: Turekian KK (ed) The Late Cenozoic Glacial Ages. Yale University, New Haven, pp 37–56

Dansgaard W, Clausen HB, Gundstrup N, Hammer CV, Johnson SF, Kristindottir PM, Rech N (1982) A new Greenland ice-core. Science 218:1273–1277

Davis DW (1982) Optimum linear regression and error estimation applied to U-Pb data. Can J Earth Sci 19:2141–2149

Davis P, Smith J, Kukla GJ, Opdyke ND (1977) Paleomagnetic study at a nuclear plant site near Bakersfield, California. Q Res 7:380–397

Davis R, Schaeffer OA (1955) Chlorine-36 in nature. Ann NY Acad Sci 62:107–121

Davis RB, Hess CT, Norton SA, Hanson DW, Hoagland KD, Anderson DS (1984) $^{137}$Cs and $^{210}$Pb dating of sediments from soft-water lakes in New England (U.S.A.) and Scandinavia, a failure of $^{137}$Cs dating. Chem Geol 44:151–185

Davis SN, De Wiest RJM (1966) Hydrogeology. Wiley, New York, 453 pp

De Atley SP (1980) Radiocarbon dating of ceramic materials: progress and prospects. Radiocarbon 22(2):987–993

Debenham NC (1983) Reliability of thermoluminescence dating of stalagmitic calcite. Nature (Lond) 304:154–156

Debenham NC, Aitken MJ (1984) TL dating of stalagmitic calcite. Archeometry 26:155–170

Debuyst R, Dejehet F, Grün R, Apers D, de Cannière P (1984) Possibility of ESR-dating without determination of the annual dose. J Radioanal Nucl Chem Lett 86:399–410

Deevey ES Jr, Gross MS, Hurchinson GE, Kraybil HL (1954) The natural $^{14}$C contents of materials from hard-water lakes. Proc Nat Acad Sci Wash 40:285–288

Degens ET, Kempe S, Spitzy A (1984) Carbon dioxide: a biogeochemical portrait. In: Hutzinger O (ed) The Handbook of Environmental Chemistry, vol I, Part C. Springer, Berlin Heidelberg New York Tokyo, pp 127–215

de Jong AFM, Mook WG (1982) An anomalous SUESS effect above Europe. Nature (Lond) 298:641–644

de Jong AFM, Mook WG, Becker B (1979) Confirmation of the SUESS wiggles: 3200–3700 BC. Nature (Lond) 280:48–49

de Jong E, Begg CBM, Kachanoski RG (1983) Estimates of soil erosion and deposition for some Saskatchewan soils. Can J Soil Sci 63:607–617

Delaloye M (1979) The total lead method. In: Jäger E, Hunziker JC (eds) Lectures in isotope geology. Springer, Berlin Heidelberg New York, pp 132–133

Delaune RD, Patrick WH Jr, Buresh RJ (1978) Sedimentation rates determined by [137]Cs dating in a rapidly accreting salt marsh. Nature (Lond) 275:532–533

Del Moro A, Puxeddu M, Radicati di Brozolo F, Villa IM (1982) Rb-Sr and K-Ar ages on minerals at temperatures of 300–400°C from deep wells in the Larderello geothermal field (Italy). Contrib Mineral Petrol 81:340–349

Deloule E, Allègre CJ, Doe B (1986) Lead and sulfur isotope microstratigraphy in galena crystals from Mississippi Valley-type deposits. Econ Geol 81:1307–1321

de Niro MJ, Epstein S (1981) Isotopic composition of cellulose from aquatic organisms. Geochim Cosmochim Acta 45:1885–1894

Denninger E (1971) The use of paper chromatography to determine the age of albuminous binders and its application to rock paintings. Suppl Afric J Sci 2:81–84

DePaolo DJ (1981a) Neodymium isotopes in the Colorado Front Range and crust-mantle evolution in the Proterozoic. Nature (Lond) 291:193–196

DePaolo DJ (1981b) Nd isotopic studies: Some new perspectives on earth structure and evolution. EOS 62:137–140

DePaolo DJ (1988) Neodymium isotope geochemistry. Springer, Berlin Heidelberg, New York, Tokyo

DePaolo DJ, Wasserburg GJ (1976a) Nd isotopic variations and petrogenetic models. Geophys Res Lett 3:249–252

DePaolo DJ, Wasserburg GJ (1976b) Inferences about magma sources and mantle structure from variations of [143]Nd/[144]Nd. Geophys Res Lett 3:743–746

DePaolo DJ, Wasserburg GJ (1979a) Petrogenetic mixing models and Nd-Sr isotopic patterns. Geochim Cosmochim Acta 43:615–627

DePaolo DJ, Wasserburg GJ (1979b) Sm-Nd age of the Stillwater complex and the mantle evolution curve for neodymium. Geochim Cosmochim Acta 43:999–1008

DePaolo DJ, Manton WI, Grew ES, Halpern M (1982) Sm-Nd, Rb-Sr and U-Th-Pb systematics of granulite facies rocks from Fyfe Hills, Enderby Land, Antarctica. Nature (Lond) 298:614–618

DePaolo DJ, Kyte FT, Marshall BD, O'Neil JR, Smit J (1983) Rb-Sr, Sm-Nd, K-Ca, O, and H isotopic study of Cretaceous-Tertiary boundary sediments, Caravaca, Spain: evidence for an oceanic impact site. Earth Planet Sci Lett 64:356–373

Dermott SF (ed) (1978) The Origin of the Solar System. Wiley, New York

Deutsch S, Hirschberg D, Picciotto E (1956) Etude quantitative des halos pleochroiques. Bull Soc Belge Geol Paleontol Hydrol 65:267–281

de Vries Hl (1958) Variation in concentration of radiocarbon with time and location on earth. Kon Ned Akad Wet Proc Ser B 61:94–102

Dickson BL, Davidson MR (1985) Interpretation of [234]U/[238]U activity ratios in groundwaters. Chem Geol 58:83–88

Dodson MH (1973) Closure temperature in cooling geochronological and petrological systems. Contrib Mineral Petrol 40:259–274

Dodson MH (1976) Kinetic processes and thermal history of slowly cooling solids. Nature (Lond) 259:551–553

Dodson MH (1979) Theory of cooling ages. In: Jäger E, Hunziker JC (eds) Lectures in isotope geology. Springer, Berlin Heidelberg New York, pp 194–202

Doe BR (1970) Lead isotopes. Minerals, rocks, and inorganic materials 3. Springer, Berlin Heidelberg New York

Doe BR (1983) The past is the key to the future. Geochim Cosmochim Acta 47:1341–1354

Doe BR, Stacey JS (1974) The application of lead isotopes to the problem of ore genesis and ore prospect evaluation: a review. Econ Geol 69:757–776

Doe BR, Zartman RE (1979) Plumbotectonics: the Phanerozoics. In: Barnes HL (ed) Geochemistry of hydrothermal ore deposits, 2nd edn. Wiley, New York, pp 22–70

Dominik J, Förstner U, Mangini A, Reineck H-E (1978) [210]Pb and [137]Cs chronology of heavy metal pollution in a sediment core from the German Bight (North Sea). Senckenbergiana Mar 10:213–227

Dorn RI, Bamforth DB, Cahill TA, Dohrenwend JC, Turrin BD, Donahue DJ, Jull AJT, Long A, Macko ME, Weil EB, Whitley DS, Zabel TH (1986) Cation-ratio and acceleration radiocarbon dating of rock varnish on Mojave artifacts and landforms. Science 231:830–833

Downes H (1984) Sr and Nd isotope geochemistry of coexisting alkaline magma series, Cantal, Massif Central, France. Earth Planet Sci Lett 69:321–334

Drozd RJ, Podosek FA (1976) Primordial $^{129}$Xe in meteorites. Earth Planet Sci Lett 31:15–30

Drozd RJ, Hohenberg CM, Morgan CJ (1974) Heavy rare gases from Rabbit Lake (Canada) and the Oklo mine (Gabon): natural spontaneous chain reactions in old uranium deposits. Earth Planet Sci Lett 23:28–32

Drozd RJ, Morgan CJ, Podosek FA, Poupeau G, Shirck JR, Taylor GJ (1977) $^{244}$Pu in the solar system. Astrophys J 212:567–580

Druffel EM (1982) Banded corals: Changes in oceanic carbon-14 during the Little Ice Age. Science 218:13–19

Druffel EM, Mok HYI (1983) Time history of human gallstones: application of the post-bomb radiocarbon signal. Radiocarbon 25 (2):629–636

Duckworth HE, Barber RC, Venkatasubramanian VS (1986) Mass Spectrometry (2nd ed). Cambridge Monographs on Physics. Cambridge University Press, Cambridge London New York New Rochell Melbourne Sydney. 337 pp

Dunbar RB, Wefer G (1984) Stable isotope fractionation in benthic foraminifera from the Peruvian continental margin. Mar Geol 59:215–225

Dungworth G (1976) Optical configuration and the racemisation of amino acids in sediments and in fossils – a review. Chem Geol 17:135–153

Dunning GR, Krogh TE, Pedersen RB (1986) U/Pb zircon ages of Appalachian-Caledonian ophiolites. Terra Cognita 6:155

Duplessy J-C, Lalou C, Vinot AL (1970) Differential isotopic fractionation in benthic foraminifera and paleotemperatures reassessed. Science 168:250–251

Dupont LM, Brenninkmeijer CAM (1984) Palaeobotanic and isotopic analysis of late Subboreal and early Subatlantic peat from Engbertsdijksveen VII, The Netherlands. Rev Palaeobot Palynol 41:241–271

Dymond J (1969) Age determinations of deep-sea sediments: a comparison of three methods. Earth Planet Sci Lett 6:9–14

Dymond J, Cobler R, Gordon L, Biscaye P, Mathieu G (1983) $^{226}$Ra and $^{226}$Rn contents of Galapagos Rift hydrothermal waters – the importance of low-temperature interactions with crustal rocks. Earth Planet Sci Lett 64:417–429

Easterbrook DJ (ed) (1988) Dating Quaternary Sediments. Geol Soc Am Special paper 227: 165pp; Boulder

Eberhardt P, Eugster O, Geiss J, Marti K (1966) Rare gas measurements in 30 stone meteorites. Z Naturforsch 21a:414–416

Eberhardt P, Geiss J, Grögler N, Krähenbühl U, Mörgeli M, Stettler A (1971) Potassium-argon age of Apollo 11 rock 10003. Earth Planet Sci Lett 11:245–247

Eckstein D, Baillie MGL, Egger H (1984) Dendrochronological Dating. Handb Archaeol 2:55. Eur Sci Found Strasbourg

Edwards RL, Chen JH, Wasserburg GJ (1986/87) $^{238}$U-$^{234}$U-$^{230}$Th-$^{232}$Th systematics and the precice measurement of time over the past 500,000 years. Earth Planet Sci Lett 81:175–192

Edwards RL, Taylor FW, Wasserburg GJ (1988) Dating earthquakes with high-precision thorium–230 ages of very young corals. Earth Planet Sci Letters 90: 371–381

Edwards TWD, Fritz P (1986) Assessing meteoric water composition and relative humidity from $^{18}$O and $^2$H in wood cellulose. Paleoclimatic implications for southern Ontario, Canada. Appl Geochem 1:715–723

Egboka BCE (1985) Appropriate monitoring techniques using bomb tritium and other geochemical parameters in hydrogeological investigations. Hydrol Sci J Sci Hydrol 30:207–224

Ehmann WD, Kohman TB (1958) Cosmic-ray induced radioactivity in meteorites - I. Chemical and radiometric procedures for aluminium, beryllium, and cobalt. II. $^{26}$Al, $^{10}$Be, $^{60}$Co in aerolites, siderites, and tektites. Geochem Cosmichim Acta 14:340–363 and 364–379

Eicher U, Siegenthaler U, Wegmüller S (1981) Pollen and oxygen isotope analyses on late and post-glacial sediments of the Tourbiere de Chirens (Dauphin, France). Q Res 15 (2):160–170

Eisenbud M, Bennett B, Blanco RE, Compere EL, Goldberg E, Jacobs DG, Koranda J, Moghissi AA, Rust J, Soldat JK (1979) Tritium in the environment – NCRP Report No. 62. In: Behaviour of Tritium in the Environment: 585–588. IAEA, Vienna

El-Daoushy F, Tolonen K (1984) Lead-210 and heavy metal contents in dated ombrotrophic peat-hummocks from Finland. Nucl Instrum Methods 223:329–399

El-Daoushy F, Tolonen K, Rosenberg R (1982) Lead-210 and moss-increment dating of two Finnish sphagnum hummocks. Nature (Lond) 296:429–431

Elitzsch C, Pernicka E, Wagner GA (1983) Thermoluminescence dating of archaeometallurgical slags. PACT 9:271–286

Ellins KK (1988) The application of $^{222}$Rn in measuring groundwater discharge to the Martha Brae River, Jamaica. Symp Tropical Hydrol 2nd Carribean Islands Water Res Congr AWRA (Am Water Res Ass) Techn Publ Ser TPS-85-1:64–68

Elmore D (1986) $^{36}$Cl and $^{129}$I geochemistry. Terra Cognita 6:121

Elmore D, Gove HE, Ferraro R, Kilius LR, Lee HW, Chang KH, Beukens RP, Litherland AE, Russo CJ, Purser KH, Murrell MT, Finkel RC (1980) Determination of $^{129}$I using tandem accelerator mass spectrometry. Nature (Lond) 286:138–139

Elmore D, Tubbs LE, Newman D, Ma XZ, Finkel R, Nishiizumi K, Beer J, Oeschger H, Andree M (1982) $^{36}$Cl bomb pulse measured in a shallow ice core from Dye 3, Greenland. Nature (Lond) 300:735–737

Elmore D, Conard NJ, Kubik PW, Gove HE, Wahlen M, Beer J, Suter M (1987) $^{36}$Cl and $^{10}$Be profiles in Greenland ice. Dating and production rate variations. Nucl Instrum Methods Phys Res B 29:207–217

Elsinger RJ, Moore WS (1983) $^{224}$Ra, $^{228}$Ra, and $^{226}$Ra in Winyah Bay and Delaware Bay. Earth Planet Sci Lett 64:430–436

Emiliani C (1954) Temperatures of Pacific bottom waters and polar superficial waters during the Tertiary. Science 119:853–855

Emiliani C (1955) Pleistocene temperatures. J Geol 63:538–573

Emiliani C (1966) Isotope palaeotemperatures. Science 154:851–857

Emmermann R (1977) A petrogenetic model for the origin and evolution of the Hercynian granite series of the Schwarzwald. N Jahrb Mineral Abh 128:219–253

Engel MH, Hare PE (1985) Gas liquid chromatographic separation of amino acids and their derivates. In: Garrett GC (ed) Chemistry and biochemistry of the amino acids. Chapman and Hall, London, pp 462–479

Engels JC, Ingamells CO (1970) Effect of sample inhomogeneity in K-Ar dating. Geochim Cosmochim Acta 34:1007–1017

Englert P, Herr W (1978) A study on exposure ages of chondrites based on spallogenic $^{53}$Mn. Geochim Cosmochim Acta 42:1635–1643

Englert P, Theis S, Michel R, Tuniz C, Moniot RK, Vajda S, Kruse TH, Pal DK, Herzog GF (1984) Production of $^7$Be, $^{22}$Na, $^{24}$Na, and $^{10}$Be from Alina $4\pi$-irradiated meteorite model. Nucl Instrum Methods B5:415–419

Ennis P, Noltmann EA, Hare PE, Slota PJ Jr, Payen LA, Prior CA, Taylor RE (1986) Use of AMS $^{14}$C analysis in the study of problems in asparctic acid racemization-deduced age estimates on bone. Radiocarbon 28 (2A):539–546

Ensley RA, Verosub KL (1982) A magnetostratigraphic study of the sediments of the Ridge Basin, southern California and its tectonic and sedimentologic implications. Earth Planet Sci Lett 59:192–207

Epstein S, Buchsbaum R, Lowenstam HA, Urey HC (1953) Revised carbonate-water isotopic temperature scale. Geol Soc Am Bull 64:1315–1325

Eriksson E (1958) The possible use of tritium for estimating groundwater storage. Tellus 10:472–478

Erlenkeuser H, Suess G, Willkomm H (1974) Industrialization affects heavy metal and carbon isotope concentrations in recent Baltic Sea sediments. Geochim Cosmochim Acta 38:823–842

Esenov SE, Egizbayeva KE, Kalinin SK, Fayn EE (1970) Radiogenic osmium in rhenium-bearing ores. Geokhimiya 5:610–615

Eugster O (1988) Cosmic-ray production rates for $^3$He, $^{21}$Ne, $^{38}$Ar, $^{83}$Kr, and $^{126}$Xe in chondrites based on $^{81}$Kr-Kr exposure ages. Geochim Cosmochim Acta 52:1649–1662

Eugster O, Eberhardt P, Geiss J (1967a) $^{81}$Kr in meteorites and $^{81}$Kr radiation ages. Earth Planet Sci Lett 2:77–82

Eugster O, Eberhardt P, Geiss J (1967b) The isotopic composition of krypton in unequilibrated and gas-rich chondrites. Earth Planet Sci Lett 3:249–257

Evans ME (1983) Do the earth's magnetic poles move? Naturwissenschaften 70:485–494

Everst DA (1964) The chemistry of beryllium. In: Robinson PL (ed) Topics in inorganic and general chemistry. A collection of Monographics. Elsevier, Amsterdam, 150 pp

Evin J (1983) Materials of terrestrial origin used for radiocarbon dating. In: Mook WG, Waterbolk HT (eds) [14]C and archaeology. PACT 8:235–275

Evin J, Marechal J, Pachiaudi C, Puissegur JJ (1980) Conditions involved in dating terrestrial shells. Radiocarbon 22 (II):545–555

Fabryka-Martin J, Bentley H, Elmore D, Airey PL (1985) Natural iodine-129 as an environmental tracer. Geochem Cocmochim Acta 49:337–347

Fabryka-Martin J, Davis SN, Elmore D (1987) Application of [129]I and [36]Cl in hydrology. Nucl Instrum Methods Phys Res B 29:361–371

Fagg BEB, Fleming SF (1970) Thermoluminescent dating of a terracotta of the NOK culture, Nigeria. Archaeometry 12:53–55

Fanale FP (1971) A case for catastrophic early degassing of the Earth. Chem Geol 8:79–105

Fanale FP, Kulp JL (1962) The helium method and the age of the Cornwall, Pennsylvania magnetite ore. Econ Geol 57:735–746

Fanale FP, Schaeffer OA (1965) Helium-uranium ratios for Pleistocene and Tertiary fossil aragonites. Science 149:312–316

Fantidis J, Ehhalt DH (1970) Variations of the carbon and oxygen isotopic composition in stalagmites and stalactites: evidence of non-equilibrium isotopic fractionation. Earth Planet Sci Lett 10:136–144

Fassett JD, Moore LJ, Travis JC, DeVoe JR (1985) Laser resonance ionization mass spectrometry. Science 230:262–267

Faure G (1977) Principles of Isotope Geology, 1st ed. Wiley, New York, 461 pp

Faure G (1982) The marine-strontium geochronometer. In: Odin GS (ed) Numerical dating in stratigraphy, part I. Wiley, Chichester, pp 73–79

Faure G (1986) Principles of Isotope Geology, 2nd edn. Wiley, New York, 589 pp

Faure G, Powell JL (1972) Strontium isotope geology. Springer, Berlin Heidelberg New York

Fehn U, Holdren GR, Elmore D, Brunelle T, Teng R, Kubik PW (1986a) Determination of natural and anthropogenic [129]I in marine sediments. Geophys Res Lett 13(1):137–139

Fehn U, Teng R, Elmore D, Kubik PW (1986b) Isotopic composition of osmium in terrestrial samples determined by accelerator mass spectrometry. Nature (Lond) 323:707–710

Fehn U, Tullai S, Teng RTD, Elmore D, Kubik PW (1987) Determination of [129]I in heavy residues of two crude oils. Nucli Instrum Methods Phys Res B 29:380–382

Feige Y, Oltman BG, Kastner J (1968) Production rates of neutrons in soils due to natural radioactivity. J Geophys Res 73:3135–3142

Ferguson CW, Huber B, Suess HE (1966) Determination of the age of Swiss lake dwellings as an example of dendrochronologically calibrated radiocarbon dating. Z Naturforsch 21a:1173–1177

Ferreira MP, Macedo R, Costa V, Reynolds JH, Riley JE Jr, Rowe MW (1975) Rare-gas dating, II. Attempted uranium-helium dating of young volcanic rocks from the Madaira Archipelago. Earth Planet Sci Lett 25:142–150

Finkel RC (1981) Uranium concentrations and [234]U/[238]U activity ratios in fault-associated groundwater as possible earthquake precursors. Geophys Res Lett 8:453–456

Finkel RC, Langway CC Jr (1985) Global and local influences on the chemical composition of snowfall at Dye 3, Greenland: the record between 10 ka B.P. and 40 ka B.P. Earth Planet Sci Lett 73:196–206

Fireman EL (1986) Uranium series dating of Allan Hills ice. J Geophys Res 91(B4):D539–544

Fischer P, Noren J, Lossing A, Odelius H (1986) Quantitative SIMS of prehistoric teeth. Archaeometry Conf, Athens

Fisher DA, Koerner RM, Paterson WSB, Dansgaard W, Gundestrup N, Reeh N (1983) Effect of wind scouring on climatic records from ice-core oxygen-isotope profiles. Nature (Lond) 301:205–209

Fisher DE (1978) Terrestrial potassium-argon abundances as limits to models of atmospheric evolution. In: Alexander EC (ed) Terrestrial rare gases. Japan Scientific Press, Tokyo, pp 173–183

Fitch FJ, Forster SC, Miller JA (1974) Geological time scale. Rep Prog Phys 37:1433–1496

Fitch WM (1977) Molecular Evolutionary Clocks. In: Ayala FJ (ed) Molecular Evolution. Sinauer Ass, Sunderland Mass, pp 160–178

Fleck RJ, Coleman RG, Cornwall HR, Greenwood WR, Hadley DG, Schmidt DL, Prinz WC,

Ratt JC (1976) Geochronology of the Arabian Shield, western Saudi Arabia: K-Ar results. Geol Soc Am Bull 87:9–21

Fleischer RL (1975) Advances in fission track dating. World Archaeol 7:136–150

Fleischer RL (1982) Alpha-recoil damage and solution effects in minerals: uranium isotopic disequilibrium and radon release. Geochim Cosmochim Acta 46:2191–2201

Fleischer RL, Price PB (1964) Decay constant for spontaneous fission of $^{238}$U. Phys Rev 133(1B):63–64

Fleischer RL, Price PB, Walker RM (1965) Effects of temperature, pressure, and ionization on the formation and stability of fission tracks in minerals and glasses. J Geophys Res 70:1497–1502

Fleischer RL, Price PB, Walker RM (1975) Nuclear Tracks in Solids. Principles and Applications. Univ California Press, Los Angeles, 605 pp

Fleishman DG, Gorin VD, Gritschenko ZG (1987) Cosmogenic $^{22}$Na and dating of natural fresh waters. In: Povinec P (ed) Low-level Counting and Spectrometry VEDA, Bratislava, pp 123–126

Fleming SJ (1970) Thermoluminescence dating: refinement of the quartz inclusion method. Archaeometry 12:53–55

Fleming SJ (1972) Thermoluminescence authenticity testing of ancient ceramics using radiation-sensitivity changes in quartz. Naturwissenschaften 59:145–151

Fleming SJ (1973) The pre-dose technique: a new thermoluminescent dating method. Archaeometry 15:13–30

Fleming SJ (1976) Dating in Archaeology. A Guide to Scientific Techniques, London

Fleming SJ, Stoneham D (1973) The subtraction technique of thermoluminescent dating. Archaeometry 15:229–238

Fleming SJ, Moss HM, Joseph A (1970) Thermoluminescence authenticity testing of some "six dynasties" figures. Archaeometry 12:57–65

Fletcher IR, Rosman KJR (1982) Precise determination of initial $\varepsilon_{Nd}$ from Sm-Nd isochron data. Geochim Cosmochim Acta 46:1983–1987

Fleyshman DG, Kanevskiy YP, Gritchenko ZG (1975) Age determination on natural waters with cosmic-ray and man-made $^{22}$Na. Geochemistry Intern 12(1):p 201

Florkowski T, Morawska L, Rozanski K (1988) Natural production of radionuclides in geological formations. Nucl Geophys 2:1–14

Förstel H, Hützen H (1983) Oxygen isotope ratios in German groundwater. Nature (Lond) 304:614–616

Foland KA, Linder JS, Laskowski TE, Grant NK (1984) $^{40}$Ar/$^{39}$Ar dating of glauconites: measured $^{39}$Ar recoil loss from well-crystallized samples. Isot Geosci 2:241–264

Folgheraiter G (1899) Sur les variations seculaires de l'inclinaison magnetique dans l'antique. Arch Sci Phys Nat 8:5–16

Folk RL, Valastro S Jr (1979) Dating of lime mortar by $^{14}$C. In: Berger R, Suess HE (eds) Radiocarbon Dating. Univ California Press, Los Angeles, pp 721–732

Fontes JC (1980) Environmental isotopes in groundwater hydrology. In: Fritz P, Fontes JC (eds) Handbook of Environmental Isotope Geochemistry, vol 1, The Terrestrial Environment A. Elsevier, Amsterdam, pp 75–140

Fontes JC (1985) Some considerations on groundwater dating using environmental isotopes. In: Hydrogeology in the Service of Man. Mem IAH, Cambridge, pp 118–154

Fontes JC, Garnier J-M (1979) Determination of the initial $^{14}$C activity of the total dissolved carbon. A review of the existing models and a new approach. Water Resour Res 15:399–413

Forman SL (1989) Applications and limitations of thermoluminescence to date Quaternary sediments. Quatern Int 1:47–59

François LM, Gerard J-C (1986) A numerical model of the evolution of ocean sulfate and sedimentary sulfur during the last 800 million years. Geochim Cosmochim Acta 50:2289–2302

Franke HW (1951) Altersbestimmung von Kalzit-Konkretionent mit radioaktivem Kohlenstoff. Naturwissenschaften 22:527

Franke HW (1966) Zur Entnahme von Sinterproben für Radiocarbondatierungen. Höhle 17:92–95

Franke T, Fröhlich K, Gellermann R, Hebert D (1986) $^{32}$Si in precipitation of Freiberg, GDR. J Radioanal Nucl Chem Lett 103(1):11–18

Frape SK, Fritz P (1982) The chemistry and isotope composition of saline groundwaters from the Sudbury Basin, Ontario. Can J Earth Sci 19:645–661

Freer R (1981) Diffusion in silicate minerals and glasses: a data digest and guide to the literature. Contrib Mineral Petrol 76:440–454

Freundel M, Schultz L, Reedy RC (1986) Terrestrial $^{81}$Kr-Kr ages of Antarctic meteorites. Geochim Cosmochim Acta 50:2663–2673

Friedman I (1983) Paleoclimate evidence from stable isotopes. In: Wright HE (ed) Late Quaternary Environments of the United States, vol I. The Late Pleistocene (Porter SC ed). Longman, London, pp 385–389

Friedman I, Long W (1976) Hydration rate of obsidian. Science 191:347–352

Friedman I, Obradovich J (1981) Obsidian hydration dating of vulcanic events. Q Res 16:37–47

Friedman I, Smith RL (1960) A new dating method using obsidian: Part I, the development of the method. Am Antiq 25:476–493

Fritz P, Fontes JC (eds) (1980) Handbook of Environmental Isotope Geochemistry, vol 1. The Terrestrial Environment A. Elsevier, Amsterdam, 546 pp

Fritz P, Fontes JC (eds) (1986) Handbook of Environmental Isotope Geochemistry, vol 2. The Terrestrial Environment B. Elsevier, Amsterdam, 557 pp

Fröhlich K, Gellermann R (1986) On the potential use of uranium isotopes for groundwater dating. Isot Geosci Sect 65:67–77

Fröhlich K, Milde G, Hebert D, Kater R (1974) Methodische und meßtechnische Erkenntnisse über die Anwendung von Tritium, $^{14}$C- und $^{32}$Si- Bestimmugen für hydrogeologische Aufgaben. Z Angew Geol 20:16–21

Fröhlich K, Gellermann R, Hebert D (1984) Uranium isotopes in a sandstone aquifer: Interpretation of data and implications for groundwater dating. In: Isotope Hydrology 1983:447–466. IAEA, Vienna

Fröhlich K, Franke T, Gellermann R, Hebert D, Jordan H (1987) Silicon-32 in different aquifer types and implications for groundwater dating. In: Isotope Techniques in Water Resources Development: 149–163. IAEA, Vienna

Fryer BJ, Taylor RP (1984) Sm-Nd direct dating of the Collins Bay hydrothermal uranium deposit, Saskatchewan. Geology 12:479–482

Fuhrmann U, Lippolt HJ (1985) Excess argon and dating of Quaternary Eifel volcanism. I. The Schellkopf phonolite. N Jahrb Geol Paläont Mh, pp 484–497

Fuhrmann U, Lippolt HJ (1986) Excess argon and dating of Quaternary Eifel volcanism: II. Phonolitic and foiditic rocks near Rieden, East Eifel, FRG. N Jahrb Geol Paläont Abh 172:1–19

Fukuchi T, Imai N, Shimokawa K (1986) ESR dating of fault movement using various defects centres in quartz; the case in the western South Fossa Magma, Japan. Earth Planet Sci Lett 78:121–128

Fullagar PD, Ragland PC (1975) Chemical weathering and Rb-Sr whole rock ages. Geochim Cosmochim Acta 39:1245–1252

Fullagar PD, Lemmon RE, Ragland PC (1971) Petrochemical and geochronological studies of plutonic rocks in the southern Appalachians: part 1. The Salisbury Pluton. Geol Soc Am Bull 82:409–416

Gabasio M, Evin J, Arnal GB, Andrieux P (1986) Origins of carbon in potsherds. Radiocarbon 28(2A):711–718

Gaffney JS, Premuzic ET, Manowitz B (1980) On the usefulness of sulfur isotope ratios in crude oil correlations. Geochim Cosmochim Acta 44:135–139

Gale NH (1972) Uranium-lead systematics in Lunar basalts. Earth Planet Sci Lett 17:65–78

Gale NH, Mussett AE (1973) Episodic uranium-lead models and the interpretation of variations in the isotopic composition of lead in rocks. Rev Geophys Space Phys 11:37–86

Gale NH, Beckinsale RD, Wadge AJ (1979) Rb-Sr whole rock dating of acid rocks. Geochim J 13:27–79

Galliker D, Hugentobler E, Hahn E (1970) Spontane Kernspaltung von 238-U und 241-Am. Helv Phys Acta 43:593–606

Ganapathy R (1982) Evidence for a major meteorite impact on the earth 34 million years ago: implication for Eocene extinctions. Science 216:885–886

Gancarz A, Tera F, Wasserburg G (1975) 3.62 AE Amitsoq gneiss from West Greenland and a 4.45 AE age of the Earth. Geol Soc Am 1975 Annual Meeting, Abstr with Programs 7:1081–1082

Ganguly J, Ruiz J (1986/87) Time-temperature relation of mineral isochrons: a thermodynamic model, and illustrative examples for the Rb-Sr system. Earth Planet Sci Lett 81:333–348

Ganssen G (1981) Isotopic analysis of foraminifera shells: interference from chemical treatment. Palaeogr Palaeoclimatol Palaeoecol 33:271–276

Ganssen G, Sarnthein M (1983) Stable-isotope composition of foraminifers: the surface and bottom water record of coastal upwelling. In: Suess E, Thiede J (eds) Coastal Upwelling. Plenum Press, New York, pp 99–121

Gariepy C, Allègre CJ, Lajoije J (1984) U-Pb systematics in single zircons from the Pontiac sediments, Abitibi greenstone belt. Can J Earth Sci 21:1296–1304

Garner EL, Murphy TJ, Gramlich JW, Paulsen PJ, Barnes IL (1976) Absolute isotopic abundance ratios and the atomic weight of a reference sample of potassium. J Res US Natl Bur Stand, Sect A, 79A:713–725

Garwin L (1984) Fission track dating comes of age. New Scientist 1418:21

Gascoyne M (1981) A simple method of uranium extraction from carbonate groundwater and its application to $^{234}U/^{238}U$ disequilibrium studies. J Geochem Explor 14:199–207

Gascoyne M, Schwarcz HP, Ford DC (1980) A palaeotemperature record for the mid-Wisconsin in Vancouver Island. Nature (Lond) 285:474–476

Gascoyne M, Schwarcz HP, Ford DC (1983) Uranium-series ages of speleothem from Northwest England: correlation with Quaternary climate. Phil Trans R Soc Lond B 301:143–164

Gat JR, Carmi I (1970) Evolution of the isotopic composition of atmospheric waters in the Mediterranean Sea area. J Geophys Res 75:3039–3048

Gat J, Gonfiantini R (eds) (1981) Stable Isotope Hydrology. Deuterium and Oxygen-18 in the Water Cycle. IAEA, Vienna, Tech Rep Ser 210:337

Gaudette HE, Hurley PM, Fairbairn HW, Lajmi T (1977) Source area for the Numidian flysch of Tunisia determined by U-Pb zircon ages. 21st progress report 1974-1976, MIT Geochronol Lab, pp 35–41

Gebauer D, Grünenfelder M (1974) Rb-Sr whole-rock dating of late diagenetic to anchimetamorphic, Palaeozoic sediments in southern France (Montagne Noire). Contrib Mineral Petrol 47:113–130

Gebauer D, Grünenfelder M (1976) U-Pb zircon and Rb-Sr whole-rock dating of low-grade metasediments. Example: Montagne Noire (Southern France). Contrib Mineral Petrol 59:13–32

Gebauer D, Grünenfelder M (1978) U-Pb zircon and Rb-Sr mineral dating of eclogites and their country rocks. Example: Münchberg Gneiss Massive, NE-Bavaria. Earth Planet Sci Lett 42:35–44

Gebauer D, Grünenfelder M (1979) U-Th-Pb dating of minerals. In: Jäger E, Hunziker JC (eds) Lectures in isotope geology. Springer, Berlin, Heidelberg, New York, pp 105–131

Gebauer D, Bernard-Griffiths J, Grünenfelder M (1981) U-Pb zircon and monazite dating of a mafic-ultramafic complex and its country rocks. Example: Sauviat-sur-Vige, French Central Massif. Contrib Mineral Petrol 76:292–300

Geiss J, Oeschger H, Schwarz U (1962) The history of cosmic radiation as revealed by isotopic changes in the meteorites and on the earth. Space Sci Rev 1:197–223

Gellermann R, Gast M (1983) Ra-Rn-Datierung der Quellwässer von Bad Brambach. Z Physiother Jg 35:129–135

Gellermann R, Börner I, Franke T, Fröhlich K (1988) Preparation of water samples for $^{32}Si$ determination. Isotopenpraxis 24:114–117

Gentner W, Zähringer J (1955) Argon-und Heliumbestimmungen in Eisenmeteoriten. Z Naturforsch 10A:498–499

Gentner W, Glass BP, Storzer D, Wagner GA (1970) Fission track ages of deposition of deep-sea microtectites. Science 168:359–361

Gerling EK, Iskanderova AD (1966) Isotopic composition of lead from carbonate rocks of different age. Akad Nauk SSSR Dokl 170/4:905–907

Gerling EK, Shukolyukov YA (1959) Isotope composition and content of xenon in uranium minerals. Radiokhimiya 1:212–222

Gerling EK, Tolstykhin IN, Shukolyukov YA, Nesmelova ZN, Azbel IY (1967) Argon isotopes and helium in natural hydrocarbon gases. Geochim Int 4:498–506

Geyh MA (1967) Experience gathered in the construction of low-level counters. In: Radioactive Dating and Low-Level Techniques: 575–589. IAEA, Vienna

Geyh MA (1969) Messungen der Tritium-Konzentration von Salzlaugen. Kali Steinsalz 5:208

Geyh MA (1970) Zeitliche Abgrenzung von Klimaänderungen mit $^{14}$C-Daten von Kalksinter und organischen Substanzen. Beih Geol Jahrb 98:15–22

Geyh MA (1972a) Basic studies in hydrology and $^{14}$C und $^3$H measurements. Proc 24th Int Geol Congr 11:227–234

Geyh MA (1972b) On the determination of the initial $^{14}$C content in groundwater. In: Rafter TA, Grant-Taylor T (eds) Proc 8th Int Conf Radiocarbon Dating: Wellington, New Zealand, pp B369–380

Geyh MA (1979) $^{14}$C Routine dating of marine sediments. In: Berger R, Suess HE (eds) Radiocarbon Dating. Univ California Press, Los Angeles, pp 470–491

Geyh MA (1980a) Holocene sea-level history: case study of the statistical evaluation of $^{14}$C dates. Radiocarbon 22 (III):695–704

Geyh MA (1980b) Einführung in die Methoden der physikalischen und chemischen Altersbestimmung. Wissenschaftliche Buchgesellschaft, Darmstadt, 276 pp

Geyh MA (1980c) Hydrogeologic interpretation of the $^{14}$C content of groundwater—a status report. Fisika 12:87–106

Geyh MA (1983) Physikalische und Chemische Datierungsmethoden in der Quartärforschung. Clausthaler Tektonische Hefte 19: pp 163. Pilger, Clausthal

Geyh MA, de Maret P (1982) Histogram evaluation of $^{14}$C dates applied to the first complete iron age sequence from West Central Africa. Archaeometry 24:158–163

Geyh MA, Franke HW (1970) Zur Wachstumsgeschwindigkeit von Stalagmiten. Atompraxis 16:1–3

Geyh MA, Hennig GJ (1986) Multiple dating of a long flowstone profile. Radiocarbon 28(2A):503–509

Geyh MA, Rohde P (1972) Weichselian chronostratigraphy, $^{14}$C dating and statistics. Proc 24th Int Geol Congr 1 Z:26–36

Geyh MA, Benzler JH, Roeschmann G (1971a) Problems of dating Pleistocene and Holocene soils by radiometric methods. In: Yaalon DH (ed) Nature, Origin and Dating of Palaeosols. Israel Univ Press, Jerusalem, pp 63–75

Geyh MA, Merkt J, Müller H (1971b) Sediment-, Pollen- und Isotopenanalysen an jahreszeitlich geschichteten Ablagerungen im zentralen Teil des Schleinsees. Arch Hydrobiol 69:366–399

Geyh MA, Krumbein WE, Kudrass H-R (1974) Unreliable $^{14}$C dating of longstored deep-sea sediments due to bacterial activity. Mar Geol 17:45–50

Geyh MA, Roeschmann G, Wijmstra TA, Middeldorp AA (1983) The unreliability of $^{14}$C dates obtained from buried sandy podzols. Radiocarbon 25:409–416

Geyh MA, Backhaus G, Andres G, Rudolph J, Rath HK (1984) Isotope study on the Keuper sandstone aquifer with a leaky cover layer. In: Isotope Hydrology 1983:499–514. IAEA, Vienna

Gill ED (1974) Carbon-14 and uranium/thorium check on suggested interstadial high sealevel around 30,000 BP. Search 5:211

Gillespie AR, Huneke JC, Wasserburg GJ (1982) An assessment of $Ar^{40}$-$Ar^{39}$ dating of incompletely degassed xenoliths. J Geophys Res 87:9247–9257

Gillespie R (1984) Radiocarbon User's Handbook. Oxford Univ Comm Archaeol, Monogr 3:36. Oxonion Rewley, Oxford

Gillot P-Y (1985) K-Ar Upper-Pleistocene dating. Terra Cognita 5:234

Gillot P-Y, Cornette Y (1986) The Cassignol technique for potassium-argon dating, precision and accuracy: examples from the late Pleistocene to recent volcanics from Southern Italy. Chem Geol (Isot Geosci Sect) 59:205–222

Glagola BG, Phillips GW, Marlow KW, Myers LT, Omohundro RJ (1984) Low level tritium detection using accelerator mass spectrometry. Nucl Instrum Methods 233 (B5):221–225

Gleadow AJW (1981) Fission-track dating methods: what are the real alternatives? Nucl Tracks 5:3–14

Godwin CI, Sinclair Aj (1982) Average lead isotope growth curves for shale-hosted zinc-lead deposits, Canadian Cordillera. Econ Geol 77:675–690

Goedicke C (1985) TL dating: a new novel of differential dating. Nucl Tracks 10:811–816

Goedicke C, Slusallek K, Kubelik M (1983) Some marginal notes on TL dating of brick structures. PACT 9:245–248

Göksu HY (1978) The TL age determination of fossil human footprints. Arch Phys 10:455–462

Göksu Y, Fremlin JH, Irvin HI, Fryxell R (1974) Age determination of burned flint by a thermoluminescence method. Science 183:651–654

Göksu-Ögelman HY (1986) Thermoluminescent dating: a review of applications to burned flint. In: Sieveking G, Hart MB (eds) The scientific study of flint and chert. Univ Press, Cambridge, pp 263–267

Göksu-Ögelman Y, Kapur S (1982) Thermoluminescence reveals weathering stages in basaltic rocks. Nature (Lond) 296:231–232

Goel PS, Kohman TP (1963) Cosmic-ray exposure history of meteorites from cosmogenic $Cl^{36}$. In: Radioactive Dating: 413–432. IAEA, Vienna

Goldberg ED (1963) Geochronology with lead-210. In: Radioactive Dating: 121–131. IAEA, Vienna

Goldberg ED, Koide M (1958) Io/Th chronology in deep-sea sediments of the Pacific. Science 128:1003

Goldberg ED, Gamble E, Griffin JJ, Koide M (1977) Pollution history of Narragansett Bay as recorded in its sediments. Estuar Coast Mar Sci 5:549–561

Goldich SS, Mudrey MG Jr (1972) Dilatancy model for discordant U-Pb zircon ages. In: Contributions to recent geochemistry and analytical chemistry (Vinogradov volume) Nauka, Moscow, pp 415–418

Goodfriend GA (1987a) Radiocarbon anomalies in shell carbonate of land snails from semi-arid areas. Radiocarbon 29:159–167

Goodfriend GA (1987b) Evaluation of amino-acid racemization/epimerization dating using radiocarbon-dated fossil land snails. Geology 15:698–700

Goodfriend GA (1987c) Chronostratigraphic studies of sediments in the Negev Desert, using amino acid epimerization analysis of land snail shells. Q Res 28:374–392

Goodfriend GA (1988) Mid-Holocene rainfall in the Negev Desert from $^{13}C$ of land snail shell organic matter. Nature 333:757–760

Goodfriend GA, Hood DG (1983) Carbon isotope analysis of land snail shells: implications for carbon sources and radiocarbon dating. Radiocarbon 25 (3):810–830

Goodman M, Weirs ML, Czelusniak J (1982) Molecular evolution above the species level branching pattern, rates, and mechanisms. Syst Zool 314:376–399

Gottfried D, Jaffe HW, Senftle Fe (1959) Evaluation of the lead-alpha (Larsen) method for determining ages of igneous rocks. US Geol Surv Bull 1097-A

Gove HE, Litherland AE, Elmore D (eds) (1987) Accelerator Mass Spectrometry. Nucl Instrum Methods Phys Res B 29 (1,2): pp 455. North-Holland, Amsterdam

Grabczak J, Zuber A, Maloszewski P, Rozanski K, Weiss W, Sliwka I (1982) New mathematical models for the interpretation of environmental tracers in groundwaters and the combined use of tritium, C-14, Kr-85, He-3, and Neon-11 for groundwater studies. Beitr Geol Schweiz Hydrol 28(II):395–406

Graf TH, Vogt S, Bonani G, Herpers U, Signer P, Suter M, Wieler R, Wölfli W (1987) Depth dependence of $^{10}Be$ and $^{26}Al$ production rates in the iron meteorite Grant. Nucl Instrum Methods Phys Res B 29:262–265

Grauert B (1974) U-Pb systematics in heterogeneous zircon populations from the Precambrian basement of the Maryland Piedmont. Earth Planet Sci Lett 23:238–248

Grauert B, Hofmann A (1973) Effects of progressive regional metamorphism and magma formation on U-Pb systems in zircon. 3rd European Colloquim of Geochronology, Cosmochronology and Isotope Geology (ECOG III), Oxford (abstract)

Grauert B, Seitz MG, Soptrajanova G (1974) Uranium and lead gain of detrital zircon studied by isotopic analyses and fission-track mapping. Earth Planet Sci Lett 21:389–399

Gray J, Se Jong Song (1984) Climatic implications of natural variations of D/H ratios in tree ring cellulose. Earth Planet Sci Lett 70:129–138

Green PF (1985a) Comparison of zeta calibration baselines of fission-track dating of apatite, zircon and sphene. Chem Geol Isot Geosci Sect 58:1–22

Green PF (1985b) In defence of the external detector method for fission track dating. Geol Mag 122:73–75

Green PF (1988) The relationship between track shortening and fission track age reduction in apatite: combined influences of inherent instability, annealing anisotropy, length bias and system calibration. Earth Planet Sci Lett 89:335–352

Grootes PM (1978) Carbon-14 time scale extended. Comparison of chronologies. Science 200:11–15

Grün R (1984) ESR dating without determination of annual dose: a first application on dating molluscs shells. Proc ESR dating and Dosimetry. IONICS, Tokyo, pp 115–123

Grün R (1988) Die ESR-Altersbestimmungsmethode. Springer Berlin Heidelberg New York Tokyo, 132 pp

Grün R (1989) Electron spin resonance (ESR) dating. Quatern Int. 1:65–109

Grün R, de Cannière P (1984) ESR dating: problems encountered in the evaluation of the naturally accumulated dose (AD) of secondary carbonates. J Radioanal Nucl Chem Lett 85:213–226

Grün R, Hentzsch B (1987) Problems encountered in ESR dating of spring-deposited travertines. 5th Specialist Seminar on TL and ESR Dating. Kings College, Cambridge 6-10 July 1987

Grün R, Invernati C (1985) Uranium accumulation in teeth and its effect on ESR dating—a detailed study of mammoth tooth. Nucl Tracks 10:869–877

Grün R, MacDonald PDM (1989) Non-linear fitting of TL/ESR dose-response curves. Appl Radiat Isot 40:1077–1080

Grün R, Schwarcz P, Chadam J (1988) ESR dating of tooth enamel: coupled correction for U-uptake and U-series disequilibrium. Nucl Tracks Radiat Meas 14:237–241

Grünenfelder M, Stern TW (1960) Das Zirkon-Alter des Bergeller Massivs. Schweiz Min Petr Mitt 40:253–259

Guérin G, Valladas G (1980) Thermoluminescence dating of volcanic plagioclases. Nature (Lond) 286:697–699

Guibert P, Bechtel F, Dubourg R, Schvoerer M (1985) Gamma-thermoluminescence dating (Γ-TL)-III: checking the homogeneity of the structure by $\gamma$-spectrometry. Nucl Tracks 10:655–662

Guichard F, Reyss J-L, Yokoyama Y (1978) Growth rate of manganese nodule measured with $^{10}$Be and $^{26}$Al. Nature (Lond) 272:155–156

Gulson BL (1977) Isotopic and geochemical studies on crustal effects in the genesis of the Woodlawn Pb-Zn-Cu deposit. Contrib Mineral Petrol 65:227–242

Gulson BL, Vaasjoki M, Carr GR (1986) Geochronology in deeply weathered terrains using lead-lead isochrons. Chem Geol Isot Geosci Sec 59:273–282

Gupta SK, Polach HA (1985) Radiocarbon dating practices at ANU. Handbook. ANU, Canberra, 173 pp

Gurfinkel DM (1987) Comparative study of the radiocarbon dating of different bone collagen preparations. Radiocarbon 29(1):45–52

Gutierrez-Negrin L, Lopez-Martinez A, Becazar-Garcia M (1984) Application of dating for searching geothermic sources. Nucl Tracks 8:385–389

Guttman J, Kronfeld J (1982) Tracing interaquifer connections in the Kefar Uriyya region (Israel), using natural uranium isotopes. J Hydrol 55:145–150

Gvirtzman H, Magaritz M (1986) Investigation of water movement in the unsaturated zone under an irrigated area using environmental tritium. Water Resour Res 22:635–642

Haeberli W, Schotterer U, Wagenbach D, Haeberli-Schwitter H, Bortenschlager S (1983) Accumulation characteristics on a cold, high-alpine firn saddle from a snow-pit study on Colle Gnifetti, Monte Rosa, Swiss Alps. J Glaciol 29:260–271

Hänny R, Grauert B, Soptrajanova G (1975) Paleozoic migmatites affected by high-grade Tertiary metamorphism in the Central Alps (Valle Bodengo, Italy), a geochronological study. Contrib Mineral Petrol 51:173–196

Hahn O, Walling E (1938) Über die Möglichkeit geologischer Altersbestimmungen rubidium-haltiger Minerale und Gesteine. Z Anorg Allg Chem 236:78–82

Hahn O, Strassman F, Mattauch J, Ewald H (1943) Geologische Altersbestimmungen mit der Strontiummethode. Chem Z 67:55–56

Hainebach K, Kazanas D, Schramm DN (1978) A consistent age for the universe. Geol Surv Open-File Rep 78–701:159–162

Hall CM, York D (1984) The applicability of $^{40}$Ar/$^{39}$Ar dating to young volcanics. In: Mahaney WC (ed) Quaternary dating methods. Elsevier, Amsterdam, pp 67–74

Hall CM, Walter RC, Westgate JA, York D (1984) Geochronology, stratigraphy, and geochemistry of Cindery Tuff in Pliocene hominid-bearing sediments of the Middle Awash, Ethiopia. Nature (Lond) 308:26–31

Halliday AN (1978) $^{40}Ar$-$^{39}Ar$ step-heating studies of clay concentrates from Irish orebodies. Geochim Cosmochim Acta 42:1851–1858

Halliday AN, Mitchell JG (1983) K-Ar ages of clay concentrates from Irish orebodies and their bearing on the timing of mineralization. Trans R Soc Edinburgh, Earth Sci 74:1–14

Hamer AN, Robbins BJ (1960) A search for variations in the natural abundance of uranium-235. Geochim Cosmochim Acta 19:143–145

Hamilton EI (1965) Applied Geochronology. Academic Press, London New York, 267 pp

Hamilton PJ, O'Nions RK, Evensen NM (1977) Sm-Nd dating of Archaean basic and ultrabasic volcanics. Earth Planet Sci Lett 36:263–268

Hamilton PJ, O'Nions RK, Evensen NM, Bridgwater D, Allaart JH (1978) SM-ND isotopic investigations of Isua supracrustals and implications for mantle evolution. Nature (Lond) 272:41–43

Hammer CU (1989) Dating by physical and chemical seasonal variations and reference horizons. In: Oeschger H, Langway Jr CC (eds) The Environmental Records in Glaciers and Ice Sheets: 85–98, Wiley, New York

Hammer CU, Clausen HB, Dansgaard W, Gundestrup N, Jihnsen SJ, Reeh N (1978) Dating of Greenland ice cores by flow models, isotopes, volcanic debris, and continental dust. J Glac 20:3–26

Hammer CU, Clausen HB, Tauber H (1986) Ice-core dating of the Pleistocene/Holocene boundary applied to a calibration of the $^{14}C$ time scale. Radiocarbon 28 (2A):284–291

Hammerschmidt K, Wagner GA, Wagner M (1984) Radiometric dating on research drill core Urach III: a contribution to its geothermal history. J. Geophys 54:97–105

Hampel W, Takagi J. Sakamoto K, Tanaka S (1975) Measurement of myon-induced $^{26}Al$ in terrestrial silicate rock. J Geophys Res 80:3757–3760

Handbook of Chemistry and Physics. CRC press, London, ca 2500 pp (annually new adition)

Hanson GN, Catanzaro EJ, Anderson DH (1971) U-Pb ages for sphene in a contact metamorphic zone. Earth Planet Sci Lett 12:231–237

Hardy EP, Volchock HL, Livingstone HD, Burke JC (1980) Time pattern of off-site plutonium deposition from rocky flats plant by lake sediment analyses. Environ Int 4:21–30

Hare PE, Mitterer RM (1967) Nonprotein amino acids in fossil shells. Carnegie Inst Wash Yearb 65:362–364

Hare PE, St John PA, Engel MH (1985) Ion exchange separation of amino acids. In: Barrett GC (ed) Chemistry and Biochemistry of the Amino Acids. Chapman and Hall, London, pp 415–425

Harkness DD, Harrison AF, Bacon B (1986) The temporal distribution of 'bomb' $^{14}C$ in a fossil soil. Radiocarbon 28 (2A):328–337

Harland WB, Smith AG, Wilcock B (eds) (1964) The Phanerozoic time-scale. Quarterly J Geol Soc London 120:458 pp

Harland WB, Cox AV, Llewellyn PG, Pickton CAG, Smith AG, Walters R (1982) A Geologic Time Scale. Cambridge Univ Press, Cambridge, 131pp

Harmon RS, Schwarcz HP (1981) Changes of $^2H$ and $^{18}O$ enrichment of meteoric water and Pleistocene glaciation. Nature (Lond) 290:125–128

Harmon RS, Thompson P, Schwarcz HP, Ford DC (1978) Late Pleistocene paleoclimates of North America as inferred from stable isotope studies from speleothems. Q. Res 9:54–70

Harmon RS, Mitterer RM, Kriausakul N, Land LS, Schwarcz HP, Garrett P, Larson GJ, Vacher HL, Rowe M (1983) U-series and amino-acid racemization geochronology of Bermuda: implications for eustatic sea-level fluctuation over the past 250,000 years. Palaeogeogr Palaeoclimatol Palaeoecol 44:41–70

Harper CT (1970) Graphical solutions to the problem of radiogenic argon-40 loss from metamorphic minerals. Eclogae Geol Helv 63:119–140

Harris WB (1982) Rubidium-strontium glaucony ages, southeastern Atlantic Coastal Plain, USA. In: Odin GS (ed) Numerical dating in stratigraphy, part I. Wiley, Chichester, pp 593–606

Harrison TM (1981) Diffusion of $^{40}Ar$ in hornblende. Contrib Mineral Petrol 78:324–331

Harrison TM, Bé K (1983) $^{40}Ar/^{39}Ar$ age spectrum analysis of detrital microclines from the southern San Joaquin Basin, California: an approach to determining the thermal evolution of sedimentary basins. Earth Planet Sci Lett 64:244–256

Harrison TM, McDougall I (1980) Investigations of an intrusive contact, northwest Nelson, New Zealand. -II. Diffusion of radiogenic and excess $^{40}$Ar in hornblende revealed by $^{40}$Ar/$^{39}$Ar age spectrum analysis. Geochim Cosmochim Acta 44:2005–2020

Hart SR (1964) The petrology and isotopic-mineral age relations of a contact zone in the Front Range, Colorado. J Geol 72:493–525

Hart SR, Davis GL, Steiger RH, Tilton GR (1968) A comparison of the isotopic mineral age variations and petrologic changes induced by contact metamorphism. In: Hamilton EI, Farquhar RM (eds) Radiometric dating for geologists. Interscience, New York, pp 73–110

Hart SR, Shimizu N, Sverjensky DA (1981) Lead isotope zoning in galena: An ion microprobe study of a galena crystal from the Buick mine, southeast Missouri. Econ Geol 76:1873–1876

Hasegawa M, Kishino H, Yano Ta (1985) Dating of human-ape splitting by a molecular clock of mitochondrial DNA. J Mol Evol 22:160–174

Haskell EH (ed) (1983) Beta dose-rate determination: preliminary results from an interlaboratory comparison of techniques. PACT 9:77–85

Haskin LA, Wildeman TR, Frey FA, Collins KA, Keedy CR, Haskin MA (1966) Rare earth in sediments. J Geophys Res 71:6091–6105

Hassan AA, Termine JD, Haynes CV Jr (1977) Mineralogical studies on bone apatite and their implications for radiocarbon dating. Radiocarbon 19 (III):364–374

Hawkesworth CJ, Vollmer R (1979) Crustal contamination versus enriched mantle: $^{143}$Nd/$^{144}$Nd and $^{87}$Sr/$^{86}$Sr evidence from the Italian Volcanics. Contrib Mineral Petrol 69:151–165

Hayatsu A, Carmichael CM (1970) K-Ar isochron method and initial argon ratios. Earth Planet Sci Lett 8:71–76

Hays JD, Imbrie, J, Shackleton NJ (1976) Variations in the earth's orbit: pacemaker of the ice ages. Science 194:1121–1132

Hearty PJ, Aharon P (1988) Amino acid chronostratigraphy of late Quaternary coral reefs: Huon Peninsula, New Guinea, and the Great Barriers Reef, Australia. Geology 16:579–583

Heath MJ (1981) The use of radon in streams as a guide to uranium distribution in south-west England. In: Proc USSHER Soc 5 (2):245

Heaton THE (1984) Rates and sources of $^4$He accumulation in groundwater. Hydrol Sci J 29:29–47

Hebeda EH, Freundel M, Schultz L (1985) Uranium and spontaneous fissiogenic xenon and krypton systematics of zircons from the Botnavatn igneous complex, SW Norway. Terra Cognita 5:280

Hedges REM (1981) Radiocarbon dating with an accelerator: Review and preview. Archaeometry 23 (1):3–18

Hedges REM, Law IA (1989) The radiocarbon dating of bone. Appl Geochem 4:249–253

Heinemeier J, Hornshøj P, Nielsen HL, Rud N, Thomsen MS (1987) Accelerator mass spectrometry applied to $^{22,24}$Na, $^{31,32}$Si, and $^{14}$C. Nucl Instrum Methods Phys Res B29:110–123

Heizer RF, Cook SF (1952) Fluorine and other chemical tests of some North American Human and fossil bones. Am J Phys Anthropol 10:289–303

Helfman PM, Bada JL (1975) Aspartic acid racemization in tooth enamel from living humans. Proc Nat Acad Sci USA 72:2891–2894

Helfman PM, Bada JL (1976) Aspartic acid racemisation in dentine as a measure of ageing. Nature (Lond) 262:279–281

Hellman KN, Lippolt HJ (1981) Calibration of the Middle Triassic time scale by conventional K-Ar and $^{40}$Ar/$^{39}$Ar dating of alkali feldspars. J Geophys. 50:73–88

Hemond C, Condomines M (1985) On the reliability of the $^{230}$Th-$^{234}$U dating method applied to young volcanic rocks–discussion. J Volcanol Geotherm Res 26:365–376

Henderson P (ed) (1984) Rare earth element geochemistry. Elsevier, Amsterdam

Hendy CH (1971) The isotope chemistry of speleothems I. Geochim Cosmochim Acta 35:801–824

Hennig GJ, Grün R (1983) ESR dating in Quaternary geology. Q Sci Rev 2:157–238

Hennig GJ, Grün R, Brunnacker K (1983) Speleothems, travertines and paleoclimates. Q. Res 20:1–29

Hennig GJ, Geyh MA, Grün R (1985) The first inter-laboratory ESR comparison project phase II: evaluation of equivalent doses (ED) or calcites. Nucl Tracks 10:945–952

Henning W, Bell WA, Billquist PJ, Glagola BG, Kutschera W, Liu Z, Lucas HF, Paul M, Rehm

KE, Yntema JL (1987) Calcium-41 concentration in terrestrial materials: prospects for dating Pleistocene samples. Science 236:725–727

Henzel N, Strebel O (1967) Modelluntersuchungen über die Tiefenverlagerung von Fallout in verschiedenen Böden. Z Geophys 33:33–47

Herpers U, Englert P (1983) $^{26}$Al production rates and $^{53}$Mn/$^{26}$Al production rate ratios in non-Antarctic chondrites and their application to bombardment histories. J Geophys Res 88 Suppl: B 312–318

Herpers U, Herr W, Wölfle R (1967) Determination of cosmic-ray produced nuclides $^{53}$Mn, $^{45}$Sc and $^{26}$Al in meteorites by neutron activation and gamma coincidence spectroscopy. In: Radioactive Dating and Methods of Low-Level Counting: 199–205. IAEA, Vienna

Herr W, Merz E (1955) Eine neue Methode zur Altersbestimmung von Rhenium-haltigen Mineralen mittels Neutronenaktivierung. Z Naturforsch 10a:613–615

Herr W, Merz E, Eberhardt P, Signer P (1958) Zur Bestimmung der $\beta$-Halbwertszeit von $^{176}$Lu durch den Nachweis von radiogenem $^{176}$Hf. Z Naturforschung 13a:268–283

Herr W, Hoffmeister W, Langhoff J (1960) Die Bestimmung von Rhenium und Osmium in Eisenmeteoriten durch Neutronenaktivierung. Z Naturforsch 15a:99–102

Herr W, Wölfe R, Eberhardt P, Koppe E (1967) Development and recent application of the Re/Os-dating method. In: Radioactive dating and methods of low-level counting: 499–508. IAEA, Vienna

Herr W, Herpers U, Englert P (1981) $^{53}$Mn and $^{26}$Al in observed chondrite falls with high exposure ages. Meteoritics 16:324–325

Herrman AG, Potts MJ, Knake D (1974) Geochemistry of the rare earth elements in spilites from the oceanic and continental crust. Contrib Mineral Petrol 44:1–16

Herron MM (1982) Glaciochemical dating techniques. In: Currie LA (ed) Nuclear and Chemical Dating Techniques. ACS, Washington DC, ACS Symp Ser 176:303–318

Herterich K, Sarnthein M (1984) Brunhes time scale: tuning by rates of calcium carbonate dissolution and cross spectral analyses with solar insolation. In: Berger A et al. (eds) Milankovitch and Climate, Part I. Reidel, Amsterdam, pp 447–466

Herweijer JC, van Luiju GA, Appelo CAJ (1985) Calibration of a mass transport model using environmental tritium. J Hydrol 78:1–17

Hess JC, Lippolt HJ, Wirth R (1987) Interpretation of $^{40}$Ar/$^{39}$Ar spectra of biotites: evidence from hydrothermal degassing experiments and TEM studies. Chem Geol Isot Geosci Sec 66:137–149

Heumann KG, Kubassek E, Schwabenbauer W, Stadler I (1979) Analytisches Verfahren zur K/Ca-Altersbestimmung an geologischen Proben. Fresenius Z Anal Chem 297:35–43

Heusser G, Ouyang Z, Kirsten T, Herpers U, Englert P (1985) Conditions of the cosmic ray exposure of the Jilin chondrite. Earth Planet Sci Lett 72:263–272

Heye D (1975) Wachstumsverhältnisse von Manganknollen. Geol Jahrb E 5:3–22

Heymann D (1977) The Inert Gases. Phys Chem Earth 10:45–55

Hickman MH, Glassley WE (1984) The role of metamorphic fluid transport in the Rb-Sr isotopic resetting of shear zones: evidence from Nordre Stromfjord, West Greenland. Contrib Mineral Petrol 87:265–281

Hillaire-Marcel C, Carro O, Casanova J (1986) $^{14}$C and U/Th dating of Pleistocene and Holocene stromatolites from East African palaeolakes. Q Res 25:312–329

Hille P (1979) An open system model for uranium series dating. Earth Planet Sci Lett 42:138–142

Hille P, Mais K, Rabeder G, Vavra N, Wild E (1981) Über Aminosäuren- und Stickstoff/Fluor-Datierung fossiler Knochen aus österreichischen Höhlen. Höhle 32:74–91

Hillman GC, Robins GV, Oduwole D, Sales KD, McNeil DAC (1983) Determination of thermal histories of archeological cereal grains with electron spin resonance spectroscopy. Science 222:1235–1236

Hinthorne JR, Andersen CA, Conrad RL, Lovering JF (1979) Single-grain $^{207}$Pb/$^{206}$Pb and U/Pb age determinations with a 10-$\mu$m spatial resolution using the ion microprobe mass analyzer (IMMA). Chem Geol 25:271–303

Hirose K, Sugimura Y (1984) Excess $^{288}$Th in the airborne dust: an indicator of continental dust from East Asian deserts. Earth Planet Sci Lett 70:110–114

Hochman HBM, Ypma PJM (1984) Thermoluminescence as a tool in uranium exploration. J Geochem Explor 22:313–331

Hofmann AW (1979) Rb-Sr dating of thin slabs: an imperfect method to determine the age of metamorphism. In: Jäger E, Hunziker JC (eds) Lectures in isotope geology. Springer, Berlin Heidelberg New York, pp 27–29

Hofmann AW, Grauert B (1973) Effect of regional metamorphism on whole-rock Rb-Sr systems in sediments. Carnegie Inst Washington Yearb 72:299–302

Hofmann A, Köhler H (1973) Whole rock Rb-Sr ages of anatectic gneisses from the Schwarzwald, SW Germany. N Jahrb Mineral Abh 119:163–187

Hofmann AW, Mahoney JW, Giletti BJ (1974) K-Ar and Rb-Sr data on detrital and postdepositional history of Pennsylvanian clay from Ohio and Pennsylvania. Geol Soc Am Bull 85:639–644

Hofmann A, White WM (1982) Mantle plumes from ancient oceanic crust. Earth Planet Sci Lett 57:421–436

Hohenberg CM (1969) Radioisotopes and the history of nucleosynthesis in the galaxy. Science 166:212–215

Holmes A (1911) The association of lead with uranium in rock minerals and its application to the measurement of geological time. Proc Soc A 85:248–256

Holmes A (1932) The origin of igneous rocks. Geol Mag 69:543–558

Holmes A (1937) The age of the Earth. Nelson Classics, London,

Holmes A (1946) An estimate of the age of the Earth. Nature (Lond) 157:680 pp

Holmes A (1947) The construction of a geological time scale. Trans Geol Soc Glasgow 21:117–152

Holmes A (1949) Lead isotopes and the age of the Earth. Nature (Lond) 163:453–456

Holmes A (1959) A revised geological time scale. Trans Edinbourgh Geol Soc 17:183–216

Holmes CW, Martin EA (1978) $^{226}$Ra chronology of Gulf of Mexico slope sediments. Geol Surv Open-File Rep 48–701:184–187

Holser WT (1977) Catastrophic chemical events in the history of the ocean. Nature (Lond) 267:403–408

Holser WT, Kaplan IR (1966) Isotope geochemistry of sedimentary sulfates. Chem Geol 1:93–135

Holser WT, Magaritz M, Wright J (1986) Chemical and isotopic variations in the world ocean during phanerozoic time. Sciences, vol 8. In: Walliser O (ed) Global Bio-Events. Lecture Notes in Earth Sciences, vol, 8 Springer, Berlin Heidelberg New York Tokyo, pp 63–74

Honda M, Kurita K, Hamano Y, Ozima M (1982) Experimental studies of He and Ar degassing during rock fracturing. Earth Planet Sci Lett 59:429–436

Hoppes DD (1984) Basic radionuclide measurements at the U.S. National Bureau of Standards. Environ Int 10:99–107

Houtermans FG (1946) The isotope ratios in natural lead and the age of uranium. Naturwissenschaften 33:185–186

Houtermans FG (1960) Die Blei-Methoden der geologischen Altersbestimmung. Geol Rundsch 49:168–196

Houtermans FG, Liener A (1966) Thermoluminescence of meteorites. J Geophys Res 71:3387–3396

Hsü KJ (1980) A scenario for the terminal Cretaceous event. Init. Reports DSDP 73:755–763

Huang WH, Walker RM (1967) Fossil alpha-particle recoil tracks; a new method of age determination. Science 155:1103–1106

Huang WH, Maurette M, Walker RM (1967) Observation of fossild-particle recoil tracks and their implications for dating measurements. In: Radioactive Dating and Methods of Low-level Counting: 415–429. IAEA, Vienna

Hudson B, Hohenberg CM, Kennedy BM, Podosek FA (1982) $^{244}$Pu in the early solar system. In: Lunar and planetary science XIII. Lunar Planet Inst, Houston, pp 346–347

Hübner H, Kowski P, Hermichen W-D, Richter W, Schütze H (1979) Regional and temporal variations of deuterium in the precipitation and atmospheric moisture of Central Europe. In: Isotope Hydrology 1978 I: 289–305. IAEA, Vienna

Hütt G, Jaek I (1989a) Infrared photoluminescence (PL) dating of sediments: modification of the physical model. Equipment and some dating results. In: Synopsis from a workshop on Long and Short Range Limits in Luminescence Dating. Occas Paper 9:18–22; Research Lab Archaeology and History of Art, Oxford University, Oxford

Hütt G, Jeck J (1989b) Dating accuracy from laboratory reconstruction of paleodose. Appl Radiat Isot 40:1057–1061

Hütt G, Jaek I, Tchonka J (1988) Optical dating: K feldspars optical response stimulation spectra. Quat Sci Rev 7 (3/4):381–385

Hughes MK, Kelly PM, Pilcher JR, La Marche VC Jr (1982) Climate from tree rings. Univ Press, Cambridge, 223 pp

Huntley DJ (1985) On the zeroing of the thermoluminescence of sediments. Phys Chem Mineral 12:122–127

Huntley DJ, Johnson HP (1976) Thermoluminescence as a potential means of dating siliceous ocean sediments. Can J Earth Sci 13:593–596

Huntley DJ, Godfrey-Smith DI, Thewalt MLW (1985) Optical dating of sediments. Nature (Lond) 313:105–107

Hunziker JC (1979) Potassium Argon Dating. In: Jäger E, Hunziker JC (eds) Lectures in isotope geology: Springer, Berlin Heidelberg New York, pp 52–76

Hurford AJ, Green PF (1982) A user's guide to fission track dating calibration. Earth Planet Sci Lett 59:343–354

Hurford AJ, Green PF (1983) The zeta age calibration of fission-track dating. Isot Geosci 1:285–317

Hurford AJ, Jäger E, Ten Cate JAM (1986) Dating Young Sediments. COOP Tech Secr, (Proc Workshop Beijing, China, Sept 1985), 393 pp

Hurst GS, Kramer SD, Lehmann BE (1982) Resonance ionization spectroscopy for low-level counting. In: Currie LA (ed) Nuclear and Chemical Dating. ACS, Washington DC, ACS Symp Ser No 176:149–158

Hurst GS, Payne HG, Kramer SD, Chen CH, Phillips RC, Allman SL, Alton GD, Dabbs JWT, Willis RD, Lehmann BE (1985) Method for counting noble gas atoms with isotopic selectivity. Rep Pros Phys 48:1333–1370

Hussain N, Krishnaswami S (1980) U-238 series radioactive disequilibrium in groundwaters: implications to the origin of excess U-234 and fate of reactive pollutants. Geochim Cosmochim Acta 44:1287–1291

Hwang FSJ (1970) Thermoluminescence dating applied to vulcanic lava. Nature (Lond) 227:940–941

Hwang FSJ, Fremlin JH (1970) A new technique using thermally stimulated current. Archaeometry 12:67–71

IAEA (1969–1986) Environmental Isotope Data. World Survey of Isotope Concentrations in Precipitation. IAEA Tech Rep Ser No 69, 117, 129, 149, 165, 192, 206, 226, 264. IAEA, Vienna

IAEA (1979) Behaviour of Tritium in the Environment. IAEA, Vienna, 711 pp

IAEA (1981a) Methods of Low-Level Counting and Spectrometry. IAEA, Vienna, 557 pp

IAEA (1981b) Migration in the terrestrial environment of long-lived radionuclides from nuclear fuel cycle. IAEA, Vienna

IAEA (1983) Guidebook on Nuclear Techniques in Hydrology. Tech Rep Ser 91:456 pp. IAEA, Vienna

IAEA (1987) Studies on sulfur isotope variations in nature. IAEA Panel proc Ser, IAEA, Vienna, 124pp

Ikeya M (1982a) A model of linear uranium accumulation for ESR age of Heidelberg (Mauer) und Tautavel Bones. Jp J Appl Phys 21:L690–692

Ikeya M (1982b) Electron-spin resonance of petrified woods for geological age assessment. Jpn J Appl Phys 20:L28–L30

Ikeya M (1983a) Electron spin resonance (ESR) dating in archaeology and geology. JEOL News 19A:26–30

Ikeya M (1983b) ESR studies of geothermal boring cores at Hachobara power station. Jpn J Appl Phys 22:L763–L765

Ikeya M (1984) Age limitations of ESR dating of carbonate fossils. Naturwissenschaften 71:421–423

Ikeya M (1985) ESR ages of bones on paleo-anthropology: uranium and fluorine accumulation. IInt Symp ESR Dating. Yamagoshi, Sept 1985

Ikeya M (1988) Dating and radiation dosimetry with electron spin resonance. Magn Reson Rev 13:91–134

Ikeya M, Miki T (1980a) Electron spin resonance dating of animal and human bones. Science 207:977–979

Ikeya M, Miki T (1980b) A new dating method with a digital ESR. Naturwissenschaften 67:191
Ikeya M, Miki T (1980c) Present status of ESR dating: reply to Nambi's comment. Jpn J Appl Phys
    19:1809–1810
Ikeya M, Miki T (eds) (1985a) ESR dating and Dosimetry. Ionics, Tokyo, 536 pp
Ikeya M, Miki T (1985b) ESR dating of organic materials: from potatochips to a dead body. Nucl
    Tracks 10:909–912
Ikeya M, Ohmura K (1983) Comparison of ESR ages of corals from marine terraces with $^{14}$C and
    $^{230}$Th/$^{234}$U ages. Earth Planet Sci Lett 65:34–38
Ikeya M, Ohmura K (1984) ESR age of Pleistocene shells measured by radiation assessment.
    Geochem J 18:11–17
Ikeya M, Miki T, Tanaka K, Sakuramoto Y, Ohmura K (1983) ESR dating of faults at Rokko and
    Atotsugawa. PACT 9:411–419
Ikeya M, Devine SD, Whitehead NE, Hedenquist JW (1986) Detection of methane in geothermic
    quartz by ESR. Chem Geol 56:185–192
Ikeya M, Kai A, Bishoff J (1987) ESR dating of lake sediments. In: Proc 5$^{th}$ Specialist Seminar on
    TL and ESR Dating. Kings College, Cambridge, July 6–10, 1977
Imamura M, Matsuda H, Horie K, Honda M (1969) Applications of neutron activation method for
    $^{53}$Mn in meteoritic iron. Earth Planet Sci Lett 6:165–172
Imamura M, Finkel RC, Wahlen M (1973) Depth profile of $^{53}$Mn in lunar surface, Earth Planet Sci
    Lett 20:107–112
Imbrie J, Imbrie JZ (1980) Modeling the climatic response to orbital variations. Science 207:943–
    953
Imbrie J, Hays JD, Martinson DG, McIntyre A, Mix AC, Morley JJ, Pisias NG, Prell WL,
    Shackleton NJ (1984) The orbital theory of Pleistocene climate: Support from a revised
    chronology of the marine $\delta^{18}$O record. In: Berger A, Imbrie J, Hays J, Kukla G. Saltzman B (eds)
    Milankovitch and climate, Part I. Reidel, Dordrecht Boston Lancaster, pp 269–305
International Study Group (1982) An inter-laboratory comparison of radiocarbon measurements
    in tree rings. Nature (Lond) 298:619–623
Ivanovich M, Harmon RS (eds) (1982) Uranium Series Disequilibrium. Applications to Environ-
    mental Problems. Clarendon, Oxford, 571 pp
Ivanovich M, Ireland P (1984) Measurement of uranium series disequilibrium in the case-hardened
    Aymamon limestone of Puerto Rico. Z Geomorph NF 28:305–319
Ivanovich M, Vita-Finzi C, Hennig GJ (1983) Uranium-series dating of molluscs of uplifted
    Holocene beaches in the Persian Gulf. Nature (Lond) 302:408–410
Ivanovich M, Ku T-L, Harmon RS, Smart PL (1984) Uranium series intercomparison project
    (USIP). Nucl Instrum Methods Phys Res 223:466–471
Jacobs JA (1984) What triggers reversals of the earth's magnetic field? Nature (Lond) 309:115
Jacobsen SB, Wasserburg GJ (1980) Sm-Nd evolution of chondrites. Earth Planet Sci Lett
    50:139–155
Jacobsen SB, Wasserburg GH (1984) Sm-Nd isotopic evolution of chondrites and achondrites, II.
    Earth Planet Sci Lett 67:137–150
Jacobsen SB, Quick JE, Wasserburg GJ (1984) A Nd and Sr isotopic study of the Trinity peridotite;
    implications for mantle evolution. Earth Planet Sci Lett 68:361–378
Jäger E (1979) The Rb-Sr method. In: Jäger E, Hunziker JC (eds) Lectures in isotope geology.
    Springer, Berlin Heidelberg New York, pp 13–26
Jäger E, Hunziker JC (1979) Lectures in Isotope Geology. Springer, Berlin Heidelberg New York,
    329 pp
Jäger E, Ji CW, Hurford AJ, Xin LR, Hunziker JC, Ming LD (1985) BB-6: A Quaternary age
    standard for K-Ar dating. Chem Geol Isot Geosci Sect 52:275–279
Jaffey AH, Flynn KF, Glendenin LE, Bentley WC, Essling AM (1971) Precision measurement of
    half-lives and specific activities of $^{235}$U and $^{238}$U. Phys Rev C 4:1889–1906
Jahn B-M, Chen PY, Yen TP (1976) Rb-Sr ages of granitic rocks in southeastern China and their
    tectonic significance. Geol Soc Am Bull 86:763–776
Jahn B-M, Vidal P, Tilton GR (1980) Archean mantle heterogeneity: evidence from chemical and
    isotopic abundances in Archean igneous rocks. Phil Trans R Soc Lond A297:353–364
Jahn B-M, Gruau G, Glickson AY (1982) Komatiites of the Onverwacht Group, S. Africa: REE
    geochemistry, Sm/Nd age and mantle evolution. Contrib Mineral Petrol 80:25–40

Jansen E (1989) The use of stable oxygen and carbon isotope stratigraphy as a dating tool. Quatern Int 1:151–161

Javoy M (1977) Stable isotopes and geothermometry. J Geol Soc Lond 133:609–636

Jenkins WJ (1980) Tritium and $^3$He in the Sargasso Sea. J Mar Res 38:533–569

Jenkins WJ, Rhines PB (1980) Tritium in the deep North Atlantic Ocean. Nature (Lond) 286:877–880

Jenkins WJ, Lott DE, Pratt MW, Boudreau RD (1983) Anthropogenic tritium in South Atlantic bottom water. Nature (Lond) 305:45–46

Jessberger EK, Ostertag R (1982) Shock-effects on the K-Ar system of plagioclase feldspar and the age of anorthosite inclusions from North-Eastern Minnesota. Geochim Cosmochim Acta 46:1465–1471

Johnsen SJ (1977) Stable isotope homogenization of polar firn and ice. In: Rodda JC (ed) Isotope and impurities in snow and ice. IAHS Publ 118, Grenoble, pp 210–219

Johnsen SJ, Dansgaard W, Clausen HB, Langway cc Jr (1972) Oxygen isotope profiles through the Antarctic and Greenland ice sheets. Nature (Lond) 235:429–434

Johnson RA, Stipp JJ, Tamers MA, Bonani G, Suter M, Wölfli W (1986) Archaeologic sherd dating: Comparison of thermoluminescence dates with radiocarbon dates by beta counting and accelerator techniques. Radiocarbon 28 (2A):719–725

Johnson WM, Maxwell JA (1981) Rock and mineral analysis. 2nd edn. Wiley, New York

Joly J (1907) Pleochroic halos. Phil Mag 13:381–383

Joly J (1908) The radioactivity of sea-water. Phil Mag J Sci Lond 6:385–393

Jones JH (1982) The geochemical coherence of Pu and Nd and the $^{244}$Pu/$^{238}$U ratio of the early solar system. Geochim Cosmochim Acta 46:1793–1804

Joshi SR (1987) Nondestructive determination of lead-210 and radium-226 in sediments by direct photon analysis. J Radioanal Nucl Chem 116:169–182

Jost DT, Marti, K, Sutter E (1981) Pu-Nd-Xe Dating: Sytematics in $^{244}$Pu fission and REE spallation components. Meteoritics 16:334

Jouzel J, Merlivat L (1981) Low-level tritium measurements in water: a complete system including liquid scintillation, gas counting and electrolysis. In: Methods of Low-level Counting and Spectrometry: 325–334. IAEA, Vienna

Jouzel J, Merlivat L, Lorius C (1982) Deuterium excess in an east Antarctic ice core suggests higher relative humidity at the ocean surface during the last glacial maximum. Nature (Lond) 299:688–691

Julg A, Lafont R, Perinet G (1987) Mechanisms of collagen racemization in fossil bones: application to absolute dating. Q Sci Rev 6:25–28

Kadlec RH, Robbins JA (1984) Sedimentation and sediment accretion in Michigan coastal wetlands (U.S.A.). Chem Geol 44:119–150

Kaipa PL, Haskell EH, Kenner GH (1988) Beta dose attenuation and calculations of effective grain size in brick samples. Nucl Tracks Radiat Meas 14:215–217

Kamen MA (1963) Early history of carbon-14. Science 140:584–590

Kaneoka I (1972) The effect of hydration on the K/Ar ages of volcanic rocks. Earth Planet Sci Lett 14:216–220

Kaneoka I, Aramaki S, Tonouchi S (1982) K-Ar ages of a basanitoid lava flow of Nanzaki Volcano and underlying Miocene andesites from the Irozaki area, Izu Peninsula, central Japan. J Geol Soc Jpn 88:919–922

Kapusta YS, Makeyev AF, Yakovleva SS (1983) Uranium-xenon geochronometry based on the crystalline material in zircon. Geochem Int 20/4:112–117

Karakostanoglou I, Schwarcz HP (1983) ESR isochron dating. PACT 9:391–398

Kasanevich ER (1968) The interpretation of lead isotopes and their geological significance. In: Hamilton EI, Farquhar RM (eds) Radiometric dating for geologists. Interscience, New York

Katz BJ, Harrison CGA, Man EH (1983) Geothermal and other effects on amino-acid racemization in selected deep-sea drilling project cores. Org Chem 5:151–156

Katzenberger O (1989) Experimente zu Grundlagen der ESR-Datierung von Molluskenschalen. Sonderveroeffentlichungen des Geologischen Instituts der Universität zu Köln 72:71 pp

Katzenberger O, Willems N (1987) Interferences encountered in the determination of AD of mollusc samples. In: Proc 5$^{th}$ Specialist Seminar on TL and ESR Dating. Kings College, Oxford, July 6–10, 1987

Kaufhold J, Herr W (1967) Influence of experimental factors on dating natural and man-made glasses by the fission track method. In: Radioactive Dating and Methods of Low-Level Counting: 403–411. IAEA, Vienna

Kaufman A (1986) The distribution of $^{230}$Th/$^{234}$U ages in corals and the number of last interglacial high-sea stands. Q Res 25:55–62

Kaufman A, Broecker WS (1965) Comparison of Th$^{230}$ and C$^{14}$ ages for carbonate materials from Lakes Lahontan and Bonneville. J Geophys Res 70:4039–4054

Kaufman A, Broecker WS, Ku T-L, Thurber DL (1971) The status of U-series methods of mollusk dating. Geochim Cosmochim Acta 35:1155–1183

Kaufman S, Libby WF (1954) The natural distribution of tritium. Phys Rev 93:1337–1344

Keevil NB (1939) The calculation of geological age. Am J Sci 237:195–214

Keir RS (1983) Reduction of thermohaline circulation during deglaciation: the effect on atmospheric radiocarbon and $CO_2$. Earth Planet Sci Lett 64:445–456

Keir RS (1984) Recent increase in Pacific $CaCO_3$ dissolution: a mechanism for generating old $^{14}$C ages. Mar Geol 59:227–250

Keisch B, Feller RL, Levine AS, Edwards RR (1967) Dating and authenticating works of art by measurement of natural alpha emitters. Science 155:1238–1242

Kelly WR, Wasserburg GJ (1978) Evidence for the existence of $^{107}$Pd in the early solar system. Geophys Res Lett 5:1079–1082

Kennet JP (ed) (1980) Magnetic stratigraphy of sediments. Dowden, Hutchinson and Ross, Stroudsburg, Penn, 448 pp (Benchmark papers in Geology Ser vol 54)

Kigoshi K (1967) Ionium dating of igneous rocks. Science 156:932–934

Kim KH, Burnett WC (1983) $\gamma$-Ray spectrometric determination of uranium-series nuclides in marine phosphorites. Anal Chem 55:1796–1800

Kim KH, Burnett WC (1985) $^{226}$Ra in phosphate nodules from Peru/Chile seafloor. Geochim Cosmochim Acta 49:1073–1081

Kimber RWL, Griffing CW (1987) Further evidence of the complexity of the racemization process in fossil shells with implications for amino acid racemization dating. Geochim Cosmochim Acta 51:829–846

Kimber RWL, Griffin CW, Milnes AR (1986) Amino acid racemization dating: evidence of apparent reversal in aspartic acid racemization with time in shells of ostrea. Geochim Cosmochim Acta 50:1159–1161

King JW, Banerjee SK, Marvin J, Lund S (1983) Use of small-amplitude paleomagnetic fluctuations for correlation and dating of continental climatic changes. Palaeogeogr Palaeoclimatol Palaeoecon 42:167–183

Kinny PD, Compston W (1986) Hafnium model ages and evidence for resetting of the U-Pb system in zircon. Terra Cognita 6:153–154

Kirsten T (1978) Time and the solar system. In: Dermott SF (ed) Origin of the Solar System. Wiley, London, pp 267–346

Klein J, Giegengack R, Middleton R, Sharma P, Underwood JR Jr, Weeks RA (1986) Revealing of exposure using in situ produced $^{26}$Al and $^{10}$Be in Lybian desert glass. Radiocarbon 28 (2A):547–555

Klopin WG, Gerling EK (1947) A new method for the determination of absolute ages of minerals. Dokl Akad Nauk SSSR 58:1415–1417 (in Russian)

Knudsen KL, Sejrup HP (1988) Amino acid geochronology of selected interglacial sites in the North Sea area. Boreas 17:347–354

Kober B (1986) Whole-grain evaporation for $^{207}$Pb/$^{206}$Pb-age-investigations on single zircons using a double-filament thermal ion source. Contrib Mineral Petrol 93:482–490

Kochenov AV, Zenev'yev VV, Lovaleva SA (1965) Some features of the accumulation of uranium in peat bogs. Geokhimiya 1:97–103

Köppel V, Grünenfelder M (1979) Isotope geochemistry of lead. In: Jäger E, Hunziker JC (eds) Lectures in isotope geology. Springer, Berlin Heidelberg New York, pp 134–153

Köppel V, Sommerauer J (1974) Trace elements and the behaviour of the U-Pb system in inherited and newly formed zircons. Contrib Mineral Petrol 43:71–82

Kohmann TP, Goel PS (1963) Terrestrial ages of meteorites from cosmogenic C$^{14}$. In: Radioactive Dating: 395–411. IAEA, Vienna

Koide M, Goldberg ED (1983) Uranium isotopes in the Greenland ice-sheet. Earth Planet Sci Lett 65:245–248

Koide M, Soutar A, Goldberg ED (1972) Marine geochronology with Pb-210. Earth Planet Sci Lett 14:442–446

Koide M, Bruland KW, Goldberg ED (1973) Th-228/Th-232 and Pb-210 geochronologies in marine and lake sediments. Geochim Cosmochim Acta 37:1171–1187

Korkisch J, Steffan I, Hille P, Vonach H, Wild E (1982) Uranium series method applied to fossil bone. J Radioanal Chem 68:107–116

Korschinek G, Morinaga H, Nolte E, Preisenberger E, Ratzinger U, Urban A, Dragovitch P, Vogt S (1987) Accelerator mass spectrometry with completely stripped $^{41}$Ca and $^{53}$Mn ions at the Munnich tandem accelerator. Nucl Instrum Methods Phys Res B 29:67–71

Kovacheva M (1982) Archaeomagnetic investigations of geomagnetic secular variations. Phil Trans R Soc Lond A 306:79–86

Kralik M, Riedmüller G (1985) Dating fault by Rb-Sr and K-Ar techniques. Terra Cognita 5:279

Kralik M, Krumm H, Schramm JM (1987) Low grade and very low grade metamorphism in the northern calcareous Alps and in the Greywacke zone: Illite-crystallinity data and isotopic ages. In: Flügel HW, Faul P (eds) Geodynamics of the eastern Alps: 164–178; Vienna (Deuticke)

Kramers JD, Smith CB (1983) A feasibility study of U-Pb and Pb-Pb dating of kimberlites using groundmass mineral fractions and whole rock samples. Isot Geosci 1:23–38

Kriausakul N, Mitterer RM (1983) Epimerization of COOH-terminal isoleucine in fossil dipeptides. Geochim Cosmochim Acta 47:963–966

Krishnaswami S, Moore WS (1973) Accretion rates of fresh-water manganese deposits. Nature Phys Sci 243:114–116

Krishnaswami S, Lal D, Martin JM, Meybeck M (1971) Geochronology of lake sediments. Earth Planet Sci Lett 11:407–414

Krishnaswami S, Graustein WC, Turekian KK, Dowd JF (1981) Chronometric applications of radium isotopes and radon in groundwater. Abstr Progr 13:491–492

Krishnaswami S, Mangini A, Thomas JH, Sharma P, Cochran JK, Turekian KK, Parker PD (1982) $^{10}$Be and Th isotopes in manganese nodules and adjacent sediments: nodule growth histories and nuclide behavior. Earth Planet Sci Lett 59:217–234

Kröll VST (1953) Vertical distribution of radium in deep-sea sediments. Nature (Lond) 171:742

Krogh TE (1973) A low-contamination method for the hydrothermal decomposition of zircon and extraction of U and Pb for isotopic age determinations. Geochim Cosmochim Acta 37:485–494

Krogh TE (1982a) Improved accuracy of U-Pb zircon dating by selection of more concordant fractions using a high gradient magnetic separation technique. Geochim Cosmochim Acta 46:631–635

Krogh TE (1982b) Improved accuracy of U-Pb zircon ages by the creation of more concordant systems using an air abrasion technique. Geochim Cosmochim Acta 46:637–649

Krogh TE, Davis GL (1973) The effect of regional metamorphism on U-Pb systems in zircons and a comparison with Rb-Sr systems in the same whole rock and its constituent minerals. Carnegie Inst Washington Yearb 72:601–610

Krogh TE, Davis GL (1975) Alteration in zircons and differential dissolution of altered and metamict zircon. Carnegie Inst Washington Yearb 73:619–623

Kromer B, Pfleiderer C, Schlosser P, Levin I, Münnich KO, Bonani G, Suter M, Wölfli W (1987) AMS $^{14}$C measurement of small volume oceanic water samples: experimental procedure and comparison with low-level counting technique. Nucl Instrum Methods Phys Res B 29:302–305

Krouse HR (1980) Chapter 11: Sulfur isotopes in our environment. In: Fritz P, Fontes JC (eds) Handbook of Environmental Isotope Geochemistry, vol 1. The Terrestrial Environment A. Elsevier, Amsterdam, pp 435–471

Ku T-L (1965) An evaluation of the $U^{234}/U^{238}$ method as a tool for dating pelagic sediments. J Geophys Res 70:3457–3474

Ku T-L (1968) Protactinium-231 method of dating corals from Barbados island. J. Geophys Res 73:2271–2276

Ku T-L (1976) The uranium-series methods of age determination. Ann Rev Earth Planet Sci 4:347–379

Ku T-L, Broecker WS (1967a) Uranium, thorium and protactinium in manganese nodule. Earth Planet Sci Lett 2:317–321

Ku T-L, Broecker WS (1967b) Rates of sedimentation in the arctic ocean. Progr Oceanogr 4:95–104

Ku T-L, Joshi LU (1981) Measurements of $^{238}$U, $^{234}$U and $^{230}$Th in impure carbonates for age determination. J Radioanal Chem 67:351–358

Ku T-L, Liang Z-C (1984) The dating of impure carbonates with decay-series isotopes. Nucl Instrum Methods Phys Res 223:563–571

Ku T-L, Omura A, Chen PS (1979a) Be$^{10}$ and U-series isotopes in manganese nodules from the Central North Pacific. In: Bischoff JL, Piper DZ (eds) Marine Geology and Oceanography of the Pacific Manganese Nodule Province. Plenum, New York, pp 791–814

Ku T-L, Buhl WB, Freeman ST, Knauss KG (1979b) Th$^{230}$-U$^{234}$ dating of pedogenic carbonates in gravelly desert soils of Vidal Valley, southeastern California. Geol Soc Am Bull I 90:1063–1073

Ku T-L, Kusakabe M, Nelson DE, Southon JR, Korteling RG, Vogel J, Nowikon I (1982) Constancy of oceanic deposition of $^{10}$Be as recorded in manganese crusts. Nature (Lond) 299:240–242

Kubik PW, Elmore D, Conard NJ, Nishiizumi K, Arnold JR (1986) Determination of cosmogenic $^{41}$Ca in a meteorite with tandem accelerator mass spectrometry. Nature (Lond) 319:568–570

Kulp JL (1961) Geological Time Scale. Science 133:1105–1114

Kulp JL, Engels J (1963) Discordances in K-Ar and Rb-Sr isotopic ages. In: Radioactive Dating: 219–238. IAEA, Vienna

Kulp JL, Volchok HL, Holland HD (1952) Age from metamict minerals. Am Mineral 37:709–718

Kulp JL, Bull WB, Freeman ST, Knauss KG (1979) Th$^{230}$-U$^{234}$ dating of pedogenic carbonates in gravelly desert soils of Vidal Valley, southeastern California. Geol Soc Am Bull 90:1063–1073

Kuroda PK (1960) Nuclear fission in the early history of the earth. Nature (Lond) 187:36–38

Kuroda PK (1963) Dating methods based on the process of nuclear fission. In: Radioactive Dating: 45–54. IAEA, Vienna

Kuroda PK (1967) Dating methods based on the extinct radionuclides iodine-129 and plutonium-244. In: Radioactive Dating and Methods of Low-Level Counting: 259–268. IAEA, Vienna

Kurz MD (1986) In situ production of terrestrial cosmogenic helium and some applications to geochronology. Geochim Cosmchim Acta 50:2855–2862

Kurz MD, O'Brien P, Garcia M, Frey FA (1985) Isotopic evolution of Haleakala volcano: primordial, radiogenic and cosmogenic helium. EOS 66:1120

Kusakabe M, KU T-L, Vogel J, Southon JR, Neslon DE, Richards G (1982) $^{10}$Be Profiles in Seawater. Nature (Lond) 299:712–714

Kutschera K, Hennig W, Paul M, Smither RK, Stephensen RJ, Yutema JL, Alburger DE, Cunning JB, Harbottle G (1980) Measurement of the $^{32}$Si half-life via accelerator mass spectrometry. Phys Rev Lett 45:592–596

Kuznetsova RI, Lavrukhina VI (1981) Aluminium-26 and proton exposure in early solar system. Meteoritics 16:344–345

Kvenvolden KA, Blunt DJ (1979) Amino acid dating of bone nuclei in manganese nodules from the North Pacific Ocean. In: Bischoff JL, Piper DZ (eds) Marine Science 9: Marine Geology and Oceanography of the Pacific Manganese Nodule Province. Plenum, New York London, pp 763–773

Kvenvolden KA, Blunt DJ, Clifton HE (1981) Age estimations based on amino acid racemization: Reply to comments of J.F. Wehmiller. Geochim Cosmochim Acta 45:265–267

Kyte FT, Zhou Z, Wasson JT (1980) Siderophile-enriched sediments from the Cretaceous-Tertiary boundary. Nature (Lond) 288:651–656

Kyte FT, Zhou Z, Wasson JT (1981) High noble metal concentrations in a late Pliocene sediment. Nature (Lond) 292:417–420

Labeyrie LD, Duplessy JC (1985) Changes in the Oceanic $^{13}$C/$^{12}$C ratio during the last 140 000 years. High-latitude surface water records. Palaeogeogr Palaeoclimatol Palaeoecol 50:217–240

Lal D (1962) Cosmic-ray-produced radionuclides in the sea. J Oceanogr Soc Jpn 20th Annivers Vol, pp 600–604

Lal D (1987a) $^{10}$Be in polar ice: data reflect changes in cosmic ray flux or polar meteorology. Geophys Res Lett 14:785–788

Lal D (1987b) Cosmogenic nuclides produced in situ in terrestrial solids. Nucl Instrum Methods Phys Res B 29:238–245

Lal D, Arnold JR (1985) Tracing quartz through the environment. Proc Indian Acad Sci 94:1–5

Lal D, Somayajulu BLK (1984) Some aspects of the geochemistry of silicon isotopes. Tectonophysics 105:383–397

Lal D, Goldberg ED, Koide M (1960) Cosmic-ray-produced silicon-32 in nature. Science 131:332–337

Lal D, Nijampurkar VN, Rama S (1970) Silicon-32 hydrology. In: Isotope Hydrology 1970:847–863. IAEA, Vienna

Lal D, Nijampurkar VN, Somayajulu BLK, Koide M, Goldberg ED (1976) Silicon-32 specific activities in coastal waters of the world oceans. Limnol Oceanogr 21:285–293

Lal D, Venkatesan RD, Davis R Jr (1986) Cosmogenic $^{37}$Ar, $^{39}$Ar in terrestrial rocks. Terra Cognita 6:250

Lal D, Nishiizumi K, Arnold JR (1987) In situ cosmogenic $^{3}$H, $^{14}$C, and $^{10}$Be for determining the net accumulation and ablation rates of ice sheets. J Geophys Res 92 (B6):4947–4952

Lambert RSJ (1971) The pre-Pleistocene Phanerozoic time-scale – further data. In: Harland WB, Francis EH (eds) The Phanerozoic Time-scale – A Supplement. Geol Soc London, Spec Publ 5: pp 9–34

Lamothe M, Huntley J (1988) Thermoluminescence dating of late Pleistocene sediments, St Lawrence Lowland, Quebec. Geogr Phys Quatern 42:33–44

Lance JC, McIntyre SC, Naney JW, Rousseva SS (1986) Measuring sediment movement at low erosion rates using cesium-127. Soil Sci Soc Am J 50:1303–1309

Lancelot JR, Vitrac A, Allègre CJ (1976) Uranium and lead isotopic dating with grain by grain zircon analysis: a study of complex geological history with a single rock. Earth Planet Sci Lett 29:357–366

Lancelot JR, Boullier AM, Maluski H, Ducrot J (1983) Deformation and related radiochronology in a Late Pan-African mylonitic shear zone, Adrar des Iforas (Mali). Contrib Mineral Petrol 82:312–326

Lanford WA (1978) $^{15}$N-hydrogen profiling scientific applications. Nucl Instrum Methods 149:1–7

Lanphere MA, Dalrymple GB (1971) A test of the $^{40}$Ar/$^{39}$Ar age spectrum technique on some terrestrial materials. Earth Planet Sci Lett 12:359–372

Larsen ES, Keevil NB (1947) Radioactivity of the rocks of the batholith of Southern California. Bull Geol Soc Am 58:483–493

Larsen ES, Keevil NB, Harrison HC (1952) Method for determining the age of igneous rocks using the accessory minerals. Bull Geol Soc Am 63:1045–1052

Latham AG, Schwarcz HP, Ford DC, Pearce GW (1979) Palaeomagnetism of stalagmite deposits. Nature (Lond) 280:383–385

Latham AG, Schwarcz HP, Ford DC, Pearce GW (1982) The palaeomagnetism and U-Th dating of the Canadian speleothems: evidence for the westward drift, 5.4–2.1 ka BP. Can J Earth Sci 19:1985–1995

Lawrence JR, White JWC (1984) Growing season precipitation from D/H ratios of eastern white pine. Nature (Lond) 311:558–560

Leavy BD, Phillips FM, Elmore D, Kubik PW, Gladney E (1987) Measurement of cosmogenic $^{36}$Cl/Cl in young volcanic rocks: an application of accelerator mass spectrometry in geochronology. Nucl Instrum Methods Phys Res B 29:246–250

Ledent D, Patterson C, Tilton GR (1964) Ages of zircon and feldspar concentrates from Northern American beach and river sands. J Geol 72:112–122

Lee C, Bada JL, Peterson E (1976) Amino acids in modern and fossil woods. Nature (Lond) 259:183–186

Lee RR (1969) Chemical temperature integration. J Appl Meteorol 8:423–430

Lee RR, Leich DA, Tombrello TA, Ericson JE, Friedman I (1974) Obsidian hydration profile measurements using a nuclear reaction technique. Nature (Lond) 250:44–47

Lee T, Schramm DN, Wefel JP, Blake JB (1978) On the apparent conflict between the time scales inferred from the cosmochronometers $^{129}$I, $^{244}$Pu, and $^{26}$Al. Geol Surv Open-File Rep 78–701:246–247

Lehmann BE, Loosli HH (1984) Use of noble gas radioisotopes for environmental research. Inst Phys Conf Ser No 71:219–226

Lehmann BE, Oeschger H, Loosli HH, Hurst GS, Allman SL, Chen CH, Kramer SD, Payne MG, Phillips RC, Willis RD, Thonnard N (1985) Counting $^{81}$Kr atoms for analysis of groundwater. J Geophys Res 90 (B13):11557–11551

Lein AY (1985) The isotope mass balance of sulfur in oceanic sediments (the Pacific ocean as an example). Mar Chem 16:249–257

Lemons KW, McAtee JL (1983) The parameters of induced thermoluminescence of some selected phyllosilicates: a crystal defect structure study. Am Mineral 68:915–923

Lerman JC, Mook WG, Vogel JC (1970) $^{14}$C in tree rings from different localities. In: Olsson IU (ed) Radiocarbon Variations and Absolute Chronology. Almqvist Wiksell, Stockholm, pp 275–301

Le Roux LJ, Glendenin LE (1963) Half-life of $^{232}$Th. Proc Nat Meet Nucl Energy, Pretoria, South Africa, pp 83–94

Levchenkov OA, Shukolyukov YA (1970) A new method for calculating age and time of metamorphism of minerals and rocks without correction for ordinary lead. Geochem Int 1970:60–65

Levchenkov OA, Makeyev AF, Shuleshko IK, Komarov AN, Ovchinnikova GN, Yakovleva SZ (1982) Uranium-lead isochron dating of heterogeneous zircons. Dokl Acad Sci USSR, Earth Sci Sec 251:51–53

Levin BY (1982) Asteroids, comets, meteor matter – their place and role in the cosmogony of the solar system. Izvestiya Earth Phys 18:414–424

Levin I, Münnich KO, Weiss W (1980) The effect of anthropogenic $CO_2$ and $^{14}$C sources on the distribution of $^{14}$C in the atmosphere. Radiocarbon 22 (II):379–391

Levy DM, Moore WS (1985) Radium-224 in continental shelf waters. Earth Planet Sci Lett 73:226–230

Lewin R (1983) Isotopes give clues to past diet. Science 220:1369

Lewis DR (1966) Exoelectron-emission phenomena and geological applications. Geol Soc Am Bull 77:761–770

Libby WF (1946) Atmospheric helium-three and radiocarbon from cosmic radiation. Phys Rev 69:671–672

Libby WF (1953) The potential usefulness of natural tritium. Proc Nat Acad Sci 39:245–247

Lindner M, Leich DA, Russ GP, Bazan JM, Borg RJ (1989) Direct determination of the half-life of $^{187}$Re. Geochim Cosmochim Acta 53:1597–1606

Linick TW (1980) Bomb-produced carbon-14 in the surface water of the Pacific ocean. Radiocarbon 22 (II):599–606

Lippolt HJ (1977) Isotopische Salz-Datierung: Deutung und Bedeutung. Aufschluβ 28:369–389

Lippolt HJ, Raczek I (1979a) Rinneite – dating of episodic events in potash salt deposits. J Geophys 46:225–228

Lippolt HJ, Raczek I (1979b) Cretaceous Rb-Sr total rock ages of Permian salt rocks. Naturwissenschaften 66:422–423

Lippolt HJ, Weigel E (1988) $^4$He diffusion in $^{40}$Ar-retentive minerals. Geochim Cosmochin Acta 52:1449–1458

Lippolt HJ, Boschmann W, Arndt H (1982) Helium und Uran in Schwarz wälder Bleiglanzen, ein Datierungsversuch. Oberrhein Geol Abh 31:31–46

Lippolt HJ, Schleicher H, Raczek I (1983) Rb-Sr systematics of Permian volcanites of the Schwarzwald (SW Germany). Part I: Space of time between plutonism and late orogenic volcanism. Contrib Mineral Petrol 84:272–280

Lippolt HJ, Fuhrmann U, Hradetzky H (1986) $^{40}$Ar/$^{39}$Ar age determinations on sanidines of the Eifel volcanic field (Federal Republic of Germany): Constraints on age and duration of a middle Pleistocene cold period. Chem Geol 59:187–204

Lister G, Kelts K, Schmid R, Bonani G, Hofmann H, Morenzoni E, Nessi M, Suter M, Wölfli W (1984) Correlation of the paleoclimatic record in lacustrine sediment sequences: $^{14}$C dating by AMS. Nucl Instr Methods 233 (B5):389–393

Livingston HD, Anderson RF (1983) Large particle transport of plutonium and other fallout radionuclides to the deep ocean. Nature (Lond) 303:228–230

Lo Bello P, Feraud G, Hall CM, York D, Lavina P, Bernat M (1987) $^{40}$Ar/$^{39}$Ar step-heating and laser fusion dating of a Quaternary pumice from Neschers, Massif Central, France: The defeat of xenocryst contamination. Chem Geol Isot Geosci Sec 66:61–71

Lodge JP, Bien GS, Suess HE (1960) The carbon-14 content of urban airborne particulate matter. Int J Air Poll 2:309–312

Löfvendahl R, Åberg G (1981) An isotope study of Swedish secondary U-Pb-minerals. Geol Fören Stockholm Förh 103:331–342

Longinelli A (1984) Oxygen isotopes in mammal bone phosphate: a new tool for paleohydrological and paleoclimatological research? Geochim Cosmochim Acta 48:385–390

Loosli HH (1983) A dating method with $^{39}$Ar. Earth Plan Sci Lett 63:51–62

Loosli HH (1988) Argon-39: a tool to investigate ocean water circulation and mixing. In: Fritz P, Fontes JCh (eds) Handbook of Environmental Isotope Geochemistry, vol 3: The Marine Environment. Elsevier, Amsterdam

Loosli HH, Oeschger HH (1968) Detection of $^{39}$Ar in atmospheric argon. Earth Planet Sci Lett 5:191–198

Loosli HH, Oeschger H (1979) Argon-39, carbon-14, and krypton-85 measurements in groundwater samples. In: Isotope Hydrology 1978 (II): 931–945. IAEA, Vienna

Lorius C, Merlivat L, Jouzel J, Pourchet M (1979) A 30,000-yr isotope climatic record from Antarctic ice. Nature (Lond) 280:644–648

Lorius C, Jouzel J, Ritz C, Merlivat L, Barkov NI, Korotkevich YS, Kotlyakov VM (1985) A 150,000-year climatic record from Antarctic ice. Nature (Lond) 316:591–596

Lovejoy CO, Burstein AH, Heiple KG (1972) Primate phylogeny and immunological distance. Science 176:803–805

Lovering JF, Hinthorne JR, Conrad RL (1976) Direct $^{207}$Pb/$^{206}$Pb dating by ion microprobe of uranium-thorium rich phases in Allende calcium-aluminium-rich clasts (cars). Lunar Sci VII:504–506

Lovlie R (1989) Paleomagnetic stratigraphy: a correlation method. Quatern Int 1:129–149

Lowe DC, Wallace G, Sparks RJ (1987) Applications of AMS in the atmospheric and oceanographic sciences. Nucl Instrum Methods Phys Res B 29:291–296

Lowe JP, Lowe DJ, Hodder APW, Wilson AT (1984) A tritium exchange method for obsidian hydration shell measurement. Isot Geosci 2:351–363

Lowrie W, Alvarez W (1981) One hundred million years of geomagnetic polarity history. Geology 9:392–397

Luck JM, Allègre CJ (1982) The study of molybdenites through the $^{187}$Re-$^{187}$Os chronometer. Earth Planet Sci Lett 61:291–296

Luck JM, Arndt NT (1985) Re/Os isochron for Archean komatiite from Alexo, Ontario. Terra Cognita 5:323

Luck JM, Turekian KK (1983) Osmium-187/Osmium-186 in manganese nodules and the Cretaceous-Tertiary boundary. Science 222:613–615

Luck JM, Birck J-L, Allègre CJ (1980) $^{187}$Re-$^{187}$Os systematics in meteorites: early chronology of the Solar System and age of the Galaxy. Nature (Lond) 283:256–259

Ludwig KR (1977) Effect of initial radioactive daughter disequilibrium on U-Pb isotope apparent ages of young minerals. J Res US Geol Surv 5:663–667

Ludwig KR (1980) Calculation of uncertainties of U-Pb isotope data. Earth Planet Sci Lett 46:212–220

Ludwig KR (1983) Plotting and regression programs for isotope geochemists, for use with HP-86/87 microcomputers. US Geol Surv Open-File Rep 83–849:94

Ludwig KR Silver LT (1977) Lead-isotope inhomogeneity in Precambrian igneous K-feldspars. Geochim Cosmochin Acta 41:1457–1471

Ludwig KR, Lindsey DA, Zielinski RA, Simmons KR (1980) U-Pb ages of uraniferous opals and implications for the history of beryllium, fluorine, and uranium mineralization at Spor Mountain, Utah. Earth Planet Sci Lett 46:221–232

Ludwig KR, Rubin B, Fishman NS, Reynolds RL (1982) U-Pb ages of uranium ores in the Church Rock uranium district, New Mexico. Econ Geol 77:1942–1945

Lugmair GW (1974) Sm-Nd ages: a new dating method. Meteoritics 9:369

Lugmair GW, Carlson RW (1978) The Sm-Nd history of KREEP. Proc 9th Lunar Planet Sci Conf, pp 689–704

Lugmair GW, Marti K (1978) Lunar initial $^{143}$Nd/$^{144}$Nd: differential evolution of the lunar crust and mantle. Earth Planet Sci Lett 39:349–357

Lugmair GW, Scheinin NB, Marti K (1975) Sm-Nd age and history of Apollo 17 basalt 75075: evidence of early differentiation of the lunar exterior. Proc Lunar Sci Conf 6th vol 2. Geochim Cosmochin Acta Suppl 6: 1419–1429

Lupton JE, Craig H (1975) Excess $^3$He in oceanic basalts, evidence for terrestrial primordial helium. Earth Planet Sci Lett 26:133–139

Luz B, Kolodny Y, Horowitz M (1984) Fractionation of oxygen isotopes between mammalian bone-phosphate and environmental drinking water. Geochim Cosmochim Acta 48:1689–1693

Lyons RG (1988) Determination of alpha effectiveness in ESR dating using nuclear accelerator techniques: methods and energy dependence. Nucl Tracks Radiat Meas 14:275–280

Lyons RG, Bowmaker GA, O'Connor CJ (1988) Dependence of accumulated dose in ESR dating on microwave power: a contra-indication to the routine use of low power levels. Nucl Tracks Radiat Meas 14:243–251

Magaritz M, Kaufman A, Paul M, Fink D (1989) The use of the chlorine isotopic composition of groundwater as a direct estimate of regional evapotranspiration. In: Black TA, Spittlehouse DL, Novak KMD, Price DT (eds) Estimation of areal evapotranspiration. IAHS Publ No. (in press)

Mahaney WC, Boyer MG (1986) Microflora distributions in paleosols: a method for calculating the validity of radiocarbon-dated surfaces. Soil Sci 142:100–107

Mahaney WC, Boyer MG, Rutter NW (1986) Evaluation of amino acid composition as a geochronometer in buried soils on Mount Kenia, East Africa. Geogr Phys Q XL:171–183

Makino T, Honda M (1977) Half life of $^{92}$Nb. Geochim Cosmochim Acta 41:1521–1523

Maloszewski P, Zuber A (1982) Determining the turnover time of groundwater systems with the aid of environmental tracers, I. Models and their applicability. J Hydrol 57:207–231

Maloszewski P, Zuber A (1983) Theoretical possibilities of the $^3$H-$^3$He method in investigations of groundwater systems. CATENA 10:189–198

Maluski H (1978) Behaviour of biotites, amphiboles, plagioclases and K-feldspars in response to tectonic events with the $^{40}$Ar-$^{39}$Ar radiometric method. Example of Corsica granite. Geochim Cosmochim Acta 42:1619–1633

Maluski H, Schaeffer OA (1982) $^{39}$Ar-$^{40}$Ar laser probe dating of terrestrial rocks. Earth Planet Sci Lett 59:21–27

Mamyrin BA, Tolstikhin IN (1984) Helium Isotopes in Nature. In: Fyfe WS (ed) Developments in Geochemistry 3. Elsevier, Amsterdam, 273 pp

Mangerud J, Gullikson S (1975) Apparent radiocarbon ages of recent marine shells from Norway, Spitsbergen, and Antarctic Canada. Q Res 5:263–273

Mangerud J, Sønstegaard E, Sejrup H-P (1979) Correlation of the Eemian (interglacial) Stage and the deep-sea oxygen-isotope stratigraphy. Nature (Lond) 277:189–192

Mangini A, Diester-Haass L (1983) Excess Th-230 in sediments off NW Africa traces upwelling in the past. In: Suess E, Thiede J (eds) Coastal Upwelling. Plenum New York, pp 455–470

Mangini A, Dominik J (1978) Th$^{230}$-excess EDTA extraction: a modification of the ionium method for high accumulation rate determinations. "Meteor" Forsch-Ergebnisse C 29:6–13

Mangini A, Key RM (1983) A $^{230}$Th profile in the Atlantic Ocean. Earth Planet Sci Lett 62:377–384

Mangini A, Sonntag C (1977) $^{231}$Pa dating of deep-sea cores via $^{227}$Th counting. Earth Planet Sci Lett 37:251–256

Mangini A, Dominik J, Müller PJ, Stoffers P (1982) Pacific deep circulation: A velocity increase at the end of the interglacial stage 5? Deep-Sea Res 29:1517–1530

Mangini A, Segl M, Bonani G, Hofmann HJ, Morenzoni E, Nessi M, Suter M, Wölfli W, Turekian KK (1984) Mass-spectrometric $^{10}$Be dating of deep-sea sediments applying the Zürich tandem accelerator. Nucl Instrum Methods 233 (B5):353–358

Mangini A, Segl M, Kudrass H, Wiedecke M, Bonani G, Hofmann HJ, Morenzoni E, Nessi M, Suter M, Wölfli W, Turekian KK (1986) Diffusion and supply rates of $^{10}$Be and $^{230}$Th radioisotopes in two manganese encrustations from the South China Sea. Geochim Cosmochim Acta 50:149–156

Manhes G, Allègre CJ, Dupré B, Hamelin B (1979) Lead-lead systematics, the "age of the earth" and the chemical evolution of our planet in a new representation space. Earth Planet Sci Lett 44:91–104

Manhes G, Allègre CJ, Dupré B, Hamelin B (1980) Lead isotope study of basic ultrabasic layered complexes: speculations about the age of the earth and primitive mantle characteristics. Earth Planet Sci Lett 47:370–382

Manhes G, Allègre CJ, Provost A (1984) U-Th-Pb systematics of the eucrite "Juvinas": Precise age determination and evidence for exotic lead. Geochim Cosmochim Acta 48:2247–2264

Mankinen EA, Dalrymple GB (1972) Electron microprobe evaluation of terrestrial basalts for whole-rock K-Ar dating. Earth Planet Sci Lett 17:89–94

Mankinen EA, Dalrymple GB (1979) Revised geomagnetic polarity time scale for the interval 0–5 my BP. J Geophys Res 84 (B2):615–626

Marshall BD, DePaolo DJ (1982) Precise age determinations and petrogenetic studies using the K-Ca method. Geochim Cosmochim Acta 46:2537–2545

Marshall BD, Woodard HH, DePaolo DJ (1986) K-Ca-Ar systematics of authigenic sanidine from Waukau, Wisconsin, and the diffusivity of argon. Geology 14:936–938

Marti K (1967) Mass-spectometric detection of cosmic-ray-produced $Kr^{81}$ in meteorites and the possibility of Kr-Kr-Dating. Phys Rev Lett 18:264–266

Marti K (1982) Krypton-81-Krypton dating by mass spectrometry. In: Currie LA (ed) Nuclear and Chemical Dating Techniques. ACS, Washington DC, ACS Conf Proc No 176:129–138

Marti K (1984) Live $^{129}I$-$^{129}Xe$ dating. In: Reedy RC, Englert P (1984) Workshop on Cosmogenic Nuclides. LPI Rep 86–06, Houston, pp 49–51

Martin WR, Sayles FL (1987) Seasonal cycles of particle and solute transport processes in nearshore sediments: $^{222}Rn/^{226}Ra$ and $^{234}Th/^{238}U$ disequlilibrium at a site in Buzzards Bay, MA. Geochim Cosmochim Acta 51:927–943

Martinez ML, York D, Hall CM, Hanes JA (1984) Oldest reliable $^{40}Ar/^{39}Ar$ ages for terrestrial rocks: Barberton Mountain komatiites. Nature (Lond) 307:352–354

Martinson DG, Pisias NG, Hays JD, Imbrie J, Moore TC Jr, Shackleton NJ (1987) Age dating and the orbital theory of the ice ages: development of a high-resolution 0 to 300,000-year chronostratigraphy. Q Res 27:1–29

Martz LW, de Jong E (1987) Using cesium-137 to assess the variability of net soil erosion and its association with topography in a Canadian prairie landscape. Catena 14:439–451

Masters PM, Bada JL (1979) Amino acid racemization dating of fossil shell from southern California. In: Berger R, Suess HE (eds) Radiocarbon Dating. Univ California Press, Los Angeles, pp 757–773

Matsuda J-I, Lewis RS, Takahashi H, Anders E (1980) Isotopic anomalies of noble gases in meteorites and their origins – VII. C3V carbonaceous chondrites. Geochim Cosmochim Acta 44:1861–1874

Matthews JA (1985) Radiocarbon dating of surface and buried soils: Principles, problems and prospects. In: Richards KS, Arnett RR, Ellis S (eds) Geomorphology and Soils. Allen and Unwin, London, pp 269–288

Mattinson JM (1986) Geochronology of high-pressure-low-temperature Franciscan metabasites: A new approach using the U-Pb system. Geol Soc Am Mem 164:95–105

Maurer C, Williams S, Riley T (1981) ESR dating of archaeological ceramics: a progress report. MASCA J 1:202–204

Mayewski PA, Lyons WB, Spencer MJ, Twickler H, Dansgaard W, Koci B, Davidson CI, Honrath RE (1986) Sulfate and nitrate concentrations from a south Greenland ice core. Science 232:975–977

Mazor E, Jaffe FC, Fluck J, Dubois JD (1986) Tritium-corrected $^{14}C$ and atmospheric noble-gas corrected $^{4}He$ applied to deduce ages of mixed groundwaters: examples from the Baden region, Switzerland. Geochim Cosmochim Acta 50:1611–1618

McConnell D (1962) Dating of fossil bones by the fluorine method. Science 136:241–244

McCulloch MT, Chappell BW (1982) Nd isotopic characteristics of S- and I-type granites. Earth Planet Sci Lett 58:51–64

McCulloch MT, Compston W (1981) Sm-Nd age of Kambalda and Kanowna greenstones and heterogeneity in the Archaean mantle. Nature (Lond) 294:322–326

McCulloch MT, Wasserburg GJ (1978) Sm-Nd and Rb-Sr chronology of continental crust formation. Science 200:1003–1011

McDougall DJ (1968) Thermoluminescence in Geological Materials. Academic Press, London New York, 311–450

McDougall DJ (1976) Fission-track dating. Sci Am 235:114–122

McDougall I, Harrison TM (1988) Geochronology and Thermochronology by the $^{40}Ar/^{39}Ar$ Method. Oxford Monographs on Geology and Geophysics No 9. Oxford University Press, Oxford. 224 pp

McDowell CA (1963) Mass spectrometry. McGraw-Hill, New York

McElhinny M, Senanayake WE (1982) Variations in the geomagnetic dipole 1: The past 50,000 years. J Geomagn Geoelectr 34:39–51

McIntyre GA, Brooks C, Compston W, Turek A (1966) The statistical assessment of Rb-Sr isochrons. J Geophys Res 71:5459–5468

McIntyre IG, Pilkey OH, Stuckenrath R (1978) Relict oysters on the United States Atlantic continental shelf: a reconsideration of their usefulness in understanding late Quaternary sea-level history. Geol Soc Am Bull 89:277–278

McKay CP, Long A, Friedman EI (1986) Radiocarbon dating of open systems with bomb effect. J Geophys Res 91 (B3):3836–3840

McKee BA, DeMaster DJ, Nittrouer CA (1984) The use of $^{234}Th/^{238}U$ disequilibrium to examine the fate of particle-reactive species on the Yangtze continental shelf. Earth Planet Sci Lett 68:431–442

McKeever SWS (1982) Dating of meteorite falls using thermoluminescence. Application to Antarctic meteorites. Earth Planet Sci Lett 58:419–429

McKeever SWS (1984) Thermoluminescence of soils. Cambridge Univ Press, Cambridge, 376 pp

McKeever SWS, Townsend PD (1982) Comment on "Thermoluminescence of meteorites and their terrestrial ages". Geochim Cosmochim Acta 46:1997–2000

McKerrow WS, Lambert RSJ, Chamberlain VE (1980) The Ordovician, Silurian and Devonian time scales. Earth Planet Sci Lett 51:1–8

Mejdahl V (1979) Thermoluminescence dating: beta dose attentuation in quartz grains. Archaeometry 21:61–72

Mejdahl V (1985) Thermoluminescence dating of partially bleached sediments. Nucl Tracks 10:711–715

Mejdahl V (1988) Long-term stability of the TL signal in alkali feldspars. Q Sci Rev 7 (357–360)

Mejdahl V (1989) How far back: life times estimated from studies of feldspars of infinite ages.-In: Synopsis from a workshop on Long and Short Range Limits in Luminescence Dating. Occas Paper 9:46–51; Research Lab Archaeology and History of Art, Oxford University, Oxford

Mejdahl V, Winther-Nielsen M (1983) TL dating based on feldspar inclusions. PACT 6:426–437

Mejdahl V, Wintle AG (1984) Thermoluminescence applied to age determination in archaelogy and geology. In: Horowitz Y (ed) Thermoluminescence and Thermoluminescent Dosimetry II. CRC, Cleveland, pp 133–190

Mejdahl V, Bowman SGE, Wintle AG, Aitken MJ (coeds) (1983) 3rd specialist seminar on TL and ESR dating, Elsenore, July 1982. PACT 9:628

Melcher CL (1981) Thermoluminscence of meteorites and their terrestrial age. Geochim Cosmochim Acta 45:615–626

Melcher CL (1982) Thermoluminscence of meteorites and their terrestrial ages: reply to comments by SWS McKeever and PD Townsend. Geochim Cosmochim Acta 46:1999–2000

Menning M (1989) A synopsis of numerical time scales 1917–1986. Episodes 12:3–5

Mensing TM, Faure G (1983) Identification and age of neoformed Paleozoic feldspar (adularia) in a Precambrian basement core from Scioto County, Ohio, USA. Contrib Mineral Petrol 82:327–333

Merlivat L, Jouzel J (1979) Global climatic interpretation of the deuterium-oxygen 18 relationship for precipitation. J Geophys Res 84:5029–5033

Merrihue C, Turner G (1976) Potassium-argon dating by activation with fast neutrons. J Geophys Res 71:2852–2857

Meshick AP, Verhovsky AB, Assonov SS, Shukolyukov YA (1987) A new method of dating young uranium minerals. Dokl Akad Nauk USSR 293 (6):1476–1478

Michard A, Gurriet P, Soudant M, Albarede F (1985) Nd isotopes in French Phanerozoic shales: external vs. internal aspects of crustal evolution. Geochim Cosmochim Acta 49:601–610

Michel R, Brinkmann G, Stück R (1982) Solar cosmic-ray-produced radionuclides in meteorites. Earth Planet Sci Lett 59:33–48

Michelot J-L, Bentley HW, Brissaud I, Elmore D, Fontes JC (1984) Progress in environmental isotope studies ($^{36}Cl$, $^{34}S$, $^{18}O$) at the Stripa site. In: Isotope Hydrology 1983:207–230. IAEA, Vienna

Michels JW (1973) Dating Methods in Archaeology. Academic Press, London New York, 240 pp

Michels JW (1986) Obsidian hydration dating. Endeavour, New Ser 10:97–100

Michels JW, Marean CW, Tsong IST, Smith GA (1982) "Invisible" hydration rims on obsidian artifacts: a test case. SAS Res Rep 2:1–4

Michels JW, Tsong IST, Smith GA (1983a) Experimentally derived hydration rates in obsidian dating. Archaeometry 25:107–117

Michels JW, Tsong IST, Nelson CM (1983b) Obsidian dating and east African archeology. Science 219:361–366

Middleton J (1844) On fluorine in bones, its source, and its application to the determination of the geological age of fossil bone. Proc Geol Soc Lond 4:432–433

Miki T, Ikeya M (1982) Physical basis of fault dating with ESR. Naturwissenschaften 69:390–391

Milankovitch M (1930) Mathematische Klimalehre und astronomische Theorie der Klima-schwankungen. In: Koppen W, Geiger R (eds) Handbuch der Klimatologie, I(A). Bornträger, Berlin

Miller DS, Wagner GA (1981) Fission-track ages applied to obsidian artifacts from South America using the plateau-annealing and the track-size age-correction techniques. Nucl Tracks 5:147–155

Miller GH, Brigham-Grette J (1989) Amino acid geochronology: resolution and precision in carbonate fossils. Quatern Int 1:111–128

Miller GH, Mangerud J (1985) Aminostratigraphy of European marine interglacial deposits. Q Sci Rev 4:215–278

Miller GH, Hollin JT, Andrews JT (1979) Aminostratigraphy of UK Pleistocene deposits. Nature (Lond) 281:539–543

Milliman PM (1969) Meteorite Research (Proc). Springer, Berlin Heidelberg New York

Milne GW (ed) (1971) Mass spectrometry: techniques and applications. Wiley, New York, 521 pp

Minster JF, Allègre CJ (1979) $^{87}$R-$^{86}$Sr chronology of H chondrites: constraint and speculations on the early evolution of their parent body. Earth Planet Sci Lett 42:333–347

Minster JF, Allègre CJ (1981) $^{87}$Rb-$^{86}$Sr dating of LL chondrites. Earth Planet Sci Lett 56:89–106

Minster JF, Allègre CJ (1982) The isotopic composition of zirconium in terrestrial and extra-terrestrial samples: implications for extinct $^{92}$Nb. Geochim Cosmochim Acta 46:565–573

Minster JF, Richard L-P, Allègre CJ (1979) $^{87}$Rb-$^{86}$Sr chronology of enstatite meteorites. Earth Planet Sci Lett 44:420–440

Mitchell JG (1968) The argon-40/argon-39 method for potassium argon age determination. Geochim Cosmochim Acta 32:781–790

Mitterer RM (1975) Ages and diagenetic temperatures of Pleistocene deposits of Florida based on isoleucine epimerization in Mercenaria. Earth Planet Sci Lett 28:275–282

Molodkov A (1988) ESR dating of subfossil mollusc shells: fading of absorbed paleodose (in Russian). Geoloogia 37:113–126

Molodkov A (1989) The problem of long-term fading of absorbed paleodose on ESR-dating of Quaternary mollusc shells. Appl Radiat Isot 40:1087–1093

Monaghan MC (1987) Greenland ice $^{10}$Be concentration and average precipitation rates north of 40°N to 45°N. Earth Planet Sci Lett 84:197–203

Monaghan MC, Krishnaswami S, Thomas JH (1983) $^{10}$Be concentrations and the long-term fate of particle-reactive nuclides in five soil profiles from California. Earth Planet Sci Lett 65:51–60

Monaghan MC, Krishnaswami S, Turekian KK (1985) The global-average production of $^{10}$Be. Earth Planet Sci Lett 76:279–287

Montag RL, Seidemann DE (1981) A test of the reliability of Rb-Sr dates for selected glauconite morphologies of the Upper Cretaceous (Navesink Formation) of New Jersey. Earth Planet Sci Lett 52:285–290

Mook WG (1970) Stable carbon and oxygen isotopes of natural waters in the Netherlands. In: Isotope Hydrology 1970:163–190. IAEA, Vienna

Mook WG (1971) Paleotemperatures and chlorinities from stable carbon and oxygen isotopes in shell carbonate. Palaeogeogr Palaeoclim Palaeoecol 9:245–263

Mook WG (1983) $^{14}$C calibration curves depending on sample time-width. In: Mook WG, Waterbolk HT (eds): $^{14}$C and Archaeology: PACT 8:517–525

Mook WG (1984) Archaeological and geological interest in applying $^{14}$C AMS to small samples. Nucl Instrum Methods 233 (B5):297–302

Mook WG, Waterbolk HT (eds) (1983) $^{14}$C and Archaeology. PACT 8:525 pp. Counc Europe, Strasbourg

Mook WG, Waterbolk HT (1985) Radiocarbon Dating. Handb Archaeol 3:65. Eur Sci Found, Strasbourg

Moorbath S (1969) Evidence for the age of deposition of the Torridonian sediments of northwest Scotland. Scott J Geol 5:154–170

Moorbath S, Welke H (1969) Lead isotope studies on igneous rocks from the Isle of Skye, northwest Scotland. Earth Planet Sci Lett 5:217–230

Moore WS (1969) Oceanic concentrations of $^{228}$Radium. Earth Planet Sci Lett 6:437–446

Moore WS, Edmond JM (1984) Radium and barium in the Amazon River system. J Geophys Res 89:2061–2065

Moore WS, Krishnaswami S (1972) Coral growth rates using $^{228}$Ra and $^{210}$Pb. Earth Plan Sci Lett 15:187–191

Morton JP (1985) Rb-Sr dating of diagenesis and source age of clays in Upper Devonian black shales of Texas. Geol Soc Am Bull 96:1043–1049

Moser H, Rauert W (eds) (1980) Isotopenmethoden in der Hydrologie. Borntraeger, Berlin, 400 pp

Mügge O (1907) Radioaktivität als Ursache der Pleochroitischen Höfe des Cordierit. Zentralbl Mineral Geol Paläontol 1907:397–399

Muhs DR, Rosholt JN, Bush CA (1989) The uranium-trend dating method: principles and application for Southern California marine terrace deposits. Quartern Int 1:19–34

Müller PJ (1984) Isoleucine epimerization in Quaternary planktonic foraminifera: effects of diagentic hydrolysis and leaching, and Atlantic-Pacific intercore correlations. "Meteor" Forsch.-Ergebnisse C38:25–47

Müller PJ, Mangini A (1980) Organic carbon decomposition rates in sediments of the Pacific manganese nodule belt dated by $^{230}$Th and $^{231}$Pa. Earth Planet Sci Lett 51:94–114

Münnich KO (1957) Messung des $^{14}$C-Gehaltes von hartem Grundwasser. Naturwissenschaften 34:32–33

Münnich KO (1968) Isotopen-Datierung von Grundwasser. Naturwissenschaften 55:158–163

Münnich KO, Vogel JC (1959) $^{14}$C-Altersbestimmung von Süßwasser-Kalkablagerungen. Naturwissenschaften 46:168–169

Muller RA (1977) Radioisotope dating with a cyclotron. Science 196:489–494

Murthy VR, Patterson CC (1962) Primary isochron of zero age for meteorites and the earth. J Geophys Res 67:1161–1167

Musset AE, McCormack AG (1978) On the use of 3-dimensional plots in K-Ar dating. Geochim Cosmochim Acta 42:1877–1883

Naeser CW (1976) Fission-track dating. US Geol Surv Open-File Rep 76–190

Naeser CW (1979) Fission-track dating and geologic annealing of fission tracks. In: Jäger E, Hunziker JC (eds) Lectures in Isotope Geology. Springer, Berlin Heidelberg New York, pp 154–169

Naeser CW (1981) The fading of fission-tracks in the geological environment – data from deep drill holes. Nucl Tracks 5:248–250

Naeser CW, Forbes RB (1976) Variation of fission-track ages with depth in two deep drill holes. Trans Am Geophys Union 57:353

Naeser CW, Izett GA, Obradovich JD (1980) Fission-track and K-Ar ages of natural glasses. US Geol Surv Bull 1489:31

Naeser ND (1986) Neogene thermal history of the northern Green river basin, Wyoming – evidence from fission-track dating. Soc Econ Paleontol Mineral 14:65–72

Naeser ND, Naeser CW (1984) Fission-track dating. In: Mahaney WC (ed) Quaternary Dating Methods. Elsevier, Amsterdam, pp 87–100

Naeser ND, Naeser CW, McCulloh (1989) The Application of Fission-Track Dating to the Depositional and Thermal History of Rocks in Sedimentary Basins. In: Naeser ND, McCulloh TH. Thermal History of Sedimentary Basins. Methods and Case Histories: 157–180; New York, Berlin Heidelberg (Springer)

Nakai S, Shimizu H, Masuda A (1986) A new geochronometer using lanthanum-138. Nature (Lond) 320:433–435

Nakamura N, Tatsumoto M, Nunes PD, Unruh DM, Schwab AP, Wildeman TR (1976) 4.4-b.y.-old clast in boulder 7, Apollo 17: A comprehensive chronological study by U-Pb, Rb-Sr and Sm-Nd methods. Proc Lunar Sci Conf 7th Geochim Cosmochim Acta Suppl 7:2309–2333

Nambi KSV (1982) ESR and TL dating studies in Quaternary marine gypsum crystals. PACT 6:314

Nambi KSV (1984) Alpha-radioactivity-related upper age limit for thermoluminescence dating? Proc Indian Acad Sci 93:47–56

Nambi KSV, Aitken MJ (1986) Annual dose conversion factors for TL and ESR dating. Archaeometry 28:202–205

Neftel A, Oeschger H, Suess HE (1981) Secular non-random variations of cosmogenic carbon-14 in the terrestrial atmosphere. Earth Planet Sci Lett 56:127–147

Neftel A, Oeschger H, Schwander J, Stauffer B, Zumbrunn R (1982) Ice core sample measurements give atmospheric $CO_2$ content during the past 40,000 yr. Nature (Lond) 295:220–223

Nelson DE, Korteling RG, Stott WR (1977) Carbon-14 direct detection of natural concentrations. Science 198:507–508

Nelson DE, Loy TH, Vogel JS, Southon JR (1986a) Radiocarbon dating blood residues on prehistoric stone tools. Radiocarbon 28 (1):170–174

Nelson DE, DeNiro MJ, Schoeninger MJ, De Paolo DJ, Hare PE (1986b) Effects of diagenesis on strontium, carbon, nitrogen, and oxygen concentration and isotopic composition of bone. Geochim Cosmochim Acta 50:1941–1949

Neretnieks I (1981) Age dating of groundwater in fissured rock: influence of water volume in micropores. Water Resour Res 17:421–422

Ness G, Levi S, Couch R (1980) Marine magnetic anomaly timescales for the Cenozoic and Late Cretaceous: a precis, critique, and synthesis. Rev Geophys Space Phys 18:753–770

Netterberg F (1978) Dating and correlation of calcretes and other pedocretes. Trans Geol Soc S Afr 81:379–391

Newman S, Finkel RC, MacDougall JD (1983) $^{230}$Th-$^{238}$U disequilibrium systematics in oceanic tholeiites from 21°N on the East Pacific Rise. Earth Planet Sci Lett 65:17–33

Newman S, MacDougall JD, Finkel RC (1984a) $^{230}$Th-$^{238}$U disequilibrium in island arcs: evidence from the Aleutians and the Marianas. Nature (Lond) 308:268–270

Newman S, Finkel RC, MacDougall JD (1984b) Comparison of $^{230}$Th-$^{238}$U disequilibrium systematics in lavas from three hot spot regions: Hawaii, Price Edward and Samoa. Geochim Cosmochim Acta 48:315–324

Newton RG (1971) The enigma of the layered crusts on some weathered glasses, a chronological account of the investigations. Archaeometry 13:1–10

Nicolaysen LO (1961) Graphic interpretation of discordant age measurements on metamorphic rocks. Ann NY Acad Sci 91:198–206

Nicolis C, Nicolis G (1984) Is there a climatic attractor? Nature (Lond) 311:529–532

Nielsen H (1965) Schwefelisotope im marinen Kreislauf und das $\delta^{34}$S der früheren Meere. Geol Rundsch 55:160–172

Nielsen H (1979) Sulfur isotopes. In: Jäger E, Hunziker JC (eds) Lectures in Isotope Geology. Springer, Berlin Heidelberg New York, pp 283–312

Nielsen H, Ricke W (1964) Schwefelisotopenverhältnisse von Evaporiten aus Deutschland, ein Beitrag zur Kenntnis von $\delta^{34}$S im Meerwasser-Sulfat. Geochim Cosmochim Acta 28:577–591

Niemeyer S (1983) I-Xe and $^{40}$Ar-$^{39}$Ar analyses of silicate from the Eagle Station pallasite and the anomalous iron meterite Enon. Geochim Cosmochim Acta 47:1007–1012

Nier AO (1938) Isotopic constitution of Sr, Ba, Bi, Tl, and Hg. Phys Rev 54:275–278

Nier AO (1950) A redetermination of the relative abundances of the isotopes of carbon, nitrogen, oxygen, argon and potassium. Phys Rev 77:789–793

Niggli E, Overweel CJ (1953) An X-ray crystallographical application of the fluorine dating method of fossil bones. Proc Koninkl Nederl Akad Wetenschappen 56 B:538–542

Nijampurkar VN, Bhandari N, Vohra CP, Krishnan V (1982) Radiometric chronology of Neh-Nar glacier, Kashmir. J Glaciol 28:91–105

Nijampurkar VN, Martin JM, Maybeck M (1983) Silicon-32 as a tool to studying silicon behaviour in estuaries. Limnol Oceanogr 28:1237–1242

Ninagawa K, Takahashi N, Wada T, Yamamoto I, Yamashita N, Yamashita Y (1988) Thermoluminescence measurements of a calcite shell for dating. Quat Sci Rev 7:367–371

Nir A (1964) On the determination of tritium "age" measurements of groundwater. J Geophys Res 69:2585–2595

Nishiizumi K (1983) Measurement of $^{53}$Mn in deep-sea iron and stony spherules. Earth Planet Sci Lett 63:223–228

Nishiizumi K, Arnold JR, Elmore D, Ma X, Newman D, Gove HE (1983a) $^{36}$Cl and $^{53}$Mn in Antarctic meteorites and $^{10}$Be-$^{36}$Cl dating of Antarctic ice. Earth Planet Sci Lett 62:407–417

Nishiizumi K, Elmore D, Honda M, Arnold JR, Gove HE (1983b) Measurements of $^{129}$I in meteorites and lunar rock by tandem accelerator mass spectrometry. Nature (Lond) 305:611–612

Nishiizumi K, Lal D, Klein J, Middleton R, Arnold JR (1986) Production of $^{10}$Be and $^{26}$Al by cosmic rays in terrestrial quartz in situ and implications for erosion rates. Nature (Lond) 319:134–136

Nkomo IT, Rosholt JN (1972) A lead-isotope age and U-Pb discordance of Precambrian Gneiss from Granite Mountains, Wyoming. US Geol Surv Prof Pap 800-C: C169-C177

Notsu K, Mabuchi H, Yoshioka O, Matsuda J, Ozima M (1973) Evidence of the extinct nuclide $^{146}$Sm in "Juvinas" achondrite. Earth Planet Sci Lett 19:29–36

Nunes PD, Tatsumoto M, Knight RJ, Unruh DM, Doe BR (1973) U-Th-Pb systematics of some Apollo 16 lunar samples. Proc Lunar Sci Conf 4th:1797–1822

Nydal R (1983) Optimal number of $^{14}$C samples and accuracy in dating problems. In: Mook WG, Waterbolk HT (eds) $^{14}$C and Archaeology. PACT 8:107–122

Nydal R, Lövseth K (1983) Tracing bomb $^{14}$C in the atmosphere 1962–1980. J Geophys Res 88:3621–3642

Nydal R, Gulliksen S, Lövseth K, Skogseth FH (1984) Bomb $^{14}$C in the ocean surface 1966–1981. Radiocarbon 26:7–45

Nyquist LE (1977) Lunar Rb-Sr chronology. Phys Chem Earth 10:103–142

Oakley KP (1949) The fluorine dating method. Yearb Phys Anthropol 5:44–52

Oakley KP (1980) Relative dating of the fossil hominids of Europe. Bull Brit Museum Nat Hist Geol Ser 34 (1):1–63

Obradovich HM, Sklash MG (1986) An isotropic and geochemical study on snowmelt runoff in a small arctic watershed. Hydrol Processes 1:15–30

Obradovich JD, Sutter JF, Kunk MJ (1986) Magnetic polarity chron tie points for the Cretaceous and early Tertiary. Terra Cognita 6:140

Odin GS (ed) (1982a) Numerical dating in stratigraphy, parts I and II. Wiley, New York

Odin GS (1982b) How to measure glaucony ages. In: Odin GS (ed) Numerical dating in stratigraphy, part I. Wiley, New York, pp 387–403

Odin GS, Dodson MH (1982) Zero isotopic age of glauconies. In: Odin GS (ed) Numerical dating in stratigraphy, part I. Wiley, New York, pp 277–305

Odom AL, Rink WJ (1988) National accumulation of Schottky-Frenkel defects: Implications for a quartz geochronometer. Geology 17:55–58

Oeschger H (1988) The Ocean System—Ocean/Climate and Ocean/$CO_2$ Interactions. In: Roswall T, Woodmansee RG, Risser PG (eds) Scales and Global Changes. Spatial and Temporal Variability in the Biospheric and Geospheric Processes. Wiley, New York, 319 pp

Oeschger H, Loosli HH (1977) New developments in sampling and low-level counting of natural radioactivity. Low-Radioactivity Measurements and Applications. Slovenske Pedagogicke Nakladatelstvo, Bratislava, pp 13–22

Oeschger H, Siegenthaler U (1978) The dynamics of the carbon cycle as revealed by isotope studies. In: Williams J (ed) Carbon Dioxide, Climate and Society. Pergamon Press, London pp 45–61

Oeschger H, Wahlen M (1975) Low level counting techniques. Ann Rev Nucl Sci 25:423–463

Oeschger H, Houtermans J, Loosli H, Wahlen M (1970) The constancy of cosmic radiation from isotopic studies in meteorites and on the earth. In: Olsson IU (ed) Radiocarbon Variations and Absolute Chronology. Almqvist Wiksell, Stockholm, pp 471–498

Oeschger H, Siegenthaler U, Schotterer U, Gugelmann A (1975) A box diffusion model to study the carbon dioxide exchange in nature. Tellus 27:168–192

Oeschger H, Stauffer B, Bucher P, Lossli HH (1977a) Extraction of gases and dissolved and particulate matter from ice in deep boreholes. In: Rodda JC (ed) Isotopes and impurities in snow and ice. IAHS PUBL. No 118, Grenoble Symp, pp 307–311

Oeschger H, Schotterer U, Stauffer B, Haeberli W, Röthlisberger H (1977b) First results from alpine core drilling project. Z Gletscherkunde Glazialgeol 13:193–208

Oestlund HG, Stuiver M (1980) GEOSECS Pacific radiocarbon. Radiocarbon 22 (1):25–53

Ohmoto H (1972) Systematics of sulfur and carbon isotopes in hydrothermal ore deposits. Econ Geol 67:551–578

Oldfield F, Appleby PG, Battarbee RW (1978) Alternative $^{210}$Pb dating: results from the new Guinea Highlands and Lough Erne. Nature (Lond) 271:339–342

Olson EA (1963) The problem of sample contamination in radiocarbon dating. PhD Thesis, Columbia University, New York, 320 pp

Olsson IU (ed) (1970) Radiocarbon Variations and Absolute Chronology. Almqvist Wiksell, Stockholm, 656 pp

Olsson IU (1979a] The radiocarbon contents of various reservoirs. In: Berger R, Suess HE (eds) Radiocarbon Dating. Univ California Press, Los Angeles, pp 613–618

Olsson IU (1979b) A warning against radiocarbon dating of samples containing little carbon. Boreas 8:203–207

Olsson IU (1980a) $^{14}$C in extractives from wood. Radiocarbon 22:515–524

Olsson IU (1980b) Content of $^{14}$C in marine mammals from northern Europe. Radiocarbon 22(3):662–675

Olsson IU (1983) Dating of non-terrestrial materials. In: Mook WG, Waterbolk HT (eds) $^{14}$C and Archaeology. PACT 8:277–294

Olsson IU, Göksu Y, Stenberg A (1968) Further investigations of storing and treatment of foraminifera and molluscs for C$^{14}$ dating. Geol Fören Stockh Förh 90:417–426

Olsson IU, Holmgren B, Skye E (1984) Questions arising when using lichen for $^{14}$C measurements in climatic studies. In: Mörner N-A, Karlèn W (eds) Climatic Changes on a Yearly to Millennial Basis. Reidel, Dordrecht, pp 303–308

Omoto K (1983) The problem and significance of radiocarbon geochronology in Antarctica. In: Oliver RL, James PR, Jago JB (eds) Antarctic Earth Science, Australian Academy of Science, Canberra, pp 450–452

O'Neil JR, King Chi-Yu (1981) Variations in stable-isotope ratios of ground waters in seismically active regions of California. Geophys Res Lett 8:429–432

O'Nions RK, Hamilton PJ, Evensen NM (1976) Variations in $^{143}$Nd/$^{144}$Nd and $^{87}$Sr/$^{86}$Sr ratios in oceanic basalts. Earth Planet Sci Lett 34:13–22

O'Nions RK, Evensen NM, Hamilton PJ (1979) Geochemical modeling of mantle differentiation and crustal growth. J Geophys Res 84:6091–6101

O'Nions RK, Hamilton PJ, Hooker PJ (1983) A Nd isotope investigation of sediment related to crustal development in the British Isles. Earth Planet Sci Lett 63:229–240

Oosthuyzen EJ, Burger AJ (1973) The suitability of apatite as an age indicator by the uranium-lead isotope method. Earth Planet Sci Lett 18:29–36

Opdyke ND (1972) Paleomagnetism of deep-sea cores. Rev Geophys Space Phys 10:213–249

Ortlam D (1983) Einsatz und Möglichkeiten von Pollution-Tracer-Verfahren am Beispiel der Weser. N Jahrb Geol Paläont Abh 165:303–325

Ortner DJ, von Endt DW, Robinson MS (1972) The effect of temperatures on protein decay in bone. Its significance in nitrogen dating of archaeological specimens. Am Antiq 37:514–520

Osmond JK, Cowart JB (1976) The theory and uses of natural uranium isotopic variations in hydrology. Atom Energy Rev 144:621–679

Osmond JK, Carpenter JR, Window HL (1965) $^{230}$U/$^{234}$U age of the pleistocene corals and oolithes of Florida. J Geophys Res 70:1843–1847

Osmond JK, Kaufman MI, Cowart JB (1974) Mixing volume, calculations, sources and aging trends for Floridan aquifer water by uranium isotopic methods. Geochim Cosmochim Acta 38:1083–1100

Osmond JK, Cowart JB, Ivanovich M (1983) Uranium isotopic disequilibrium in ground water as an indicator of anomalies. Int J Appl Radiat Isot 34:283–308

Ostic RG, Russel RD, Reynolds PH (1963) A new calculation for the age of the earth from abundances of lead isotopes. Nature (Lond) 199:1150–1152

Otlet RL, Huxtable G, Evans GV, Humphreys DC, Short TD, Conchie SJ (1983a) Development and operation of the Harwell small counter facility for the measurement of $^{14}$C in very small samples. Radiocarbon 25 (2):565–576

Otlet RL, Walker AJ, Longley H (1983b) The use of $^{14}$C in natural materials to establish the average gaseous dispersion patterns of releases from nuclear installations. Radiocarbon 25:593–602

Ottaway BS (1973) Dispersion diagrams: a new approach to the display of $^{14}$C dates. Archaeometry 15:5–12

Ottaway BS (ed) (1983) Archaeology, Dendrochronology and the Radiocarbon Calibration Curve. Occass Pap 9:100. Univ Edinbourgh

Ovchinnikova GV, Levchenkov OA, Varshavskaya ES, Kutyavin EP, Yakovleva SZ (1980) Comparative study of K-Ca, Rb-Sr and K-Ar systems of lepidolites. Geokhimya 8:1166–1173

Oversby VM (1974) A new look at the lead isotope growth curve. Nature (Lond) 248:132–133

Oversby VM (1975) Lead isotopic systematics and ages of Archaean acid intrusives in the Kalgoorlie-Norseman area, Western Australia. Geochim Cosmochim Acta 39:1107–1125

Oversby VM (1976) Isotopic ages and geochemistry of Archaean acid igneous rocks from the Pilbara, Western Australia. Geochim Cosmochim Acta 40:817–829

Owen LB (1974) Age determinations by the lutetium-176/hafnium-176 method. PhD thesis, Dept Geol Mineral, Ohio State Univ, Columbus, Ohio (unpublished)

Özer AM, Wieser A, Göksu HY, Müller P, Regulla DF, Erol O (1989) ESR and TL age determination of caliche nodules. Appl Rad Isot 40: 1159–1162

Ozima M (1975) Ar isotopes and Earth-atmosphere evolution models. Geochim Cosmochim Acta 39:1127–1134

Ozima M, Zashu S (1983) Noble gases in submarine pillow volcanic glasses. Earth Planet Sci Lett 62:24–40

Ozima M, Podosek FA, Igarashi G (1985) Terrestrial xenon isotope constraints on the early history of the Earth. Nature (Lond) 315:471–474

Pal DK, Tuniz C, Moniot RK, Kruse TH, Herzog GF (1982) Beryllium-10 in Australasian tektites: evidence for a sedimentary precursor. Science 218:787–789

Palmer AR (1983) The decade of North American geology: 1983 geologic time scale. Geology 11:503–504

Pankhurst RJ, O'Nions RK (1973) Determination of Rb/Sr and $^{87}$Sr/$^{86}$Sr ratios of some standard rocks and evolution of X-ray fluoresence spectrometry in Rb-Sr geochronology. Chem Geol 12:127–136

Pankhurst RJ, Smellie JL (1983) K-Ar geochronology of the South Shetland Island, Lesser Antarctica: apparent lateral migration of Jurassic to Quaternary island arc volcanism. Earth Planet Sci Lett 66:214–222

Papanastassiou DA, Wasserburg GJ (1971) Rb-Sr ages of igneous rocks from the Apollo 14 mission and the age of the Fra Mauro formation. Earth Planet Sci Lett 12:36–48

Papanastassiou DA, Wasserburg GJ (1973) Rb-Sr ages and initial strontium in basalts from Apollo 15. Earth Planet Sci Lett 17:324–337

Papanastassiou DA, Wasserburg GJ (1981) Microchrons: the $^{87}$Rb-$^{87}$Sr dating of microscopic samples. Proc 12$^{th}$ Lunar Planet Sci Conf, 1027 pp

Papanastassiou DA, DePaolo DJ, Wasserburg GJ (1977) Rb-Sr and Sm-Nd chronology and genealogy of basalts from the Sea of Tranquility. Proc 8th Lunar Sci Conf, pp 1639–1672

Paquette JL, Peucat J-J, Bernard-Griddiths J, Marchand J (1985) Evidence for old Precambrian relics shown by U-Pb zircon dating of ecologites and associated rocks in the Hercynian belt of south Brittany, France. Chem Geol Isot Geosci Sec 52:203–216

Parrish R, Krogh T (1987) Synthesis and purification of $^{205}$Pb for U-Pb geochronology. Chem Geol Isot Geosci Sect 66:103–110

Patchett PJ (1983a) Importance of the Lu-Hf isotopic system in studies of planetary chronology and chemical evolution. Geochim Cosmochim Acta 47:81–91

Patchett PJ (1983b) Hafnium isotope results from mid-ocean ridges and Kerguelen. Lithos 16:47–51

Patchett PJ, Bridgwater D (1984) Origin of continental crust of 1.9–1.7 Ga age defined by Nd isotopes in the Kitilidian terrain of South Greenland. Contrib Mineral Petrol 87:311–318

Patchett PJ, Tatsumoto M (1980a) Lu-Hf total-rock isochron for the eucrite meteorites. Nature (Lond) 288:571–574

Patchett PJ, Tatsumoto M (1980b) A routine high-precision method for Lu-Hf isotope geochemistry and chronology. Contrib Mineral Petrol 75:263–267

Patchett PJ, Tatsumoto M (1981) Lu/Hf in chondrites and definition of a chondritic hafnium growth curve. Lunar Planet Sci XII:822–824

Patchett PJ, Kouvo O, Hedge CE, Tatsumoto M (1981) Evolution of continental crust and mantle heterogeneity: evidence from Hf isotopes. Contrib Mineral Petrol 78:279–297

Patchett PJ, White WM, Feldmann H, Kielinczuk S, Hofmann AW (1984) Hafnium/rare earth element fractionation in the sedimentary system and crustal recycling into the Earth's mantle. Earth Planet Sci Lett 69:365–378

Paterne M, Guichard F, Labeyrie J, Gillot PY, Duplessy JC (1986) Tyrrhenian Sea tephrochronology of the oxygen isotope record for the past 60,000 years. Mar Geol 72:269–285

Patterson CC (1955) The $Pb^{207}/Pb^{206}$ ages of some stone meteorites. Geochim Cosmochim Acta 7:151–153

Patterson CC (1956) Age of meteorites and the earth. Geochim Cosmochim Acta 10:230–237

Paul M, Kaufman A, Magaritz M, Fink D, Hennig W, Kaim R, Kutschera W, Meirav O (1986) A new $^{36}Cl$ hydrological model and $^{36}Cl$ systematics in the Jordan River/Dead Sea system. Nature (Lond) 321:511–515

Paul M, Fink D, Hollos G, Kaufman A, Kutschera W, Magaritz M (1987) Measurement of $^{129}I$ concentrations in the environement after the Chernobyl reactor accident. Nucl Instrum Methods Phys Rev B 29:341–345

Pavich MJ, Brown L, Harden J, Klein J, Middleton R (1986) $^{10}Be$ distribution in soils from Merced River terraces, California. Geochim Cosmochim Acta 50:1727–1735

Pavlov AN (1970) On the determination of the ages of subterranean waters by means of the helium-argon method. Sowj Geol 13:140–148 (in Russian)

Pazdur A (1988) The relations between carbon isotope composition and apparent age of freshwater tuffaceous sediments. Radiocarbon 30:7–18

Pearson FJ Jr, Rightmire CT (1980) Sulfur and oxygen isotopes in aqueous sulfur compounds. In: Fritz P, Fontes JCH (eds) Handbook of Environmental Isotope Geochemistry, vol 1. The Terrestrial Environment A. Elsevier, Amsterdam, pp 227–258

Pearson FJ Jr, Swarzenki WV (1974) $^{14}C$ Evidence for the origin of arid region groundwater, northern province, Kenya. In: Isotope Techniques in Groundwater Hydrology 1974 (II):95–108. IAEA, Vienna

Pearson FJ Jr, Noronha CJ, Andrews RW (1983) Mathematical modeling of the distributing of natural 14-C, 234-U in a regional groundwater system. Radiocarbon 25:291–300

Pearson GW (1986) Precise calendar dating of known growth-period samples using a 'curve fitting' technique. Radiocarbon 28 (2A):292–299

Peng T-H, Broecker WS (1984) The impact of bioturbation on the age difference between benthic and planktonic foraminifers in deep sea sediments. Nucl Instrum Methods 233 (B5):346–352

Pepin R, Bradley J, Dragon J, Nyquist L (1972) K-Ar dating of lunar fines: Apollo 12, Apollo 14, and Luna 16. Proc Lunar Sci Conf 3:1569–1588

Pernicka E, Wagner GA (1982) Radioactive equilibrium and dose rate determination in TL dating. PACT 6:132–144

Peterman ZE, Hildreth RA, Nkomo IT (1971) Precambrian geology and geochronology of the Granite Mountains, Central Wyoming (abs.). Geol Soc Am Abstr with Programs 3, 6:403–404

Petit J-C, Langevin Y, Dran J-C (1985) $^{234}U/^{238}U$ disequilibrium in nature. Theoretical reassessment of the various proposed models. Bull Mineral 108:745–753

Pettersson H (1937) Das Verhältnis Thorium zu Uran in dem Gestein und im Meer. Anz Akad Wiss Wien Math Naturwiss Kl:127

Pettingill HS, Patchett PJ (1981) Lu-Hf total-rock age for the Amitsoq gneisses, West Greenland. Earth Planet Sci Lett 55:150–156

Pettingill HS, Sinha AK, Tatsumoto M (1984) Age and origin of anorthosites, charnockites, and granulites in the Central Virginia Blue Ridge: Nd and Sr isotopic evidence. Contrib Mineral Petrol 85:279–291

Peucat JJ, Tisserant D, Caby R, Clauer N (1985) Resistance of zircons to U-Pb resetting in a prograde metamorphic sequence of Caledonian age in east Greenland. Can J Earth Sci 22:330–338

Phillips FM, Smith GI, Bentley HW, Elmore D, Grove HE (1983) Chlorine-36 dating of saline sediments: preliminary results from Searles Lake, California. Science 222:925–927

Phillips FM, Goff F, Vuataz F, Bentley HW, Elmore D, Gove HE (1984a) $^{36}Cl$ as a tracer in geothermal systems: example from Valles Caldera, New Mexico. Geophys Res Lett 11, 12:1227–1230

Phillips FM, Trotman KN, Bentley HW, Davis SN, Elmore D (1984b) Chlorine-36 from atmospheric nuclear weapons testing as a hydrologic tracer in the zone of aeration in arid climates. Proc RIZA Symp, Munich

Phillips FM, Leavy BD, Jannik NO, Elmore D, Kubik PW (1986a) The accumulation of cosmogenic chlorine-36 in rocks: a method for surface exposure dating. Science 231:41–43

Phillips FM, Bentley HW, Davis SN, Elmore D, Swanick GB (1986b) Chlorine-36 dating of very old groundwater 2. Milk River aquifer, Alberta, Canada. Water Resour Res 22, 13:2003–2016

Phillips FM, Mattick JL, Duval TA, Elmore D, Kubik PW (1988) Chlorine 36 and tritium from nuclear weapons fallout as tracers for long-term liquid and vapor movement in desert soil. Water Res Res 24:1877–1891

Phinney D (1972) 36/Ar, Kr, and Xe in terrestrial materials. Earth Planet Sci Lett 16:413–420

Piasecki MAJ (1985) Isotopic dating of the time of movements in ductile shear zones. Terra Cognita 5:237–238

Picciotto E, Wilgain S (1954) Thorium determination in deep-sea sediments. Nature (Lond) 173:632

Pidgeon RT, O'Neil RJ, Silver LT (1973) Observations on the crystallinity and the U-Pb system of a metamict Ceylon zircon under experimental hydrothermal conditions. Fortschr Mineral 50:118

Pidgeon RT, Compston W, Wilde SA, Baxter JL, Collins LB (1986) Archaean evolution of the Jack Hills metasedimentary belt, Yilgarn Block, Western Australia. Terra Cognita 6:146

Piggot CS, Urry WMD (1939) The radium content of an ocean bottom core. Washington Acad Sci J 29:405–415

Piggot CS, Urry WMD (1942) Time relations in ocean sediments. Bull Geol Soc Am 53:1187–1210

Pilcher JR, Baillie MGL, Schmidt B, Becker B (1984) A 7,272-year tree-ring chronology for western Europe. Nature (Lond) 312:150–152

Pilot J, Rösler HJ (1967) Altersbestimmungen von Kalisalzmineralen. Naturwissenschaften 54:490

Piper JDA (1982) The Precambrian palaeomagnetic record: the case for the Proterozoic Supercontinent. Earth Planet Sci Lett 59:61–89

Pisias NG (1983) Geologic time series from deep-sea sediments: time scales and distortion by bioturbation. Mar Geol 51:99–113

Pisias NG, Shackleton NJ (1984) Modelling the global climate response to orbital forcing and atmospheric carbon dioxide changes: a frequency domain approach. Nature (Lond) 310:757–759

Podosek FA, Honda M, Ozima M (1980) Sedimentary noble gases. Geochim Cosmochim Acta 44:1875–1884

Polach D (1988) Radiocarbon Dating Literature. Academic Press, London. 370 pp

Polach HA (1987a) Perspectives in radiocarbon dating by radiometry. Nucl Instrum Methods Phys Res B 29:415–423

Polach HA (1987b) Evaluation and status of liquid scintillation counting for radiocarbon dating. Radiocarbon 29(1):1–11

Polach HA, Kojola H, Nurmi J, Soini E (1984) Multiparameter liquid scintillation spectrometry. Nucl Instrum Methods Phys Res B5:439–442

Poupeau G (1981) Precision, accuracy and meaning of fission track ages. Proc Earth Planet Sci 90:403–436

Preuszer F (1978) Differential thermal analysis of paint samples. ICOM Comm for Conservation (V Triennial Meeting), Zagreb, 20 Feb 1978

Preuszer F (1979) Untersuchung von Werken der Kunst- und Kulturgeschichte mit Hilfe der Differentialthermoanalyse I. Altersbestimmung von Ölgemälden. J Thermal Anal 16:277–283

Prevot M, Mankinen EA, Gromme CS, Coe RS (1985) How the geomagnetic field reverses. Nature (Lond) 316:230–234

Price PB, Walker RM (1962) Observation of fossil particle tracks in natural micas. Nature (Lond) 196:732–734

Price PB, Walker RM (1963) Fossil tracks of charged particles in mica and the age of minerals. J Geophys Res 68:4847–4862

Price WJ (1972) Analytical atomic absorption spectrometry. Heyden, London

Przewlocki K, Yurtsever Y (1974) Some conceptual mathematical models and digital simulation approach in the use of tracers in hydrological systems. In: Isotope Techniques in Groundwater Hydrology 1974 (II):425–448. IAEA, Vienna

Pye K (1982) Thermoluminescence dating of sand dunes. Nature (Lond) 299:376

Radicati di Brozolo F, Huneke JC, Papanastassiou PA, Wasserburg GJ (1981) $^{40}$Ar-$^{39}$Ar and Rb-Sr age determinations on Quaternary volcanic rocks. Earth Planet Sci Lett 53:445–456

Radtke U, Grün R (1988) ESR dating of corals. Quat Sci Rev 7:465–470

Radtke U, Mangini A, Grün R (1985) ESR Dating of marine fossil shells. Nucl Tracks 10:879–884

Rae A, Hedges REM (1989) Further studies for uranium-series dating of fossil bone. Appl Geochem 4:331-337

Rae AM, Ivanovich M (1986) Successful application of uranium series dating of fossil bone. Appl Geochem 1:419–426

Raisbeck GM, Yiou F (1979) Possible use of $^{41}$Ca for radioactive dating. Nature (Lond) 277:42–44

Raisbeck GM, Yiou F, Fruneau M, Loiseaux JM (1978) Beryllium-10 mass spectrometry with a cyclotron. Science 202:215–217

Raisbeck GM, Yiou F, Stephan C (1979) $^{26}$Al measurement with a cyclotron. J Phys Lett 40:241–244

Raisbeck GM, Yiou F, Fruneau M, Loiseaux JM, Lieuvin M, Ravel JC, Lorius J (1981) Cosmogenic $^{10}$Be concentrations in Antarctic ice during the past 30,000 years. Nature (Lond) 292:825–826

Raisbeck GM, Yiou F, Klein J, Middleton R (1983) Accelerator mass spectrometry measurement of cosmogenic $^{26}$Al in terrestrial and extraterrestrial matter. Nature (Lond) 301:690–692

Raisbeck GM, Yiou F, Bourles D, Kent DV (1985) Evidence for an increase in cosmogenic $^{10}$Be during a geomagnetic reversal. Nature (Lond) 315:315–317

Rajan RS, Huneke JC, Smith SP, Wasserburg GJ (1979) Argon$^{40}$-argon$^{39}$ chronology of lithic clasts from the Kapoeta howardite. Geochim Cosmochim Acta 43:957–971

Ramdohr P (1960) Neue Beobachtungen an radioaktiven Höfen in verschiedenen Mineralien mit kritischen Bemerkungen zur Auswertung der Höfe zur Altersbestimmung. Geol Rundsch 49:253–263

Rampino MR (1979) Possible relationships between changes in global ice volume, geomagnetic excursions, and the eccentricity of the Earth's orbit. Geology 7:584–587

Reardon EJ, Fritz PE (1978) Computer modelling of groundwater $^{13}$C and $^{14}$C isotope compositions. J Hydrol 36:201–224

Reedy RC, Arnold JR, Lal D (1983) Cosmic-ray record in solar system matter. Ann Rev Nucl Part Sci 33:505–537

Reeh N (1989) Dating of ice flow modeling: a useful tool or an exercise in applied mathematics? In: Oeschger H, Langway Jr CC (eds) The Environmental Records in Glaciers and Ice Sheets. Wiley, New York, pp 141–160

Regulla DE, Wieser A, Göksu HY (1985) Effects of sample preparation on the ESR spectra of calcite, bone and volcanic material. Nucl Tracks 10:825–830

Rendell HM, Mann SJ, Townsend PD (1988) Spectral measurements of loess TL. Nucl Tracks Radiat Meas 14:63–72

Reymer APS (1982) Indirect dating of tectonic events by Rb-Sr analysis of syntectonic garnets: an example from schists of the Seve nappe, central Scandinavian Caledonides. Geol Mag 119:599–604

Reynolds JH (1963) Xenology. J Geophys Res 68:2939–2956

Reynolds PH, Muecke GK (1978) Age studies on slates: applicability of the $^{40}$Ar/$^{39}$Ar stepwise outgassing method. Earth Planet Sci Lett 40:111–118

Reyss JL, Yokoyama Y, Tanaka S (1976) Aluminum-26 in deep-sea sediment. Science 193:1119–1121

Reyss JL, Yokoyama Y, Duplessy JC (1978) A rapid determination of oceanic sedimentation rates by non-destructive gamma-gamma coincidence spectrometry. Deep-Sea Res 25:491–498

Reyss JL, Yokoyama Y, Guichard F (1981) Production of cross sections of $^{26}$Al, $^{22}$Na, $^{7}$Be from argon and of $^{10}$Be, $^{7}$Be from nitrogen: implications for production rates of $^{26}$Al and $^{10}$Be in the atmosphere. Earth Planet Sci Lett 53:203–210

Rhodes EG, Polach HA, Thom BG, Wilson SR (1980) Age structure of Holocene coastal sediments. Gulf of Carpentaria, Australia. Radiocarbon 22 (III):718–727

Rhodes EJ (1988) Methodical considerations on the optical dating of quartz. Q Sci Rev 7:395–400

Richards JR, Compston W, Paterson RG (1982) Isotopic information on the Ardlethan tinfield, New South Wales. Proc Austr Inst Mining Metallurgy 284:11–16

Rittmann K (1984) Argon in Hornblende, Biotit und Muskovit bei der geologischen Abkühlung— $^{40}$Ar/$^{39}$Ar-Untersuchungen. PhD thesis, Univ Heidelberg (unpublished)

Robbins JA (ed) (1984) Geochronology of recent deposits. Chem Geol 44:1–340

Robbins JA, Edgington DN (1975) Determination of recent sedimentation rates in Lake Michigan using Pb-210 and Cs-317. Geochim Cosmochim Acta 39:285–304

Robbins JA, Seeley NJ, MacNeil DAC, Symons MCR (1978) Identification of ancient heat treatment in flint artifacts by ESR spectroscopy. Archaeometry 23:103–107

Roberts JH, Gold R, Armani RJ (1968) Spontaneous fission decay constant of $^{238}$U. Phys Rev 174:1482–1484

Robin GdeQ (ed) (1983) The climatic record in polar ice sheets. Univ Press, Cambridge, pp 180–195

Robinson SW (1981) Natural and man-made radiocarbon as a tracer for coastal upwelling process. Coast Upwelling Coast Estuar Sci 1:298–302

Roddick JC (1978) The application of isochron diagramms in $^{40}$Ar-$^{39}$Ar dating: a discussion. Earth Planet Sci Lett 41:233–244

Roddick JC (1983) High-precision intercalibration of $^{40}$Ar-$^{39}$Ar standards. Geochim Cosmochim Acta 47:887–898

Roddick JC, Loveridge DW, Parrish R (1987) Precise U/Pb dating of zircon at the sub-nanogram Pb level. Chem Geol Isot Geosci Sect 66:111–121

Roe KK, Burnett WC (1985) Uranium geochemistry and dating of Pacific island apatite. Geochim Cosmchim Acta 49:1581–1592

Roe KK, Burnett WC, Lee AIN (1983) Uranium disequilibrium dating of phosphate deposits from the Lau Group, Fiji. Nature (Lond) 302:603–606

Roether W, Münnich KO, Schoch H (1980) On the $^{14}$C to tritium relationship in the North Atlantic ocean. Radiocarbon 22 (I):636–646

Roman D, Airey PL (1981) The application of environmental chlorine-36 to hydrology-I. Liquid scintillation counting. Int J Appl Radiat Isot 32:287–290

Rosen AA, Rubin M (1965) Discriminating between natural and industrial pollution through carbon dating. J Water Poll Centr Fed Wash 37:1302–1307

Rosholt JN Jr (1957) Quantitative radiochemical methods for the determination of the source of natural activity. Anal Chem 29:1398–1408

Rosholt JN (1967) Open system model for uranium-series dating of Pleistocene samples. In: Radioactive Dating and Methods of Low-Level Counting: 299–311. IAEA, Vienna

Rosholt JN (1978) Uranium-trend dating of alluvial deposits. Geol Surv Open-File Rep 78–701:360–362

Rosholt JN (1980) Uranium-trend dating of Quaternary sediments. US Geol Surv Open-File Rep 80–1087

Rosholt JN, Bartel AJ (1969) Uranium, thorium, and lead systematics in Granite Mountains, Wyoming. Earth Planet Sci Lett 7:141–147

Rosholt JN Jr, Emiliani C, Geis J, Kozy FF, Wangersky JP (1961) Absolute dating of deep-sea core by the $^{231}$Pa/$^{230}$Th method. J Geol 69:162–185

Roth E, Poty B (1985) Méthodes de datation par les phenomènes nucleaires naturels: Applications. Collection CEA. Masson, Paris, 642 pp

Roth E, Poty B (eds) (1989) Nuclear Methods of Dating. Dordrecht (Kluwer)

Rozanski K (1985) Deuterium and oxygen-18 content in European groundwaters—links to atmospheric circulation in the past. Chem Geol 52:349–363

Rozanski K, Dulinski M (1987) Deuterium content of European paleowaters as inferred from isotopic composition of fluid inclusions trapped in carbonate cave deposits. In: Isotope Techniques in Water Resources Development: 565–578; IAEA Vienna

Rozanski K, Florkowski T (1979) Krypton-85 dating of groundwater. In: Isotope Hydrology 1978 (II):949–959. IAEA, Vienna

Ruddiman WF, McIntyre A (1981) The North Atlantic ocean during the last deglaciation. Palaeogeogr Palaeoclim Palaeoecol 35:145–214

Ruddiman WF, McIntyre A (1984) Ice-age thermal response and climatic role of the surface atlantic ocean 40°N to 63°N. Bull Geol Soc Am 95:381–396

Ruiz J, Jones LM, Kelly WC (1984) Rubidium-strontium dating of ore deposits hosted by Rb-rich rocks, using calcite and other common Sr-bearing minerals. Geology 12:259–262

Russ GP, Bazan JM (1986) $^{187}$Re-$^{187}$Os chronology using ICP-MS. Terra Cognita 6:150

Russel WA, Papanastassiou DA, Tombrello TA (1978) Ca isotope fractionation on the Earth and other solar system materials. Geochim Cosmochim Acta 42:1075–1090

Rutherford E (1906) Radioactive Transformations. Scribner, New York

Rutter NW, Crawford RJ (1984) Utilizing wood in amino acid dating. In: Mahaney WC: Quaternary Dating Methods. Elsevier, Amsterdam, 195–202

Rutter N, Brigham-Grette J, Catto N (eds) (1989) Applied Quaternary Geochronology. Quatern Int 1:1–166

Sackett WM (1960) The protactinium-231 content of ocean water and sediments. Science 132:1761–1762

Sakanoue M, Konishi K, Komura K (1967) Stepwise determinations of thorium, protactinium and uranium isotopes and their applications in geochronological studies. In: Radioactive Dating and Methods of Low-Level Counting: 313–329. IAEA, Vienna

Sales KD, Oduwole AD, Robins GV, Olsen S (1985) The radiation and thermal dependence of ESR signals in ancient and modern bones. Nucl Tracks 10:845–851

Saliege JF, Fontes JC (1984) Essai de determination expérimentale du fractionnement des isotopes $^{13}C$ and $^{14}C$ du carbone au cours de processus naturels. Int J Appl Radiat Isot 35:55–62

Salomons W, Mook WG (1976) Isotope geochemistry of carbonate dissolution and reprecipitation in soils. Soil Sci 122:15–24

Salvamoser J (1986) Quantitative separation of admixed young groundwater and surface water with the $^3H$-$^{85}Kr$ method. Conjunctive Water Use, IAHS 156:355–363

Sammet F, Herr W (1963) Studies on the cosmic-ray produced nuclides $Be^{10}$, $Al^{26}$ and $Cl^{36}$ in iron meteorites. In: Radioactive Dating: 343–354. IAEA, Vienna

Sano Y (1986) Helium flux from the solid earth. Geochem J 20:227–232

Santos ES, Ludwig KR (1983) Age of uranium mineralization at the Highland mine, Powder River basin, Wyoming, as indicated by U-Pb isotope analyses. Econ Geol 78:498–501

Santschi PH, Li Y-H, Alder DM, Amdurer M, Bell J, Nyfeller UP (1983) The relative mobility of natural (Th, Pb and Po) and fallout (Pu, Am, Cs) radionuclides in the coastal marine environment: result from model ecosystems (MERL) and Narragansett Bay. Geochim Cosmochim Acta 47:201–210

Sarafin R, Bonani G, Herpers U, Signer P, Hofmann H, Nessi M, Morenzoni E, Suter M, Wieler R, Wölfli W (1984) Spallogenic nuclides in meteorites by conventional and accelerator mass spectrometry. Nucl Instrum Methods Phys Res B 29:411–414

Sarda P, Staudacher T, Allègre CJ (1985) $^{40}Ar/^{36}Ar$ in MORB glasses: constraints on atmosphere and mantle evolution. Earth Planet Sci Lett 72:357–375

Sarich VH, Wilson AC (1967) Immunological time scale for hominoid evolution. Science 158:1200–1203

Sato T (1981) ESR dating of calcareous fossils in deep-sea sediment. Rock Magnetism and Paleogeophysics 8:85–88

Sato T (1982) ESR dating of planktonic foraminifera. Nature (Lond) 300:518–521

Saupe F, Strappa O, Coppens R, Guillet B, Jaegy R (1980) A possible source of error in $^{14}C$ dates: volcanic emanations (examples from the Monte Amiata district, provinces of Grosseto and Sienna, Italy). Radiocarbon 22 (II): 525–531

Sayre EV, Harbottle GY, Stoenner RW, Washburn W, Olin JS, Fitzhugh W (1982) The carbon-14 dating of an iron bloom associated with the voyages of Sir Martin Frobisher. In: Currie LA (ed) Nuclear and Chemical Dating Techniques. ACS, Washington DC, pp 441–451

Schaeffer GA, Schaeffer OA (1977) $^{39}Ar$-$^{40}Ar$ ages of lunar rocks. Geochim Cosmochim Acta Suppl 10:2253–2300

Schaeffer OA (1967) Direct dating of fossils by the helium-uranium method. In: Radioactive Dating and Methods of Low-Level Counting: 395–402. IAEA, Vienna

Schaeffer OA, Zähringer J (eds) (1966) Potassium-Argon Dating. Springer, Berlin Heidelberg New York

Schärer U (1984) The effect of initial $^{230}Th$ disequilibrium on young U-Pb ages: the Makalu case, Himalaya. Earth Planet Sci Lett 67:191–204

Scharpenseel HW (1979) Soil fraction dating. In: Berger R, Suess HE (eds) Radiocarbon Dating. Univ California Press, Los Angeles, pp 277–283

Scharpenseel HW, Schiffmann H (1977) Radiocarbon dating of soils. A review. Z Pflanzenernähr Bodenkde 140:159–174

Schell WR, Swanson JR, Currie LA (1983) Anthropogenic changes in organic carbon and trace metal input to Lake Washington. Radiocarbon 25:621–628

Schiegl WE (1972) Deuterium content of peat as a paleoclimatic recorder. Science 175:512–513

Schlax M, Oldenburg DW (1984) Age bounds from lead isotope data. Earth Planet Sci Lett 68:413–421

Schleicher H, Baumann A, Keller J (1990) Pb isotopic systematics of alkaline volcanic rocks and carbonatites from the Kaiserstuhl, Upper Rhine rift valley, FRG (submitted to Chem Geol)

Schleicher H, Lippolt HJ, Raczek I (1983) Rb-Sr systematics of Permian volcanites in the Schwarzwald (SW Germany). Part II: Age of eruption and the mechanism of Rb-Sr whole rock age distortions. Contrib Mineral Petrol 84:281–291

Schleicher H, Keller J, Kramm U (1990) Isotope studies on alkaline volcanics and carbonatites from the Kaiserstuhl, Fed. Rep. Germany. Lithos spec issue. "Alkaline Rocks and Carbonatites" (in press)

Schlesinger WH (1985) The formation of caliche in soils of the Mojave Desert, California. Geochim Cosmchim Acta 49:57–66

Schlosser P, Stute M, Dörr H, Sonntag CH, Münnich KO (1988) Tritium/$^3$He dating of shallow groundwater. Earth Planet Sci Lett 89:353–362

Schoch-Fischer H, Rozanski K, Jacob H, Sonntag C, Jouzel I, Östlund G, Geyh M (1984) Hydrometeorological factors controlling the time variation of D, $^{18}$O and $^3$H in atmospheric water vapour and precipitation in the northern westwind belt. In: Isotope Hydrology 1983:3–30. IAEA, Vienna

Schoell M (1983) Genetic characterisation of natural gases. Am Ass Petrol Geol Bull 67:2225–2238

Schoell M, Faber E (1978) New isotopic evidence for the origin of Red Sea brines. Nature (Lond) 275:436–438

Schoeninger MJ, DeNiro MJ (1982) Carbon isotope ratios of apatite from fossil bone cannot be used to reconstruct diets of animals. Nature (Lond) 297:577–578

Schotterer U, Finkel R, Oeschger H, Siegenthaler U, Wahlen M, Bart G, Gäggeler H, von Gunten HR (1977) Isotope measurements on firn and ice cores from alpine glaciers. In: Rodda JC (ed) Isotopes and impurities in snow and ice. IAHS-AISH Publ 118:232–236

Schroeder RA, Bada JL (1978) Aspartic acid racemization in Late Wisconsin Lake Ontario sediments. Q Res 9:193–204

Schroll E (1975) Analytische Geochemie. I: Methodik. Enke Stuttgart

Schultz L (1986) Terrestrial $^{81}$Kr-ages of four Yamato meteorites. Mem Nat Inst Polar Res Spec Issue 41:319–327

Schultz L, Kruse H (1981) Light noble gases in stony meteorites – a compilation. Nucl Track Detect 2:65–103

Schultz L, Signer P (1976) Depth dependence of spallogenic helium, neon, argon in the St. Severin chrondrite. Earth Planet Sci Lett 30:191–199

Schvoerer M, Delavergne M-C, Chapoulie R (1988) The thermoluminescence (TL) of Egyptian Blue. Nucl Tracks Radiat Meas 14:321–327

Schwarcz HP (1980) Absolute age determination of archaeological sites by uranium series dating of travertines. Archaeometry 22:3–29

Schwarcz HP (1985) ESR dating of tooth enamel. Nucl Tracks 10:865–867

Schwarcz H, Gascoyne M (1984) Uranium-series dating of Quaternary deposits. In: Mahaney WC (ed) Quaternary Dating Methods. Elsevier, Amsterdam, pp 33–51

Schwarcz H, Gascoyne M, Ford DC (1982) Uranium-series disequilibrium studies of granitic rocks. Chem Geol 36:87–106

Scott WE, McCoy WD, Shroba RR, Rubin M (1983) Reinterpretation of the exposed record of the last two cycles of lake Bonneville, Western United State. Q Res 20:261–285

Sears DWG, Hasan FA (1986) Thermoluminescence and Antarctic Meteorites. In: Annexstad JO, Schultz L, Wänke H (eds) Antarctic Meteorites. LPI Tec Rep 86–01:83–100

Segl M, Levin I, Schoch-Fischer H, Münnich M, Kromer B, Tschiersch J, Münich KO (1983) Anthropogenic $^{14}$C variations. Radiocarbon 25:583–592

Segl M, Mangini A, Bonani G, Hofmann HJ, Morenzoni E, Nessi M, Suter M, Wölfli W (1984) $^{10}$Be dating of the inner structure of Mn-encrustations applying the Zürich tandem accelerator. Nucl Instrum Methods 233 (B5):359–364

Seidemann DE, Masterson WD, Dowling MP, Turekian KK (1984) K-Ar dates and $^{40}$Ar/$^{39}$Ar age spectra for Mesozoic basalt flows of the Hartford Basin, Connecticut, and the Newark Basin, New Jersey. Geol Soc Am Bull 95:594–598

Seifert A (1978) Methodischer Beitrag zur He-, Ne-, Ar-Isotopenuntersuchung an Grundwässern. Z Angew Geol 24:97–100

Sguigna AP, Larabee AJ, Waddington JC (1982) The half-life of $^{176}$Lu by $\gamma$–$\gamma$ coincidence measurement. Can J Phys 60:361–364

Shackleton N (1967) Oxygen isotope analyses and Pleistocene temperatures re-assessed. Nature (Lond) 215:15–17

Shackleton N (1970) Stable isotope study of the palaeoenvironment of the neolithic site of Nea Nikomedeia, Greece. Nature (Lond) 227:943–944

Shackleton NJ (1973) Oxygen isotope analysis as a means of determining season of occupation of prehistoric midden sites. Archaeometry 15:133–141

Shackleton NJ (1982) The deep-sea sediment record of climate variability. Prog Oceanogr 11:199–218

Shackleton NJ (1987) Oxygen isotopes, ice volume, and sea level. Q Sci Rev 6:183–190

Shackleton NJ, Opdyke ND (1973) Oxygen isotope and palaeomagnetic stratigraphy of equatorial Pacific core V28–238: oxygen isotope temperatures and ice volumes on a $10^5$ year and $10^6$ year scale. Q Res 3:39–55

Shackleton NJ, Imbrie J, Hall MA (1983a) Oxygen and carbon isotope record of East Pacific core V19-30: implications for the formation of deep water in the late Pleistocene North Atlantic. Earth Planet Sci Lett 65:233–244

Shackleton NJ, Hall MA, Line J, Shuki Chan (1983b) Carbon isotope data in core V19-30 confirm reduced carbon dioxide concentration in the ice age atmosphere. Nature (Lond) 306:319–322

Shafiqullah M, Damon PE (1974) Evaluation of K-Ar isochron methods. Geochim Cosmochim Acta 38:1341–1358

Sharma P, Middleton R (1989) Radiogenic production of $^{10}$Be and $^{26}$Al in uranium and thorium ores: implications for studying terrestrial samples containing low levels of $^{10}$Be and $^{26}$Al. Geochim Cosmochim Acta 53:709–716

Sharma P, Rama, Moore WS (1984) Spatial variation of U-Th series radionuclides and trace metals in deep-sea manganese encrustations. Earth Planet Sci Lett 67:319–326

Sharma P, Somayajulu BLK (1984) Implications of precise $^{10}$Be measurements in deep-sea sediments. Isot Geosci 2:89–96

Shaw J (1983) Recent advances in archaeomagnetism. J Geomagn Geoelectr 37:119–127

Shestakov GI (1972) Diffusion of lead in monazite, zircon, sphene and apatite. Trans Geokhimiya 10:1197–1202

Shields WR (1960) Comparison of Belgian Congo and synthetic "normal" samples. Table 6 in Appendix A, Rep 8, Nat Bureau Stand Meeting of the Advisory Committee for Standard Materials and Methods of Measurement, May 17 and 18, 1960, 37 pp

Shimizu H, Makishima A, Nakai S, Masuda A (1986) $^{138}$La-$^{138}$Ce geochronology of an Amitsoq gneiss, Greenland, and a Mustikkamaki pegmatite, Finland. Terra Cognita 6:145

Shimokawa K, Imai N (1987) Simultaneous determination of alteration and eruption ages of volcanic rocks by electron-spin resonance. Geochim Cosmochim Acata 51:115–119

Shukolyukov YA (1983) Radiogenic-isotope geochmistry. Geokhimiya 3:333–347

Shukolyukov YA, Meshick AP (1987) Application of xenon isotopes for dating pitchblendes. Chem Geol Isot Geosci Sect 66:123–136

Shukolyukov YA, Mirkina SL (1963) Determination of the age of monazites by the xenon method. Geochemistry 729–731

Shukolyukov YA, Ashkinadze GS, Komarov AN (1974a) A new Xe-Xe neutron-activated method of mineral dating. Dokl Akad Nauk USSR 219 (4):952–954

Shukolyukov J, Kirsten T, Jessberger EK (1974b) The Xe-Xe spectrum technique, a new dating method. Earth Planet Sci Lett 24:271–281

Shukolyukov J, Ashkinadze GS, Kirsten T, Jessberger EK (1976) The $Kr_s$-$Kr_n$ dating method for radioactive minerals. Geochem Int 13:182–185

Shukolyukov YA, Kapusta YS, Vekhovskiy AB, Vaasjoki M (1979) Neutron-induced fission xenon dating of zircon. Geochem Int 16:122–133

Siegele R, Mangini A (1985) ESR studies on foraminifera in deep-sea sediments. Terra Cognita 5:85–86

Siegenthaler U (1988) Causes and effects of natural $CO_2$ variations during the glacial-interglacial cycles. In: Wanner H., Siegenthaler U (eds): Long and Short Term Variability of Climate. Lecture Notes in Earth Sciences 16:153–171; Berlin Heidelberg New York (Springer)

Siegenthaler U, Eicher U, Oeschger H, Dansgaard W (1984) Lake sediments as continental $\delta^{18}$O records from the glacial/post-glacial transition. Ann Glac 5:149–152

Sigurgeirsson T (1962) Dating recent basalt by the potassium-argon method. Rep Phys Lab, Univ Iceland, 9 pp

Silver LT, Deutsch S (1963) Uranium-lead isotopic variations in zircons: A case study. J Geol 71:721–758

Silver LT, Schultz PH (1982) Geological implications of impacts of large asteroids and comets on the earth. Spec Pap 190: 528. Geol Soc Am, Boulder

Simonsen A (1983) Procedures for sampling, sorting and pretreatment of charcoal for $^{14}$C dating. In: Mook WG, Waterbolk HT (eds) $^{14}$C and Archaeology. PACT 8:313–318

Sims PK, Peterman ZE, Prinz WC, Benedict FC (1984) Geology, geochemistry, and age of Archean and early Proterozoic rocks in the Marenisco-Watersmeet area, northern Michigan. US Geol Surv Prof Pap 1292-A

Singhvi AK, Mejdahl V (1985) Thermoluminescence dating of sediments. Nucl Tracks 10:137–161

Singhvi AK, Wagner GA (1986) Thermoluminescence dating and its applications to young sedimentary deposits. In: Hurford AJ, Jäger E, Then Cate JAM (eds) Dating Young Sediments. CCOP Tech Secr, Bangkok, pp 159–197

Singhvi AK, Sharma YP, Agrawal DP (1982) Thermoluminescence dating of sand dunes in Rajasthan, India. Nature (Lond) 295:313–315

Singhvi AK, Nambi KSV, Sunta CM, Durranni SA, Mejdahl V (1985) Theory and practice of thermally stimulated luminescence and related phenomena. Nucl Tracks 10:1–2

Sinha AK, Tilton GR (1973) Isotopic evolution of common lead. Geochim. Cosmochim. Acta 37:1823–1849

Smethie WM Jr, Østlund HG, Loosli HH (1986) Ventilation of the deep Greenland and Norwegian seas: evidence from krypton-85, tritium, carbon-14 and argon-39. Deep-Sea Res 33:675–703

Smewing JD, Potts PJ (1976) Rare earth abundances in basalts and metabasalts from the Troodos Massif, Cyprus. Contrib Mineral Petrol 57:245–257

Smith BW (1988) Zircon from sediments: a combined OSL and TL auto-regenerative dating technique. Q Sci Rev 7:401–406

Smith BW, Smart PL, Symons MCR (1985) ESR signals in a variety of speleothem calcites and their suitability for dating. Nucl Tracks 10:837–844

Smith BW, Aitken MJ, Rhodes EJ, Robinson PD, Geldard DM (1986) Optical dating: methodical aspects. Radiat Protect Dosim 17:229–233

Smith DB, Wearn PL, Richards HJ, Rowe PC (1970) Water movement in the unsaturated zone of high and low permeability strata by measuring natural tritium. In: Isotope Hydrology 1970:73–88. IAEA, Vienna

Smits F, Gentner W (1950) Argonbestimmungen in Kalium-Mineralen. I. Bestimmungen an tertiären Kalisalzen. Geochim Cosmochim Acta 1:22–27

Snelling NJ (ed) (1985) The chronology of the geological record. Geol Soc Lond Mem 10, 343 p

Sobotovich EV (1961) Possibility of determining the absolute age of the granites of the Terskey Ala-Tau by the lead method included in them. Akad Nauk SSSR Kom Opredeleniyu Absolyut Vozrasta Geol Formatsii Trudy Sess 9:269–280

Sobotovich EV, Grashchenko SM, Lovtsyus AV (1963a) Rock age of Taromskoe quarry according to data of lead isochronous method. Akad Nauk SSSR Kom Opredeleniyu Absolyut Vozrasta Geol Formatsii Trudy Sess 11:353–356

Sobotovich EV, Grashchenko SM, Aleksandruk VM, Shats MM (1963b) Determination of the age of the most ancient rocks by the lead isochronous and the strontium isotope-spectral methods. Akad Nauk SSSR Izv Ser Geol 28:3–14

Somayajulu BLK, Tatsumoto M, Rosholt JN, Knight RJ (1966) Disequilibrium of the $^{238}$U series in basalt. Earth Planet Sci Lett 1:387–391

Somayajulu BLK, Lal D, Craig H (1973) Silicon-32 profiles in the south Pacific. Earth Planet Sci Lett 18:181–188

Somayajulu BLK, Sharma P, Berger WH (1983/1984) $^{10}$Be, $^{14}$C and U-Th decay series nuclides and $\delta^{18}$O in a box core from the central north Atlantic. Mar Geol 54:169–180

Somayajulu BLK, Rengarajan R, Lal D, Weiss RF, Craig H (1987) GEOSECS Atlantic $^{32}$Si profiles. Earth Planet Sci Lett 85:329–342

Sonntag C, Thorweihe U, Rudolf J, Löhnert EP, Junghans C, Münnich KO, Klitzsch E, El Shazly EM, Swailem FM (1980) Paleoclimatic evidence in apparent $^{14}$C ages of Saharian groundwaters. Radiocarbon 22 (III):871–878

Sprenkel EL (1959) Produced Cl$^{36}$ and Ar$^{39}$ in iron meteorites. Bull Am Phys Soc 4:223

Srdoc D, Obelic B, Sliepcevic A (1983) Precise Radiocarbon dating of wooden beams from St.

Donat's church in Zadar. In: Mook WG, Waterbolk HT (eds) [14]C and Archaeology. PACT 8:319–328

Srivastava AP, Rajagopalan G, Ambwani K (1986) Fission-track dating of fossil palm wood from Shahpura, Mandla district, Madhya Pradesh. Geophytology 16:136–137

Stacey JS, Hedlund DC (1983) Lead-isotopic compositions of diverse igneous rocks and ore deposits from southwestern New Mexico and their implications for early Proterozoic crustal evolution in the western United States. Geol Soc Am Bull 94:43–57

Stacey J, Kramers J (1975) Approximation of terrestrial lead isotope evaluation by a two-stage model. Earth Planet Sci Lett 26:207–221

Stacey JS, Stern TW (1973) Revised tables for the calculation of lead isotope ages. US Dept Commerce, Natl Tech Inf Service, Springfield, Virginia, 22151, PB–20919

Stacey JS, Stoeser DB (1983) Distribution of oceanic and continental leads in the Arabian-Nubian Shield. Contrib Mineral Petrol 84:91–105

Stacey JS, Zartman RE, Nkomo IT (1968) A lead isotope study of galenas and selected feldspars from mining districts in Utah. Econ Geol 63:796–814

Stafford ThW Jr, Jull AJT, Brendel K, Duhamel RC, Donahue D (1987) Study of bone radiocarbon dating accuracy at the University of Arizona NSF accelerator facility for radioisotope analysis. Radiocarbon 29(1):24–44

Stafford ThW Jr, Brendel K, Duhamel RC (1988) Radiocarbon, [13]C and [15]N analysis of fossil bone: removal of humates with XAD-2 resin. Geochim Cosmochim Acta 52:2257–2267

Stahl W, Faber E (1984) Carbon isotopes as a petroleum exploration tool. Proc 11th World Petrol Congr Chichester 2:147–159

Stahl W, Jordan R (1969) General considerations on isotopic paleotemperature determinations and analyses on Jurassic ammonites. Earth Planet Sci Lett 6:173–178

Stanton RL, Russel RD (1959) Anomalous leads and the emplacement of lead sulfide ores. Econ Geol 54:588–607

Staudacher T, Allègre CJ (1982) Terrestrial xenology. Earth Planet Sci Lett 60:389–406

Staudigel H, Gillis K, Duncan R (1968) K/Ar and Rb/Sr ages of celadonites from the Troodos ophiolite, Cyprus. Geology 14:72–75

Stauffer BR (1989) Dating of ice by radioactive isotopes. In: Oeschger H, Landway Jr CC (eds) The environmental records in glaciers and ice sheets. Wiley, New York, pp 123–140

Steiger RH, Jäger E (1977) Subcommission on geochronology: convention on the use of decay constants in geo- and cosmochronology. Earth Planet Sci Lett 36:359–362

Steiger RH, Wasserburg GJ (1966) Systematics in the $Pb^{208}$-$Th^{232}$, $Pb^{207}$-$U^{235}$, $Pb^{206}$-$U^{238}$ systems. J Geophys Res 71:6065–6090

Steiger RH, Wasserburg GJ (1969) Comparative U-Th-Pb systematics in $2.7 \times 10^9$ yr plutons of different geological histories. Geochim Cosmochim Acta 33:1213–1232

Steinhof A, Henning W, Müller M, Roeckl E, Schüll D, Korschinek G, Nolte E, Paul M (1987) Accelerator mass spectrometry of $^{41}$Ca with a positive-ion source and the UNILAC accelerator. Nucl Instrum Methods Phys Res 29:59–62

Stenhouse MJ, Baxter MS (1983) [14]C dating reproducibility: evidence from routine dating of archaeological samples. In: Mook WG, Waterbolk HT (eds) [14]C and Archaeology. PACT 8:147–161

Stern TW, Goldich SS, Newell MF (1966) Effects of weathering on the U-Pb ages of zircon from the Morton gneiss, Minnesota. Earth Planet Sci Lett 1:369–371

Stettler A, Albarède F (1978) $^{39}$Ar-$^{40}$Ar systematics of two millimeter-sized rock fragments from Mare Crisium. Earth Planet Sci Lett 38:401–406

Stichler W, Herrmann A (1983) Application of environmental isotope techniques in water balance studies of small basins. In: van der Beken A, Herrmann A (eds) New approaches in water balance computations 148. IAHS Press, Wallingford, pp 93–112

Stichler W, Maloszewski P, Moser H (1986) Modelling of river water infiltration using oxygen-18 data. J Hydrol 83:355–365

Stiller M, Nissenbaum A (1980) Variations of stable hydrogen isotopes in plankton from a freshwater lake. Geochim Cosmochim Acta 44:1099–1101

Stoneham D, Winter MB (1988) The removal of anomalous fading in authenticity samples. Nucl Tracks Radiat Meas 14:127–130

Storetvedt KM (1970) On remagnetization problems in palaeomagnetism: further considerations. Earth Planet Sci Lett 9:407–415

Storzer D, Poupeau G (1973) Plateau ages of minerals and glasses by the fission track method. CR Acad Sci Paris 276:137–139

Storzer D, Wagner GA (1969) Correction of thermally lowered fission track ages of tektites. Earth Planet Sci Lett 5:463–468

Storzer D, Wagner GA (1982) The application of fission track dating in stratigraphy: a critical review. In: Odin GS (ed) Numerical Dating in Stratigraphy. Wiley, New York, pp 199–221

Strutt RJ (1908) On the accumulation of helium in geological time. Proc R Soc A 81:272–277

Stuiver M (1970) Oxygen and carbon isotope ratios of fresh water carbonates as climatic indicators. J Geophys Res 75:5247–5257

Stuiver M (1978) Radiocarbon timescale tested against magnetic and other dating methods. Nature (Lond) 273:271–274

Stuiver M (1982) A high-precision calibration of the AD radiocarbon time scale. Radiocarbon 24 (1):1–26

Stuiver M, Braziunas TF (1987) Tree cellulose $^{13}C/^{12}C$ isotope ratios and climatic change. Nature (Lond) 328:58–60

Stuiver M, Kra R (1986) Calibration Issue. Radiocarbon 28(2B):805–1030

Stuiver M, Oestlund HG (1980) GEOSECS Atlantic Radiocarbon. Radiocarbon 22 (1):1–24

Stuiver M, Polach HA (1977) Discussion reporting of $^{14}C$ data. Radiocarbon 19 (3):355–363

Stuiver M, Quay PD (1980a) Patterns of atmospheric $^{14}C$ changes. Radiocarbon 22 (II):166–176

Stuiver M, Quay PD (1980b) Changes in atmospheric carbon-14 attributed to a variable sun. Science 207:11–19

Stuiver M, Heuser CJ, Yong C (1978) North American glacial history extended to 75 000 yr ago. Science 200:16–21

Stuiver M, Quay PD, Oestlund HG (1983) Abyssal water carbon-14 distribution and the age of the world oceans. Science 219:849–851

Suess HE (1955) Radiocarbon concentration in modern wood. Science 122:415–417

Suess HE (1986) Secular variations of cosmogenic $^{14}C$ on earth: their discovery and interpretation. Radiocarbon 28 (2A):259–265

Sullivan CH, Krueger HW (1981) Carbon isotope analysis of separate chemical phases in modern and fossil bone. Nature (Lond) 292:333–335

Sun SS, Hanson G (1975) Evolution of the mantle. Geology 3:297–302

Sun SS, Tatsumoto M, Schilling J-G (1975) Mantle plume mixing along the Reykjanes ridge axis: Lead isotopic evidence. Science 190:143–147

Sunin LV, Malyshev VI (1983) The thermoisochron method of determining Pb-Pb ages. Geochem Int 20/3:34–45

Sunta CM, Unnikrishnan K, Kathuria SP (1983) Retrapping and intertrap charge migration during the decay of the glow peaks—effect on the order of kinetics. PACT 9:97–108

Suter M, Balzer R, Bonani G, Wölfli W (1984) A fast beam pulsing system for isotope ratio measurements. Nucl Instrum Methods 233 (B5):242–246

Sutter JF, Husain L, Schaeffer OA (1971) $^{40}Ar/^{39}Ar$ ages from Fra Mauro. Earth Planet Sci Lett 11:249–253

Sutton SR, Walker RM (1986) Thermoluminescence of antarctic meteorites: a rapid screening techniques for terrestrial age estimation, pairing studies, and identification of specimens with unusual pre-fall histories. In: Annexstad JO, Schultz L, Wänke H (eds) Antarctic Meteorites. LPI Tech Rep 86–01:104–106

Szabo BJ (1979) $^{230}Th$, $^{231}Pa$, and open system dating of fossil corals and shells. J Geophys Res 84:4927–4930

Szabo BJ, Butzer KW (1979) Uranium-series dating of lacustrine limestones from pan deposits with final Acheulian assemblage at Rooidam, Kimberley district, South Africa. Q Res 11:257–260

Szabo BJ, Rosholt JN (1969) Uranium series dating of Pleistocene molluscan shells from southern California—an open system model. J Geophys Res 74:3254–3260

Szabo BJ, Miller GH, Andrews JT, Stuiver M (1981) Comparison of uranium-series, radiocarbon, and amino acid data from marine molluscs, Baffin Island, Arctic Canada. Geology 9:451–457

Taddeucci A, Broecker WS, Thurber DL (1967) $^{230}Th$ dating of volcanic rocks. Earth Planet Sci Lett 3:338–342

Takagi J, Hampel W, Kristen T (1974) Cosmic ray muon-induced $^{129}I$ in tellurium ores. Earth Planet Sci Lett 24:141–150

Takayanagi M, Ozima M (1987) Temporal variation of $^3$He/$^4$He ratio recorded in deep-sea sediment cores. J Geophys Res 92(B12):531–538

Talbot MR, Kelts K (1986) Primary and diagenetic carbonates in the anoxic sediments of Lake Bosumtwi, Ghana. Geology 14:912–916

Tamers MA (1969) Dating of recent events. Atompraxis 15:271–276

Tamers MA (1970) Validity of radiocarbon dates on terrestrial snail shells. Am Antiq 35:94–100

Tamers MA (1975) Validity of radiocarbon dates of groundwater. Geophys Surv 2:217–239

Tamers MA (1979) Radiocarbon transmutation mechanism for spontaneous somatic cellular mutations. In: Berger R, Suess HE (eds) Radiocarbon Dating. Univ California Press, Los Angeles, pp 355–364

Tamers MA, Ronzani C, Scharpenseel HW (1969) Observation of naturally occurring chlorine-36. Atompraxis 15:1–5

Tanaka S, Sakamoto K, Takagi J, Tschuchimoto M (1964) Search for $^{26}$Al induced by cosmic-ray muons in terrestrial rocks. Inst Nuclear Study, Univ Tokio, INS-TCS-19

Tanaka T, Masuda A (1982) The La-Ce geochronometer: a new dating method. Nature (Lond) 300:515–518

Tans PP, de Jong AFM, Mook WG (1978) Chemical pretreatment and radial flow of $^{14}$C in tree rings. Nature (Lond) 271:234–235

Tans PP, de Jong AFM, Mook WG (1979) Natural atmospheric $^{14}$C variation and the SUESS effect. Nature (Lond) 280:826–828

Tarling DH (1975) Archaeomagnetism: the dating of archaeological materials by their magnetic properties. World Archaeol 7:185–197

Tatsumoto M (1970) U-Th-Pb age of Apollo 12 rock 12013. Earth Planet Sci Lett 9:193–200

Tatsumoto M (1978) Isotopic composition of lead in oceanic basalt and its implication to mantle evolution. Earth Planet Sci Lett 38:63–87

Tatsumoto M, Knight RJ, Allègre CJ (1973) Time differences in the formation of meteorites as determined from the ratio of lead-207 to lead-206. Science 180:1279–1283

Tatsumoto M, Unruh D, Desborough G (1976) U-Th-Pb and Rb-Sr systematics of Allende and U-Th-Pb systematics of Orgueil. Geochim Cosmochim Acta 40:617–634

Tauber H (1981) $^{13}$C Evidence for dietary habits of prehistoric man in Denmark. Nature (Lond) 292:332–333

Tauber H (1983) $^{14}$C dating of human beings in relation to dietary habits. In: Mook WG, Waterbolk HT (eds) $^{14}$C and Archaeology. PACT 8:365–375

Taylor HP Jr, Turi B (1976) High $^{18}$O igneous rocks from the Tuscan magmatic province, Italy. Contrib Mineral Petrol 55:33–54

Taylor LA, Shervais JW, Hunter RH, Shih C-Y, Bansal BM, Wooden J, Nyquist LE, Laul LC (1983) Pre-4.2 AE mare-basalt volcanism in the lunar highlands. Earth Planet Sci Lett 66:33–47

Taylor PN, Moorbath S (1986) Dating of carbonate rocks by the Pb/Pb method. Terra Cognita 6:157

Taylor RE (1975) Flourine diffusion: a new dating method for chipped lithic materials. World Archaeol 7:125–135

Taylor RE (1982) Problems in the radiocarbon dating of bone. In: Currie LA (ed) Nuclear and Chemical Dating Techniques. ACS, Washington DC, pp 453–473

Taylor RE (1987a) Dating techniques in archeology and paleoanthropology. Anal Chem 59(4):317A–331A

Taylor RE (1987b) Radiocarbon Dating: An Archaeological Perspective. Academic Press, London New York, 212 pp

Taylor SR (1982) Planetary Science: a lunar perspective. Lunar Planet Inst, Houston, 481 pp

Taylor SR, McLennan SM (1981) The composition and evolution of the continental crust: rare earth element evidence from sedimentary rocks. Phil Trans R Soc Lond Ser A 301:381–399

Teitsma A, Clarke WB (1978) Fission xenon isotope dating. J Geophys Res 83:5443–5453

Teitsma A, Clarke WB, Allègre CJ (1975) Spontaneous fission—neutron fission xenon: a new technique for dating geological events. Science 189:878–880

Templer RH (1986a) The localized transition model of anomalous fading. Radiat Protect Dosim 17:493–497

Templer RH (1986b) Auto-regenerative TL dating of zircon inclusions. Radiat Protect Dosim 17:235–239

Templer RH, Smith BW (1988) Auto-regenerative TL dating with zircon inclusions from fired materials. Nucl Tracks Radiat Meas 14:329–332

Tera F, Wasserburg GJ (1972) U-Th-Pb systematics in three Apollo 14 basalts and the problem of initial Pb in lunar rocks. Earth Planet Sci Lett 14:281–304

Tera F, Wasserburg GJ (1974) U-Th-Pb systematics in lunar rocks and inferences about lunar evolution and the age of the moon. Proc Lunar Sci Conf 5th 2:1571–1599

Tetzlaff U, Milde G, Hagendorf U (1973) Erfahrungen bei der Anwendung der He-Ar-Methode zur Bestimmung des physikalischen Grundwasseralters und der Klärung praktischer hydrogeologischer Probleme. Z Angew Geol 19:142–147

Thellier E, Thellier O (1959) Sur l'intensité du champ magnétique terrestre dans le passé historique et geologique. Ann Geophys 15:285–376

Thiel K, Herr W (1976) The $^{238}$U spontaneous fission decay constant redetermined by fission tracks. Earth Planet Sci Lett 30:50–56

Thiel K, Vorwerk R, Saager R, Stupp HD (1983) $^{235}$U fission tracks and $^{238}$U-series disequilibria as a means to study recent mobilization of uranium in Archaean pyritic conglomerates. Earth Plan Sci Lett 65:249–262

Thode HG, Rees CE (1970) Sulfur isotope geochemistry and Middle East oil studies. Endeavour 29:24–28

Thompson KC, Reynolds RJ (1978) Atomic absorption, fluorescence and flame emission spectroscopy. Griffin, London, 319 pp

Thompson M, Walsh JN (1983) A Handbook of Inductively Coupled Plasma Spectrometry. Chapman and Hall, New York, 273 pp

Thompson P, Ford DC, Schwarcz HP (1975) $U^{234}/U^{238}$ ratios in limestone cave seepage waters and speleothem from West Virginia. Geochim Cosmochim Acta 39:661–669

Thompson R (1983) $^{14}$C Dating and Magnetostratigraphy. Radiocarbon 25:229–238

Thonnard N, Willis RD, Wright MC, Davis WA, Lehmann BE (1987) Resonance ionization spectroscopy and the detection of $^{81}$Kr. Nucl Instrum Methods Phys Res B 29:398–406

Thorpe RI, Goodz MD, Jonasson IR, Blenkinsop J (1986) Lead-isotope study of mineralization in the Cobalt district, Ontario. Can J Earth Sci 23:1568–1575

Thurber DL (1962) Anomalous $U^{234}/U^{238}$ in nature. J Geophys Res 67:4518–4520

Thurber DL (1963) Natural variations in the ratio of $U^{234}$ to $U^{238}$. In: Radioactive Dating: 113–120. IAEA, Vienna

Thurber DL, Broecker WS, Blanchard RL, Potratz HA (1965) Uranium-series ages of Pacific Atoll coral. Science 149:55–58

Tilton GR (1960) Volume diffusion as a mechanism for discordant lead ages. J Geophys Res 65:2933–2945

Tilton GR (1973) Isotopic lead ages of chondritic meteorites. Earth Planet Sci Lett 19:321–329

Tilton GR (1983) Evolution of depleted mantle: The lead perspective. Geochim Cosmochim Acta 47:1191–1197

Tilton GR, Steiger RH (1965) Lead isotopes and the age of the Earth. Science 150:1805–1808

Titayeva NA (1966) Possibility of absolute dating of organic sediments by the ionium method. Geokhimiya 10:1183–1191

Tite MS (1981) Methods of Physical Examinations in Archaeology. Seminar Press, London, 385 pp

Todt W (1976) Zirkon-U/Pb-Alter des Malsburg-Granits vom Süd-Schwarzwald. N Jahrb Mineral Mh: 532–544

Todt W, Büsch W (1981) U-Pb investigations on zircons from pre-Variscan gneisses - I. A study from Schwarzwald, West Germany. Geochimica Cosmochimica Acta 45:1789–1801

Tolstikhin IN (1975) Helium isotopes in the earth's interior and in the atmosphere: A degassing model of the earth. Earth Planet Sci Lett 26:88–96

Tolstikhin IN, Kamenskiy IL (1969) Determination of ground-water ages by the T-$^3$He method. Geokhimiya 8:1027–1029

Top Z, Clarke WB (1981) Dissolved helium isotopes and tritium in lakes: further results for uranium prospecting in Central Labrador. Econ Geol 76:2018–2031

Torgersen T (1980) Controls on pore-fluid concentration of $^4$He and $^{222}$Rn and the calculation of $^4$He/$^{222}$Rn ages. J Geochem Explor 13:57–75

Torgersen T, Clarke WB (1985) Helium accumulation in groundwater I: An evaluation of sources and the continental flux of crustal $^4$He in the Great Artesian Basin, Australia. Geochim Cosmochim Acta 49:1211–1218

Torgersen T, Clarke WB (1987) Helium accumulation in groundwater, III. Limits on helium transfer across the mantle-crust boundary beneath Australia and the magnitude of mantle degassing. Earth Planet Sci Lett 84:345–355

Torgersen T, Ivey GN (1985) Helium accumulation in groundwater. II: A model for the accumulation of the crystal $^4$He degassing flux. Geochim Cosmochim Acta 49:2445–2452

Torgersen T, Jenkins WJ (1982) Helium isotopes in geothermal systems: Iceland, The Geysers, Raft River and Steamboat Springs. Geochim Cosmochim Acta 46:739–748

Torgersen T, Clarke WB, Jenkins WJ (1979) The tritium/helium-3 method in hydrology. In: Isotope Hydrology 1978 (II):917–929. IAEA, Vienna

Torgersen T, Lupton JE, Sheppard DS, Giggenbach WF (1982) Helium isotope variations in the thermal areas of New Zealand. J Volcanol Geotherm Res 12:283–298

Torgersen T, Clarke WB, Habermehl MA (1987) Helium isotopic evidence from recent subcrustal volcanism in eastern Australia. Geophys Res Lett 14:1215–1218

Tremba EL, Faure G, Katsikatsos GC, Summerson CH (1975) Strontium-isotope composition in the Tethys Sea, Euboea, Greece. Chem Geol 16:109–120

Trembour F, Friedman I (1984) The present status of obsidian hydration dating. In: Mahaney WC (ed) Quaternary Dating Methods. Elesevier, Amsterdam, pp 141–151

Trivedi BMP (1977) A new approach to nucleocosmochronology. Astrophys J 215:877–884

Tsong IST, Houser CA, Yusef NA, Messier RF, White WB, Michels JW (1978) Obsidian hydration profiles measured by sputter-induced optical emission. Science 201:339–341

Tsoulfanidis N (1983) Measurement and Detection of Radiation. McGraw-Hill, New York, 571pp

Tuniz C, Pal DK, Moniot RK, Savin W, Kruse TH, Herzog GF, Evans JC (1983) Recent cosmic ray exposure history of ALPHA 81005. Geophys Res Lett 10:804–806

Tuniz C, Smith CM, Moniot RK, Kruse TH, Savin W, Pal DK, Herzog GF, Reedy RC (1984) Beryllium-10 contents of core samples from the St Severin meteorite. Geochim Cosmochim Acta 48:1867–1872

Turekian KK, Nelson E (1976) Uranium-series dating of travertines of Caune de l'Arago (France). In: Labeyrie J, Lalou C (eds) Colloque I. Datations absolues et analyses isotopiques en Prèhistoire. Mèthods et limites. Union des Sci Prehist Protohist, IX Congres, Nice, pp 172–179

Turekian KK, Kharkar DP, Funkhouser J, Schaeffer OA (1970) An evaluation of the uranium-helium method of dating fossil bones. Earth Planet Sci Lett 7:420–424

Turner G (1971) $^{40}$Ar/$^{39}$Ar ages from the lunar maria. Earth Planet Sci Lett 11:169–191

Turner G (1977) Potassium-argon chronology of the Moon. Phys Chem Earth 10:145–195

Tuross N, Fogel ML, Hare PE (1988) Variability in the preservation of the isotopic composition of collagen from fossil bones. Geochim Cosmochim Acta 52:929–935

Turpin L (1985) Rb-Sr dating of hydrothermal alteration in St. Sylvestre leucogranite (French Massif Central). Terra Cognita 5:280–281

Ulrych TJ (1967) Oceanic basalt leads: a new interpretation and an independent age for the Earth. Science 158:252–256

Unruh DM, Tatsumoto M (1976) Lead isotopic composition and uranium, thorium, and lead concentrations in sediments and basalts from the Nazca Plate. Initial Rep Deep-Sea Drilling Project 34:431–437

Unruh DM, Stille P, Patchett PJ, Tatsumoto M (1984) Lu-Hf and Sm-Nd evolution in lunar mare basalts. Proc 14th Lunar Planet Sci Conf, J Geophys Res Suppl 89:B459-B477

Urey HC (1947) The thermodynamic properties of isotopic substances. J Chem Soc 1947:562–581

Urey HC (1973) Cometary collisions and geological periods. Nature (Lond) 242:32–33

Valette-Silver JN, Brown L, Pavich M, Klein J, Middleton R (1986) Detection of erosion events using $^{10}$Be profiles: example of the impact of agriculture on soil erosion in the Chesapeake Bay area (USA). Earth Planet Sci Lett 80:82–90

van den Bergh S (1981) Size and age of the universe. Science 213:825–830

van den Bogaard P, Hall CM, Schmincke H-U, York D (1987) $^{40}$Ar/$^{39}$Ar laser dating of single grains: ages of Quaternary tephra from the East Eifel volcanic field, FRG. Geophys Res Lett 14:1211–1214

van den Boom GP (1987) Helium distribution pattern of measured and corrected data around the "Eldingen" oil field, NW Germany. J Geophys Res 91 (B12):547–555

van der Merwe NJ (1982) Carbon isotopes, photosynthesis, and archaeology. Am Sci 70:596–606

van der Pflicht J, Mook WG (1989) Calibration of radiocarbon dates by computer. 13th Int Radiocarbon Conf, Dubrovnik, June 1988. Radiocarbon (in press)

van der Wijk A, El-Daoushy F, Arends AR, Mook WG (1986) Dating peat with U/Th disequilibrium: some geochemical consideratations. Chem Geol Isot Geosci Sect 59:283–292

van Everdingen RO, Shakur MA, Krouse HR (1985) Role of corrosion by $H_2SO_4$ fallout in cave development in a travertine deposit-evidence from sulfur and oxygen isotopes. Chem Geol 49:205–211

van Strydonck M, Dupas M, Dauchot-Dehon M, Pachiaudi C, Marechal J (1986) The influence of contaminating (fossil) carbonate and the variations of $\delta^{13}C$ in mortar dating. Radiocarbon 28 (2A):702–710

Veeh HH (1966) $Th^{230}/U^{238}$ and $U^{234}/U^{238}$ ages of Pleistocene high sea level stand. J Geophys Res 71:3379–3386

Veeh HH (1982) Concordant $^{230}$Th and $^{231}$Pa ages of marine phosphorites. Earth Planet Sci Lett 57:278–284

Veizer J, Compston W (1976) $^{87}Sr/^{86}Sr$ in Precambrian carbonates as an index of crustal evolution. Geochim Cosmochim Acta 40:905–914

Veizer J, Holser WT, Wilgus CK (1980) Correlation of $^{13}C/^{12}C$ and $^{34}S/^{32}S$ secular variations. Geochim Cosmochim Acta 44:579–587

Venkatesan MI, Kaplan IR, Mankiewicz P, How K, Sweeney RE (1982) Determination of petroleum contamination in marine sediments by organic geochemical and stable sulfur isotope analyses. Environ Deep-Sea II:93–104, Rubey

Verosub KL (1977) Depositional and post-depositional processes in the magnetization of sediments. Rev Geophys Space Phys 15:129–143

Verosub KL (1982) Geomagnetic excursions: a critical assessment of the evidence as recorded in sediments of the Bruhnes epoch. Phil Trans R Soc Lond A306:161–168

Vilcsek E, Wänke H (1963) Cosmic-ray exposure ages and terrestrial ages of stone and iron meteorites derived from $Cl^{36}$ and $Ar^{39}$ measurements. In: Radioactive Dating: 381–393. IAEA, Vienna

Villa IM, Huneke JC, Wasserburg GJ (1983) $^{39}Ar$ recoil losses and presolar ages in Allende inclusions. Earth Planet Sci Lett 63:1–8

Vincent D, Clocchiatti R, Langevin Y (1984a) Fission-track dating of glass inclusions in volcanic quartz. Earth Planet Sci Lett 71:340–348

Vincent J-S, Morris WA, Occhietti S (1984b) Glacial and nonglacial sediments of Matuyama paleomagnetic age of Banks Island, Canadian Arctic Archipelago. Geology 12:139–142

Vitrac AM, Albarède F, Allègre CJ (1981) Lead isotopic composition of Hercynian granitic K-feldspars constraints continental genesis. Nature (Lond) 291:460–464

Vogel JC (1970) Carbon-14 dating of groundwater In: Isotope Hydrology 1970:225–240. IAEA, Vienna

Vogel JC (1982) Ionium dating of peat: a note. In: Goetze JA, van Zinderen-Bakker EM (eds) Palaeoecology of Africa and the Surrounding Islands. Balkema, Rotterdam, pp 161–162

Vogel JC (1983) 14-C Variations during the upper Pleistocene. Radiocarbon 25:213–218

Vogel JC, Ehhalt D (1963) The use of the carbon isotopes in groundwater studies. In: Radioisotopes in Hydrology: 383–395. IAEA, Vienna

Vogel JC, Kronfeld J (1980) A new method for dating peat. S Afr J Sci 76:557–558

Vogel JC, Uhlitzsch I (1975) Carbon-14 as an indicator of $CO_2$ pollution in cities. In: Isotope Ratios as Pollutant Source and Behaviour Indicators: 143–150. IAEA, Vienna

Vogel JS, Southon JR, Nelson DE, Brown TA (1984) Performance of catalytically condensed carbon for use in accelerator mass spectrometry. Nucl Instrum Methods 233 (B5):289–293

Vollmer R, Ogden P, Schilling J-G, Kingsley RH, Waggoner DG (1984) Nd and Sr isotopes in ultrapotassic volcanic rocks from the Leucite Hills, Wyoming. Contrib Mineral Petrol 87:359–368

von Weizsäcker CF (1937) Über die Möglichkeit eines dualen Betazerfalls von Kalium. Phys Z 38:623–624

Voshage H (1984) Investigations on the cosmic-ray produced nuclides in iron meteorites, 6. The Signer-Nier model and the history of the cosmic radiation. Earth Planet Sci Lett 71:181–194

Voshage H, Hintenberger H (1963) The cosmic-ray exposure ages of iron meteorites as derived from the isotopic composition of potassium and the production rates of cosmogenic nuclides in the past. In: Radioactive Dating: 367–379. IAEA, Vienna

Waddell C, Fountain JC (1984) Calcium diffusion: a new dating method for archeological materials. Geology 12:24–26

Wagner GA (1978) Archaeological applications of fission-track dating. Nucl Track Detect 2:51–64

Wagner GA (1979) Correction and interpretation of fission track ages. In: Jäger E, Hunziker JC (eds) Lectures in Isotope Geology. Springer, Berlin Heidelberg New York, pp 170–177

Wagner GA (1981) Fission-track ages and their geological interpretation. Nucl Tracks 5:15–25

Wagner GA (1983) Thermoluminescence Dating. Handb Archaeol 1:47. Eur Sci Found, Strasbourg

Wagner GA, Reimer GA, Jäger E (1977) Cooling ages derived by apatite fission track, mica Rb-Sr and K-Ar dating the uplift and cooling history of the Central Alps. Memoir 1st Geol Min Univ Padova 30:1–27

Wagner GA, Aitken MJ, Singhvi AK, Mangini A, Mejdahl V, Pernicka E, Durrani SA (1985) Thermoluminescence and Electron-Spin-Resonance Dating. Nucl Tracks 10:485–955

Wahl W (1941) Die Bedeutung der Isotopenforschung für die Geologie. Geol Rundsch 32:550–562

Wakshal E, Nielsen H (1982) Variations of $\delta^{34}S$ ($SO_4$), $\delta^{18}O$ ($H_2O$) and $Cl/SO_4$ ratio in rainwater over northern Israel, from the Mediterranean Coast to Jordan Rift Valley and Golan Heights. Earth Planet Sci Lett 61:272–282

Walliser OH (1983/1984) Geologic processes and global events. Terra Cognita 4:17–20

Walther HW, Förster H, Harre W, Kreuzer H, Lenz H, Müller P, Raschka H (1981) Early Cretaceous porphyry copper mineralization on Cebu Island, Philippines, dated with K-Ar and Rb-Sr methods. Geol Jahrb Ser D 48:21–35

Walton A, Baxter MS, Callow WJ, Baker MJ (1967) Carbon-14 concentrations in environmental materials and their temporal fluctuations. In: Radioactive Dating and Methods of Low-Level Counting: 41–47. IAEA, Vienna

Walton D (1977) Archaeomagnetic intensity measurements using a SQUID magnetometer. Archaeometry 19:192–200

Wang S, McDougall I, Tetley N, Harrison TM (1980) $^{40}Ar/^{39}Ar$ age and thermal history of the Kirin chondrite. Earth Planet Sci Lett 49:117–131

Wardlaw NC (1968) Carnalite-sylvite relationships in the Middle Devonian Prairie evaporite formation, Saskatchewan. Bull Geol Soc Am 79:1273–1294

Waring CL, Worthing H (1953) A spectrographic method for determining trace amounts of lead in zircon and other minerals. Am Mineral 38:827–833

Wasserburg GJ (1963) Diffusion processes in lead-uranium systems. J Geophys Res 68:4823–4846

Wasserburg GJ (1987) Isotopic abundances: inferences on solar system and planetary evolution. Earth Planet Sci Lett 86:129–173

Wasserburg GJ, Wen T, Aronson J (1964) Strontium contamination in mineral analysis. Geochim Cosmochim Acta 28:407–410

Wasserburg GJ, Schramm DN, Huneke JC (1969) Nuclear chronologies for the galaxy. Ap J 157:L91-96

Wasserburg GJ, Jakobsen SB, DePaolo DJ, McCulloch MT, Wen T (1981) Precise determination of Sm/Nd ratios, Sm and Nd isotopic abundances in standard solutions. Geochim Cosmochim Acta 45:2311–2324

Wasson JT (1963) Radioactivity in interplanetary dust. Icarus 2:54–87

Wasson JT (1985) Meteorites: Their Record of Early Solar System History. Freeman New York, 286 pp

Waterbolk HT (1983) Ten guidelines for the archaeological interpretation of radiocarbon dates. In: Mook WG, Waterbolk HT (eds) $^{14}C$ and Archaeology. PACT 8:57–70

Watkins ND (1976) Polarity subcommission sets up some guide-lines. Geotimes 21:18–20

Webster RK (1960) Mass spectrometric isotope dilution analysis. In: Smales AA, Wager LR (ed) Methods in Geochemistry. Interscience, New York, pp 202–246

Wehmiller JF (1982) A review of amino acid racemization studies in Quaternary mollusks: Stratigraphic and chronologic applications in coastal and interglacial sites, Pacific and Atlantic coasts, United States, United Kingdom, Baffin Island, and Tropical Islands. Q Sci Rev 1:83–120

Wehmiller JF (1984a) Relative and absolute dating of Quaternary mollusks with amino acid racemization: evaluation, applications and questions. In: Mahaney WC (ed) Quaternary Dating Methods. Elsevier, Amsterdam, pp 171–193

Wehmiller JF (1984b) Interlaboratory comparison of amino acid enantiomeric ratios in fossil Pleistocene mollusks. Q Res 22:109–120

Wehmiller JF (1986) Amino acid racemization geochronology. In: Hurford AJ, Jäger E, Ten Cate JAM (eds) Dating Young Sediments. COOP Tech Secr, Bangkok, pp 139–158

Wehmiller JF, Belknap DF (1982) Amino acid age estimates. Quaternary Atlantic coastal plain. Comparison with U-series dates, biostratigraphy, and paleomagnetic control. Q Res 18:311–366

Weis D (1981) Lead isotopic composition in whole rocks: methodology. Bull Soc Chim Belg 90:1127–1140

Weiss W, Jenkins WJ (1980) Tritium-helium-3 dating on natural waters. ZFI Mitt 29:292–301

Weiss W, Roether W (1980) The rates of tritium input to the world oceans. Earth Planet Sci Lett 49:435–446

Weiss W, Sittkus A, Stockburger H, Sartorius H, Münnich KO (1983) Large-scale atmospheric mixing derived from meridional profiles of krypton-85. J Geophys Res 88 (C 13):8574–8578

Weiss W, Stockburger H, Sartorius H, Rozanski K, Heras C, Östlund HG (1986) Mesoscale transport of $^{85}$Kr originating from European sources. Nucl Instrum Methods Phys Res B 17:571–580

Welz B (1983) Atomabsorptionsspektrometrie. Verlag Chemie, Weinheim, 527 p

Wendt I (1984) A three-dimensional U-Pb discordia plane to evaluate samples with common lead of unknown isotopic composition. Isot Geosci 2:1–12

Wendt I, Carl C (1985) U/Pb dating of discordant 0.1 Ma old secondary U minerals. Earth Planet Sci Lett 73:278–284

Wendt I, Stahl W, Geyh M, Fauth F (1967) Model experiments for $^{14}$C water-age determinations. In: Isotopes in Hydrology: 321–337. IAEA, Vienna

Weninger B (1986) High-precision calibration of archaeological radiocarbon dates. Acta Interdisciplinaria Archaeologica IV:11–53

Wetherill GW (1956) Discordant uranium-lead ages. Trans Am Geophys Union 37:320–326

Wetherill GW (1963) Discordant uranium-lead ages – Pt 2, discordant ages resulting from diffusion of lead and uranium. J Geophys Res 68:2957–2965

Wetherill GW (1971) Of Time and the Moon. Science 173:383–392

Wickmann FE, Aberg G, Levi B (1983) Rb-Sr dating of alteration events in granitoids. Contrib Mineral Petrol 83:358–362

Wigley TML (1977) Carbon-14 dating of groundwater from closed and open systems. Water Resour Res 11:324–328

Wigley TML, Muller AB (1981) Fractionation corrections in radiocarbon dating. Radiocarbon 23 (II):173–190

Wigley TML, Plummer LN, Pearson FJ Jr (1978) Mass transfer and carbon isotope evolution in natural water systems. Geochim Cosmochim Acta 42:1117–1139

Wilgain S, Picciotto E, de Brenck W (1965) Strontium-90 fall-out in Antarctic. J Geophys Res 70:6023–6032

Wilhelm HG, Ackermann W (1972) Altersbestimmung nach der K-Ca-Methode an Sylvin des Oberen Zechsteines des Werragebietes. Z Naturforsch 27a:1256–1259

Williams DF, Lerche I, Full WE (1988) Isotope Chronostratigraphy. Theory and Methods. Academic Press, San Diego New York Berkeley Boston. 345 pp

Williams GE, Polach HA (1969) The evaluation of $^{14}$C ages for soil carbonate from the arid zone. Earth Planet Sci Lett 7:240–242

Williams HM, Smith GG (1977) A critical evaluation of the application of amino acid racemization to geochronology and geothermometry. Origins Life 8:91–144

Williams IS, Compston W, Black LP, Ireland TR, Foster JJ (1984) Unsupported radiogenic Pb in zircon: a cause of anomalously high Pb-Pb, U-Pb and Th-Pb ages. Contrib Mineral Petrol 88:322–327

Williams KL (1987) Introduction to X-ray spectrometry. Allen Unwin, London

Williamson JH (1968) Least-square fitting of a straight line. Can J Phys 46:1845–1848

Willkomm H (1983) The reliability of archaeologic interpretation of $^{14}$C dates. Radiocarbon 25:645–646

Winograd IJ, Szabo BJ, Coplen TB, Riggs AC (1988) A 250,000-Year climatic record from Great Basin vein calcite: implications for Milankovitch theory. Science 242:1275–1280

Wintle AG (1978) A thermoluminescence dating study of some Quaternary calcite: potential and problems. Can J Earth Sci 15:1977–1986

Wintle AG (1982) Thermoluminescence properties of fine-grain minerals in loess. Soil Sci 134:164–170

Wintle AG (1987) Thermoluminescence dating of loess. Catena Suppl 9:103–115

Wintle AG, Catt JA (1985) Thermoluminescence dating of Dimlington Stadial deposits in eastern England. Boreas 14:231–234

Wintle AG, Huntley DJ (1979) Thermoluminescence dating of a deep sea sediment core. Nature (Lond) 279:710

Wintle AG, Huntley DJ (1980) Thermoluminescence dating of a ocean sediments. Can J Earth Sci 17:348–360

Wintle AG, Huntley DJ (1982) Thermoluminescence dating of sediments. Q Sci Rev 1:31–53

Wintle AG, Huntley DJ (1983) ESR studies of planktonic foraminifera. Nature (Lond) 305:161–162

Wintle AG, Shackleton NJ, Lautridou JP (1984) Thermoluminescence dating of periods of loess deposition and soil formation in Normandy. Nature (Lond) 310:491–493

Whiticar MJ, Faber E, Schoell M (1986) Biogenic methane formation in marine and freshwater environments: $CO_2$ reduction vs. acetate formation – isotope evidence. Geochim Cosmochim Acta 50:693–709

Wölfli W (1987) Advances in accelerator mass spectrometery. Nucl Instrum Methods Phys Res B 29:1–13

Wölfli W, Polach HA, Andersen HH (1984) Accelerator Mass Spectrometry. Nucl Instrum Methods 233 (B5):448

Wörner G, Zindler A, Staudigel H, Schmincke H-U (1986) Sr, Nd, and Pb isotope geochemistry of Tertiary and Quaternary alkaline volcanics from West Germany. Earth Planet Sci Lett 79:107–119

Wood BJ (1975) Influence of pressure, temperature and bulk composition on the appearance of garnet in orthogneiss – an example from South Harris, Scotland. Earth Planet Sci Lett 26:299–311

Wright RJ, Simms LA, Reynolds MA, Bogard DD (1973) Depth variation of cosmogenic noble gases in the 120-kg Keyes chondrite. J Geophys Res 78:1308–1318

Yamada M, Tsunogai S (1984) Postdepositional enrichment of uranium in sediment from Bering Sea. Mar Geol 54:263–276

Yang H-S, Noraki Y, Sakai H (1986) The distribution of $^{230}$Th and $^{231}$Pa in the deep-sea surface sediments of the Pacific ocean. Geochim Cosmochim Acta 50:81–89

Yapp CJ, Epstein S (1982a) A reexamination of cellulose-bound hydrogen D measurements and some factors affecting plant-water D/H relationships. Geochim Cosmochim Acta 46:955–965

Yapp CJ, Epstein S (1982b) Climatic significance of the hydrogen isotope ratios in tree cellulose. Nature (Lond) 297:636–639

Yengingil Z, Göksu Y (1982) Fission-track dating of obsidians. Nucl Tracks 6:43–48

Yiou F, Raisbeck G, Bourlès D, Lestringuez J, Deboffle D (1986) Measurement of $^{10}$Be and $^{26}$Al with a tandetron accelerator mass spectrometer facility. Radiocarbon 28 (2A):198–203

Yokoyama Y, Guichard F, Reyss J-L, Van NH (1978) Oceanic residence times of dissolved beryllium and aluminum deduced from cosmogenic tracers $^{10}$Be and $^{26}$Al. Science 201:1016–1017

Yokoyama Y, Reyss J-L, Guichard F (1977) Production of radionuclides by cosmic rays at mountain altitudes. Earth Planet Sci Lett 36:44–50

York D (1966) Least-squares fitting of a straight line. Can J Phys 44:1079–1086

York D (1969) Least-squares fitting of a straight line with correlated errors. Earth Planet Sci Lett 5:320–324

York D, Farquar RM (1972) The earth's age and geochronology. Pergamon Press, New York, 178 pp

York D, Hall CM, Yanase Y, Hane JA, Kenyon WJ (1981) $^{40}$Ar/$^{39}$Ar dating of terrestrial minerals with a continuous laser. Geophys Res Lett 8:1136–1138

York D, Masliwec A, Kuybida P, Hanes JA, Hall CM, Kenyon WJ, Spooner ETC, Scott SD (1982) $^{40}$Ar/$^{39}$Ar dating of pyrite. Nature (Lond) 300:52–53

Zaikowski A, Roberts EH (1981) Importance of resolution for helium detectors used in uranium exploration. AAPG Bull 65:1011

Zartman RE (1974) Lead isotope provinces in the Cordillera of the western United States and their geologic significance. Econ Geol 69:792–805

Zartman RE, Doe BR (1980) A Hewlett-Packard 9830A BASIC language program for Plumbotectonics. US Geol Surv Open File Rep 80–1088:33 pp

Zartman RE, Doe BR (1981) Plumbotectonics – the model. Tectonophysics 75:135–162

Zashu S, Ozima M, Nitoh O (1986) K-Ar isochron dating of Zaire cubic diamonds. Nature (Lond) 323:710–712

Zaun PE (1982) Einflüsse der Bodenlargerung auf antike Keramik. Mineralneu und-rückbildungen als mögliche Grundlagen für neue Datierungshilfen. Ein Beitrag zur archäometrischen Forschung. N Jahrb Mineral Abh 3:106–118

Zaun PE, Wagner GA (1985) The potential of zircon fission track studies for deep-drilling projects. Naturwissenschaften 72:143–144

Zeller EJ, Levy PW, Mattern PL (1967) Geologic dating by electron spin resonance. In: Radioactive Dating and Methods of Low-Level Counting: 531–540. IAEA, Vienna

Zimmerman DW (1967) Thermoluminescence from fine grains from ancient pottery. Archaeometry 10:26–28

Zimmerman DW (1971) Uranium distribution in archaeological ceramics: dating of radioactive inclusions. Science 174:818–819

Zimmermann U, Ehhalt D, Münnich KO (1967) Soil-water movement and evatranspiration: changes in the isotopic composition of water. In: Isotopes in Hydrology: 567–585. IAEA, Vienna

Zindler A, Hart S (1986) Helium: problematic primordial signals. Earth Planet Sci Lett 79:1–8

Zindler A, Staudigel H, Hart SR, Endres R, Goldstein S (1983) Nd and Sr isotope study of a mafic layer from Ronda ultramafic complex. Nature (Lond) 304:226–230

Zito R, Donahue DJ, Davis SN, Bentley HW, Fritz P (1980) Possible subsurface production of carbon-14. Geophys Res Lett 7 (4):235–238

Zöller L, Wagner GA (1989) Strong or partial thermal washing in Tl dating of sediments? Workshop Long & Short Term Limits in Luminescence Dating. Research Lab of Archaeology & History of Art, Oxford University, Oxford

Zuber A (1983) On the environmental isotope method for determining the water balance components of some lakes. J Hydrol 61:409–427

Zuber A (1986) Mathematical models for the interpretation of environmental radioisotopes in groundwater systems. In: Fritz P, Fontes JC (eds) Handbook of Environmental Isotope Geochemistry, vol 2. The Terrestrial Environment B. Elsevier, Amsterdam, pp 1–59

Zuber A, Grabczak J, Kolonko M (1979) Environmental and artificial tracers for investigating leakages into salt mines. In: Isotope Hydrology 1978 (I): 45–62. IAEA, Vienna

Zuckerkandl E, Pauling L (1962) Molecular disease, evolution, and genetic heterogeneity. In: Kasha M, Pullman B (eds) Horizons of Biochemistry. Academic Press, New York, pp 189–225

# Acknowledgments

A textbook of this breadth could only have been written with the active and implicit contributions of many. Over the years we have received advice and suggestions as well as valuable criticism from such a number of specialists in the various disciplines so that we could not begin to name them all. Here we include friends and colleagues in our immediate disciplines, also the many geoscientists and archeologists who over the past few decades have collaborated with the Hannover Radiocarbon Laboratory on geochronological problems. Indispensable were the specialists whom we requested to review the various sections of the book. Among these there is a group to whom we are particularly thankful; the present structure, the presentation of the individual methods as well as the selection of references, were much enhanced by their criticism, corrections, and suggestions for improvement:

Alvarez, L.W., Berkeley, CA, USA
Clauer N., Strasbourg, France and
  Luxemburg
Elmore D., Rochester, NY, USA
Englert P., San Jose, CA, USA
Grauert B., Münster, FRG
Hofmann A.W., Mainz, FRG
Holser W.T., Eugene, OR, USA
Ivanovich M., Oxford, UK
Lanphere M.A., Menlo Park, CA,
  USA
Latham A., Waterloo, ON, Canada
Marshall B.D., Denver, CO, USA

Michels J.W., St. College, PA, USA
Naeser C.W., Denver, CO, USA
Naeser N.D., Denver, CO, USA
Olsson I.U., Uppsala, Sweden
Patchett, J., Tucson, AR, USA
Scott M., Glasgow, UK
Shackleton N.J., Oxford, UK
Shukolyukov Yu.A., Moscow, USSR
Smith B.W., Oxford, UK
Stacey J.A., Menlo Park, CA, USA
Verhagen B.Th., Johannesburg, SA
Wahlen M., Albany, NY, USA
Wehmiller J.F., Newark, NJ, USA

Reviewing the work of others is a thankless task. We are well aware of this and are no less appreciative of the contribution of the other reviewers. For example, expressions were improved, or attention was drawn to papers to be cited which we would otherwise have missed and to inadvertent errors. Here we wish to thank:

Aitken M.J., Oxford, Uk
Bada J.L., La Jolla, CA, USA
Bell K., Ottawa, Canada
Compston W., Canberra, Australia
Delaloye M., Geneva, Switzerland

Kirsten T., Heidelberg, FRG
Lehmann B.E., Bern, Switzerland
Loosli H.H., Bern, Switzerland
Luck J.M., Paris, France
Mangini U., Heidelberg, FRG

Fehn U, New York, NY, USA         Menzies M.A., London, UK
Friedman I., Denver, CO, USA       Nijampurkar V.N., Ahmedabad, India
Göksu-Ögelman Y., Neuherberg, FRG  Nishiizumi K., San Diego, CA, USA
                                Polach H., Canberra, Australia
Hedges R.E.M., Oxford, UK          Pucher R., Hanover, FRG
Heumann K.G., Regensburg, FRG     Schleicher K., Woods Hole, MA, USA
Hsü K.J., Zurich, Switzerland        Schotterer U., Berne, Switzerland
Huntley D.J., Burnaby, BC, Canada    Siegenthaler U., Berne, Switzerland
                                Wagner G., Heidelberg, FRG
Lkeya M, Yamaguchi, Japan          Wild E, Vienna, Austria

In particular, we extend our appreciation to Dr. Clark Newcomb. He did more than simply translate, he contributed his scientific knowledge to eliminate vague and superficial formulations, as well as laboratory jargon, in order to produce a text which is clear and accessible to readers from a wide range of fields.

We wish to acknowledge Jens Pielawa for his patience in making corrections, when they became apparent, in the figures he had drawn. Also, our employers and closest colleagues for their support through their patience and consideration are acknowledged.

Our wives, children, and personal friends need to be especially mentioned for their understanding and forbearance. No longer will *Physical and Chemical Dating Methods* have to be considered in the planning of weekends and vacations, the evening hours and at those times when we were particularly needed. To this large circle, including those we have not mentioned by name, who have contributed in so many ways to the completion of this book, we wish to express our sincere appreciation.

Figures 5.13, 6.92, 6.93, 6.94, and 8, 10 have been taken from Mahaney (1984) with the kind permission of Elsevier, Amsterdam. Springer-Verlag followed the preparation of this book with patience. Finally, thanks are due to Drs. D. Hohm and W. Engel for their valuable advice.

After all the years of work on this book, we did not expect anyone could stimulate us to willingly do the endless correction necessary for any book. With her friendly manner and suggestions Ms. S. Fink of the Springer publishing company set a constructive and warm atmosphere for the concluding stages of the preparation of this book, for which we wish to express our special thanks.

Bennetze and Freiburg                                Mebus A. Geyh
May 15, 1990                                    Helmut Schleicher

# Appendix A:  Geochronology Glossary

AA analysis: atomic absorption analysis; substances are analyzed on the basis of the wavelengths they absorb (*see* Sect. 5.2.4.3)

absolute age: *see* age

accelerator, particle: device in which charged particles are accelerated in an electrical field to high energies; in geochronology accelerators are used for high resolution mass spectrometry (*see* Sect. 5.3.2.2)

accessory mineral: minerals present in a rock in amounts too small to be considered for the classification of the rock; although they occur only in minor amounts in a rock, they often have a wide distribution (e.g., zircon, apatite, magnetite)

accuracy: nearness of a measurement to its accepted (e.g., as represented by a standard) or true value (*see also* precision)

achondrite: stoney meteorite that contains no chondrules (see also chondrite)

acidic rock: igneous rock with a relatively high silica content (more that 60%)

activation: *see* neutron-activation analysis

activity: *see* decay rate

AD: (L. anno domini) years after Christ

age, absolute: age expressed in units of time, especially sidereal years (Sect. 2.1)

age, apparent: age determined according to a model assumption that in some cases may not apply to the substance being dated; reasons for this could be, for example, the reservoir effect or contamination; apparent ages may differ from the true ones in either direction by an amount that is often unknown

age, conventional: age determined by any of the advanced geochronological methods for which the half-life, the year of reference, the standard, etc. have been defined according to international convention (*see* Sect. 2.3)

age, radiocarbon: age determined from the $^{14}C$ content of terrestrial organic matter on the basis of the Libby model using a conventional half-life of 5568 years and 1950 AD as reference year and corrected by the $\delta^{13}C$ value referred to $-25‰$ (*see* Sect. 6.2.1)

age: *see also* cooling age, crustal residence age, crystallization age, gas retention age, cosmic ray exposure age, isochron age, mean residence time, model age

aliquot: a fraction (e.g., 1/2, 1/3, 1/4, 1/5, ...) of a homogeneous mixture

allochthonous: originating elsewhere, not native to a place (in the context of this book: not the same age as that which is to be dated, e.g., the stratum in which the object was found); see autochthonous

alpha decay: radioactive decay of an isotope with emission of a helium nucleus with a mass of 4 amu and a charge of $+2$ (alpha particle); in this process the atomic number of the parent isotope decreases by two and the atomic mass decreases by four (e.g., $^{238}_{92}U \rightarrow {}^{234}_{90}Th$, *see* Sect. 5.1)

alpha-recoil tracks: *see* recoil tracks

alpha-spectrometry: *see* spectrometry

alteration (in the geological sense of the word):
   a) any change in the mineralogical composition of a rock by chemical or physical means, e.g., by hydrothermal solutions or by diagenesis;
   b) any change in the mineralogical and/or chemical composition of a rock by weathering

anatexis: melting of pre-existing rock

anion: negatively charged particle that moves to the anode in an electrical field; *see* cation

annealing: the repair of lattice defects (e.g., fission tracks) by recrystallization at elevated temperature; the annealing temperature differ from mineral to mineral

annihilation radiation: two 0.51-MeV photons (gamma radiation) emitted in diametrically opposed directions when an electron and a positron combine

anode: positive pole of a source of direct current, e.g., a battery

anomalous lead: lead with isotope ratios that indicate a higher or lower model age than is possible on the basis of the stratigraphic age

anthropogenic: man-made; arising from or changed by the influence of man

anticoincidence: non-simultaneous occurrence of events (in the context of this book: a pulse that occurs within a given time interval in only one of two or more detectors connected in parallel; *see* Sect. 5.2.2

ANU standard (for Australian National University): secondary standard for the $^{14}C$ method (Sect. 6.2.1); *see* NBS oxalic acid

apparent age: *see* age

apatite dating: $^{14}C$ age determination (Sect. 6.2.1) of the carbon in the apatite in bones, which is highly susceptible to isotopic exchange; apatite $^{14}C$ ages are, therefore, less reliable than those done on collagen

aragonite: orthorhombic calcium carbonate; can be transformed into calcite by recrystallization

Archean: rocks of the Archeozoic

Archeozoic: earlier part of Precambrian time (*see* Table 9.1)

archeological dose: *see* equivalent dose

argon, inherited: *see* inherited argon

atomic mass: mass of an atom, usually given in atomic mass units (amu); it is the atomic weight multiplied by the atomic mass unit, which is equal to exactly 1/12 the mass of a neutral atom of the most abundant isotope of carbon, $^{12}C$; it is approximated by the number of protons and neutrons in the nucleus

atomic number: number representing the position of an element in the periodic table, it is equal to the number of protons in the nucleus of an atom of that element

authigenic: formed or generated in place; said of a mineral that crystallized at the spot where it is now found; often used for a mineral (e.g., quartz and feldspar) formed after deposition of the sediment

autochthonous: indigenous, native, originating at the place where found (in the context of this book: the same age as that which is to be dated); *see* allochthonous

autoclave: in geochronological analyses, container (usually of stainless steel coated with teflon) used for acid digestion of samples under pressure and elevated temperature

background sample: standard whose activity is zero; the counting rate for such a sample (background counting rate) by a detector is then due to the radiation from the surroundings and from the detector itself (*see* Sect. 5.2.2)

basement: crust of the Earth below the sediment cover

basic rock: igneous rock with a relatively low silica content (between 44 (45)% and 51 (52)% $SiO_2$)

bell curve: frequency distribution of data in the shape of a bell (Sect. 4.2, Fig. 4.2); *see* normal distribution

beta decay: radioactive decay of an isotope with emission of an electron resulting in a daughter isotope with the same mass but with an atomic number increased by one

BC: years before Christ

bioluminescence: *see* luminescence

biosphere: zone of the earth that contains living organisms (including decomposing organic matter); all the living organisms of the earth

biostratigraphy: branch of the geosciences that deals with the succession and classification of stratified rocks on the basis of their fossil content: *see* stratigraphy

bioturbation: disturbance of sediments by the burrowing or digging activity of animals or by plant roots; can lead to complete disappearance of the layering; deposits in which bioturbation is present are often unsuitable for dating

blocking temperature: *see* closure temperature

BP: before present; the years before 1950, the reference year for all geochronological time-scales

branching: the occurrence of two or more modes by which a radionuclide can decay; the different types of decay always occur in a definite proportion, called the branching ratio; branching occurs at least once in each of the three decay series, another example is $^{40}K$ (Sect. 6.1.1)

Bremsstrahlung (lit. "braking radiation"): a fast-moving electron approaching the nucleus of an atom is deflected, resulting in the emission of energy as radiation, slowing or "braking" the electron

calibration: the fixing, checking, or correction of a measuring instrument or the adjustment of a time-scale

Calvin cycle ($C_3$ cycle): main path of $CO_2$ incorporation in plants native to humid climates; leads to depletion of $\delta^{13}C$ by $-17‰$ (Sect. 6.2.1); see Hatch-Slack cycle

Cambrian: earliest period of the Paleozoic era (*see* Fig. 9.1; Table 9.1)

Carboniferous: geological period of the Paleozoic era (*see* Fig. 9.1; Table 9.1)

cation: positively charged particle that moves to the cathode in an electrical field; see anion

Cenozoic: geologic era from the beginning of the Tertiary period to the present; follows the Mesozoic era (*see* Table 9.1)

chalcophile: term for an element that tends to concentrate in sulfide minerals and ores; cf. lithophile, siderophile

chemoluminescence: *see* luminescence

chi-square test: statistical test for values for a random sample (or samples) to determine whether they belong to a common frequency distribution with a given probability (Sect. 4.2.1)

chondrite: stony meteorite that contains chondrules (rounded aggregates about 1 mm and smaller) of olivine and/or orthopyroxene in a fine crystalline matrix

chromatography: method for separation of mixtures based on differences in the adsorption properties of the components

chronology: science of measuring time in fixed periods and of dating events and epochs and arranging them in the order of occurrence; arrangement of events, etc. in the order of occurrence

chronometer: instrument (or method) for measuring time precisely

chronostratigraphy: science involving the correlation and classification of strata on the basis of their relative or absolute ages; see chronology, stratigraphy

CHUR: Chondritic Uniform Reservoir

class (statistics): interval in a frequency distribution

clastic rock: rock or sediment composed principally of fragments produced by mechanical degradation of other rocks

closed system: *see* system

closure temperature (also called blocking temperature): narrow, mineral-specific temperature range below which closed conditions prevail for a given parent/daughter isotope system, i.e., no loss of radiogenic gases and/or no isotope exchange with the surroundings occur (Fig. 6.3) (cf. Curie temperature)

coccolith: microscopic calcareous plate covering the body of some floating marine organisms

coeval: of the same age

cogenetic: of the same origin or the same formation

coincidence: simultaneous occurrence of events (in the context of this book: simultaneous pulses within a given time interval in two or more parallel detectors); *see* anticoincidence (Sect. 5.2.2)

collagen dating: relatively reliable radiocarbon dating of the fibrous protein found in connective tissue, bone, and cartilage (Sect. 6.2.1); will become substituted by the more reliable radiocarbon dating of specific amino acids from the collagen

collimator: part of a spectrometer for confining the particles of a beam within a given solid angle

common lead: lead in a mineral or rock that was not produced by the in-situ decay of a radioactive parent isotope since the formation of the mineral or rock; the isotopic composition of common lead is the same as the average of the geochemical reservoir at the time of formation of the mineral or rock; this definition of "common" is also applied to other elements

concordant: a) radiometric ages determined by more than one method that are in agreement within experimental error (*see* Sect. 6.1.9); b) structurally conformable; said of strata showing parallelism of bedding or structure

concordia curve: curve on which the points all represent the same (i.e., concordant) age (a term from the U/Pb age dating method); rocks or minerals with concordant $^{238}U/^{206}Pb$ and $^{235}U/^{207}Pb$ ages have isotope concentrations that lie on a curve in a plot of $^{206}Pb/^{238}U$ versus $^{207}Pb/^{235}U$ (Sect. 6.1.9, Fig. 6.28)

confidence interval: range given by the mean $\pm$ a multiple of the standard deviation; if the factor is one (corresponding to the $1\sigma$ criterion), this range includes 68% of data with a normal distribution (Fig. 4.2)

confidence probability: probability with which a decision made on the basis of a statistical test is right

conformable lead: lead whose isotope ratios yield model ages in a two-stage lead development model which agree with the stratigraphic age of the sample (Sect. 6.1.10)

contamination: (in the context of this book) allochthonous material with an age different from the sample in which it occurs; in the case of organic samples, roots and humic acids are common contaminants (see Sect. 6.2.1)

contemporary: (a) living or happening in the same period of time; (b) of about the same age; (c) of the present or recent times. The emphasis of this word is on a comparison of two or more things

continental effect: increase in the tritium content, as well as the $\delta^2$H and $\delta^{18}$O values in precipitation with increasing distance from the coast

conventional age: see age

cooling age: time since the last cooling of a mineral, rock, or meteorite below a narrow, substance-specific temperature range below which radiation damage (e.g., fission tracks) is no longer annealed (Sect. 6.4.7), gases no longer diffuse away (Sects. 4.1.2 and 6.1.1; see gas-retention age), or isotopic homogenization no longer occurs

cooling model: model for the cooling history of a rock unit or an extensive geological area

cosmic-ray exposure age: period of time during which a meteoroid or lunar rock was exposed to extraterrestrial radiation (Sect. 6.5.6)

cosmochronology: science of the dating of events in the universe; time-scale for these events

cosmogenic: produced by the action of cosmic rays

counter: radiation detector which delivers electrical pulses whose frequency is proportional to the activity; in proportional counters the size of the pulse is proportional to the energy of the detected radiation (see Sect. 5.2.2.1); see also Geiger-Müller counter, liquid scintillation counter

Cretaceous: final period of the Mesozoic era (see Fig. 9.1; Table 9.1)

criterion, $1\sigma$ ($2\sigma$, etc.): see confidence interval

crustal residence age: time that a specific material has been in the Earth's crust

crystal lattice: three-dimensional repeating pattern in which the atoms, ions, or molecules of a crystalline solid are arranged

crystallization age: time since the crystallization of a specific mineral in a rock

Curie temperature: substance-specific temperature above which ferromagnetic substances acquire paramagnetic properties (Sect 7.1); synonym: Curie point (cf. closure temperature)

dating range: range of age that can be covered with a specific dating method (see Foldout Table at the end of this book)

daughter isotope: product of the decay of a radionuclide (which is called the parent isotope) (Sect. 5.1); synonym for decay product decay constant: isotope-specific constant that corresponds to the probability with which a radionuclide will decay per time unit; is inversely proportional to the half-life (Sect. 5.1)

decay rate (activity): number of atoms that decay per time unit

decay series: *see* radioactive decay series

declination, magnetic: deviation of the Earth's magnetic north pole from the geographic one (Sect. 7.1)

dehydration: removal of water from organic or inorganic compounds

delta notation: usual form in which isotope abundances (e.g., of hydrogen, oxygen, and carbon) relative to a standard substance (PDB, SMOW) are given (in ‰ deviation from the abundance of the standard) (Sect. 6.2.1)

dendrochronological correction of $^{14}C$ ages: conversion of conventional $^{14}C$ ages into absolute (calibrated $^{14}C$) ages using tables or curves (Stuiver and Kra 1986) (*see* Sect. 6.2.1)

dendrochronology: branch of geochronology dealing with the absolute dating of pieces of wood by tree-ring analysis (Sect. 2.1)

dendro-year: age based on the annual rings of a tree not necessarily correlated with a usual reference year

depleted mantle: part of the Earth's mantle that is depleted in certain elements due to formation of crustal material by partial melting

detection threshold: minimum level that something being measured (e.g., activity) can be distinguished from the background level

detritus: fragments of rock produced by mechanical disintegration; any accumulation of such disintegrated material or debris

deuterium ($^2H$): stable isotope of hydrogen with a mass of 2

Devonian: geological period of the Paleozoic era (*see* Fig. 9.1; Table 9.1)

De Vries effect: changes in the initial $^{14}C$ concentration of the atmospheric carbon dioxide during the past, caused by variation in radiocarbon production and by changes due to climate in the distribution of carbon in the global reservoirs (biosphere, hydrosphere, atmosphere) (Sect. 6.2.1)

diagenesis: physical, chemical, and biological changes occurring in sediments during and after the period of initial deposition, including lithification, but excluding surficial alteration (weathering) and metamorphism

dielectric: term for a material that does not conduct electricity; a property of electrical insulators

differentiation, magmatic: process by which various types of igneous rocks are formed from a source (or parent) magma

diffusion: intermingling of gaseous, liquid, or solid substances in contact with each other without external energy, but due to Brownian motion; diffusion finally results in equalization of primary concentration differences

diffusivity: capacity of a material to let substances diffuse through it

dilatancy: increase in volume during a geological process caused by a change from a close-packed structure to a more open one; is associated with an increase in the pore volume and accompanied by microfracturing and grain boundary slippage

dipole field: magnetic or electric field of a rod magnet or two electrical charges (*see* Sect. 7.1) dipole moment of the Earth's magnetic field: the magnetic field of the Earth corresponds, in first approximation, to that of a rod magnet (dipole field) whose magnetic moment is the product of magnetic charge and the shortest distance between the poles; the vector of the dipole moment points from the magnetic south pole to the north pole (Sect. 7.1)

discordant: a) radiometric ages determined by more than one method that are in disagreement beyond experimental error (*see* Sect. 6.1.9); b) structurally un-conformable; said of strata lacking conformity or parallelism of bedding or structure

discordia: line in a plot of $^{206}Pb/^{238}U$ versus $^{207}Pb/^{235}U$ formed by the data from phases that have lost lead or gained uranium as the result of a secondary (i.e., more recent) geological event (e.g., metamorphism) (*see* Sect. 6.1.9)

discrimination, mass: supression or amplification of certain ranges of mass during a mass spectrometric measurement by the use of certain instrument parameters (*see* Sect. 5.2.3.1)

disequilibrium, radioactive: radioactive equilibrium disturbed by a chemical or physical process (*see also* radioactive equilibrium) (Sect. 6.3)

dose: *see* radiation dose

dpm: disintegrations per minute

dps: disintegrations per second

dual decay: *see* branching

Earth's crust: outermost layer of the Earth, divided into upper and lower continental crust and oceanic crust; the boundary between the crust and the mantle is called the Mohorovičic (Moho) discontinuity

Earth's mantle: zone of the Earth below the crust and above the core; divided into upper and lower mantle

Eemian: interglacial stage before the Holocene, beginning 125,000 years ago

electron capture (EC): mode of radioactive decay involving the capture and inclusion in the nucleus of an electron from the innermost electron shell (the K-shell) of an atom with emission of $\gamma$-radiation (Sect. 5.1, Fig. 5.1); the daughter isotope has the same mass with an atomic number one less than the parent isotope, e.g., $^{40}K \rightarrow ^{40}Ar$ (also called K-electron capture)

electron microprobe: *see* microprobe

electron spin resonance analysis: *see* ESR spectroscopy

enantiometric ratio: ratio of the number of mirror-image molecules of a substance (Sect. 8.1)

energy band: quantum mechanical concept to describe the energy distribution of electrons in atomic or molecular bonding; according to this concept, electrons in atoms move only in discrete paths (orbitals) of differing energy levels; the higher the energy level, the further the orbital from the atomic nucleus; the spaces between orbitals is forbidden for electrons; this simplified concept requires a correction inasmuch as bands exist instead of discrete energy levels because of thermal movement of the electrons themselves and because the outer orbitals lie close to one another due to interaction between neighboring atoms; each band, which can have several energy levels, may contain no more than two electrons with opposite spin; electrons in the outermost band are quasi free to move within the crystal lattice; those in the lower luminescence band are bound to a single atom; a characteristic luminescence occurs if electrons are raised from an inner orbital to the luminescence band by addition of energy (excited atom) and other electrons drop back to the vacated inner orbital (Sect. 6.4)

enrichment, isotopic: increase in the concentration of a specific isotope in a substance or phase by physical or chemical isotope fractionation

environmental isotope: natural or anthropogenic isotope present in nature and whose concentration is determined by natural processes

environmental radiation: radiation from environmental isotopes and cosmic radiation

epigenetic: term for minerals and rocks that are younger than their surroundings

epithermal: term for a hydrothermal mineral deposit formed in the 50–200°C temperature range at depths of up to about 1 km

equivalent dose: laboratory dose needed to obtain a TL signal of the same size as that of the natural thermoluminescence of a sample; also called total dose (Sect. 6.4)

equilibrium, radioactive: steady-state condition of a system in which the number of atoms of the members of a natural radioactive decay series does not change and the half-lives are directly proportional (secular equilibrium); if the half-life of a parent isotope is relatively small, but still larger than that of the daughter isotope, then an equilibrium is present in which although the ratio of the atomic numbers of the parent and daughter isotopes remains constant, the absolute numbers decrease with the half-life of the parent isotope (Sect. 5.1)

equilibrium, steady-state: continuous state of a closed, dynamic system in which specific conditions do not change, e.g., when the rate of decay and the rate of formation of a radioactive material are equal

error: *see* standard deviation

error, a priori: assumed on the basis of previous experience, includes all possible chemical and physical sources of error

errorchron: *see* pseudoisochron

ESR spectroscopy [electron-spin resonance spectroscopy, also called electron paramagnetic resonance (EPR) spectroscopy]: sensitive analytical method for measuring the paramagnetic properties of atoms and molecules. Depending on their energy levels, electrons absorb differing amounts of energy when placed in an external, alternating magnetic field causing a transition between spin states (Sect. 6.4.3)

eutrophic: designating or of a lake, pond, etc., rich in plant nutrient minerals and organisms but often deficient in oxygen in midsummer

evolution diagram: graph in which changes in isotopic composition in a closed system that result from the formation of radiogenic isotopes are plotted as a function of time

evolution, single-stage (two-stage, multi-stage): mineral, rock, or geochemical reservoir in which the isotopic composition of a specific element has been determined by one, two, or more events

excess isotope: isotope incorporated in a rock or mineral by processes other than in-situ radioactive decay of the parent nuclide (*see* Sect. 6.1.1)

exchange, ionic and isotopic: reversible process in which ions or isotopes are exchanged between two or more compounds or phases; the concentrations of the different ions or isotopes change, but not the number of ions or nuclides involved since atoms, ions, or radicals of a chemical compound are only replaced by others; in molecular exchange, molecules go from one phase to another; this process ends in a steady-state equilibrium; *see also* equilibrium, fractionation, diffusion

excited atom: an atom in which an electron is in a higher than normal orbit after energy has been supplied (e.g., irradiation or heat)

excursion of the Earth's magnetic field: a change in the direction of the Earth's magnetic poles by more than 40° (*see* Sect. 7.1)

exothermic: term for a chemical change occurring with liberation of heat

expectation (expected) value: the values for random samples with a Poisson or normal distribution lie quasi symmetrically around the expectation value; in this case, it is identical with the median for an infinitely large number of random samples (Sect. 4.2)

exponential model: hydrogeological: model for calculating the mean residence time of long-term karst water or groundwater in open aquifers that consist of components of differing age whose proportions increase exponentially with increasing age (Sect. 7.4)

exposure age: *see* cosmic ray exposure age

extraneous isotope: sum of the inherited and excess isotopes in a sample (*see* excess isotope)

extrusion age: time since the extrusion of a rock

extinct radionuclides: radionuclides produced during nucleosynthesis that have completely decayed due to their relatively short half-lives; their existence is demonstrated by radiation damage and stable daughter nuclides produced by their decay (Sect. 6.5.3)

extraterrestrial: outside the Earth and its atmosphere

facies: sum of the lithological, stratigraphic, paleontological, sedimentary, or petrographic features of a rock unit that reflect the environment in which the rock was formed

factor of merit: ratio of the square of the standard net counting rate and the background counting rate of a detector or the square root of this ratio (*see* Sect. 5.2.2)

fading: annealing of radiation damage signal (e.g. fission tracks) during heating; anomalous fading occurs at ambient temperatures (*see* Sect. 6.4)

fallout: radioactive precipitation resulting from the explosion of nuclear weapons or from nuclear power plants (Sect. 7.4)

felsic: term for light-colored minerals (e.g., quartz, feldspar, muscovite) and for rocks composed chiefly of such minerals; *see* mafic

ferromagnetism: ability of iron, nickel and cobalt to assume a high degree of magnetism independent of the strength of the magnetizing field; this ability is dependent on the crystalline state and tends to saturation; returns to a lower level if the temperature rises above the Curie point (*see* Sect. 7.1); cf. paramagnetism

field: spatial distribution of the physical effect of a substance

field strength: vector giving the direction and strength of a field at any particular point

fission, nuclear: the splitting of a nucleus into two fragments of similar mass together with the conversion of part of the mass into energy; some nuclides can undergo spontaneous fission, but most nuclides fission only when bombarded with neutrons or protons (*see* Sect. 6.4.7)

fission track: path left by particulate (e.g., fragments from the spontaneous fission of $^{238}U$) radiation (called radiation damage) made microscopically visible by etching of the surface of minerals and glasses (*see* Sect. 6.4.7)

flame photometer: instrument in which metal atoms are thermally excited to emit light; the frequencies of the emitted light are characteristic for each element, the intensities are proportional to concentration (Sect. 5.2.4.3)

fluid inclusions: tiny cavity in a mineral containing liquid and/or gas; formed by entrapment of fluid, commonly that from which the rock crystallized

fluorescence: light of a specific wavelength emitted by a substance when irradiated with UV rays, X-rays, or electrons; fluorescence can be utilized for quantitative analysis (*see* Sect. 6.4.1)

foraminifer: any of an order (Foraminifera) of small, one-celled sea animals with calcareous shells full of tiny holes through which slender filaments project; they form the main component of chalk and many deep-sea oozes (Sect. 7.2)

formation age: time since the formation of a rock or mineral

formation interval: the time between the termination of nucleosynthesis associated with the creation of the universe ("Big Bang") and the accretion of the Earth (*see* Sect. 6.5.3)

fossil: (in the context of Sect. 6.2) lying outside the dating range of a method, e.g., for radiocarbon: older than about 50,000 years (*see* Sect. 6.2.1)

fractionation, isotopic: change in the isotopic composition during a chemical or physical process as a result of differences in the rates of the different isotopes present in the molecules involved. The difference in the rates is due to the slight difference in the masses of the isotopes. Hence, isotopic fractionation is greater for the lighter elements because the difference between the masses of their isotopes is greater

frequency: the number of times an event, value, or characteristic occurs in a given period; the ratio of the number of times a characteristic occurs to the number of trials in which it can potentially occur

frequency distribution: a function that gives the frequency of the values for a given sample in the individual frequency intervals (called classes) (*see* Sect. 4.2, Fig. 4.2)

gamma-rays: high-energy electromagnetic radiation of extremely short wavelength (about $10^{-11} - 10^{-14}$ m) emitted by the decay of many radionuclides (Sect. 5.1)

gamma-spectrometry: *see* spectrometry

gas chromatography: the process of separating constituents of a gas or vapor mixture based on differences in adsorbtion on a column followed by qualitative or quantitative analysis of the separated components

gas retention age: time since the rock, mineral, or meteorite cooled to a temperature at which it became a closed system, from which occluded gas or gas formed by radioactive decay could no longer escape (e.g., Sect. 6.1.1)

Geiger region (also called the Geiger plateau): the range of applied voltage in a Geiger–Müller counter in which the size of the electrical pulse produced by ionizing radiation is independent of the radiation energy (*see* Sect. 5.2.2.1) but the counting rate is proportional to the decay rate

Geiger–Müller counter (usually abbreviated to Geiger counter): *see* counter

geobarometry: branch of geosciences dealing with the determination of the pressure conditions under which a rock or mineral was formed

geochron: the isochron in a graph of $^{207}Pb/^{204}Pb$ vs. $^{206}Pb/^{204}Pb$ (Sect. 6.1.10) representing the present time

geochronology: branch of the geosciences dealing with the age of the Earth and its materials, the dating of geological events and evolutionary stages in plant and animal development, etc.

geothermometry: branch of the geosciences dealing, in part, with the temperature(s) existing during the formation of rocks and minerals

getter: material introduced into a vacuum to chemically remove any residual gases

glacial stage: subdivision of a glacial period (e.g., the Pleistocene) in which large, non-polar regions were covered with glaciers and inland ice (Sect. 7.2)

glow curve: graph of thermoluminescent emission produced by heating certain, previously irradiated substances (Sect. 6.4.1, Fig. 6.81)

Gregorian calendar: a corrected form of the Julian calendar, introduced by Pope Gregory XIII in 1582 and now used in most countries of the world; it provides for an ordinary year of 365 days and a leap year of 366 days in years divisible by four, except for century years, which are leap years only if divisible by 400; year zero is the year of the birth of Jesus (*see* Sect. 2.1)

half-life: time required for the disintegration of half of the atoms of a specific radioisotope originally present in a sample (Sect. 5.1); in geochronology, a distinction is made between the physical (actual) half-life and the conventional half-life, the latter has been fixed by international convention for calculating conventional ages: for example, $^{14}C$ ages (Sect. 6.2.1) are calculated with a half-life of 5568 years by convention (thus called the conventional half-life), although the physical half-life is 5730 years

hard-water effect (special case of the reservoir effect): phenomenon that organic matter formed in fresh-water lakes or groundwater have a lower initial $^{14}C$ concentration than, for example, wood, which is formed by plants that assimilate their carbon from the atmosphere; samples affected by the hard-water effect yield apparent $^{14}C$ ages (Sect. 6.2.1) that are larger than wood of the same age; see reservoir effect

Hatch-Slack ($C_4$) cycle: main path of $CO_2$ incorporation in plants native to semiarid climates; leads to depletion of $\delta^{13}C$ by $-7‰$ (*see* Sect. 6.2.1); cf. Calvin cycle

histogram: graphic representation in which frequency of values is shown by columns with areas whose widths correspond to a definite range of frequencies and whose heights correspond to the number of frequencies occurring within the range (Sect. 4.2.3); with the advent of computers, more complex representations are now being used (Fig. 4.4); *see also* frequency distribution

Holocene: present epoch of the Quaternary period beginning about 10,000 years ago at the end of the Weichselian glacial stage (Table 9.1)

humic acids: amorphous compounds produced by decomposition of organic substances in soil, coal, and other decayed plant materials (*see* Sect. 6.2.1)

hydration: chemical reaction of water with another substance in a definite molecular ratio to form a hydrate (*see* Sect. 8.6)

hydroisochrons: contour lines on a map connecting groundwater of the same age

hydrosphere: all the water on the surface of the Earth, including oceans, lakes, rivers, and groundwater, as well as snow and ice and the water in the atmosphere

hydrothermal: term for hot water, with or without association with igneous processes, and materials (e.g., a mineral deposit) precipitated from a hot aqueous solution

ICP analysis: inductive coupled plasma analysis; *see* Sect. 5.2.4.3

igneous: rock that solidified from molten or partly molten material, i.e., magma; igneous rocks make up one of three main classes of rocks, the others being sedimentary and metamorphic rocks

immunological distance (ID): relative concentration of antiserum required to produce a peak as high as that given by the homologous protein (which is the protein used to produce the antibody for this test); also called the index of dissimilarity

industry effect: *see* Suess effect

inherited argon: $^{40}Ar$ formed by radioactive decay of $^{40}K$ in a mineral or rock before the event (e.g., formation of a rock or mineral) being dated and not completely driven off by that event

initial concentration (or ratio): isotopic concentration (or ratio) in a substance when it begins to age, for the $^{14}C$ method for example, the death of the living organism (Sect. 6.2.1)

input curve: changes in the concentration of certain isotopes (e.g., $^{14}C$, $^{3}H$, and $^{85}Kr$) in a reservoir (e.g., the atmosphere or hydrosphere) since the beginning of the 1950's (*see* Sect. 7.4, Fig. 7.8)

in situ: in its original place

interglacial stage: time between successive glacial stages (e.g., Riss and Weichselian) with warm climate and the corresponding flora and fauna (e.g., the Eemian) (*see* Sect. 7.2)

interstade: relatively short period of time within a glacial stage in which the climate becomes warmer and the ice recedes but the characteristic flora and fauna of the glacial stage remain; adj. interstadial

intrusion age: time since the intrusion of a magma

ion: an electrically charged atom or group of atoms; the electrical charge of which results when a neutral atom or group of atoms loses or gains one or more electrons during a chemical reaction, by the action of certain forms of radiant energy, etc.

ion exchange: *see* exchange

ionium: earlier term for the thorium isotope with a mass number of 230; decays by alpha emission; parent isotope is $^{234}U$ in the $^{238}U$ decay series; half-life is 75,200 years (Sect. 6.3.1, Fig. 6.3.1)

ionization: formation of an ion by removal or addition of an electron from or to the outer shell of an atom or a molecule, for example as in a gas under the influence of radiation; dissociation into ions of opposite charge, for example during solution in water ion microprobe: *see* microprobe

ion substitution: replacement of an ion in a crystal lattice with one of another element with the same charge, of similar size, and similar electronegativity

irradiation age: *see* cosmic ray exposure age

isobar: any of two or more atoms with the same mass number but different atomic number

isochron: a) line on a map or graph through points with the same age; b) a graph of D/N vs. P/N will yield a straight line ( = isochron) for samples of the same age and same initial isotope ratio D/N (D = radiogenic isotope, N = nonradiogenic isotope of the same element, P = parent isotope of D); the slope of an isochron increases with increasing age of the sample (Sect. 6.1.3)

isochron age: age determined by the construction of an isochron through points for several cogenetic samples

isomer: chemical compound that has the same molecular formula as another compound, but has a different structural formula; isomers that are mirror images of each other (called enantiomers) differ in their optical properties (Sect. 8.1)

isotherm: of the same temperature

isotope: a nuclide of a chemical element that has a different mass than the other isotopes of that element (i.e., same number of protons, different number of neutrons); there are stable and radioactive isotopes, the latter decay with a half-life specific to each isotope by emission of alpha or beta particles or electromagnetic radiation (Sect. 5.1)

isotope dilution analysis: method of quantitative analysis for determining extremely small quantities of an element by adding a known quantity of tracer with a known isotopic composition (called a spike); Sect. 5.2.4.1

isotope geochemistry (syn. isotope geology): branch of the geosciences dealing with the study and application of stable and radioactive isotopes in geology and petrology

isotopic depletion: opposite of isotopic enrichment

isotopic enrichment: *see* enrichment

isotopic exchange: *see* exchange

isotopic fractionation: *see* fractionation

isotropic: having one or more chemical or physical properties that is (are) the same in all directions

IUGS: International Union of Geological Sciences

Jurassic: second period of the Mesozoic era (*see* Fig. 9.1; Table 9.1)

K-electron capture: see electron capture

lacustrine: pertaining to, produced by, or formed in a lake, e.g., sediments

late glacial: Late Pleistocene after the end of the last ice age until the beginning of the Holocene

lattice: *see* crystal lattice

lattice defect: an imperfection in an otherwise ideal crystal lattice; formed during crystal growth, caused by ionizing radiation, or by mechanical deformation

law of radioactive decay: the decay rate is proportional to the number of radionuclides (Sect. 5.1)

layered intrusion: intrusive body with layers of varying mineralogical composition layer structure: type of crystal structure formed by distinct layer units (e.g., mica)

least squares fit: *see* regression analysis

leucocratic: term for a light-colored igneous rock (poor in mafic minerals)

LIL: large ion lithophile

limnic: originating in a fresh water body, e.g., coals, also used for flora and fauna

liquid scintillation counter (LSC): detector for measuring radioactive radiation of a liquid sample or suspension that has been mixed with a scintillator liquid (called a cocktail); radiation produces flashes of light, which are then detected and counted with a photomultiplier (Sect. 5.2.2.2, Fig. 5.8)

lithophile: term for elements that tend to concentrate in silicate/oxide minerals and ores; cf. chalcophile, siderophile

lithosphere: solid rocky part of the Earth, includes the crust and the upper part of the mantle

lithostratigraphy: branch of the geosciences dealing with the nature, distribution and relations of rock strata on the basis of their lithologic character

low-level techniques: radiometric procedures for preparing and measuring very low concentrations of rare, radioactive, usually cosmogenic isotopes (Sect. 6.2)

LREE: light rare earth elements; *see* REE

luminescence: emission of light at other than high temperatures: e.g., fluorescence, caused by absorption of electromagnetic or particulate radiation; bioluminescence, light given off by living organisms; chemoluminescence, light given off during certain chemical reactions; thermoluminescence, release of stored energy in the form of light when a substance is heated; triboluminescence, luminescence resulting from friction, observed at the surface of certain crystalline materials

mafic: term for igneous rocks composed chiefly of ferromagnetic (dark-colored) minerals; it is the complement of felsic

magnetic separator: apparatus for separating paramagnetic minerals from each other and from non-magmatic ones by a magnetic field (Sect. 5.2.1.1)

magnetization, remanent: magnetization of a ferromagnetic substance that remains after the material is removed from a magnetic field (*see* Sect. 7.1)

magnetometer: instrument for measuring the magnetic dipole moment or magnetic field strength (*see* Sect. 7.1)

mass number: number of nucleons (i.e., protons and neutrons) in an atomic nucleus (*see also* atomic number)

mass spectrometer: apparatus for determining relative abundances of ionized atoms or molecular fragments that have different masses (Sect. 5.2.3.1)

master curve, dendrochronological: a generalized change in the tree growth rings valid for limited regions and specific tree species in the past; the age of a sample can be determined usually to within 1 to 2 years by comparing the growth spectrum of a piece of wood that has at least 100 growth rings with the dendrochronological master curve (Sect. 2.1); *see* tree-ring analysis

maximum age: a) radiometrically determined age that is obviously too high and thus represents only the upper limit of the unknown true age; b) highest age that can be determined with a specific method, corresponds to about ten times the half-life of the radionuclide used for a radiometric method

mean life: half-life divided by ln(2)

mean residence time: mean age value calculated from an isotope concentration according to a model, e.g., an exponential model (Sect. 7.4)

median: the value below which 50% of the sample values fall; the value of the middle item (or the mean of the values of the two middle items) when the items are arranged in an increasing or decreasing order of magnitude; the median is the same as the mean in a symmetrical frequency distribution

memory effect: the effect on analytical results caused by contamination of a sample by remnants of the previous sample(s) in apparatus for sample preparation or measurement

Mesozoic: geologic era between the Paleozoic and Cenozoic eras; it is divided into the Triassic, Jurassic, and Cretaceous periods (Fig. 9.1; Table 9.1)

metamict: term for a mineral whose lattice has been disrupted and in which changes have taken place as a result of radiation damage caused by radioactive elements contained in the mineral without changing the morphology of the mineral

metamorphism: mineralogical, chemical, and structural changes in rocks resulting from physical and chemical conditions at depth below the Earth's surface which differ from the conditions under which the rocks originated

metastable: in a state or phase that does not correspond to the present P, T conditions

meteor: luminous extraterrestrial body when passing through the Earth's atmosphere

meteorite: fragment of a meteoroid that survives passage through the atmosphere as it falls to the Earth

meteoroid: small solid body traveling through space; it is seen (as a meteor) only when it passes through the Earth's atmosphere

microprobe (ion or electron): device using ions or electrons to determine the isotopic or chemical composition of microsamples

migmatite: rock consisting of macroscopically distinguishable igneous (or apparently igneous) and metamorphic components

minimum age: a) radiometrically determined age that is obviously too low and thus represents only the lower limit of the unknown true age; b) lowest age that can be determined with a specific method

mixing line: linear fit of the points in a plot of measured values of two parameters of the system being studied; the linearity of this plot indicates the mixing of two (and only two) components in the system (e.g., a magma or groundwater)

model: illustrative description of a system; in a box model, for example, complex processes are reduced, broken up into spacially and temporally coupled units; each box is described by the place and process that occurs in it; the parameters of a box model, coupling and time constants, spacial boundary values, and distributions, are varied until the calculated results correspond satisfactorily with the measured values

model age: age determined on the basis of a model concept

model concept: concept on which every method for age determination is based; a model concept sets forth the conditions that must be fulfilled so that certain properties (e.g., the isotopic concentration) correspond to the age of a substance

monitor mineral: mineral of reliably known age that is irradiated and analyzed together with the sample of unknown age in order to determine the radiation parameters for methods that utilize neutron radiation (see Sects. 6.1.1.2 & 6.4.7)

MORB: Mid-Ocean Ridge basalts

monochromator: device for selecting particles of the same energy or radiation of only one wavelength

multi-channel analyzer: electronic apparatus that sorts and stores pulses according to many amplitude classes; used, for example, in gammaspectrometry

NBS oxalic acid standard (National Bureau of Standards, Washington, D.C.): internationally accepted primary standard for the $^{14}C$ method (Sect. 6.2.1); 95% of its activity corresponds to the $^{14}C$ concentration of a tree ring that grew in 1950 to which a Suess correction has been applied; a new standard (NOX) has replaced the NBS standard

neutron activation analysis: method of quantitative determination of trace elements: the sample is irradiated with neutrons and the activity of the radioactive isotopes thus produced is measured (see Sect. 5.2.4.2)

noble gases: the elements He, Ne, Ar, Kr, Xe, and Rn, which are gases at normal temperatures and pressure; they have a stable configuration of electrons (sometimes called the "noble-gas configuration"), are almost completely inert and thus occur in nature only in elemental form

non-dipole magnetic field of the earth: magnetic field remaining when the theoretical dipole magnetic field of the earth is subtracted from the actual one (Sect. 7.1)

normal distribution: symmetric, bell-shaped frequency distribution of observed data from random events (Sect. 4.2)

nuclear weapons effect: change in the abundance of long-lived environmental radioisotopes, e.g., radiocarbon in the carbon dioxide of the atmosphere and the tritium in precipitation, due to the addition of anthropogenic $^{14}C$ and $^{3}H$, produced, for example, by atomic weapons tests (Sect. 7.4; Fig. 7.8); see input curve

nucleon: name for the constituent particles of the nucleus of an atom, i.e., a neutron or proton

nucleotide: basic repeating unit of DNA and RNA consisting of a phosphate ester of a carbohydrate bound to a purine or pyrimidine base

nuclide: term for a species of atom characterized by the constitution of its nucleus, i.e., the numbers of neutrons and protons it contains

nucleosynthesis: formation of heavier chemical elements from the nuclei of hydrogen or other lighter elements, e.g., in the interior of a star or immediately after the "Big Bang"

obsidian: black or grey glass rock formed by the solidification of acid volcanic magma; obsidian was worked into knives and arrowheads during the neolithic period (late Stone Age) (Sect. 8.6)

ooid (also oolith): a tiny, spherical or ellipsoid particle with concentric layers, usually of calcium carbonate, formed around a nucleus (e.g., a shell fragment or sand grain) in wave-agitated sea water

oolite: a rock composed chiefly of ooids

open system: see system

Ordovician: geological period of the Paleozoic era (*see* Fig. 9.1; Table 9.1)

orogenesis: process of mountain formation

orthologous proteins: proteins that differ only as a result of divergence of the genes of the species (i.e., speciation) from which they were taken

paleomagnetism: remanent magnetism in a rock that is proportional to the dipole moment of the Earth's magnetic field at the time it last cooled below the Curie temperature; under certain conditions, this remanent magnetism can be used to determine the age of rocks (Sect. 7.1)

paleontology: branch of the geosciences dealing with life in the geological past

Paleozoic: era of geologic time between the Precambrian and the Mesozoic (Fig. 9.1; Table 9.1)

paramagnetism: property of materials to become magnetized in a magnetic field proportionally to the strength and direction of the field; *see* ferromagnetism (Sect. 7.1)

parent isotope: radionuclide that decays (either directly or as a later member of a radioactive series) into another isotope, called the daughter isotope (Sect. 5.1)

particle accelerator: *see* accelerator

path, mean free: average distance a particle travels between collisions

PDB standard: internationally recognized carbonate reference for $\delta^{13}C$ and $\delta^{18}O$ determinations on organic matter and on carbonates; the standard was taken from a Cretaceous belemnite in the Peedee Formation in South Carolina; its isotopic composition represented a good average of that of marine limestone; because this standard has been exhausted a new standard NBS-19 is supplied by the NBS and IAEA, but all data are still referred to PDB (Sect. 7.2; *see also* SMOW standard)

pelagic: originating at oceanic depths of more than about 800 m

Permian: last geological period of the Paleozoic era (*see* Fig. 9.1; Table 9.1)

Phanerozoic: term for the geologic time represented by rocks containing fossils, i.e., the time since the Precambrian

phase: homogeneous form of a substance existing in a heterogeneous system, bounded by sharp boundaries, optically distinguishable, and mechanically separable

phototube: vacuum tube used to convert electromagnetic radiation (light, etc.) into electrical pulses, which can be measured; an essential component of a liquid scintillation counter, for example

phylogeny: origin and evolution of a division, group, or race of animals or plants

plankton: animal (zooplankton) and plant (phytoplankton) life, usually microscopic, that floats or drifts with no motion of its own in the ocean or in bodies of fresh water

plateau age: the mean of concordant apparent ages determined, for example, during successive degassing steps of rock or mineral samples in the $^{39}Ar/^{40}Ar$ (Sect. 6.1.1.2) and xenon/xenon (Sect. 6.1.15.2) methods; in TL dating (Sect. 6.4.1), it is obtained from a certain part of the glow curve; in FT dating (Sect. 6.4.7), it is determined using a special correction technique

Pleistocene: first subdivision of the Quaternary; it began less than 4 million years ago and extended to 10,000 years ago

pleochroic halo: minute zone of color or darkening surrounding and produced by a radioactive mineral or inclusion (*see* Sect. 6.4.9)

plumbotectonics: combined concepts of common-lead isotope geochemistry and plate tectonics

pmc (also % modern or percent modern carbon): unit of $^{14}C$ concentration; 100 pmc corresponds to the $^{14}C$ concentration of wood that grew in 1950 to which a Suess correction has been applied (Sect. 6.2.1); *see* NBS oxalic acid standard

Poisson distribution: statistical distribution of events in which the classes of interest within a long series of observations are seldom occupied; an example is the decay of radionuclides within a given time span (Sect. 4.2)

polymetamorphic: term for metamorphic rocks that have been subjected to more than one phase of metamorphism

positron: positive elementary particle with the same mass as an electron

post-emplacement disturbance: disturbance in a rock system after its formation, often in connection with tectonic events and geochemical and geochronologically relevant changes

post-glacial: pertaining to the time since the last glacial stage; Holocene

Precambrium: term for the geologic time represented by rocks containing no fossils, the time before the Paleozoic (*see* Fig. 9.1; Table 9.1)

precision: the agreement between the numerical values of two or more measurements made in an identical fashion; these values may or may not be close to the true value (the difference is called the bias); the standard deviation is a measure of precision (cf. accuracy)

pretreatment: treatment of the sample before dating to prepare suitable fractions (e.g., mineral phases) or to remove contamination by physical and chemical procedures

primary element concentration: element concentration at the time of rock or mineral formation

primordial: (in the sense of this book) existing since the nucleosynthesis

probability: the number of times something will probably occur over the range of possible occurrences, expressed as a ratio; for certain experiments (e.g., rolling dice), the probability is known theoretically (*see* Poisson distribution)

progressive metamorphism: progressive change from low-grade to high-grade metamorphism

proportional counter: *see* counter

proton-induced decay: disintegration of a nucleus due to bombardment with high-energy protons (*see* neutron activation analysis)

pseudoisochron (syn. errorchron): invalid isochron calculated from a number of isotope ratios; can sometimes, under suitable circumstances, be recognized as such because the points for which the line was calculated have a greater scatter than indicated by the standard deviation

pulse discrimination: selection of pulses according to amplitude, rise-time, coincidence, etc., to increase the signal-to-noise ratio, i.e., to diminish the influence of pulses from contaminants and other background radiation (Sect. 5.2.2)

pyroclastic: term for a clastic rock consisting of material ejected during a volcanic explosion or from a volcanic vent (e.g., ash, tuff, tephra)

Quaternary: most recent geologic period, beginning 2 to 4 million years ago (*see* Fig. 9.1; Table 9.1)

racemization: conversion of an optically active substance into an optically inactive one (racemate), which contains an equal number of molecules of isomers that rotate light in opposite directions (Sect. 8.1)

radiation, corpuscular (or particulate): radiation consisting of matter, e.g., helium-4 nuclei (i.e., $\alpha$-particles), as opposed to electromagnetic radiation, which consists of photons

radiation damage: change in the physical or chemical properties of a substance due to irradiation (Sect. 6.4)

radiation detector: instrument for the measurement of, for example, radioactive radiation by conversion into measurable and countable electrical or optical pulses (Sect. 5.2.2); *see* counter, liquid scintillation counter, semiconductor detector

radiation dose: radiation energy absorbed by matter (unit is the gray = Gy); the dosage (unit is Gy/a) is the dose absorbed per time unit (*see* Sect. 6.4)

radiation exposure age: see cosmic ray exposure age

radical: group of atoms that behave chemically as a unit

radioactive: giving off of radiant energy in the form of particles or rays by the spontaneous disintegration of radionuclides; *see* law of radioactive decay

radioactive decay series (or chain): series of radionuclides in which the daughter isotope of the decay of one member of the series (chain) is in turn the parent isotope of the next member of the series; there are three naturally occurring radioactive decay series, the $^{238}$U, $^{235}$U, and $^{232}$Th series (Fig. 6.64); *see* equilibrium, radioactive

radiocarbon: $^{14}$C (*see* Sect. 6.2.1)

radiogenic: produced by radioactive decay

radiometric age: age calculated on the basis of a quantitative analysis of radioactive isotopes and their decay products; expressed in years (*see* Sect. 5.1)

radiometry: discipline dealing with the measurement of the intensity and energy of radiation from radioactive substances

radionuclide: radioactive atom

radon: radioactive nobel gas with the atomic number 86; is produced by alpha decay of radium-226 and is thus a member of the $^{238}$U decay series (Fig. 6.64); *see* (Sect. 6.3.15)

random experiment: experiment in which the results are due to random events

random sample: sample taken at random used to draw conclusions about the population as a whole

rare earth element: any of the elements with atomic numbers 57 to 71 inclusive

recent: in geology, Recent corresponds to the Holocene; in the sense of this book, either marked by anthropogenic effects (Sect. 7.4) or so young that the initial concentration is still present (Sect. 6.2); cf. contemporary

recoil tracks: traces of lattice defects in non-conductors created by alpha-emitting nuclides (involves the emission of at least four alpha particles); *see* Sect. 6.4.8

recombination: return of an ionized atom or molecule to the neutral state

REE: rare earth element

reference year: for the geochronological time scale, AD 1950 is used as reference year; for the Gregorian calendar, the year of Christ's birth is used (*see* Chap. 2)

regression analysis: mathematical method using the principle of the least squares to determine the parameters of a given function (the simplest being a straight line) which give the best fit (least squares fit) to a set of data

regression curve (line): curve (line) determined by regression analysis

remanent magnetization: *see* magnetization, remanent

reservoir effect: falsification of the radiometric age resulting from reservoir-specific, initial concentrations that do not correspond to the model (giving an apparent age) (Sect. 6.2.1)

reservoir, geochemical: part of the Earth, usually very large, that is a closed system over a long period of time and can thus be distinguished on the basis of its isotopic evolution; examples are the atmosphere, biosphere, hydrosphere, lithosphere, and parts of them

resonant radiation: characteristic radiation emitted by an atom when acted upon by electromagnetic radiation of a specific wavelength (energy) (Sect. 6.4.3)

retention age: *see* gas retention age

retentivity: capability of crystalline matter to resist diffusion; opposite of diffusivity

retrograde metamorphism: change from high-grade to low-grade metamorphism; opposite to progressive metamorphism

saturation concentration: steady-state concentration, e.g., of a member of a radioactive decay series; in a closed system it is established after a period of time corresponding to a number of half-lives of a radionuclide produced at a constant rate in a decay chain (Sects. 5.1 and 6.2.1)

scintillation counter: *see* liquid scintillation counter

secondary isochron: isochron indicating a secondary geochronological event, i.e., one following and not associated with the initial event

secular variation: change in the resultant of the Earth's magnetic moment by wandering and variation in the Earth's magnetic non-dipole field (Sect. 7.1)

semiconductor detector: a detector made of a semiconductor of which two opposite sides serve as electrodes; if a potential is placed across the detector, a current pulse will be produced by ionizing radiation; semiconductor detectors are characterized by very high energy resolution and sensitivity (Sect. 5.2.2.3)

shield: a body of material used to reduce the passage of corpuscular or electromagnetic radiation, often of lead or iron and neutron-absorbing materials

sidereal year: *see* year

siderophile: term for elements that have a weak affinity for oxygen and sulfur but readily soluble in molten iron

Silurium: geological period of the Paleozoic era (*see* Fig. 9.1)

SMOW standard (s̲tandard m̲ean o̲cean w̲ater): standard for determination of the $\delta^{18}O$ value of water with an isotopic composition close to that of ocean water; by definition its $\delta^{18}O$ and $\delta^{2}H$ values $= 0$; SMOW-V standard is distributed by IAEA, Vienna (*see* Sect. 7.2); *see also* PDB standard

solidification age: the time since a rock or mineral solidified

solid-state track recorder (SSTR): term for solid substances that can preserve fission tracks within certain pressure and temperature ranges

solidus: curve in a temperature-phase plot below which the system is completely solid

space erosion: erosion of meteoroids by extraterrestrial corpuscular radiation and cosmic dust; space erosion is still in debate (Sect. 6.5)

spallation: term for a nuclear reaction induced by high-energy bombardment resulting in the emission of many nucleons (as many as 20, 30, or more) leaving a series of products of lower atomic mass and number; this is in contrast to fission, which results in two, more or less equal parts

spectrometry: discipline dealing with the theory and measurement of the frequencies and intensities of electromagnetic or particulate radiation and thus determine the components (molecules, isotopes) and concentrations of a mixture

spike: artificial substance with exactly known isotopic composition that is significantly different from natural conditions; an exactly known amount of it is added to an aliquot of the sample; used in isotope dilution analysis (Sect. 5.2.4.1)

spin: the intrinsic angular momentum of an elementary particle or photon produced by rotation about its own axis

spinner magnetometer: a cubic sample is positioned on a holder and rotated with a given velocity. This induces a voltage in a detector coil that is proportional to the vector of the remanent magnetization. Both amplitude and phase of the voltage are measured at six positions at least, yielding the $x$, $y$, $z$ components of the remanent magnetization (Sect. 7.1)

spontaneous decay: mode of radioactive decay in which the nucleus of a heavy atom fragments into two nuclides of more or less equal mass and several neutrons

standard: reference substance: substance of known (isotopic) composition or known or defined age; see SMOW, PDB, and NBS standards

standard deviation: measure of scatter in a statistical population; the expectation value $\pm$ the standard deviation is the range in which 68% of the observed values of a random experiment occur (Sect. 4.2)

steady-state equilibrium: see equilibrium, steady-state

STP: standard temperature and pressure

strata-bound: term for a mineral deposit confined to a single stratigraphic unit

stratigraphy: branch of the geosciences dealing with the study of the nature, distribution, and relations of the strata in rocks of the Earth's crust; see biostratigraphy, chronostratigraphy and lithostratigraphy

stratosphere: atmospheric zone above the troposphere up to about 80 km above the Earth's surface

Suess correction: conversion of the $^{14}$C ages determined in the 1950s into conventional ages when a piece of wood affected by the Suess effect was used as a standard (Sect. 6.2.1); see NBS oxalic acid standard

Suess effect (also called industry effect): decrease in the initial $^{14}$C concentration of atmospheric carbon dioxide since the middle of the nineteenth century caused by the constantly increasing ejection of fossil $CO_2$ into the atmosphere from the burning of fossil fuels (Sect. 6.2.1)

superlinearity, supralinearity: nonlinear rise of the growth curve of radiation damage with increasing dose (Sect. 6.4, Fig. 6.77)

syngenetic: formed at the same time as the surrounding rock

synsedimentary: formed at the same time as the deposition of the sediment (e.g., ore deposits)

systems, open and closed: things or processes related or connected so that they form a whole; in an open system there is exchange with its surroundings, in a closed system there is none

tephrochronology: relative dating on the basis of tephra (which is a general term for all of the pyroclastics of a volcano)

terrestrial: a) said of a sediment deposited on land; b) pertaining to the Earth, as opposed to extraterrestrial

terrestrial age (of meteorites): the time since the meteorite in question fell to the Earth (*see* Sect. 6.5.7)

terrigenous: derived from the continents; cf. lacustrine

Tertiary: first geological period of the Cenozoic era; divided into five epochs: the Paleocene, Eocene, Oligocene, Miocene, and Pliocene (Fig. 9.1; Table 9.1)

thermal neutrons: slow neutrons with an energy corresponding to environmental temperatures

thermoluminescence (TL): the release in the form of light of stored radiant energy from a substance when it is heated (Sect. 6.4.1)

thermoremanence magnetization: magnetization of ferromagnetic and paramagnetic substances which remains after heating above the Curie temperature and recooling in an external magnetic field (Sect. 7.1)

TL: thermoluminescence

trace element: element occurring in only minute amounts in a sample

tracer: a substance, either naturally in the environment or anthropogenic that can be used to trace the course of a geological process (*see*, for example, Sects. 6.1.6 & 7.4)

track fading: annealing of fission tracks at environmental temperatures

track retention age: *see* cooling age

tree-ring analysis: measurement of the widths of the annual growth rings of a piece of wood in the direction of growth and assignment of the pattern to a dendrochronological time-scale as given by the master curve that is valid for the area of origin

Triassic: first geological period of the Mesozoic era (Fig. 9.1; Table 9.1)

triboluminescence: *see* luminescence

tritium: cosmogenic radionuclide of hydrogen, half-life = 12.43 years; used as a tracer in hydrological studies due to its presence in the hydrological cycle (Sects. 6.2.2 and 7.4); *see also* TU

troposphere: the lowest zone of the atmosphere, up to 12 km thick, in which the weather takes place; the thickness of the troposphere depends on the season and latitude

two-stage evolution: *see* evolution

TU: tritium unit = 1 tritium atom per $10^{18}$ hydrogen atoms; used for tritium concentration in isotope hydrology studies, it corresponds to $1.182 \times 10^{-4}$ Bq/mL (*see* Sect. 6.2.2)

tuff: general term for consolidated pyroclastic (i.e., volcanic) material turbidity currents: underwater mud slides that occur on the edge of the continental slopes, leading to disturbances of the shelf and deep-sea sediments

ultrabasic: term for igneous rocks with less silica than basic rocks (i.e., less than about 44%)

ultramafic: term for igneous rocks composed chiefly of mafic minerals; note that ultramafic and ultrabasic can apply to the same rock but that they are based on two different aspects of the composition of the rock

varve: an annual layer of sedimentary material deposited in lakes and fiords by glacial meltwater, consists of two distinct bands of sediment deposited in summer and winter, respectively

vector: a quantity with both magnitude and direction, is represented by an arrow

velocity, Darcy: velocity of a mass of groundwater given by the product of hydraulic conductivity ($k_f$) and the gradient of the groundwater table

velocity, tracer: velocity of a groundwater tracer, e.g., as determined by environmental isotope analysis; the tracer velocity is the Darcy velocity divided by the total pore volume; cf. velocity, Darcy

westward drift: westward drift of magnetic anomalies as a result of the movement of the non-dipole field (Sect. 7.1)

whole-rock sample: a rock sample large enough so that it may be considered as a closed system for dating purposes

xenolith: foreign inclusion in an igneous rock

X-ray: electromagnetic radiation of very short wavelength (about $10^{-11}$–$10^{-8}$ m)

X-ray fluorescence analysis (XRF): rapid method for the quantitative analysis of elements with an atomic number greater than 9. When a substance is subjected to irradiation with primary X-rays, the emitted X-rays (fluorescence) permit identification of the elements it contains and their concentrations (*see* Sect. 5.2.4.5)

year: period of one revolution of the earth around the sun; the sideral year is the period spent by the sun in its apparent passage from a fixed star and back to the same position again; the reference for the solar or tropical year is the vernal equinox, it is insignificantly shorter than the sidereal year; the calender year is the unit of the Gregorian calender (Sect. 2.1)

---

During the preparation of this glossary, extensive reference was made to the Glossary of Geology, 2nd Ed., edited by Bates and Jackson (1980), and Webster's New World Dictionary, 2nd College Edition

# Appendix B: Radioactive and Stable Isotopes in Geochronology

(Half-lives are known with an accuracy of ca 1% apart of the long-lived $\beta^-$ emitting isotopes as $^{87}$Rb and $^{187}$Re)

| Z Element | Symbol | Ref. Nuclide | Radiation type | Energy keV | Half-life |
|-----------|--------|--------------|----------------|------------|-----------|
| 1 Hydrogen | $^3$H | $^{1,2}$H | $\beta^-$ | 18.6 | 12.43 a |
| 4 Beryllium | $^7$Be | $^9$Be | EC | 478 | 53 d |
| | $^{10}$Be | " | $\beta^-$ | 560 | 1.51 Ma |
| 6 Carbon | $^{14}$C | $^{12}$C | $\beta^-$ | 158 | 5568 a |
| 11 Sodium | $^{22}$Na | $^{23}$Na | $\beta^+$, EC | 540 | 2.602 a |
| 13 Aluminum | $^{26}$Al | $^{27}$Al | $\beta^+$, EC | 1170 | 716 ka |
| 14 Silicon | $^{32}$Si | $^{28}$Si | $\beta^-$ | 200 | 105 a |
| 15 Phosphorus | $^{32}$P | $^{31}$P | $\beta^-$ | 1710 | 14.3 d |
| 17 Chlorine | $^{36}$Cl | $^{35}$Cl | $\beta^-$ | 714 | 301 ka |
| 18 Argon | $^{37}$Ar | $^{40}$Ar | EC | | 35 d |
| | $^{39}$Ar | Cl | $\beta^-$ | 565 | 269 a |
| 19 Potassium | $^{40}$K | $^{39}$K | $\beta^-$ | 1330 | 1397 Ma |
| | " | " | EC | 1460 | 11,930 Ma |
| | $^{42}$K | " | $\beta^-$ | 3500 | 12.36 h |
| | $^{43}$K | " | $\beta^-$ | 800 | 22.2 h |
| 20 Calcium | $^{41}$Ca | $^{40}$Ca | EC | 427 | 103 ka |
| 25 Manganese | $^{53}$Mn | $^{55}$Mn | EC | 598 | 3.7 Ma |
| | $^{54}$Mn | " | EC | 835 | 312.2 d |
| 26 Iron | $^{55}$Fe | $^{56}$Fe | EC | 232 | 2.7 a |
| 27 Cobalt | $^{60}$Co | $^{59}$Co | $\beta^-$ | 1500 | 5.272 a |
| 36 Krypton | $^{81}$Kr | Kr | EC | 276 | 210 ka |
| | $^{85}$Kr | " | $\beta^-$ | 687 | 10.76 a |
| 37 Rubidium | $^{87}$Rb | $^{85}$Rb | $\beta^-$ | 275 | 48.8 Ga |
| 38 Strontium | $^{90}$Sr | $^{88}$Sr | $\beta^-$ | 540 | 28.5 a |
| 41 Niobium | $^{92}$Nb | $^{93}$Nb | EC, $\beta^+$ | 561 | 36 Ma |
| 43 Technetium | $^{97}$Tc | Tc | EC | | 2.6 Ma |
| | $^{98}$Tc | " | $\beta^-$ | 4000 | 4.2 Ma |
| 46 Palladium | $^{107}$Pd | $^{106}$Pd | $\beta^-$ | 30 | 6.5 Ma |
| 53 Iodine | $^{129}$I | $^{127}$I | $\beta^-$ | 150 | 15.7 Ma |
| | $^{130}$I | " | $\beta^-$ | 536 | 12.36 h |
| 55 Cesium | $^{137}$Cs | $^{133}$Cs | $\beta^-$ | 1180 | 30.17 a |
| 57 Lanthanum | $^{138}$La | $^{139}$La | $\beta^-$ | 789 | 270 Ga |
| | " | " | EC | 1436 | 156 Ga |
| 62 Samarium | $^{146}$Sm | $^{149}$Sm | $\alpha$ | 2550 | 103 Ma |
| | $^{147}$Sm | " | $\alpha$ | 2234 | 106 Ga |
| | $^{148}$Sm | " | $\alpha$ | 1960 | 7000 Ga |
| 71 Luthetium | $^{176}$Lu | $^{173}$Lu | $\beta^-$ | 430 | 35.7 Ga |
| | " | " | EC | | |
| | $^{177}$Lu | " | $\beta^-$ | 500 | 6.71 d |
| 72 Hafnium | $^{174}$Hf | $^{178}$Hf | $\alpha$ | 2500 | 2000 Ta |
| | $^{180}$Hf | " | $\gamma$ | 443 | 5.5 h |
| 75 Rhenium | $^{187}$Re | $^{185}$Re | $\beta^-$ | 8 | 42.3 Ga |
| | $^{188}$Re | " | $\beta^-$ | 2120 | 17 h |

| Z Element | Desc. Nuclide | Abundancy | Origin, $\tau_p$ Remarks |
|---|---|---|---|
| 1 Hydrogen | $^3$He | $10^{-16}$ | $\cos_{N,O,Mg,Si}$ |
| 4 Beryllium | $^7$Li | | $\cos_{O,S,Mg,Si}$ |
| | $^{10}$B | $10^{-9}$ | $\cos_{N,O}$ |
| 6 Carbon | $^{14}$N | $1.2 \times 10^{-12}$ | $\cos_{N,O,Mg,Si}$ $T_p = 5730\,a$ |
| 11 Sodium | $^{22}$Ne | | $\cos_{Ar_{sp}}$ |
| 13 Aluminum | $^{26}$Mg | $2 \times 10^{-12}$ | $\cos_{Ar_{sp},Si,Al}$ |
| 14 Silicon | $^{32}$P | $4 \times 10^{-14}$ | $\cos_{Ar_{sp}}$ $T_p = 105\text{–}171\,a$ |
| 15 Phosphorus | $^{32}$S | | dau $^{32}$Si |
| 17 Chlorine | $^{36}$Ar | $2.2 \times 10^{-11}$ | $\cos_{Cl,Ca,K,Ar_{sp}}$ |
| 18 Argon | $^{37}$Cl | | $\cos_{K,Ca,Ar_{sp}}$ |
| | $^{39}$K | $0 \times 10^{-16}$ | $\cos_{K,Ar,Ca}$ |
| 19 Potassium | $^{40}$Ca | .01167 | $\gamma = 89\%$; $T_p = 1{,}469$ Ma |
| | $^{40}$Ar | | $\gamma = 11\%$; $T_p = 11{,}848$ Ma |
| | $^{42}$Ca | | NAA |
| | $^{43}$Ca | | NAA |
| 20 Calcium | $^{41}$K | $2 \times 10^{-14}$ | $\cos_{Kr,Ar,Ca}$ |
| 25 Manganese | $^{53}$Cr | | $\cos_{Fe}$; $T_p = 3.9 \pm .6$ Ma |
| | $^{54}$Cr | | $\cos_{Fe}$, NAA |
| 26 Iron | $^{55}$Mn | | ant |
| 27 Cobalt | $^{60}$Ni | | ant |
| 36 Krypton | $^{81}$Br | $5 \times 10^{-13}$ | $\cos_{Kr,Rb,Sr}$ |
| | $^{85}$Rb | $10^{-11}$ | ant |
| 37 Rubidium | $^{87}$Sr | 27.83% | $T_p = 48{,}813$ Ga |
| 38 Strontium | $^{90}$Y | | ant |
| 41 Niobium | $^{92}$Zr | | |
| 43 Technetium | $^{97}$Mo | | |
| | $^{98}$Ru | | |
| 46 Palladium | $^{107}$Ag | | ex |
| 53 Iodine | $^{129}$Xe | $3 \times 10^{-13}$ | ex, $\cos_{Xe_{sp},Te,Ba}$ |
| | $^{130}$Xe | | NAA |
| 55 Cesium | $^{137}$Ba | | ant |
| 57 Lanthanum | $^{138}$Ce | 0.09% | nat |
| | $^{138}$Ba | | |
| 62 Samarium | $^{142}$Nd | | ex; $T_p = 50\text{–}130$ Ma |
| | $^{143}$Nd | 15.0% | nat |
| | $^{144}$Nd | 11.3% | nat |
| 71 Luthetium | $^{176}$Hf | 2.6% | pri; $T_p = 33 \pm 6$ Ga |
| | $^{176}$Yb | | |
| | $^{177}$Hf | | NAA |
| 72 Hafnium | $^{170}$Yb | 0.16% | nat |
| | $^{180}$Hf | 35.2% | NAA |
| 75 Rhenium | $^{187}$Os | 62.6% | nat |
| | $^{188}$Os | | NAA |

**Table** (*Continued*)

| Z Element | Symbol | Ref. Nuclide | Radiation type | Energy keV | Half-life |
|---|---|---|---|---|---|
| 81 Thallium | $^{206}$Tl | $^{205}$Tl | $\beta^-$ | 1500 | 4.3 m |
| | $^{207}$Tl | " | $\beta^-$ | 1400 | 4.8 m |
| | $^{208}$Tl | " | $\beta^-$ | 1800 | 3.05 m |
| | $^{210}$Tl | " | $\beta^-$ | 1900 | 1.3 m |
| 82 Lead | $^{205}$Pb | $^{208}$Pb | EC | 35 | 15 Ma |
| | $^{210}$Pb | " | $\beta^-$, $(\alpha)$ | 18 | 22.3 a |
| | $^{211}$Pb | " | $\beta^-$ | 1400 | 36.1 m |
| | $^{212}$Pb | " | $\beta^-$ | 300 | 10.64 h |
| | $^{214}$Pb | " | $\beta^-$ | 700 | 26.8 m |
| 83 Bismuth | $^{210}$Bi | $^{209}$Bi | $\beta^-$ | 1170 | 5 d |
| | " | " | $\alpha$ | 4946 | 3 Ma |
| | $^{211}$Bi | " | $\alpha$, $(\beta^-)$ | 6623 | 2.15 m |
| | $^{212}$Bi | " | $\beta^-$ | 2300 | 60.6 m |
| | " | " | $\alpha$ | 6340 | 25 m |
| | $^{214}$Bi | " | $\beta^-$, $(\alpha)$ | 1500 | 19.9 m |
| 84 Polonium | $^{210}$Po | Po | $\alpha$ | 5304 | 138.38 d |
| 86 Radon | $^{219}$Rn | Rn | $\alpha$ | 6819 | 3.96 s |
| | $^{220}$Rn | " | $\alpha$ | 6288 | 55.6 s |
| | $^{222}$Rn | " | $\alpha$ | 5489 | 3.825 d |
| 88 Radium | $^{223}$Ra | Ra | $\alpha$ | 5716 | 11.4 d |
| | $^{224}$Ra | " | $\alpha$ | 5685 | 3.66 d |
| | $^{226}$Ra | " | $\alpha$ | 4784 | 1600 a |
| | $^{228}$Ra | " | $\beta^-$ | 41 | 5.75 a |
| 89 Actinium | $^{227}$Ac | Ac | $\beta^-$ | 46 | 21.77 a |
| | $^{228}$Ac | " | $\beta^-$ | 1200 | 6.13 h |
| 90 Thorium | $^{227}$Th | Th | $\alpha$ | 6038 | 18.72 d |
| | $^{228}$Th | " | $\alpha$ | 5423 | 1.913 a |
| | $^{230}$Th | " | $\alpha$ | 4688 | 75.4 ka |
| | $^{231}$Th | Th | $\beta^-$ | 300 | 25.5 h |
| | $^{232}$Th | " | $\alpha$ | 4013 | 14.05 Ga |
| | $^{234}$Th | " | $\beta^-$ | 190 | 24.1 d |
| 91 Protactinium | $^{231}$Pa | Pa | $\alpha$ | 5014 | 32.760 ka |
| | $^{232}$Pa | " | $\beta^-$ | 270 | 1.31 d |
| | $^{234}$Pa | " | $\beta^-$ | 2300 | 1.17 d |
| 92 Uranium | $^{232}$U | $^{238}$U | $\alpha$ | 5320 | 70 a |
| | " | " | sf | | |
| | $^{234}$U | " | $\alpha$ | 4775 | 244.6 ka |
| | $^{235}$U | " | $\alpha$ | 4400 | 703.8 Ma |
| | " | " | sf | | $2 \times 10^{17}$ a |
| | $^{238}$U | " | $\alpha$ | 4197 | 4.468 Ga |
| | " | " | sf | | $81 \times 10^{14}$ a |
| 94 Plutonium | $^{244}$Pu | Pu | $\alpha$ | 4589 | 82.6 Ma |
| | " | " | sf | | 66.1 Ga |
| 95 Americium | $^{241}$Am | Am | $\alpha$ | 5486 | 432.6 a |
| 96 Curium | $^{247}$Cm | Cm | $\alpha$ | 4869 | 15.6 Ma |
| | $^{248}$Cm | " | $\alpha$ | 5078 | 339.7 ka |
| | " | " | sf | | |
| | $^{250}$Cm | " | sf | | < 11.3 ka |

*Abbreviations*: cos$_{N,O \cdot C, Ar}$, cosmogenic from N, O, C, and Ar; Ar$_{sp}$, spallogenic from argon; $\tau_c$ and $\tau_p$, conventional and physical half-lives; ex, extinct; dau, daughter of; ant, anthropogenic; ▶

| Z Element | Desc. Nuclide | Abundancy | Origin, $\tau_p$ Remarks |
|---|---|---|---|
| 81 Thallium | $^{206}$Pb | | uds |
| | $^{207}$Pb | | uds |
| | $^{208}$Pb | | NAA, thds |
| | $^{210}$Pb | | uds |
| 82 Lead | $^{205}$Tl | | ex |
| | $^{210}$Bi | | uds; $5 \times 10^{-5}\%\alpha$ |
| | $^{211}$Bi | | uds |
| | $^{212}$Bi | | thds |
| | $^{214}$Bi | | uds |
| 83 Bismuth | $^{210}$Po | | uds |
| | $^{206}$Tl | | uds; $5 \times 10^{-5}\%\alpha$ |
| | $^{207}$Tl | | uds; $.32\%\beta^-$ |
| | $^{212}$Po | 63.8% | thds; $63.8\%\gamma$ |
| | $^{208}$Tl | 36.2% | uds; $36.2\%\gamma$ |
| | $^{214}$Po | | uds; $.04\%\alpha$ |
| 84 Polonium | $^{206}$Pb | | dau $^{228}$Ra; $1.2\%\alpha$ |
| 86 Radon | $^{215}$Po | | rad |
| | $^{216}$Po | | rad |
| | $^{218}$Po | | dau $^{226}$Ra |
| 88 Radium | $^{219}$Rn | | rad |
| | $^{220}$Rn | | dau $^{230}$Th |
| | $^{222}$Rn | | desc $^{230}$Th |
| | $^{228}$Ac | | desc $^{232}$Th |
| 89 Actinium | $^{227}$Th | | desc $^{231}$Pa |
| | $^{228}$Th | | dau $^{228}$Ra; $1.2\%\alpha$ |
| 90 Thorium | $^{223}$Ra | | rad; uds |
| | $^{224}$Ra | | rad |
| | $^{226}$Ra | | rad; uds; syn ionium |
| | $^{231}$Pa | | rad; uds |
| | $^{228}$Ra | | par; thds |
| | $^{234}$Pa | | rad; uds |
| 91 Protactinium | $^{227}$Ac | | rad; uds |
| | $^{232}$U | | rad |
| | $^{234}$U | | rad; uds |
| 92 Uranium | $^{228}$Th | | rad |
| | $^{230}$Th | 0.005% | nat; uds |
| | $^{231}$Th | 0.72% | par; uds |
| | $^{234}$Th | 99.275% | par; uds |
| 94 Plutonium | $^{240}$U, $^{131-136}$Xe, | | ex |
| | $^{131-136}$Xe, | | ex; sf/$\alpha = 1.25 \times 10^{-3}$ |
| 95 Americium | $^{237}$Np | | ex |
| 96 Curium | $^{131-136}$Xe, $^{243}$Pu | | ex |
| | $^{131-136}$Xe, $^{244}$Pu | | ex |
| | | | ex |
| | $^{131-136}$Xe | | ex |

pri, primordial; rad, radiogenic; uds, uranium decay series; thds, thorium decay series; desc, descendent nuclide; nat, natural; sf, spontaneous fission; NAA, neutron absorption analysis

# Appendix C: List of Addresses

The field of geochronology has developed so rapidly in the last several decades – and the number of laboratories working in the field – that any list of addresses that presumes to be complete would be too long to be included in this book. Instead, we have included those laboratories that are known through their large number of publications. Thus, some laboratories, particularly commercial ones, may not be mentioned. In addition, the attempt was made to include for each group of methods at least one laboratory in North America, Europe, and the Eastern World countries.

The laboratories are listed according to chapters in this book and alphabetically according to the cities in which they are located. Contact persons, taken from their publications, may have changed.

## Section 6.1: Parent/Daughter Isotope Ratios as a Chronometer

Amsterdam, The Netherlands: Laboratorium voor Isotopen-Geologie, De Boelelaan 1085, Amsterdam 11: 6.1.1.1, 6.1.3, 6.1.6 (Boelrijk NAIM, Hebeda EH)

Athens, USA: Department of Geology, University of Georgia, Athens, GA 30602: 6.1.1.1, 6.1.1.2 (Dallmeyer RD)

Berne, Switzerland: Laboratory for Isotope Geology, University of Berne, Erlachstr. 9a, CH-3012 Berne: 6.1.1.1, 6.1.3, 6.1.6, 6.1.9, (Jäger E) 6.4.7 (Huford AJ)

Brussels, Belgium (a): Laboratories Associés de Géologie-Pétrologie-Géochronologie, Université de Bruxelles, Avenue F.D. Roosevelt 50, B-1050 Brussels: 6.1.3 (Andre L, Deutsch S)

Brussels, Belgium (b): Laboratoire de Géochronologie, Vrije Universiteit Brussel, Pleinlaan 2, B-1050 Brussels: – (Pasteels P)

Cambridge, UK: Department of Earth Sciences, Cambridge University, Downing Street, Cambridge: 6.1.3, 6.1.6 (Hamilton PJ, O'Nions RK)

Cambridge, USA: Massachusetts Institute of Technology, 77 Massachusetts Avenue, Cambridge, MA 02139: 6.1.1, 6.1.3, 6.1.7 (Hart SR)

Cambridge, USA → 6.2

Canberra, Australia: Research School of Earth Sciences. The Australian National University, G.O.P. Box 4, Canberra, A.C.T. 2601: 6.1.1.1, 6.1.1.2, 6.1.3, 6.1.6, 6.1.9, 6.1.10, 6.1.11 (Compston W, McCulloch MT, McDougall I, Oversby VM, Williams IS), 6.2.1 (Polach H), 7.1 (Barton CE, Merril RT)

Clermont-Ferrand, France: Laboratoire de Géologie, Centre de Recherches Volcanologiques, Faculté des Sciences, F-63038 Clermont-Ferrand: 6.1.1.1, 6.1.1.2

Columbus, USA: Laboratory for Isotope Geology and Geochemistry (Isotopia), Department of Geology and Mineralogy, The Ohio State University, Columbus, Ohio 43210: 6.1.1.1, 6.1.1.2, 6.1.3 (Faure G)

Copenhagen, Denmark: Institute of Petrology, University of Copenhagen, Øster Voldgade 10, DK-1350 Copenhagen K: 6.1.1.1, 6.1.3 (Bailey JC, Munksgaard NC)

Debrecen, Hungary: Institute of Nuclear Research 18/C Bem Ter, H-4001 Debrecen PF 51: 6.1.1.1 (Balogh K)

Denver, USA: US Geological Survey, Federal Center, Building 21, MS 963, PO Box 25046, Denver, Colorado 80225: 6.1.2, 6.1.3, 6.1.6, 6.1.7, 6.1.9, 6.3.1, 6.4.7, 8.3 (Friedman I, Gleason JD, Marshall BD, Smith RL, Obradovich JD, Naeser CN, Naeser ND, Rosholt JN, Szabo BJ, Tatsumoto M, Zartman RE)

Edmonton, Canada: Department of Geology, University of Alberta, Edmonton, T6G 2E3: 6.1.2, 6.1.3, 6.1.6, 6.1.9, 6.1.11 (Baadsgaard H, Cumming GL)

Egham, UK: University of London, Royal Holloway and Bedford New College, Egham Hill, Egham, Surrey TW20 OEX: 6.1.3, 6.1.6 (Menzies MA)

Gif-sur-Yvette, France: Centre des Faibles Radioactivités, Laboratoire mixte C.N.R.S.-C.E.A., Place de l'Eglise, F-91190 Gif-sur-Yvette: 6.1.1.1, 6.1.3, 6.1.6, 6.1.9 (Cassignol C, Gillot P-Y), 6.2.3, 6.2.5, 6.2.10 (Duplessy DC, Labeyrie LD, Reyss LD, Reyss JL), 6.3 (Laleou C, 6.4.1, 6.4.3, 6.4.7, 7.2 (Poupeau G, Messelet R, Regas Jl, Valladas G, Valladas H, Yokoyama Y)

Glasgow, UK: Isotope Geology Unit, Scottish Universities Research and Reactor Centre, East Kilbridge, Glasgow, G75 OQU: 6.1.1.2, 6.1.3, 6.1.6, 6.1.9 (Aftalion M, Dickin AP, Hamilton PJ), 6.2.1 (Harkness DD)

Halifax, Canada: Department of Geology, Dalhousie University, Halifax, N.S.: 6.1.1.2 (Reynolds PH)

Hannover, Fed. Rep. of Germany: Federal Institute for Geosciences and Natural Resources, Stilleweg 2, D-3000 Hannover 51: 6.1.1.1, 6.1.1.2, 6.1.3, 6.1.9, 6.3.1 (Carl C, Höhndorf A, Kreutzer H, Wendt I)

Heidelberg, Fed. Rep. of Germany: Laboratory for Geochronology, University of Heidelberg, Im Neuenheimer Feld 234, D-6900 Heidelberg: 6.1.1.1, 6.1.1.2, 6.1.3, 6.1.10, 6.1.11 (Hess JC, Kober B, Lippolt HJ)

Hyderabad, India: National Geophysical Research Institute, Uppal Road, Hyderabad 500007: 6.1.1.1 (Baksi AK), 6.2.1, 6.2.2.2 (Sukhija BS), 7.1, 7.5 (Athavale RN)

Johannesburg, South Africa: Bernard Price Institute of Geophysical Research, University of Witwatersrand, 1 Jan Smuts Avenue, Johannesburg 2001: 6.1.3, 6.1.9, 6.1.11

Lawrence, USA: Isotope Geochemistry Laboratory, Department of Geology, University of Kansas, Lawrence, Kansas 66045: 6.1.3, 6.1.9 (Bickford ME)

Leeds, UK: Department of Earth Sciences, University of Leeds, Leeds, LS2 9JT: 6.1.1.1, 6.1.1.2, 6.1.3, 6.1.6, 6.1.9 (Cliff RA, Vollmer R)

Lower Hutt, New Zealand: Institute of Nuclear Sciences, Lower Hutt: 6.1.3 (Graham IJ), 6.2.1, 6.2.2, 7.4 (Taylor C)

Mainz, Fed. Rep. of Germany: Max-Planck-Institut für Chemie, Saarstr. 23, Postfach 3060, D-6500 Mainz; 6.1.3, 6.1.6, 6.1.7, 6.1.9, 6.5 (Hofmann AW, Kröner A, Schultz L, Todt W. Wänke H, Weber HW)

Menlo Park, California, USA: Branch of Isotope Geology, U.S. Geological survey, MS 345 Middlefield Road, Menlo Park, CA 94025: 6.1.1.1, 6.1.3 (Dalrymple GB, Lanphere M, Stacey J)

Milton Keynes, UK: Department of Earth Sciences, The Open University, Walton Hall, Milton Keynes, MK7 6AA: 6.1.3, 6.1.6, 6.1.10 (Hawkesworth CJ)

Montpellier Cedex, France: Laboratoire de Géochimie Isotopique L.P. 361 et U.E.R. 9 des Sciences de la Terre, Université de Montpellier II, Place E. Bataillon, F-34060 Montpellier Cedex: 6.1.1.1, 6.1.1.2, 6.1.3, 6.1.9 (Lancelot JR)

Moscow, USSR (a): Vernadsky Institute of Geochemistry and Analytical Chemistry USSR Academy of Sciences, Moscow: 6.1.1, 6.1.1.2, 6.1.3, 6.1.15, 6.5 (Meshick AP, Shukolyokov YuA), 6.3 (Kuznetsova L, Titayeva NA, Usinova GK)

Moscow, USSR (b): Laboratory of Geochronology, Institute of Geology, Ac Staromenetny pereulok 35 IGEM, Moscow 109017: 6.1

Munnich, Fed. Rep. of Germany: Mineralogisch-Petrographisches Institut, Universität München, D-8000 München: 6.1.3, 6.1.6 (Köhler H, Müller-Sohnius D)

Münster, Fed. Rep. of Germany: Zentrallaboratorium für Geochronologie, Mineralogisches Institut, Universität Münster, Corrensstr. 24, D-4400 Münster: 6.1.3, 6.1.6, 6.1.9, 6.1.10 (Baumann A, Grauert B, Hansen B, Kramm U)

Oslo, Norway: Mineralogisk-Geologisk Museum, Sarsgate 1, N-0562 Oslo 5: 6.1.3, 6.1.6, 6.1.9 (Griffin WL)

Ottawa, Canada (a): Geochronology Section, Lithosphere and Canadian Shield Division, Geological Survey of Canada, Ottawa, Ont. K1A OE8: 6.1.9 (Parrish RR, Roddick JC)

Ottawa, Canada (b): Ottawa-Carleton Centre for Geoscience Studies, Department of Geology, Carleton University, Ottawa, Ont.: 6.1.3, 6.1.6 (Bell K, Blenkinshop J)

Oxford, UK: Department of Geology and Mineralogy, University of Oxford, Parks Road, Oxford, OX1 3PR: 6.1.3, 6.1.6, 6.1.10, 6.1.11 (Jones NW, Moorbath S, Taylor PN)

Palisades, USA: Lamont-Doherty Geological Observatory, Colombia University, Palisades, NY 10964: 6.1.3, 6.1.6, 6.1.10 (Zindler A), 6.2.1 (Thurber TL), 6.3.1, 6.3.2, 6.3.3, 6.3.4, 6.3.11 (Kulp JL, Volchock HL), 7.1.1, 7.1.2 (Broecker WC, Opdyke ND, Ruddiman WF, Hays JD)

Paris, France (a): Laboratoire de Géochimie et Cosmochimie, Institut de Physique du Globe, Department des Sciences de la Terre, Universités de Paris VI et VII, 4, Place Jussieu, F-75230 Paris Cedex 05: 6.1.3, 6.1.6, 6.1.7, 6.1.8, 6.1.9, 6.1.10, 6.1.11, 6.5, 7.2 (Allègre CJ), 6.3 (Minister J-F, Condomines M)

Paris, France (b): Department de Géologie Dynamique, Université Pierre et Marie Curie, 4 Place Jussieu, F-75230 Paris Cedex 05: 6.1.1.1, 6.1.1.2 (Odin GS)

Pasadena, USA: Lunaticum Asylum, Division of Geological and Planetary Sciences, California Institute of Technology, Pasadena, CA 91125: 6.1.3, 6.1.6, 6.1.9, 6.1.10, 6.3, 6.5 (Chen JH, Huneke JC, Lee RR, Silver LT, Taylor HP, Wasserburg GJ), 7.2 (Epstein S), 8.6

Pisa, Italy: Istituto di Geochronologia e Geochimica, CNR, Via Cardinale Maffi, 36, I-56100 Pisa: 6.1.1.1, 6.1.3, 6.5 (Del Moro A, Villa IM)

Rehovot, Israel: The Weizman Institute of Science, Isotope Department, 76100 Rehovot: 6.1.1.1, 6.1.10 (Stiller M), 6.2.1, 6.2.2, 6.3.1, 6.3.4 (Kaufman a, Carmi I), 7.2 (Gat J. Magaritz M, Nissenbaum A), 8.1 (Goodfriend GA)

Rennes Cedex, France: Centre Armoricain d'Etude Structurale des Socles, Université de Rennes, Institut de Géologie, Campus de Beaulieu, F-35042 Rennes Cedex: 6.1.1.1, 6.1.3, 6.1.6, 6.1.9, 6.1.11 (Bernhard-Griffiths J, Jahn BM, Peucat JJ)

Rhode Island, USA: University of Rhode Island, Graduate School of Oceanography, Kingston, Rhode Island 02881: 6.1.3, 6.1.6, 6.1.9 (Schilling JG)

Santa Barbara, USA: Department of Geological Sciences, University of Geosciences, University of California, Santa Barbara, CA 93106: 6.1.9, 6.1.10 (Mattinson JM, Tilton GR)

Sao Paulo, Brasil: Instituto de Geochronologia, Department of Geology, University of Sao Paulo, Caixa Postal 20.899, Sao Paulo CEP 01498: 6.1.1.1, 6.1.3 (Cordani UG, Kawashita K)

Sichuan, China: Division of Isotope Geology, Chengdu College of Geology, Chendu, Sichuan 610059: 6.1.1.1 (Dong YB)

Stockholm, Sweden (a): Naturhistoriska Riksmuseet, Laboratoriet för Isotopgeologi, Box 50007, S-104 05 Stockholm 50: 6.1.1.1, 6.1.3, 6.1.9 (Christianson K, Claesson S)

Stockholm, Sweden (b): Laboratory för Isotopgeologi, Geologiska Institutionen, S-10691 Stockholm: 6.1.1.1, 6.1.3, 6.1.9 (Aberg G)

Stony Brook, USA: Department of Earth and Space Sciences, State University of New York, Stony Brook, NY 11794: 6.1.1.1, 6.1.1.2, 6.3.14 (Schaeffer OA)

Strasbourg, France: Institut de Géologie, 1, rue Blessig, F-67084 Strasbourg Cedex: 6.1.1.1, 6.1.3 (Bonhomme MG, Clauer N)

Tokyo, Japan: Laboratory for Rare Earth Element Microanalysis, Department of Chemistry, University of Tokyo, Hongo, Tokyo, 113: 6.1.4, 6.1.5, 6.1.6 (Masuda A)

Toronto, Canada (a): Department of Geology and Mineralogy, Royal Ontario Museum, 100 Queen's Park, Toronto, Ont. M5S 2C6: 6.1.3, 6.1.9 (Dunning GR, Krogh TE)

Toronto, Canada (b): Geophysics Division, Department of Physics, University of Toronto, Toronto, Ont. M5S 1A7: 6.1.1.1, 6.1.1.2 (Farquar RM, Hall CM, York D)

Tuscon, USA: Department of Geosciences, The University of Arizona, Tuscon, Arizona 85721: 6.1.3, 6.1.6 (Patchett PJ, Ruiz J)

Vancouver, Canada: Department of Geological Sciences, Geochronology Laboratory, The University of British Columbia, Vancouver, Br Col V6T 2B4: 6.1.3, 6.1.9, 6.1.10, 6.1.11 (Armstrong RL)

Vandoeuvre-les-Nancy, Francy: Centre de Recherches Pétrographiques et Géochimique and Ecole Nationale Superieure de Géologie, 15 rue Norte Dame les Pauvres, F-54501 Vandoeuvre-les-Nancy: 6.1.3, 6.1.6, 6.1.9, 6.1.10 (Albarède, F, Juteau M, Michard A)

Vienna, Austria → 6.3

Washington, USA: Department of Terrestrial Magnetism, Carnegie Institution of Washington, D.C.: 6.1.3, 6.1.6, 6.1.9, 7.1

Westwood, USA: Teledyne Isotopes, Age Determination Laboratory, 50 Van Buren Avenue, Westwood, N.J. 07675 (commercial): 6.1.1.1, 6.1.3, 6.1.9, 6.2.1

Yerevan, USSR: Institut de Géologie, rue Barekamountain 24a, 375019 Yerevan, Armenia: 6.1.1.1 (Bagdsaryan GP)

Zurich, Switzerland: Laboratory of Isotope Geochemistry, IKP, Swiss Federal Institute of Technology, Sonneggstr. 5, CH-8092 Zurich: 6.1.3, 6.1.6, 6.1.9, 6.1.10 (Gebauer D, Grünenfelder M, Steiger RH)

## Section 6.2: Dating with Cosmogenic Radionuclides

A Complete list of addresses of $^{14}$C laboratories is published annually in the journal RADIOCARBON.

Ahmedabad, India: → 6.3
Berkeley, USA: → 7
Berne, Switzerland: Physikalisches Institut, Universität Bern, Sidlerstraße 5, CH 3012 Bern: 6.2.1, 6.2.2, 6.2.3, 6.2.7, 6.2.10, 6.5, 7.2, 7.5, 7.6 (Andree M, Beer J, Lehmann BE, Loosli HH, Oeschger H, Schotterer U, Siegenthaler U), 6.5 (Eberhardt P, Engster O, Geiss J)
Brussels, Belgium: Service de Géologie et Géochimie Nucleaires, Université Libre de Bruxelles, 50 Av. FD Roosevelt, B-1000 Brussels 5: 6.1.1, 6.2.1 (Boudin A, Crozaz G, Picciotto E)
Budapest → 6.3
Cambridge, USA: Krueger Enterprises, Inc., Geochron Laboratories Division, 24 Blackstone Street, Cambridge, MA 02139 (commercial): 6.1.1.1, 6.1.3, 6.1.11, 6.2.1 (Krueger DA)
Cambridge, UK → 7
Cambridge, USA → 6.1
Canberra → 6.1
Chalk River, Canada: Atomic Energy of Canada Limited, Chalk River Nuclear Laboratories, Chalk River, Ontaria KOJ, 1JO: 6.2.1, 6.2.3, 6.2.6 (Andrews HR, Brown RM, Milton JCD)
Coral Gables, USA: Beta Analytic Inc., P.O. Box 248113, Coral Gables, Fl 33124 (commercial): 6.2.1, 6.2.2, 6.4.1 (Stipp JJ, Tamers MA)
Cracow → 6.3
Frankfurt → 8
Freiberg → 6.3
Gif-sur-Yvette, France → 6.1
Glasgow, UK → 6.1
Groningen, The Netherlands: Centrum voor Isotopen Onderzoek, Rijksu University of Groningen, Westersingel 34, Groningen: 6.2.1, 6.3.1, 7.2, 7.5 (Mook WG)
Hannover, Fed. Rep. of Germany: Nieders. Landesamt f. Bodenforschung, Postfach 510153, D-3000 Hannover 51: 6.2.1, 6.2.2, 6.3.1, 6.1.13, 7.1, 7.2, 7.5 (Geyh MA, Hennig GJ)
Heidelberg, Fed. Rep. of Germany: Institut für Umweltphysik, Im Neuenheimer Feld 366, D-6900 Heidelberg: 6.2.1, 6.2.2, 6.2.3, 6.2.5, 6.2 7, 6.3.1, 6.3.5, 6.3.6, 6.3.13, 6.4.3, 7.2, 7.5 (Cromer B, Mangini A, Münnich KO, Roether W, Sonntag C)
Helsinki, Finland: Radiocarbon Dating Laboratory, University of Helsinki, Snellmaninkatu 5, SF-00170 Helsinki 17: 6.2.1, 6.4.1 (Jungner H)
Hyderabad, India → 6.1
Johannesburg, South Africa: Schonland Research Centre, The Witwatersrand University, Jan Smuts AV. 1, Johannesburg 2001: 6.2.1, 6.2.2, 7.5 (Verhagen BTh)
Kiel, Fed. Rep. of Germany: Institut für Reine und Angewandte Kernphysik and Geologisch-Paläontologisches Institut, Universität Kiel, Olshausenstr. 40–60, D-2300 Kiel 1:6.2.1, (Erlenkeuser H. Willkomm H), 7.2 (Sarnthein M)
La Jolla, USA: Scripps Institution of Oceanography, Department of Chemistry and California Space Institute, B-017, University of California at San Diego, La Jolla, Cal 92 093: 6.1.3, 6.1.15, 6.2.1, 6.2.3, 6.2.5, 6.2.9, 6.3.12, 6.5 (Arnold JR, Goldberg ED, Griffin JS, Killingly JS, Koide M, Lugmair GW, McDougall JD, Marti K, Newman S, Nishiizumi K, Somayajulu BLK), 7.2, 7.6 (Berger HW, Craig H, Finkel RC)
Leningrad, USSR → 6.3
London, UK (a) → 6.4
Los Angeles, USA → 6.3
Lower Hutt → 6.1
Menlo Park → 6.1

Miami, USA: Rosenstiel School of Marine and Atmospheric Science, University of Miami, 4600 Rickenbacker Causeway, Miami, Florida 33149: 6.2.1, 6.2.2, 7.5 (Östlund G)

Neuherberg, Fed. Rep. of Germany: Institut für Hydrologie and Institut Gür Dosimetrie, Ingolstädter Landstr. 1, D-8042 Neuherberg 6.2.1, 6.2.2, 7.2, 7.5 (Fritz P, Rauert W), 6.4.1, 6.4.3 (Göksu-Ögelman Y, Regulla DF)

Newburyport, USA: General Ionex Corporation, New Buryport, Mass (Commercial): 6.2 (Purser KH, Schneider RJ)

New Haven, USA → 6.3

Oak Ridge, USA: Atomic Sciences, Inc., 114 Ridgeway Center, P.O. Box 138, Oak Ridge, TN 37830 (commercial): 6.2 (Schmitt HW)

Orsay, France: Laboratoire René Bernas, Centre de Spectrométrie Nucleaire de Spectrométrie de Masse, Bat. 108, and Laboratoire d'Hydrologie et de Géochemie Isotopique, Université de Paris-Sud, F-91406 Orsay: 6.2.1, 6.2.2 (Fontes J-Ch), 6.2.3, 6.2.5, 6.2.7, 6.2.9, 6.3.1 (Raisbeck GM, Yiou F, Bourles D, Fruneau M)

Oxford, UK → 6.4

Palisades → 6.3

Philadelphia, USA: Dept. of Geophysics and Geochemistry, College of Mineral Industries and Tandem Accelerator Laboratory, The Pennsylvania State University, Philadelphia, PE 19104: 6.2.1, 6.2.3, 6.2.5 (Herzog F, Klein J), 8.3 (Middleton R)

Pretoria → 6.3

Rehovot → 6.1

Reston, USA: Branch of Isotope Geology, U.S. Geological Survey, National Center, MS 981 Reston, Virginia 22092 (Rubin M, Pavich MJ): 6.2.1, 6.2.3, 7.5

Rochester, USA: Nuclear Structure Research Laboratory, University of Rochester, Rochester, NY 14627: 6.2.1, 6.2.3, 6.2.5, 6.2.7, 6.2.8, 6.2.10, 6.2.12, 6.2.13, 6.2.14, 7.5 (Elmore D, Gove HE)

Seattle, USA: Department of Zoology and Geology, Quaternary Research Center and Nuclear Physics Laboratory, University of Washington, Seattle, WA 98 195: 6.2.1, 7.2 (Grootes PM, Smith FH, Stuiver M)

Toronto, Canada: Department of Physics and Isotrace Laboratory, University of Toronto, Toronto: 6.2.1, 6.2.3, 6.2.6, 6.2.7, 6.2.10, 6.2.12, 6.2.13, 6.2.14 (Beukens RP, Chang KH, Kilius RL, Lee HW, Litherland AE, Wilson GC)

Trondhem → 7

Tuscon, USA (A): Department of Hydrology and Water Resources and NSF Accelerator Facility for Radioisotope Analysis, University of Arizona, Tuscon, AR 85 721: 6.2.1, 6.2.7, 6.5 (Bentley HW, Damon PE, Donahue DJ, Douglas J, Klein, J. Long, A, Zabel TH)

Uppsala, Sweden: Institute of Physics, Uppsala University, Box 530, S-75121 Uppsala: 6.2.1, 6.3.13 (El-Daoushy F, Olsson IU)

Upton, USA: Brookhaven National Laboratory, Upton, NY 11973: 6.2.1, 6.2.2, 6.2.6 (Alburger DE, Gaffney JS, Harbottle G, Levy PW, Manowitz B) 6.3.14, 6.4.1, 7.3

Vienna → 6.3

Washington DC, USA: Center for Analytical Chemistry, National Bureau of Standards, Washington DC 20234: 6.2.1, 7.5, 7.6 (Curie LA, Klouda GA)

Westwood → 6.1

Woods Hole → 6.3

Zurich, Switzerland: Institut für Mittelenergiephysik, ETH Zurich, Hönggerberg, CH-8093 Zurich: 6.2.1, 6.2.3, 6.2.5, 6.2.7, 6.2.12, 6.2.13, 6.2.14 (Balzer R, Bonani G, Suter M, Wölfli W)

## Section 6.3: Uranium Series Dating

Ahmedabad, India: Physical Research Laboratory, Navrangpura, Ahmedabad 380009: 6.2.1, 6.2.2, 6.2.3, 6.2.6, 6.3.1 (Bhandari N. Goswami JN, Hussain N, Krishnaswami S, Lal D, Nijampurkar VN, Potchar MB. Sharma YP, Somayajulu BLK), 6.4.1 (Singhvi AK), 6.5, 7.4

Ann Arbor, USA: Department of Chemical Engineering, The University of Michigan, Ann Arbor, MI 48109: 6.3.1, 6.3.3 (Kadlec RH), 7.4 (Robbins JA)

Bath, UK: School of Chemistry, University of Bath, Bath BA2 7AY: 6.3.14 (Andrews JN)

Bedford Park, Australia: School of Physical Sciences, The Flinders University, Bedford Park, SA 5042: 6.3.1, 6.3.2, 6.3.4, 6.3.5, 6.3.6 (Veeh HH)

Bombay, India: Tata Institute of Fundamental Research, Bombay 5: 6.3.12, 6.3.13 (Krishnaswami S)

Budapest, Hungary: Central Research Institute for Physics, Hungarian Academy of Science, Budapest: 6.2.1, 6.3, 6.4.1, 8.1 (Szabo PP)

Cambridge, USA: Smithsonian Astrophysical Observatory, Cambridge, MA 021138: 6.3.1, 6.5 (Fireman EL)

Chapel Hill, USA: Marine Science Program, University of North Carolina, Chapel Hill, NC 27514: 6.3.13 (Chanton JP)

Chicago, USA: Department of Geophysical Sciences, University of Chicago, 5734 S Ellis AV., Chicago, IL 60637: 6.3 (Aller RC)

College Station, USA: Dept. of Oceanography, Texas A & M University, College Station, Texas 77343: 6.3 (Scott M)

Cologne → 6.4

Columbia, USA: Department of Geology, University of South Carolina, Columbia, South Carolina 29208: 6.3.11, 6.3.12, 6.3.15 (Levy DM, Moore WS)

Corvallis, USA: School of Oceanography, Oregon State University, Corvallis, OR 97331: 6.3.11, 6.3.15, 7.1, 7.2 (Couch R, Dymond J, Pisias NG)

Cracow, Poland: Institute of Physics and Nuclear Techniques, University of Mining and Metallurgy, Pl-30.059 Cracow: 6.2.1, 6.2.2, 7.2, 7.4 (Dulinski M, Rozanski K, Zuber A)

Denver, USA: US Geological Survey, Federal Center, MS 963, PO Box 25046, Denver, Colorado 80225: 6.3.1, 6.4.7, 8.6 (Friedman I, Naeser CN, Naeser ND, Obradovich J, Rosholt JN, Smith RL, Szabo BJ, Tatsumoto M)

East Kilbride, UK: Isotope Geology Unit, Scottish Universities Research and Reactor Centre, East Kilbride, Scottland G75 OQU: 6.3 (Harmon RS)

Freiberg, German Democratic Republic: Sektion Physik, Bergakademie Freiberg, von-Cotta-Str. 4, DDR-9200 Freiberg: 6.2.1, 6.2.2, 6.2.6, 6.3.1, 6.3.2, 6.3.3, 6.3.4, 7.4 (Fröhlich K, Gellermann R, Hebert D)

Frunze, USSR: Institute of Physics and Mathematics, Academy of Sciences, Kirghiz SSR, Frunze: 6.3.4 (Chalov PL, Tozuva TV)

Gif-sur-Yvette, France → 6.1

Glasgow → 6.1

Groningen → 6.2

Hakodate, Japan: Dept. of Chemistry, Graduate School of Fisheries Science, Hokkaido University, Hakodate 041: 6.3.13 (Takedo Y, Tanaka N, Tsunogai S)

Hamilton, Canada: Dept of Physics, Dept of Geology, Dept of Geography, McMaster University, 1280 Main Street West, Hamilton, Ontario L8S 4MI: 6.3.1, 6.3.2, 6.3.3, 6.3.4, 6.4.3, 7.2 (Ford DC, Gascoyne M, Schwarcz HP), 7.1 (Latham AG)

Hannover → 6.1, 6.2

Heidelberg → 6.2

Johannesburg, South Africa → 6.2

La Jolla, USA → 6.2

Leningrad, USSR: Laboratory of Precambrian Geology, Leningrad Branch of the Mathematical Institute, USSR Academy of the Sciences, Leningrad: 6.2.2, 6.3.14 (Gerling EK, Tolstykhin IN, Shukolyukov YuA)

Los Angeles, USA: Dept. of Geological Sciences, University of Southern California, Los Angeles, California 90007: 6.2.3, 6.3.1, 7.1, 8.1 (Joshi, IU, Ku T-L, Kusakabe M, Lund S)

Montreal, Canada: Laboratoire de Géochimie Isotopique et de Géochronologie, Université du Quebec a Montreal, PB 8888, Succ-A, Montreal H3C 3P8: 6.3.1 (Hillaire-Marcel C)

Moscow, USSR (a) → 6.1

Moscow, USSR (b): Geological Institute, Academy of Sciences, Moscow: 6.3 (Cherdyntsev VV)

Naples, Italy: Observatorio Vesuviano, Ercolano, Naples: 6.3.1, 6.3.4, 6.3.7 (Capaldi G, Pece R)

New Haven, USA: Dept. of Geology and Geophysics, Yale University, PO Box 6666, New Haven Connecticut 065511: 6.2.3, 6.3.15 (Cochran JK, Krishnaswami S, Turekian KK)

Orsay → 6.2

Oxon, UK: Nuclear Physics Division, Building 7, AERE Harwell, Oxon OX11 ORA: 6.3 (Ivanovich M)

Palisades, USA: Lamont-Doherty Geological University, Colombia University, Palisades, NY 10964: 6.2.1 (Thurber TL), 6.3.1, 6.3.2, 6.3.3, 6.3.4, 6.3.11 (Kulp JL, Volchock HL), 6.4.7, 7.1, 7.2 (Opdyke ND, Ruddiman WF, Hays JD, Broecker WS)

Paris (a) → 6.1

Pretoria, South Africa: National Isotopes Division, National Physical Research Institute, P.O. Box 395, Pretoria 0001: 6.1, 6.2.1, 6.2.2, 6.3.1, 6.3.15 (Heaton THE, Vogel JC)

Rehovot, Israel → 6.1

Seattle (a) → 7

Tallahassee, USA: Geological Department, Florida State University, 600 West College Av., Tallahassee, Florida 32306: 6.3.1, 6.3.2, 6.3.3, 6.3.4 (Burnett WC, Cowart JB, Kim KH, Osmond JK)

Tokyo, Japan: Ocean Research Institute and Department of Chemistry, Faculty of Sciences, University of Tokyo, Nakano-ku and Hongo, Bunkyo-ku, 164 Tokyo: 6.3.1, 6.3.2, 6.3.5, 6.3.6 (Masuda A, Nozaki Y, Sakai H, Yang H-S)

Trondhem → 7

Tsukuba, Japan: Geochemical Laboratory, Meteorological Research Institute, Nagamine I-1, Yatabe, Tsukuba, Ibaraki 305: 6.3.12 (Hirose K, Sugimura Y)

Uppsala → 6.2

Upton → 6.2

Vienna, Austria: Institut für Radiumforschung und Kernphysik, Österr. Akademie der Wissenschaften, Boltzmanngasse 3, A-1090 Wien: 6.1, 6.2.1, 6.3.1, 6.3.2, 6.3.3, 6.3.4 (Eisenbarth P, Hille P), 8.3, 8.9 (Korkisch J, Pake, Wild E)

Woods Hole, USA: Oceanographic Institution, Woods Hole, Massachusetts 02534: 6.2.1 (Druffel C), 6.2.2 (Jenkins WJ, Torgersen TH), 6.3.1, 6.3.2, 6.3.5, 6.3.6, 6.3.15, 7.4 (Anderson RF, Bacon MP, Brewer PG, Cochran JK, Keigwin LD Jr, Kulp TL, Livingston HD, Rosholt JN)

## Section 6.4:    Radiation Damage Dating

Adana, Turkey: Department of Physics and Space Science, Cukurova University, Basic Science Faculty, Adana: 6.4.1 (Göksu-Ögelmann Y, Türetken N)

Adelaide, Australia: Physics Department, University of Adelaide, Adelaide, SA 5001: 6.4.1 (Jensen HE, Prescott JR)

Ahmedabad – 6.3

Athens, Greece: Ministry for Culture and Science, Department of Underground Antiquities, 58 Omiro Street, Athens: 6.4.1, 6.4.3 (Liritzis Y)

Auckland, New Zealand: Dept. of Geography, University of Auckland: 6.4.1, 6.4.3 (Lyons RG)

Berlin, Fed. Rep. Germany: Rathgen-Forschungslabor, Staatliche Museen, Stiftung Preussischer Kulturbesitz, Schlosstr. 1a, D-1000 Berlin 19: 6.4.1 (Goedicke C)

Berne, Switzerland – 6.1

Birmingham, UK: Dept. of Physics, University of Birmingham, P.O. Box 363, Birmingham B15 2TT: 6.4.1, 6.4.3, 6.4.7, 6.5 (Durrani SA, Green PF, McKeever SWS)

Bombay, India: Health Physics Division, Bhabha Atomic Research Centre, Bombay 400 085: 6.4.1, 6.4.3 (Nambi KSV, Sakaran AV)

Bordeaux, France: Universite' de Bordeaux III-CNRS, Lab de Physique de Solide (CRIAA ERA CNRS 584), Esplanade des Antilles, MSA Domaine Universitaire, F-33405 Talence Cedex: 6.4.1 (Schvoerer M)

Bradford, UK: School of Archaeological Sciences, University of Bradford, Bradford: 6.4.1 (Warren SE)

Brighton, UK: School of Engineering and Applied Sciences and School of Mathematical and Physical Sciences, University of Sussex, Falmer, Brighton BN1 9QH: 6.4.1, 6.5 (McKeever SWS, Townsend PD)

Bristol, UK: Department of Geography, University of Bristol, University Road, Bristol BS8 1SS: 6.4.3 (Smart PL, Symons MCR)

Brussels → 6.2

Bupadest → 6.3

Burnaby, Canada: Department of Physics, Simon Fraser University, Burnaby, British Columbia V5A 1S6: 6.4.1, 6.4.2 (Berger CW, Huntley DJ)

Cambridge, UK: Thermoluminescence Dating Research Laboratory, Sub-Dept. of Quaternary Research, Cambridge University, The Godwin Laboratory, Free School Lane, Cambridge CB2 3RS (Grün R): 6.4.1, 6.4.3, 6.4.4

Cologne, Fed. Rep. Germany: Institut für Kernchemie, Universität Köln, Zülpicherstr. 49, D-5000
    Köln: 6.4.7, 6.5 (Englert P, Thiel K, Herpers U)
Columbia, USA: Department of Physics, University of Missouri, Columbia, Missouri 65211:
    6.4.3 (Cowan DL, Rowlett RM)
Coral Gables, USA → 6.2
Denver → 6.3
Durham, UK: DURTL Thermoluminescence Dating & Research Service, Department of Archae-
    ology, Durham University, South Road, Durham DHI 3LE: 6.4.1 (Bailiff IK, Watson I)
Edinburgh, UK: National Museum of Antiquities, West Granton Road, Edinburgh: 6.4.1 (Tate J)
    Gif-sur-Yvette, France → 6.1
Hamilton, Canada → 6.3
Heidelberg (a) → 6.2
Heidelberg, Fed. Rep. Germany (B): Max-Planck-Institut, Postfach 1248, 6900 Heidelberg: 6.4.1,
    6.4.3, 6.4.7 (Wagner GA, Zöller L) 6.5 (Kirsten T, Wagner G, Heusser GJ, Hampel W)
Helsinki → 6.2
Ibaraki, Japan: Institute of Geoscience, University of Tsukuba, 1-I Tennodai, Sakura-mura,
    Ibaraki, 305: 6.4.3 (Fukuchi T)
Leicester, UK: Department of Chemistry, University of Leicester, Leicester LE1 7RH: 6.4.3
    (Symons MCR)
London, UK (a): British Museum Research Laboratory, Great Russell St., London WC1B 3DG:
    6.2.1, 6.4.1 (Bowman SGE)
London, UK (b): Institute of Archaeology, University of London, 31-34 Gordon Square, London
    WC 1H OPY: 6.4.3 (Griffiths D)
London, UK (c): Waltham Technical College, London: 6.4.4 (Dalton P)
Louvain-la-Neuve, Belgium: Laboratoire de Chimie Inorganique et Nucleaire, Université de
    Louvain, 2, Chemin du Cyclotron, B-1348 Louvain la-Neuve: 6.4.3 (Apers D, De Canniere
    P, Debuyst R)
Milano, Italy: Laboratorio di Thermoluminescenza, Istituto di Fisiche del'Universita, Via Celoria
    16, I-20133, Milano: 6.4.1 (di Caprio NC, Spinola G)
Neuherberg → 6.2
Osaka, Japan: Dept. of Physics, Faculty of Engineering Science, Osaka University, Toyonaka,
    Osaka: 6.4.3 (Sato T)
Oxford, UK: Research Laboratory of Archaeology and History of Arts, Oxford University, 6
    Keble Road, Oxford OX1 3QJ:6.2.1 (Hedges REM) 6.4.1, 6.4.3, 7.1 (Aitken MJ, Debenham
    N, Huxtable J, Smith BW, Stoneham D, Templer RH, Walton D)
Palisades, USA → 6.3
Paris, France (a): Laboratoire de Recherches des Musées de France, Palais du Louvre, F-7000
    Paris: 6.4.1 (Gautier J)
Paris, France (b): Laboratoire de Minéralogie, Museum National d'Histoire Naturelle, F-75005
    Paris: 6.4.7, 6.5.7, 6.5 (Bourot-Denise M, Pellas P, Storzer D)
Parkville, Australia: Department of Geology, University of Melbourne, Parkville, Victoria 3052:
    6.4.7 (Gleadow AJW)
Philadelphia, USA (B): The University Museum, University of Pennsylvania, 33 & Spruce Streets
    F1, Philadelphia, PE 19174: 6.4.1 (Fleming SJ, Han HC)
Rennes, France: Laboratoire d'Archèométrie, Université de Rennes, Rennes: 6.4.1 (Langouët M)
    Roskilde, Danemark: The Nordic Laboratory for Thermoluminescence Dating, Risø National
    Laboratory, DK-4000 Roskilde: 6.4.1, 6.4.2 (Meydahl V, Miller A)
Salt Lake City, USA: Radiobiology Division, Bldg 351, University of Utah, Salt Lake City,
    Utah 84112: 6.4.1 (Haskell EH)
Schenectady, USA: General Electric Research & Development Center, Schenecatdy, New York
    12302: 6.4.7, 6.4.8 (Fleischer RL)
St. Louis, USA: Dept. of Earth and Planetary Sciences, Box 1105, and McDonnel Center for
    Space Science, The Washington University, St. Louis MO 63130: 6.4.7, 6.4.8, 6.5 (Crozaz G,
    Hohenberg CM, Huang WH, Lindström MM, Walker RM)
Tallin, Estonian SSR: Institute of Geology, Academy of Sciences of the Estonian SSR, Estonia
    Puistee 7, Tallinn: 200101: 6.4.1, 6.4.2, 6.4.3 (Hütt G, Molodkow A, Vaikmäe R)
Tessaloniki, Greece: Nuclear Physics Department, University of Thessaloniki, Thessaloniki: 6.4.1,
    6.4.2 (Charalambous S)

Tokyo, Japan: INA Civil Engineering Consultant, Suido-cho, Sinjuku-ku, Tokyo 112: 6.4.3
   (Sawada S, Tanaka T)
Toyonaka, Japan: Dept. of Physics, Osaka University, Machikaneyama J-560, Toyonaka, Osaka:
   6.4.3, 8.4 (Ikeya M)
Ube, Japan: Technical College, Yamaguchi University, Tokiwadai, J-755 Ube: 6.4.3, 8.4 (Miki T)
   Upton → 6.2
Warsaw, Poland: Laboratorium Wieku Thermoluminescencji, Uniwersytet Warszawaski, Madal-
   inskiego 42/62, Warsaw: 6.4.1 (Proszynska H)
West Lafayette, USA: Department of Geosciences, Purdue University, West Lafayette, Indiana
   47907: 6.4.7 (Crough STh)
Williamstown, USA: Chemistry Dept., Williams College, Williamstown, MA 01267: 6.4.3 (Skinner
   AF)
Zurich, Switzerland: Schweizerisches Landesmuseum, Museumstr., Zurich, 6.4.1 (Voute K)


## Section 6.5:   Dating Meteorites and Lunar Rocks

Ahmedabad → 6.3
Berkeley → 7
Berne → 6.2
Birmingham → 6.4
Brighton → 6.4
Cambridge, USA (b) → 6.3
Chigaco, USA: Enrico Fermi Institute and Department of Chemistry, University of Chicago,
   Chicago, IL 60637: 6.5 (Alaerts L, Anders E, Lewis RS)
Colonge → 6.4
Heidelberg (b) → 6.4
Houston, USA: Department of Geology and Space Science, Rice University, Houston, Texas:
   6.5 (Adams JAS, Fryer GE, Heymann D)
La Jolla → 6.2
London, UK: School of Applied Physics, The Polytechnic of North London, London N7 8DB:
   6.5 (Barton JC, Watson AH)
Los Alamos, USA: Nuclear Chemistry Group, Los Alamos National Laboratory, Mail Stop
   J514, Los Alamos NM 87545: 6.5 (Miller CM, Reedy RC)
Mainz → 6.1
Moscow (a) → 6.1
New Brunswick, USA: Department of Chemistry, Douglass College, Rutgers University, New
   Brunswick, NJ 08903: 6.5 (Hall GS, Herzog GF, Pal D)
Paris (a) → 6.1
Paris (b) → 6.4
Pasadena → 6.1
Pisa → 6.1
St. Louis → 6.4
Tuscon, USA: Lunar and Planetary Laboratory, University of Arizona, Tuscon AZ 85721: 6.5
   (Jones JH)
Victoria, Canada: Dominion Astrophysical Observatory, Herzberg Institute of Astrophysics,
   National Research Council of Canada, Victoria, British Colombia V8X 4M6: 6.5 (van den
   Bergh S)


## Section 7:   Chronostratigraphic Methods

Ahmedabad → 6.3
Amherst, USA: Ice Core Laboratory, SUNY Buffalo, Amherst, NY: 7.2, 7.6 (Finkel RC, Langway
   CC Jr)
Ann Arbor → 6.3
Bergen, Norway: Department of Geology, Sect. B, Bergen Universitetet, Allegaten 41, N-5007
   Bergen: 7.1, 7.2, 8.1 (Mangerud J, Sejrup HP)

Berkeley, USA: Dept. of Geology and Geophysics: 7.4 (Alvarez LW, Alvarez W, Asaro F), 8.6
    (Michel HV); and Lawrence Berkeley Laboratory and Space Sciences Laboratory: 6.2.1, 6.5
    (Muller RA, Reynolds JH), University of California, Berkeley, CA 94720
Berne → 6.2
Bloomington, USA: Corporate Technology Center, Honeywell, Inc., 10701 Lyndale Avenue
    South, Bloomington, Minnesota 55420: 7.1 (Johnson RG)
Cambridge, UK: Godwin Laboratory for Quaternary Research, University of Cambridge, Free
    School Lane, Cambridge CB2 3RS: 7.2 (Shackleton NJ)
Canberra → 6.1
Copenhagen, Danemark: Geophysical Isotope Laboratory, University of Copenhagen,
    Copenhagen: 7.2 (Dansgaard W, Johnsen SJ)
Corvallis → 6.3
Cracow → 6.3
Davis, USA: Department of Geology, University of California, Davis, CA 95616: 7.1 (Ensley RA,
    Verosub KL)
Edinburgh, UK: Dept. of Geophysics, University of Edinburgh, Mayfield Road, Edinburgh EH9
    3JZ: 7.1 (Creer KM, Thompson R, Tucholka P)
Eugene, USA: Department of Geology, University of Oregon, Eugene, Oregon 97493: 7.3 (Holser
    WT)
Flushing, USA: Queens College of the City of New York, Flushing, NY 11367: 7.2 (McIntyre A)
Gif-sur-Yvette → 6.1
Gif-sur-Yvette, France (a): Centre d'Etudes Nucleaires de Saclay, Dept. of Physico-Chemistry,
    Boite Postale 2, F-91190 Gif-sur-Yvette: 7.2 (Duplessy JC, Jouzel I, Merlivat L)
Grenoble, France: Laboratoire de Glaciologie du CNRS, 2 rue Tres Colîtres, F-38031 Grenoble
    Cedex 7.2 (Jouzel J, Lorius C, Merlivat L)
Freiberg → 6.3
Groningen → 6.2
Hamilton → 6.3
Hannover → 6.2
Heidelberg (a) → 6.2
Hyderabad → 6.1
Johannesburg → 6.2
Kiel → 6.2
Kingston, USA: Graduate School of Oceanography, University of Rhode Island, Kingston, RI
    02881: 7.1, 7.2 (Kennett JP, Kominz MA, Pisias NG)
La Jolla → 6.2
Liverpool, UK: Sub-Department of Geophysics, University of Liverpool, PO Box 147, Liverpool
    L69 3BX: 7.1 (Piper JDA)
Los Angeles (a) → 6.3
Los Angeles, USA (B): Institute of Geophysics and Planetary Science, University of California,
    Los Angeles, CAL 90024: 7.4 (Berger R, Kyte FT, Wasson JT, Zhou Z)
Lower Hutt → 6.1
Miami (a) → 6.2
Miami, USA (b): Department of Geology, University of Miami, Miami, Florida 33124:7.2
    (Emiliani C)
Minneapolis, USA: Department of Geology and Geophysics, University of Minnesota,
    Minneapolis, MN 55454: 7.1 (Banerjee SK, King JW , Marvin J)
Moscow, USSR (a): Geology Department, Moscow State University, Moscow: 7.4 (Chizova NI,
    Vhizhov AB)
Moscow, USSR (b): Institute of Biochemistry and Physiology of Microorganisms, USSR Academy
    of Sciences, Pushchino, Moscow 142292: 7.3 (Lein AYu)
Neuherberg → 6.2
Oxford → 6.4
Palisades → 6.3
Pasadena → 6.1
Prague, CSSR: Geophysical Institute, Czechoslovak Academy of Science, Pragua: 7.1 (Bucha V)
Rehovot → 6.1
Reston → 6.2

Rochester → 6.2
Sofia, Bulgaria: Geophysical Institute, Bulgarian Academy of Science, Sofia: 7.1 (Kovacheva M)
Seattle → 6.2
Seattle, USA (a): Laboratory of Radiation Ecology, College of Fisheries, University of Washington, WH-10, Seattle, WA 98195: 6.3.13, 7.5 (Schell WR)
Stanford, USA: School of Earth Sciences, Stanford University, Stanford, California 94305: 7.1 (Cox AV)
Thunder Bay, Canada: Lakehead University, Thunder Bay, Ont. P7B 5E1: 7.1 (Mothersill JS)
Trondheim, Norway: Radiological Dating Laboratory, The Norwegian Institute of Technology, Trondheim: 6.2.1, 6.3.1, 7.4 (Gulliksen S, Nydal R, Skogseth FH)
University Park → 8
Upton → 6.2
Washington → 6.1
Woods Hole → 6.3
Zurich, Switzerland (C): Geologisches Institut, ETH Zürich, CH-8092 Zürich: 7.1, 7.4 (Hsü KJ)


## Section 8:   Chemical Dating Methods

Amherst, New York, USA: Dept. of Geological Sciences, State University of New York at Buffalo, Amherst, New York 14226: 8.8 (Fountain JC, Waddell AC)
Bergen → 7
Budapest → 6.3
Boulder, USA: Amino Acid Geochronology Laboratory, University of Colorado, Boulder, Colorado 80309 (commercial): 8.1 (Andrews JT, Brigham JK, Miller GH)
Denver → 6.3
Frankfurt, Fed. of Rep. of Germany: Institut für Anthropologie, Humangenetik und Humanbiologie, Goethe Universität, Siesmayerstr. 70, D 6000 Frankfurt/Main: 6.2.1, 8.1 (Protsch R)
Hamburg, Fed. Rep. of Germany: Archäologisches Institut, Universität Hamburg, Johnsallee 35, D-2000 Hamburg 13: 8.7 (Zaun PE)
London, UK: University of London, Intercollegiate Research Service, London: 8.4 (Robins GV)
Los Angeles → 6.3
Los Angeles, USA: UCLA Obsidian Hydration Laboratory, Archaeological Survey, Dept. of Anthropology, University of California, Los Angeles, California 90024 (commercial): 8.6 (Clark CL, Meighan CW)
Menlo Park, USA: U.S. Geological Survey, Pacific-Arctic Branch of Marine Geology, 345 Middlefield Road, Menlo Park, CA 94025: 8.1 (Kvenvolden KA, Blunt DJ)
New Haven, USA: Department of Physics, Yale University, New Haven, Connecticut 06520: 8.6 (Landford WA)
Newark, USA: Department of Geology, University of Delaware, Newark, Delaware 19711: 8.1. (Wehmiller JF)
Nijmegen, The Netherlands: Dept. of Exobiology, Faculty of Science, University of Nijmegen, Nijmegen: 8.1 (Dungworth G)
Pasadena → 6.1
Philadelphia → 6.2
Pulman, USA: Department of Geology, The Washington State University, Pulman, WA 99164: 8.3, 8.9 (Hassan AA)
Richardson, USA: Geosciences Program, University of Texas at Dallas, P.O. Box 688, Richardson, Texas 75080: 8.1 (Mitterer RM, Kriausakul N)
Riverside, USA: Department of Anthropology, Institute of Geophysics and Planetary Physics, UCLA, Riverside: 8.3, 8.9 (Taylor RE)
San Diego, USA: Amino Acid Dating Laboratory, Scripps Institution of Oceanography, UCLA, La Jolla, California 92093: 8.1, 8.2 (Bada JL, Darling D, Hoopes E)
State College, USA: MOHLAB, 1188 Smithfield Street, State College, PA 16801 (commercial): 8.6 (Michels JW)
Tempel, USA: Department of Physics, Arizona State University, Tempe, Arizona 85281: 8.6 (Tsong IST)

Tokiwadai Ube → 6.4

University Park, USA: Pennsylvania State Obsidian Dating Laboratory, The Pennsylvania State University, University Park, Pennsylvania 16802: 7.3, 8.6 (Michels JW)

Vienna → 6.3

Washington, DC, USA: Geophysical Laboratory, Carnegie Institution of Washington, Washington, D.C. 20008: 8.1, 8.2 (Hare PE, Mitterer RM)

# Subject Index

The numbers in italics refer to the main discussion of the subject in each case.

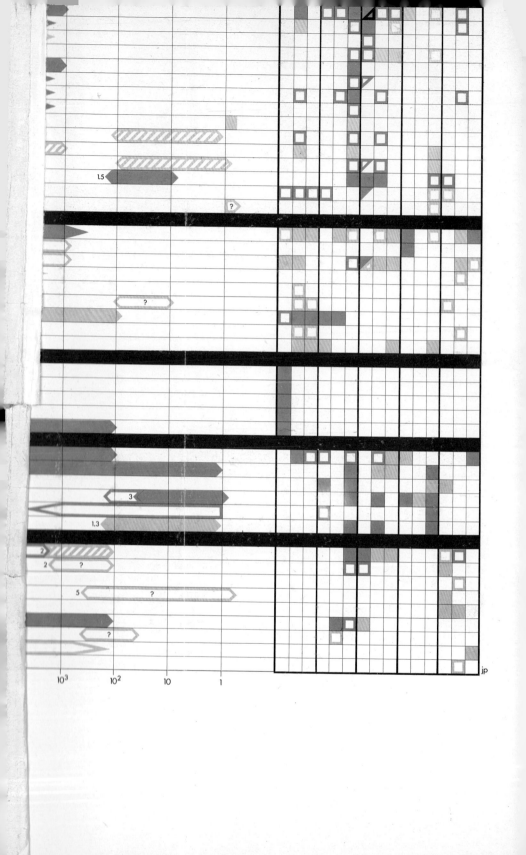

1.5

?

?

3

1.3

2

2

5

?

?

$10^3$      $10^2$      10      1

jp